Nature-Based Solutions

Nature-Based Solutions

Editors

Panayiotis Dimitrakopoulos
Mario A. Pagnotta
Miklas Scholz
Arshiya Noorani

MDPI

Basel • Beijing • Wuhan • Barcelona • Belgrade • Novi Sad • Cluj • Manchester

Editors
Panayiotis Dimitrakopoulos
University of the Aegean
Mytilene
Greece

Mario A. Pagnotta
Tuscia University
Viterbo
Italy

Miklas Scholz
atene KOM
Berlin
Germany

Arshiya Noorani
FAO
Rome
Italy

Editorial Office
MDPI
St. Alban-Anlage 66
4052 Basel, Switzerland

This is a reprint of articles from the Topic published online in the open access journals *Sustainability* (ISSN 2071-1050), *Forests* (ISSN 1999-4907), *Land* (ISSN 2073-445X), and *Plants* (ISSN 2223-7747) (available at: https://www.mdpi.com/topics/Nat_Sol).

For citation purposes, cite each article independently as indicated on the article page online and as indicated below:

Lastname, A.A.; Lastname, B.B. Article Title. *Journal Name* **Year**, *Volume Number*, Page Range.

ISBN 978-3-0365-9206-0 (Hbk)
ISBN 978-3-0365-9207-7 (PDF)
doi.org/10.3390/books978-3-0365-9207-7

Cover image courtesy of Panayiotis Dimitrakopoulos

Contents

About the Editors

Panayiotis Dimitrakopoulos

Panayiotis Dimitrakopoulos is Professor of Functional Ecology at the Department of Environment of the University of the Aegean, Greece. He obtained his Ph.D. degree from the University of the Aegean in 2001. His postdoctoral research was carried out at the University of Zurich, Switzerland, and was funded by the European Science Foundation (LINKECOL program). His research focuses on functional plant ecology, community ecology, biodiversity conservation, and conservation policy. He has served as President of the Hellenic Ecological Society (2012–2014), member of the "Natura 2000" National Committee (2010–2020), and Honorary Visiting Senior Fellow at Anglia Ruskin University, Cambridge (2019–2022). He serves as Section Editor-in-Chief of Ecology in Biology (Basel) and is a member of the editorial board of six journals and referee for more than 60 journals. He has published more than 90 papers in peer-reviewed journals.

Mario A. Pagnotta

Mario A. Pagnotta is Associate Professor of Agricultural genetics at the University of Tuscia. His activities have included project management, student supervising, lecturing, and participation in several national and international projects. The research activity focuses mainly on the assessment of genetic variability in natural populations, landraces, and the characterization of plant germplasm of several species, and the relationship between genetic variation and edaphic factors, as well as on topics related with the biodiversity conservation, adaptation and variety identification. The analyses are conducted using traditional and advanced methodologies, based on different kinds of molecular markers. Particular attention is paid to the resistance to abiotic stress, especially drought stress, looking also at the distribution and identification of SNP markers for genes related to drought and salt tolerance. He has reviewed several projects and edited scientific papers for several journals. He is on the Editorial Board of *Diversity* (MDPI), serving as Editor-in-Chief of the "Plant Diversity" section. He also acts as Associate Editor for *Frontiers in Plant Science*, in the "Plant Abiotic Stress" section.

Miklas Scholz

Miklas Scholz, cand ing, BEng (equiv), PgC, MSc, PhD, DSc, CWEM, CEnv, CSci, CEng, FHEA, FIEMA, FCIWEM, FICE, Fellow of IWA, Fellow of IETI is a Distinguished Professor at Johannesburg University, South Africa. Miklas holds the title of Chair in Civil Engineering as a Professor at The University of Salford, UK, and he is also a Senior Expert in Water Management for atene KOM, Germany, a Technical Specialist for Nexus by Sweden, Sweden, and a Hydraulic Engineer for Adams Kunststofftechnik, Germany. He has published 6 books and 311 journal articles. Prof. Scholz's total citations amount to approximately 13660 (above 8510 citations since 2018), resulting in an h-index of 57 and an i10-Index of 227. He belongs to the top 2% of academics regarding the i10-index in the past five years. Miklas also belongs to the World's Top 2% of scientists according to Stanford University. A bibliometric analysis of all constructed wetland-related publications and corresponding authors with a minimum number of 20 publications and 100 citations indicates that Miklas is 5th in the world out of approximately 70 authors (including those who have sadly passed away). In 2019, Prof. Scholz was awarded EURO 7M for the EU H2020 REA project Water Retention and Nutrient Recycling in Soils and Streams for Improved Agricultural Production (WATERAGRI). He received EURO 1.52M for the JPI Water 2018 project Research-based Assessment of Integrated approaches to Nature-based Solutions (RAINSOLUTIONS).

Arshiya Noorani

Arshiya Noorani is an Agricultural Officer based at at FAO headquarters in Rome, and is a member of the Seeds and Plant Genetic Resources Team, Plant Production and Protection Division (AGP). Arshiya has been with FAO since 2011 but has been working since 2000 in the conservation, assessment and management of biodiversity, including protected areas. She holds a PhD in Plant Conservation Genetics, an MSc in Environmental Protection and Management, and BSc (Hons) in Zoology. Prior to joining FAO, she worked on in situ conservation and on-farm management of plant genetic resources for food and agriculture at Bioversity International. She previously taught ecological surveying and protected area management planning in both terrestrial and freshwater habitats at the University of Edinburgh. At FAO, Arshiya works on developing and implementing projects, developing technical guidelines as well as assessing data submitted by Members when monitoring the conservation and use of plant genetic resources. She supports Members in the implementation of the rolling Global of Action for Plant Genetic Resources for Food and Agriculture and the associated reporting, which feed into the preparation of global periodic assessments. She is contributes to FAO's normative work, especially through the Commission on Genetic Resources for Food and Agriculture and the International Treaty on Plant Genetic Resources for Food and Agriculture.

Preface

'Nature-based solutions' (NbS) is an umbrella term for ecosystem-based approaches. It includes several actions in harmony with natural principles to achieve development goals and addressing challenges while benefitting human livelihoods and the environment. The NbS approach encompasses actions that are carried out in a participatory manner, in collaboration with diverse stakeholders, such as local communities and policy makers, focusing on key elements such as protection, restoration or management of ecosystems and ecosystem services; sustainable management of water systems, coastal areas, arable, grassland, and forest areas; and the creation of green zones in and around urban and peri-urban areas. In this multidisciplinary topic, 18 review, research and data papers have been published covering a variety of topics including, inter alia, (a) natural and green infrastructures, (b) ecosystem-based management approaches, (c) ecosystem and biodiversity conservation, (d) ecosystem-based adaptation and climate change adaptation, and (e) nature-based recreational activities. As humanity faces two interdependent crises, i.e., the climate-biodiversity nexus, nature-based solutions are seen as the way forward in transformative change, implemented at the local, national and international level.

Panayiotis Dimitrakopoulos, Mario A. Pagnotta, Miklas Scholz, and Arshiya Noorani
Editors

sustainability

MDPI

view

ibliometric Analysis and Comprehensive Review of tormwater Treatment Wetlands: Global Research Trends and xisting Knowledge Gaps

ash Jett D. G. Reyes, Franz Kevin F. Geronimo, Heidi B. Guerra and Lee-Hyung Kim *

Civil and Environmental Engineering Department, Kongju National University, Cheonan City 31080, Republic of Korea
* Correspondence: leehyung@kongju.ac.kr

Abstract: Stormwater treatment wetlands are widely recognized as efficient and cost-effective solutions to growing stormwater problems. This study presented a new approach to evaluating the current status and trends in stormwater treatment wetlands research. The annual scientific productivity of different states was identified using a bibliometric analysis approach. The number of publications related to stormwater treatment wetlands has exhibited an increasing trend since the earliest record of publication. USA and China were among the states that had the most number of stormwater treatment wetlands-related publications and international collaborations. In terms of the population-to-publication ratio, Australia, Canada, and South Korea were found to have a higher level of scientific productivity. Analysis of frequently used keywords and terms in scientific publications revealed that the efficiency of stormwater treatment wetlands and the processes involved in the removal of nutrients and trace elements were adequately investigated; however, inquiries on the removal of organic micropollutants and emerging pollutants, such as pharmaceuticals and personal care products, microplastics, and industrial compounds, among others, are still lacking. Through the comprehensive review of related scientific works, the design, components, and primary factors affecting the performance of stormwater treatment wetlands were also identified. Future works that address the aforementioned knowledge gaps are recommended to optimize the benefits of stormwater treatment wetlands.

Keywords: bibliometrics; constructed wetland; nature-based solution; stormwater management

ation: Reyes, N.J.D.G.; Geronimo, .F.; Guerra, H.B.; Kim, L.-H. liometric Analysis and nprehensive Review of rmwater Treatment Wetlands: bal Research Trends and Existing owledge Gaps. *Sustainability* **2023**, 2332. https://doi.org/10.3390/ 5032332

demic Editors: ayiotis Dimitrakopoulos, Mario Pagnotta, Miklas Scholz and hiya Noorani

eived: 6 December 2022
ised: 20 January 2023
epted: 24 January 2023
lished: 27 January 2023

1. Introduction

Stormwater management is an essential component of land-use planning and development. Over the past decades, stormwater-related disasters have continued to aggravate due to the conversion of natural land cover into impermeable areas and other forms of land-use changes [1]. Flooding is a natural disaster that affects the largest number of people. In the year 2014, it was estimated that approximately 16 billion USD and 1500 casualties were recorded due to flooding incidents [2]. Despite being a natural phenomenon, flood risks may also be influenced by land use and land-use changes. Urban areas experience higher flood risks since urban areas are characterized by a high percentage of impervious land cover or built-up zones that facilitate increased rainfall-to-runoff conversion and greater overland flow. Moreover, deforestation and the loss of vegetative cover stimulate runoff formation in headwater locations [3].

Stormwater is also an imminent threat to the environment and human health. Stormwater runoff contains a complex mixture of pollutants from natural and anthropogenic sources. A number of studies have indicated that stormwater runoff serves as a major source of water pollution and toxicity. Heavy metals are among the most toxic compounds that can be found in stormwater. Ma et al. (2016) developed a hazard index to identify the heavy metal species present in stormwater that can pose the highest risk to human health. The results indicated that chromium (Cr), manganese (Mn), and lead (Pb) were among the m toxic heavy metals that pose adverse effects to human health despite their relatively l concentrations in stormwater [4]. Apart from heavy metals, xenobiotic organic carbo including pesticides and polycyclic aromatic hydrocarbons (PAHs), may also have de mental effects on the environment [5]. Pesticides are associated with aquatic toxicity a biodiversity loss, whereas PAHs are known to possess human health and environmen toxicity [6,7].

The characteristics and amounts of pollutants in stormwater greatly depend on t land use and potential sources of pollution within the catchment. In the study conduct by Yang and Toor (2018), non-point phosphorus pollution in urban catchments originat from stormwater runoff [8]. Lee et al. (2020) reported the presence of pathogenic bacter protists, and fungi in stormwater that can potentially trigger an increase in water-bor disease outbreaks [9]. In urban areas, high vehicular traffic and anthropogenic activiti may increase the pollutant concentrations in stormwater. Variable concentrations of to compounds (i.e., heavy metals, PAHs, pesticides, etc.) were observed by Zgheib et al. (20 in the stormwater samples collected from densely populated urban areas [10]. In a mc recent study, Pramanik et al. (2020) identified different emerging pollutants, such as p and polyfluorinated substances and microplastics, in urban runoff [11]. The stormwa from agricultural areas also contains a considerable amount of pollutants that can pot tially degrade the quality of receiving water bodies. Specifically, agrichemicals and orga materials that are incorporated in stormwater can be the primary drivers of water quali degradation. Generally, insufficient or inappropriate stormwater management schem can lead to severe economic, environmental, and human health consequences.

Constructed wetlands (CWs) are increasingly utilized as efficient and effective storm ter treatment technologies. These engineered systems were designed as nature-based cilities that utilize biological and physico-chemical mechanisms to improve water quali Additionally, CWs also provide provisioning, cultural, supporting, and other ecosyste services that benefit humans and the environment [12]. The effectiveness of CWs in treati stormwater runoff was widely reported in different scientific publications. Stefanakis (20 highlighted the importance of CWs in urban stormwater and wastewater management [1 The performance of CWs as stormwater runoff intervention facilities in the upstream gions of a watershed was investigated by Kabenge et al. (2018). It was found that vegetat microcosm wetlands effectively reduced pollutant concentrations in the stormwater up to 76% [14]. Alihan et al. (2017) recommended the use of CWs for treating urb stormwater runoff. The CWs, which treat runoff from impervious roads and parking lo effectively reduced the total runoff volume and pollutant concentrations by up to 37% a 81%, respectively [15]. Studies regarding the applicability of CWs in agricultural run treatment were also conducted by several authors. The horizontal subsurface flow C evaluated by Grinberga and Lagzdins (2017) exhibited a considerably reduced conce tration of nutrients in the stormwater runoff from an agricultural farmyard [16]. Apa from nutrients, McMaine et al. (2019) reported that constructed wetlands were capable reducing the concentration of pesticides in plant nursery runoff by more than 79% [17]. T valuable contribution of constructed wetlands as sediment traps was highlighted in t study conducted by Ockenden et al. (2014). It was estimated that small CWs can interce up to 70 tons of sediment from agricultural runoff over the period of three years, th significantly reducing the sediment loads to waterways [18].

Review papers also provided considerable information regarding treatment efficien design, and other factors that affect the performance of constructed wetlands in mana ing stormwater. The removal pathways of heavy metals in constructed wetlands we reviewed by Headley and Tanner (2006), whereas Sharma et al. (2021) summarized t constructed wetland mechanisms involved in the effective removal of heavy metals a nutrients in stormwater [19,20]. Ingrao et al. (2020) compiled different studies related the environmental and operational issues of constructed wetlands [21]. In the synthes conducted by Li et al. (2018), several scientific publications regarding the performance constructed wetlands in treating non-point source (NPS) pollution were highlighted [22]. Despite a large number of studies and review papers, there are currently no bibliometric analyses conducted to determine the current status of publication about the application of constructed wetlands in stormwater management. Therefore, this study was conducted to establish the trends of scientific publications focused on constructed wetlands used for stormwater management through a bibliometric analysis approach. The data from the extensive collection of published research works were used to determine the scientific productivity of various authors, institutions, and states in terms of the number of publications and citations. A comprehensive review was also conducted to create a detailed summary of the facility and catchment area characteristics, pollutant concentrations, the effectiveness of stormwater treatment wetlands in improving water quality, and other pertinent information contained in relevant scientific publications. Ultimately, the major research hotspots, knowledge gaps, and future research directions were also identified in this review.

2. Materials and Methods

2.1. Data Collection and Bibliometric Analysis

The Web of Science (WoS) platform is one of the most reliable and comprehensive scientific databases that is commonly used for bibliometric analysis [23]. A standard document search was conducted in the Science Citation Index Expanded (SCI-Expanded) and Emerging Sources Citation Index (ESCI) collections of the WoS platform. The terms ("constructed wetland*" OR "treatment wetland*" OR "engineered wetland*" OR "artificial wetland*") AND ("stormwater*" or "storm water*") were used to retrieve all documents related to the search term. In the analysis field, all documents containing the search terms in the title, abstract, keywords, keywords plus, and publications from the earliest records available in WoS (1 January 1990) up to 31 December 2021 were considered. The bibliographic information exported from the database includes authors, titles, sources, abstracts, author keywords, affiliations, and cited references. Initial query results returned a total of 452 documents. Among the list, 414 documents were classified as articles, 34 were review papers, seven were proceedings papers, and seven documents were classified as editorial materials, corrections, and news items. The list was further downsized to articles written in English, thus resulting in a total of 413 documents used in the analysis. The detailed process of data collection and the processes associated with the bibliometric analysis were illustrated in Figure 1.

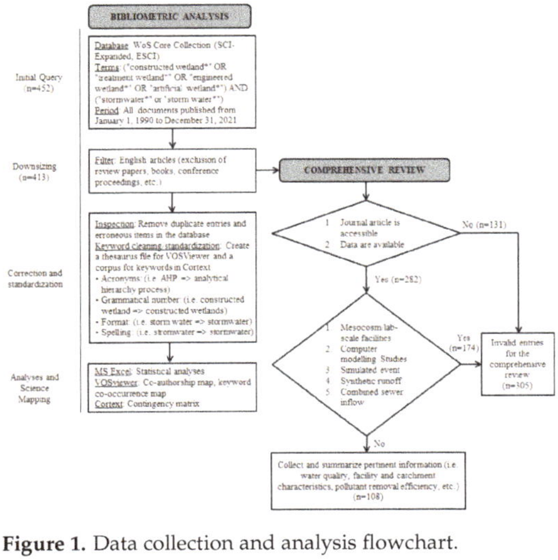

Figure 1. Data collection and analysis flowchart.

2.2. *Data Visualization and Science Mapping*

Aside from determining the trends and status of research, the bibliometric analysi approach also provides an overview of the interrelationships among the terms, authors and other pertinent information that can be obtained from large volumes of bibliographi information [24]. Science mapping tools were utilized to generate accurate visualization of bibliographic information. Co-authorship networks and keyword co-occurrence map were created using the VOSviewer software, whereas contingency matrices were generated in the Cortext platform (www.cortext.net). VOSviewer is commonly used as a tool fo generating maps or networks from an extensive collection of bibliographic information from different scientific databases. This software can be used to create visualization of co-authorships, keyword co-occurrences, and co-citations, among others, in order to analyze the relationships or interrelationships among different variables. Cortext is an online platform that provides tools for textual analysis, text mining, and science mapping It can also be used to generate diagrams that effectively show the frequency, correlation and evolution of terms used in different scientific publications.

Prior to mapping, bibliographic entries were inspected to remove duplicates and erroneous items. Manual inspection and correction were also conducted to standardize th terms used in the analysis. The network map generated through VOSviewer provides infor mation regarding the frequency, relatedness, and co-occurrence of keywords. Larger circle and labels indicate that the terms were used more frequently in different publications. Th distance between the terms and the curved lines that connect the keywords indicate thei relatedness based on co-occurrences. Terms are also grouped into clusters, as represented by the color groups, to signify a relatively higher degree of relations. The Cortext platform enables users to generate contingency matrices that show interrelationships among th different variables contained in a bibliographic dataset. A contingency matrix expresses th degree of correlations and anti-correlations between two selected parameters. Cells that are highlighted in red describe a positive relationship or a high degree of co-occurrence be tween the two parameters. On the other hand, blue-colored cells indicate anti-correlation or lower co-occurrences as compared to the expected number of co-occurrences. The whit cells represent items that do not show any correlation. The Chi^2 score expresses the rati between the square of the difference between the number of co-occurrences between A(and B(j) and its expected number of observations and the expected number of observa tions [25,26]. The data inputs and methods of analysis employed in the science mapping software were summarized in Table 1.

Table 1. Data inputs and methods of analyses used in VOSviewer and Cortext.

Software	Parameter	Inputs/Method of Analysis
VOSviewer (Network map)	Bibliographic database file	WoS plaintext file
	Type of analysis	Co-occurrence
	Unit of analysis	Author keywords
	Counting method	Full counting (terms have the same weight)
	Minimum number of co-occurrences	5
Cortext (Contingency matrix)	Bibliographic database file	WOS plaintext file
	Field values	Author keywords—Country/State Author keywords—Year of publication
	Contingency analysis measure	Chi^2 score
	Number of nodes	10 (Default software value)

2.3. *Comprehensive Review*

A comprehensive review was performed to summarize important information pro vided in relevant scientific publications. From the downsized list of 413 research articles only 282 documents were accessible or contained pertinent data. In order to limit the review to real-world application scenarios, studies that utilize computer models or simulate runoff events were excluded from the review. Inquiries that investigated combined sewer inflows or utilized mesocosms, lab-scale facilities, or synthetic stormwater runoff were also excluded to reflect the actual performance of stormwater treatment wetlands in reducing runoff pollutant concentrations. A total of 108 scientific publications were considered for the comprehensive review. Essential information, such as the size and characteristics of study areas, the types and components (i.e., filter media and plants) of stormwater treatment wetlands, and the investigated water quality parameters, were compiled from each of the valid documents.

3. Results and Discussion

3.1. Bibliometric Analysis and Science Mapping

3.1.1. Trend of Scientific Productivity and Characteristics of Published Literature

The annual scientific productivity or number of published documents related to stormwater treatment wetlands was exhibited in Figure 2. Based on the records of the WoS database, papers that focused on constructed wetlands used for stormwater management were first published in the year 2012. The annual scientific productivity exhibited an increasing trend until the year 2017, when the maximum number of articles was reached (57 articles). Despite the relatively lower number of publications from the year 2018 to 2021, the number of documents published within this period was 10% to 55% higher than the number of publications in the year 2012. In terms of the number of citations, a continuously increasing trend was observed from 2012 to 2021. The initial number of citations in the year 2012 only amounted to 11, whereas the maximum number of citations, amounting to 1268, was observed in 2021. The number of citations proportionally increased alongside the surge in the number of publications since the network of researchers specializing in constructed wetlands and stormwater management also expanded over the years. Generally, the observed growth in the number of scientific publications and citations indicated that constructed wetlands have become an important component of stormwater management strategies.

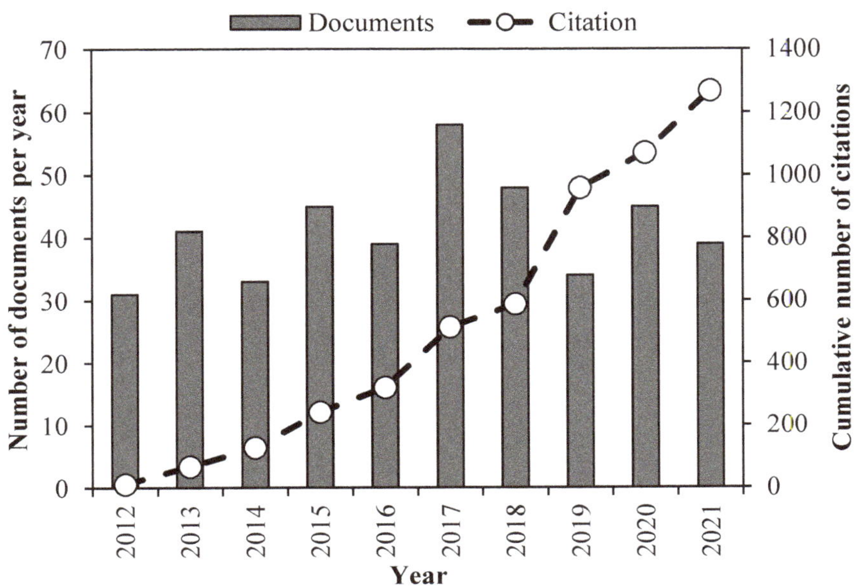

Figure 2. Annual number of published documents and cumulative number of citations.

The contribution of various states to the total number of publications was illustrat in Figure 3. Researchers from 51 different states contributed to the total number of index publications. Authors from the USA had the largest contribution to scientific publicati with a total of 165 articles and an average annual scientific productivity of 17 articl The second-most productive state in terms of the number of publications was Austral with a total of 71 documents. China (62 articles) and South Korea (31 articles) were t leading Asian states in stormwater treatment wetlands research, whereas France was fou to be the most active European state in terms of the number of published papers in th particular field of research (31 articles) in this area of study. The scientific productivity different states was also evaluated in terms of the population-to-publication ratio. Th ratio is indicative of the state's contribution to a specific research field in relation to total population. Australia, Canada, and South Korea were found to have the highest lev of scientific productivity, with corresponding population-to-publication ratios of 36,00(1,530,000:1, and 1,669,000:1. Despite having the largest number of scientific publications the Asian Region, China had the lowest population-to-publication ratio, with approximat 23,000,000 people per article in the WoS database.

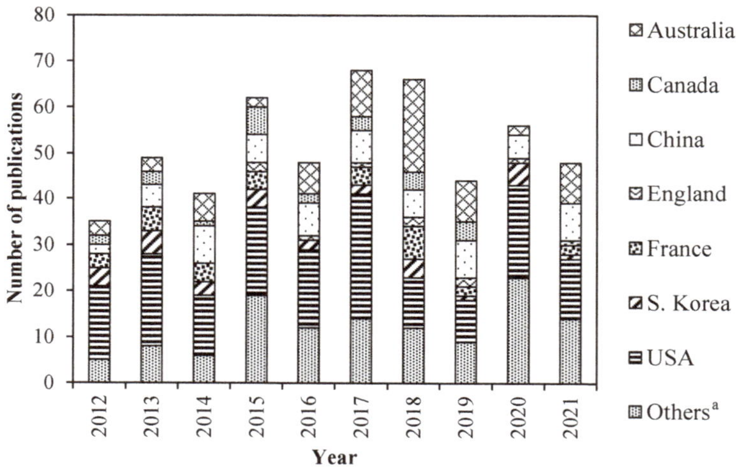

[a] States with less than ten publications

Figure 3. Annual scientific productivity of different states.

3.1.2. Journal Publications and Related Subject Areas

The research on constructed wetlands used for stormwater management covers a wi range of subject areas and specializations. The major subject areas related to stormwa treatment wetlands research was listed in Table 2. A total of 32 subject areas were identifi among the list of publications. This indicated that studies on constructed wetlands ar stormwater management are multi-faceted and established through complex collabo tions among professionals and researchers from various fields of expertise. Among t 32 subject areas identified, approximately 81% of the published literature was direct relevant to "Environmental Sciences Ecology." Constructed wetlands are widely-used te nologies for environmental preservation and restoration, and thus, most of the analyz published literature is related to "Environmental Sciences Ecology." Publications that f under "Engineering" (49%) and "Water Resources" (35%) were also found to be abunda Despite being considered nature-based facilities, constructed wetlands are built syster that integrate advanced engineering techniques and natural processes to achieve wa quality goals, mitigate flooding conditions, and support water resource conservation effo

Generally, it was observed that individual documents can be classified into multiple subject areas, implying that most scientific publications are multi-disciplinary in nature.

Table 2. Major subject areas relevant to constructed wetlands and stormwater management research.

Subject Area	Number of Documents	% of Total
Environmental Sciences Ecology	334	81%
Engineering	204	49%
Water Resources	146	35%
Geology	19	5%
Marine Freshwater Biology	19	5%
Science Technology Other Topics	16	4%
Meteorology Atmospheric Sciences	11	3%
Others [a]	68	16%

[a] Subject areas with less than 10 documents.

The journals having the most publications related to stormwater treatment wetlands were summarized in Table 3. The 413 articles retrieved in the query were published in 116 journals. The Ecological Engineering journal contains the largest number of publications, amounting to 83, followed by Science of the Total Environment with 26 and Water with 22 articles. Journal impact factors are commonly used to evaluate the quality of journals or the research articles published in a specific journal. Despite the opposition from several researchers and institutions, this metric is still widely used for journal rankings due to the lack of alternatives and ease of use [27,28]. Four of the most productive journals in stormwater treatment wetlands research have impact factors greater than three. Moreover, three of the journals on the list belong to Q1 (top 25%) in the ranking of Environmental Engineering journals.

Table 3. List of most productive journals in terms of the number of publications.

Journal	Impact Factor (as of 2020)	Number of Documents	% of Total
Ecological Engineering [a]	4.035	83	20%
Science of the Total Environment [a]	7.963	26	6%
Water	3.103	22	5%
Water Science and Technology	1.915	21	5%
Desalination and Water Treatment	1.254	17	4%
Water Research [a]	11.24	14	3%
Journal of Environmental Engineering	1.746	11	3%
Others [b]	-	219	53%

[a] Q1 journals in the field of Environmental Engineering. [b] Journals with less than 10 publications.

3.1.3. Frequently Cited Research Works on Constructed Wetlands

Four out of the five most-cited articles presented results on the effectiveness of constructed wetlands in treating nutrients and heavy metals in stormwater. As listed in Table 4, the study conducted by Xu et al. (2017) received the highest number of citations. This study explored the mechanisms involved in the removal of nitrogen and phosphorus in stormwater. Specifically, the nutrient uptake capabilities of two wetland macrophytes (i.e., *Iris pseudacorus* and *Thalia dealbata*) were evaluated to determine their contribution to the nutrient removal process. A systematic plant harvesting strategy was also recommended to optimize the performance of the constructed wetland [29]. The second-highest number of citations was noted in the paper by Winston et al. (2013) about the conversion of stormwater ponds into floating treatment wetlands. While there were no significant statistical differences in pollutant removal observed after retrofitting the ponds, results suggested that the addition of plants to the system contributed to beneficial treatment mechanisms [30]. The article published by White and Cousins (2013) received a total of 80 citations within a period of eight years. This inquiry focused on the evaluation of the season-long nutrient removal efficiency of a floating treatment wetland. The nutrient assimilation rates of two macrophytes (i.e., *Canna flaccida* and *Juncus effuses*) and the changes in physico-chemical characteristics of the wetland effluent were also documented in this study [31]. The paper published by Borne et al. (2013) and Payne et al. (2014) were also among the most cited literature related to constructed wetlands and stormwater. Borne et al. (2013) explored the heavy metal treatment performance of floating treatment wetlands, whereas Payne (2014) highlighted the fate and dominant processes involved in the removal of nitrogen from stormwater [32,33].

Table 4. List of highly cited publications.

Title	Author/s	Journal (Year)	Number of Citations
Improving Urban Stormwater Runoff Quality by Nutrient Removal through Floating Treatment Wetlands and Vegetation Harvest [29]	Xu, Bing, Xue Wang, Jia Liu, Jiaqiang Wu, Yongjun Zhao, and Weixing Cao	Scientific Reports (2017)	101
Evaluation of floating treatment wetlands as retrofits to existing stormwater retention ponds [30]	Winston, R. J., Hunt, W. F., Kennedy, S. G., Merriman, L. S., Chandler, J., & Brown, D.	Ecological Engineering (2013)	97
Floating treatment wetland aided remediation of nitrogen and phosphorus from simulated stormwater runoff [31]	White, S. A., & Cousins, M. M.	Ecological Engineering (2013)	80
Floating treatment wetland retrofit to improve stormwater pond performance for suspended solids, copper and zinc [32]	Borne, K. E., Fassman, E. A., & Tanner, C. C.	Ecological Engineering (2013)	77
Temporary Storage or Permanent Removal? The Division of Nitrogen between Biotic Assimilation and Denitrification in Stormwater Biofiltration Systems [33]	Payne, G., Fletcher, T., Russel, D., Grace, M., Cavagnaro, T., Evrard, V., Deletic, A., Hatt, B., & Cook, P.	PloS ONE (2014)	67

3.1.4. Co-Occurrence of Keywords and International Research Collaborations

Keywords are one of the most essential pieces of bibliographic information since they represent the important contents of a scientific publication [34]. Overall, 1120 unique keywords were identified from the 413 articles considered in this study. Keywords that were used at least two times in different documents only amounted to 268. Increasing the minimum number of keyword occurrences to three further limited the list to 127 unique terms. The considerable decrease in the number of unique keywords as the minimum number of occurrences increased implied that most studies were focused on a specific topic. In order to extract the terms most relevant to stormwater treatment wetlands, the minimum number of occurrences of a unique keyword should be increased. The network map shown in Figure 4a represents the most frequently used terms in publications related to stormwater treatment wetlands. The keywords exhibited in the figure were derived by setting the minimum number of occurrences to five. Among the initial 1120 unique keywords, only 67 terms met the threshold.

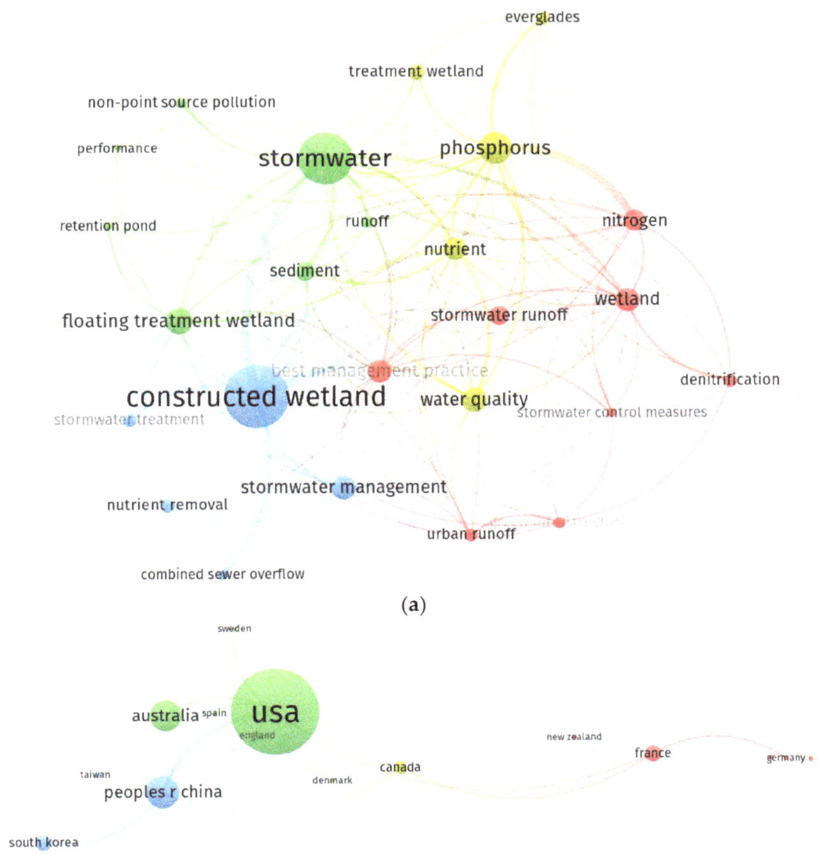

(a)

(b)

Figure 4. (**a**) Network map of keyword co-occurrence; (**b**) collaboration network among different states.

Excluding the terms used in the query, the most frequently used keywords in related documents were phosphorus (n = 37), water quality (n = 26), best management practice (n = 24), nutrient (n = 23), and nitrogen (n = 22). The frequently used keywords reflected the primary role of constructed wetlands in water quality improvement. Specifically, the ability of constructed wetlands to treat nutrients (i.e., nitrogen and phosphorus) incorporated in stormwater runoff was extensively studied. The application of constructed wetlands as a valuable stormwater best management practice can also be deduced from the map. In terms of co-occurrence, constructed wetlands had high relations with best management practices and stormwater management, whereas water quality exhibited considerable relations with stormwater control measures and nutrients. Phosphorus and nitrogen also had strong links, indicating that the two nutrient compounds are commonly studied jointly in scientific publications. There were seven clusters identified based on the relatedness of terms. The clusters represent the main topics associated with the research on stormwater treatment wetlands. Moreover, these groups also suggest topics that have been extensively investigated by past inquiries or publications.

Aside from the co-occurrence of keywords, the status of international collaboration can also be visualized through a network map. As illustrated in Figure 4b, the USA had the highest number of networks or co-authorships with other states. Aside from being the top publishing state, the USA established international collaboration networks that resulted in high scientific productivity. The highest link strength was observed between the USA a China, indicating that the two states had the most co-authored publications. The numb of co-authored papers among European states, including France and Germany, was al relatively high; however, the collaboration among European and Asian states was found be limited.

3.1.5. Shifts in Research Interest and State-Specific Research Trends

Research trends change over time due to the changes in the policy of a specific state, t degree of saturation or number of current studies for a certain topic, and the advancemer in analysis procedures that can significantly affect data acquisition, among others. It is al important to note the shifts in research interests to determine the existing knowledge ga or potential research directions. The contingency matrix exhibited in Figure 5a shows the nual trend of keyword use. In the early year of research on stormwater treatment wetlan (2012), "stormwater runoff" and "phosphorus" were the most dominant keywords used scientific publications. The terms "stormwater" and "floating treatment wetlands" emerg as the most commonly used keywords in the succeeding years, alongside the growth in t number of publications from different states. "Constructed wetland" and "water quali became the most relevant terms in the years 2018 and 2019. Except for the year 2016, was observed that there was a shift from using the term "floating treatment wetland" to more general term, "constructed wetland." This also highlighted the developments in t design of engineered systems used for stormwater management. Aside from free-wat surface wetlands, other variants of engineered wetlands (i.e., subsurface flow construct wetlands and hybrid constructed wetlands) became available, thereby resulting in chang in the frequency of keywords used in publications. More recent studies published in t years 2020 and 2021 utilized the terms "nitrogen" and "stormwater management" mo frequently. Generally, shifts or changes in the pattern of keyword use created a general id of the current status of research in a particular subject area.

State-specific use of terms also provides significant information that can be useful the development of policies and environmental management strategies. The contingen matrix of the most relevant keywords used by the researchers from the top publishi states is shown in Figure 5b. The term "phosphorus" had the highest correlation wi the USA, implying that stormwater treatment wetlands in the USA were primarily us for treating phosphorus in stormwater. "Stormwater" and "stormwater runoff" were t dominant terms used in the publications from South Korea, China, Spain, and Austral This suggested that constructed wetlands were extensively used as a tool for stormwat management in the aforementioned states. Publications from Canada and France ha increased usage of the term "sediment", whereas Germany and India frequently us the keyword "constructed wetland." The term "nutrient" was highly associated wi publications from the United Kingdom (UK), indicating that the function of stormwat treatment wetlands as nutrient sinks was of particular interest in the state.

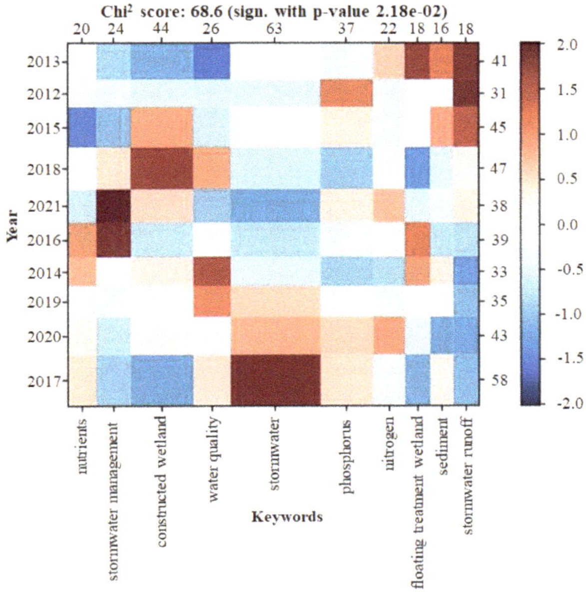

(**a**)

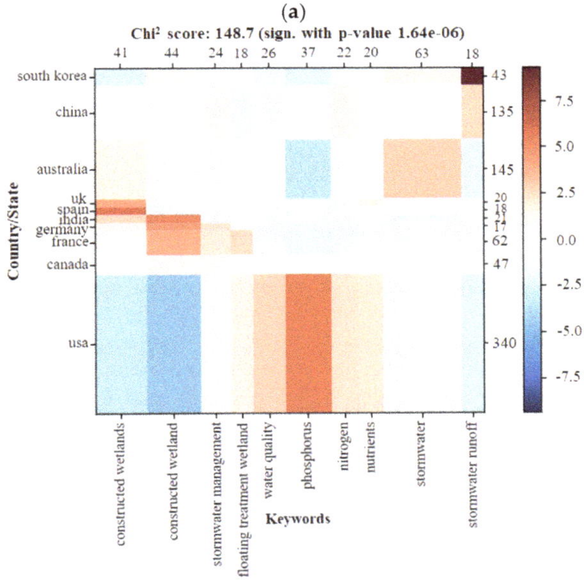

(**b**)

Figure 5. Contingency matrix of (**a**) keywords and year of publication and (**b**) keywords and state.

3.2. Data Synthesis and Comprehensive Review

3.2.1. Land Use Types and Catchment Area Characteristics

Land use and catchment area characteristics are among the primary factors that affect the quality of stormwater runoff. The extent of anthropogenic activities and the intensity of development may also exhibit direct and indirect relationships with the distribution of pollutants within the catchment area. The list of typical land use types frequently investigated in different studies is summarized in Table 5. Most scientific inquiries focused on the application of stormwater wetlands to treating runoff from urban catchments. Cor structed wetlands emerged as cost-effective and socially acceptable stormwater treatmer technologies due to their multiple benefits fit for an urban setting [35]. Wetlands designe for agricultural pollution mitigation were also extensively investigated. Treatment we lands are commonly used as low-energy and low-cost alternatives for abating pollute agricultural runoff. Specific wetland components, such as plants and microorganisms, als contribute to efficient nutrient cycling to prevent the excessive deposition of pollutants i natural waterways [36]. The review of published literature also suggested that wetland are extensively utilized to treat stormwater runoff from residential areas, parking lot highways, and mixed land use catchments.

Table 5. Drainage areas (in ha) of dominant land use types in reviewed scientific publications.

Land Use Type	Agricultural	Highway	Mixed	Parking lot	Residential	Urban	Others [a]
Frequency, n	41	10	25	7	12	54	5
Minimum	0.81	0.13	3.60	2.31	2.00	0.04	0.09
Maximum	86000	13.07	3139	2.37	572	2060	2.30
Median	42.70	1.70	320	2.37	5.40	95	0.45
Average	7876.51	3.01	781.30	2.35	76.51	271.86	0.82
Standard Deviation	24704.92	4.14	1059.91	0.03	176.99	465.53	0.90

[a] Land use types with a frequency of less than five (i.e., suburban, grassland, municipal, etc.).

The size of catchment areas investigated in previous studies varied greatly. As ex hibited in Table 5, treatment wetlands were applied to site-specific or catchment-scal treatment of stormwater runoff. Parking lots, highways, and residential areas const tuted relatively small catchment areas, with mean values ranging from 3.01 ha, 2.35 ha, an 76.61 ha, respectively. One major factor limiting the application of CWs is spatial constraint however, recent developments have allowed the installation of CWs despite the limite space availability. The concept of "pocket wetlands" can be applied to small drainag basins to provide additional stormwater treatment or achieve water quality goals. The siz of pocket wetlands is more restricted as compared to their catchment-scale counterpart and thus, these systems are often used for stormwater runoff polishing [37,38]. Among the identified land uses where stormwater treatment wetlands were applied, the highes drainage area was recorded in an agricultural catchment (86,000 ha) located in a sectio of the Everglades agricultural area, South Florida, United States. With an approximat treatment area exceeding 27,000 ha, the Everglades Stormwater Treatment Areas (STAs) ar considered the world's largest and most complex constructed wetlands. The constructio of STAs was implemented through the Everglades Forever Act, which aims to reduce th TP loads discharged from predominantly agricultural areas upstream of the Everglade Protection Area [39–41].

3.2.2. Types and Sizes of Stormwater Treatment Wetlands

The choice of design and components of stormwater treatment wetlands can be in fluenced by the climate, availability of materials, influent pollutant concentrations anc water quality targets, public perception, and existing environmental regulations, among others [42–45]. Based on the reviewed literature, free water surfaces (FWS) (n = 40) anc floating treatment wetlands (FTW) (n = 26) were commonly employed for stormwate treatment. FWS wetlands are commonly used as stormwater management facilities du to their high volume capacities. Apart from the intrinsic capability of FWS wetland to attenuate flooding, these systems can also provide efficient treatment of stormwate due to the prolonged retention time in the system [46,47]. FTW systems were also ex tensively studied due to their potential for increasing the ecosystem benefits of existing stormwater infrastructure. Unlike other types of treatment wetlands with considerabl space requirements for field-scale applications, FTW systems can be incorporated int existing stormwater ponds to improve their general function. In the study conducted by

Borne et al. (2013), retrofitted stormwater ponds with FTWs showed higher treatment efficiencies as compared to conventional detention ponds [32]. Winston et al. (2013) and Tirpak et al. (2022) also reported improvements in the pollutant removal performance of stormwater ponds retrofitted with FTWs; however, the treatment contributions provided by FTWs were found to be limited or highly influenced by the pond design [30,48]. Studies regarding other wetland designs, including hybrid, horizontal subsurface flow (HSSF), and vertical subsurface flow (VSSF) stormwater treatment wetlands, were relatively scarce, with reporting frequencies of five, four, and seven times, respectively.

The size of stormwater treatment wetlands from the compiled data ranged from 7 m^2 to more than 270 km^2. Small facilities were specifically designed to treat runoff from site-specific sources, whereas treatment wetlands with relatively large surface areas were designated for catchment-scale runoff management. The surface area-to-catchment area (SA/CA) ratios of stormwater treatment wetlands applied in various land use types were illustrated in Figure 6. The typical SA/CA ratios of FWS and hybrid treatment wetlands were approximately 0.5% to 3%, whereas HSSF treatment wetlands exhibited higher SA/CA ratios ranging from 1.65% to 12.78%. It can be noted that FTWs had low SA/CA ratios (0.07% to 0.38%), since these are compact and modular facilities that are used as additional features to a stormwater detention pond. Nature-based facilities, including stormwater treatment wetlands, usually have small surface areas relative to the drainage area. In the study conducted by Hong et al. (2016), low-impact development facilities with SA/CA ratios of 1% to 5% were capable of reducing runoff volume by more than 40% [49]. Choi et al. (2018) also indicated green stormwater infrastructures with SA/CA ratios of 1% to 2% can provide adequate pollutant removal from stormwater [50].

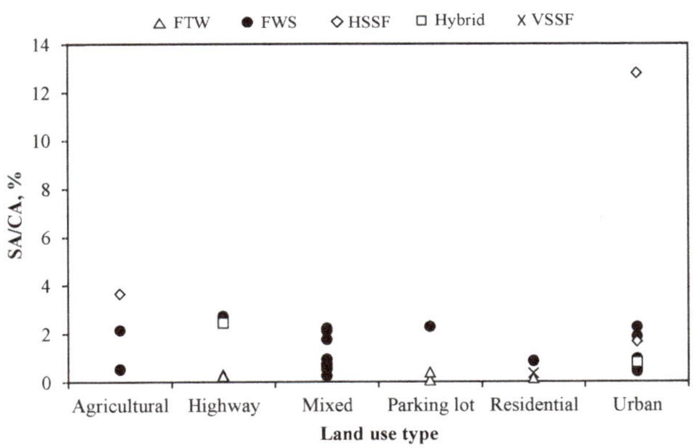

Figure 6. SA/CA ratio of stormwater treatment wetlands in different land uses.

3.2.3. Choice of Filter Media and Substrates

Treatment wetlands that contain subsurface structures are equipped with filter media that help enhance the facilities' filtration and pollutant retention functions. In the case of FTWs, the platforms or modules where the treatment units are mounted also contain substrates that support the growth of plants. The summary of the usage rates of various filter media and substrates reported in relevant scientific publications is shown in Figure 7. Sand and gravel were the most common filter media used for HSSF, hybrid, and VSSF stormwater treatment wetlands. The high usage rate of sand (18% to 36%) and gravel (27% to 50%) can be attributed to their relative abundance as construction materials and effective pollutant removal performance [51,52]. Marine-grade foam (48%) and recycled plastic fibers (31%) were the most common compositions of FTWs. These materials provide buoyancy and bond the platform carrying the substrates and plants [19,53]. Other types of filter media, such as laterite, rubber mulch, pebbles, woodchips, volcanic rock, and bioceramic, were also utilized as filter media for stormwater treatment wetlands. The treatment wetland presented in the study conducted by Adyel et al. (2016) utilized later aggregates to enhance phosphorus removal in stormwater. It was found that laterite act as an important phosphorus sink due to the ligand exchange reaction that prompt effective phosphorus adsorption [54]. Packed rubber mulch and pebbles were primar used to enhance the physico-chemical and biological processes in the filter bed. Han a Tao (2014) attributed the enhanced industrial runoff treatment to the effective biosorpti and adsorption of pollutants in the packed rubber mulch and pebbles [55]. The trade of using woodchip filter materials for treating stormwater was reported by Niu et (2018) [56]. Zhang et al. (2020) cited high adsorption capacity, porous properties, a resistance to degradation as the major advantages of using volcanic rock and bioceram as filter media [57]. Woodchips serve as additional carbon sources for enhancing t denitrification process in a filtration system but may also lead to elevated chemical oxyg demand (COD) concentrations in the effluent. Based on the information presented in t reviewed articles, the choice of filter media or substrates was found to be influenced by t availability and properties of the materials.

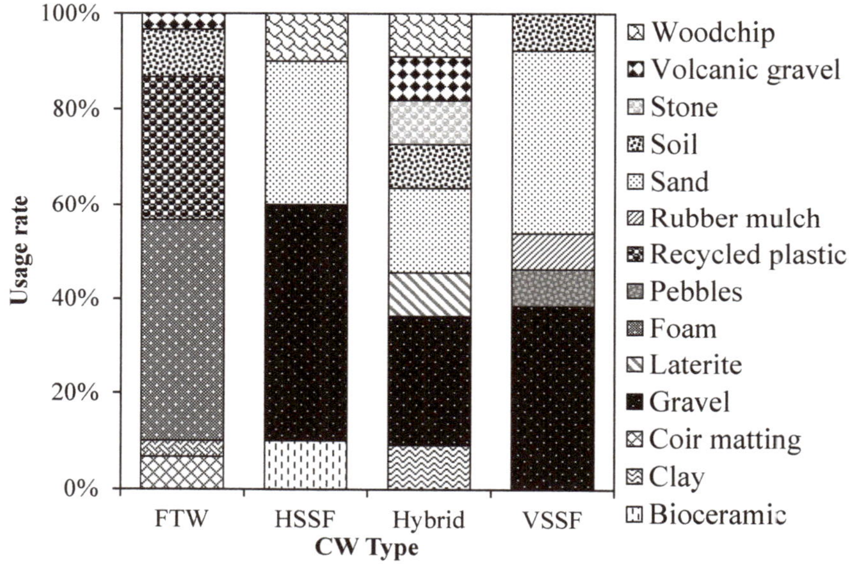

Figure 7. Usage rate of filter media and substrates used in stormwater treatment wetlands.

3.2.4. Plants in Stormwater Treatment Wetlands

Plants are considered essential components of stormwater treatment wetlands. T vegetative components of wetlands directly contribute to the efficient management nutrients, the removal of toxic pollutants through plant uptake and assimilation, and t carbon storage functions of the system [58–60]. Plants can also enhance the treatme properties of wetlands by prompting efficient sedimentation of particulates in the wetla bed, providing favorable conditions for the growth of microorganisms, and preventi internal algal blooms through shading. A total of 94 genera and 160 species of wetla plants were identified from the reviewed articles (see Table S1). As exhibited in Figure the five most common plant genera used in stormwater treatment wetlands include *Typ*, *Juncus*, *Carex*, *Phragmites*, and *Schoenoplectus*, with reporting frequencies of 45, 38, 29, and 16 times, respectively.

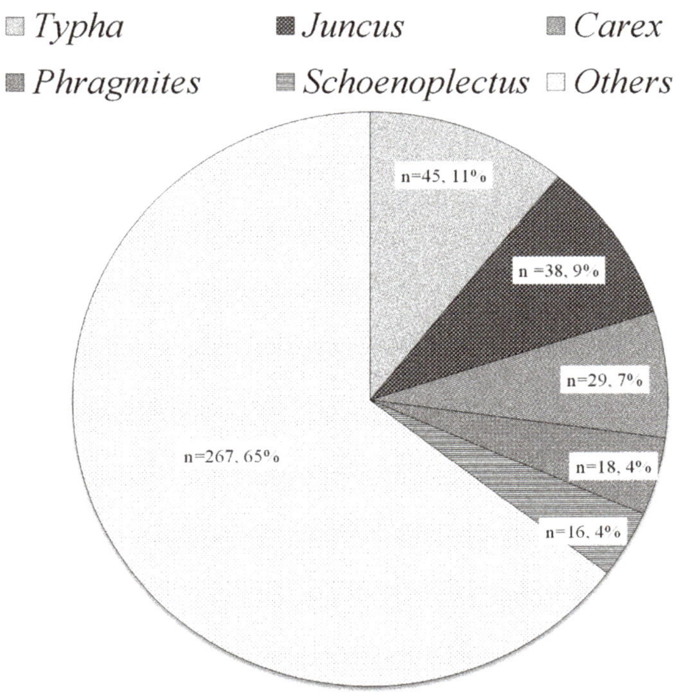

Figure 8. Distribution of the most commonly used plants in stormwater treatment wetlands.

Typha is widely used in treatment wetlands due to its high pollutant uptake capabilities and extremely resilient properties. *Typha* is known to be an important nutrient sink, making it suitable for managing eutrophication [61]. Chandra and Yadav (2011) emphasized that *Typha* can be used for heavy metal phytoremediation since these plants can accumulate Cd, Cr, Cu, Fe, Ni, and Pb in their roots [62]. These plants also have a high phytotoxic tolerance, making them suitable components of wetlands receiving highly contaminated runoff [63]. *Juncus* is also a typical wetland plant known for its high ecosystem value. It was previously reported that different *Juncus* species have promising phytoremediation potential and high biomass yield [64–66]. Plants of the genus *Carex* are commonly applied on FTWs due to their adaptability, ability to uptake heavy metals, and ease of harvesting [20,67]. *Phragmites* are wetland plants known for being potent hyperaccumulators of trace elements [68–70]. *Phragmites* species are often planted in treatment wetlands to increase pollutant remediation; however, these plants are also considered ubiquitous, highly invasive, and phytotoxic due to the production of allelochemicals detrimental to the growth of other plant species [71]. Wetland plants belonging to the genus *Schoenoplectus* are leafless species with large underwater surface areas. These plants are commonly utilized in treatment wetlands due to their remarkable tolerance to physicochemical changes in water quality (i.e., pH, temperature, and salinity) and high nutrient-regulating properties [72,73].

3.2.5. Runoff Water Quality and Treatment Performance of Stormwater Wetlands

A complex mixture of chemicals and compounds can be incorporated into stormwater as a factor of land use, geomorphological characteristics, and the patterns of pollutant deposition in the catchment area. Stormwater treatment wetlands are usually employed in different catchments due to their versatility in treating a wide range of pollutant compounds with substantially varying concentrations. A total of 91 unique water quality parameters and constituents were identified from the collection of reviewed articles concerning stormwater treatment wetlands. As shown in Figure 9, total phosphorus (TP) an total nitrogen (TN) were the most commonly investigated runoff constituents. Variou nitrogen forms, including total Kjeldahl nitrogen (TKN) and nitrate-nitrogen (NO_3-N), wer also typically included in inquiries related to stormwater treatment wetlands. Nutrients ar primary stormwater pollutants that can trigger eutrophication and algal blooms. Treatmer wetlands are associated with the treatment of nutrients in stormwater runoff since these sy tems perform mechanisms that effectively remove different nitrogen and phosphorus form in stormwater. Wetlands remove nitrogen from stormwater through the combination c physico-chemical and biological processes (i.e., adsorption, plant uptake, ammonificatio etc.) involved in the transformation of nitrogen compounds. The removal of phosphorus i CWs can also be influenced by biological processes, such as biodegradation and plant up take, but sedimentation and soil retention are considered the main pathways for long-terr phosphorus removal in treatment wetlands [74]. Suspended solids (SS) concentration i also a well-represented water quality parameter in the reviewed articles since stormwate is a major transport route of sediments and particulates to waterways. The high reportin frequencies of SS in published literature can also be attributed to the relative simplicit of experimental methods for quantifying sediments or particulates in stormwater [75,76 Heavy metals, such as calcium (Ca), copper (Cu), and zinc (Zn), and dissolved organi carbon (DOC) were also widely investigated, with reporting frequencies of more than 10 The presence of emerging stormwater pollutants, including pesticides, polycyclic aromati hydrocarbons, and microplastics, was also documented in some studies, but the reporting frequencies of these compounds were relatively low [77–79].

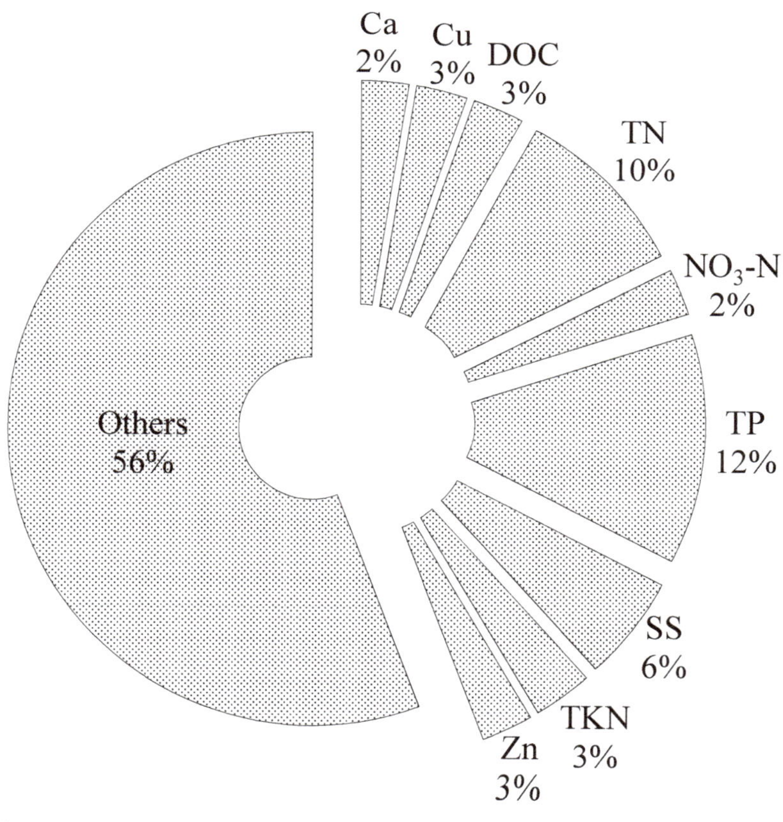

Figure 9. Typical water quality parameters investigated in stormwater treatment wetlands research

The concentrations of the three most frequently reported water quality parameters were summarized in Figure 10. The highest concentration of TP, amounting to 700 mg/L, was reported in the runoff from an agricultural catchment, whereas maximum TN (54 mg/L) and SS (1953 mg/L) concentrations were observed from a predominantly urban catchment. Phosphorus loads may originate from various natural or anthropogenic-related processes; however, agricultural activities are known as the principal sources of excessive phosphorus loads in water bodies [80]. In urban areas, SS and TN concentrations are mostly influenced by the disturbance of natural features and the accumulation of pollutants on impermeable surfaces. Elevated SS concentrations in urban areas usually originate from construction activities, road and highway maintenance, traffic-related activities, and wet and dry atmospheric deposition [81,82]. High nitrogen concentrations in urban runoff were also reported, considering its diverse sources. The dominant sources of nitrogen in urban runoff include fertilizers applied to lawns, wastewater, atmospheric deposition, and combustion [83]. Among the different land use types identified in the review, parking lot and highway runoff had the lowest mean TP (0.17 mg/L), TN (1.64 mg/L), and SS (42 mg/L) concentrations. Highways and parking lots are usually maintained through sweeping or the removal of accumulated detritus. Since various pollutants are bound to particles, reducing the sediment build-up also resulted in a significant decrease in stormwater pollutant concentrations [84,85].

Stormwater treatment wetlands are capable of reducing pollutant concentrations to a certain extent. As seen in Figure 9, effluent pollutant concentrations were considerably lower than the observed concentrations in the inflow. The treatment wetland investigated by Byeon and Nam (2020) exhibited a pollutant removal performance of up to 99%, indicating that the facility is fit for mitigating the negative impacts of NPS pollution [86]. Li et al. (2020) also reported removal efficiencies exceeding 90%, citing the contribution of effective microorganisms in improving the overall performance of wetlands in treating nutrients in runoff [87]. The stormwater treatment wetlands monitored by Grinberga et al. (2021) showed relatively lower mean pollutant removal efficiencies, ranging from 17% to 80%. It was highlighted that the poor pollutant removal performance of the system can be attributed to the low influent pollutant concentrations caused by stormwater dilution [88]. Some studies also reported negative removal efficiencies or higher outflow concentrations after receiving treatment from stormwater treatment wetlands. Walaszek et al. (2018) recorded a negative removal of PAHs and heavy metals in a stormwater treatment wetland due to the resuspension of solids containing particulate heavy metal fractions in the system [78]. Howitt et al. (2014) emphasized the effect of external factors, including wind mixing, fine sediment resuspension, and external pollutant loading, on the performance of treatment wetlands [89]. The information compiled from the reviewed scientific literature clearly indicated that stormwater runoff management is essential in mitigating environmental degradation. Studies also suggested that treatment wetlands are effective green stormwater infrastructures, but the design of facilities should also be adapted to the stormwater characteristics and catchment area conditions.

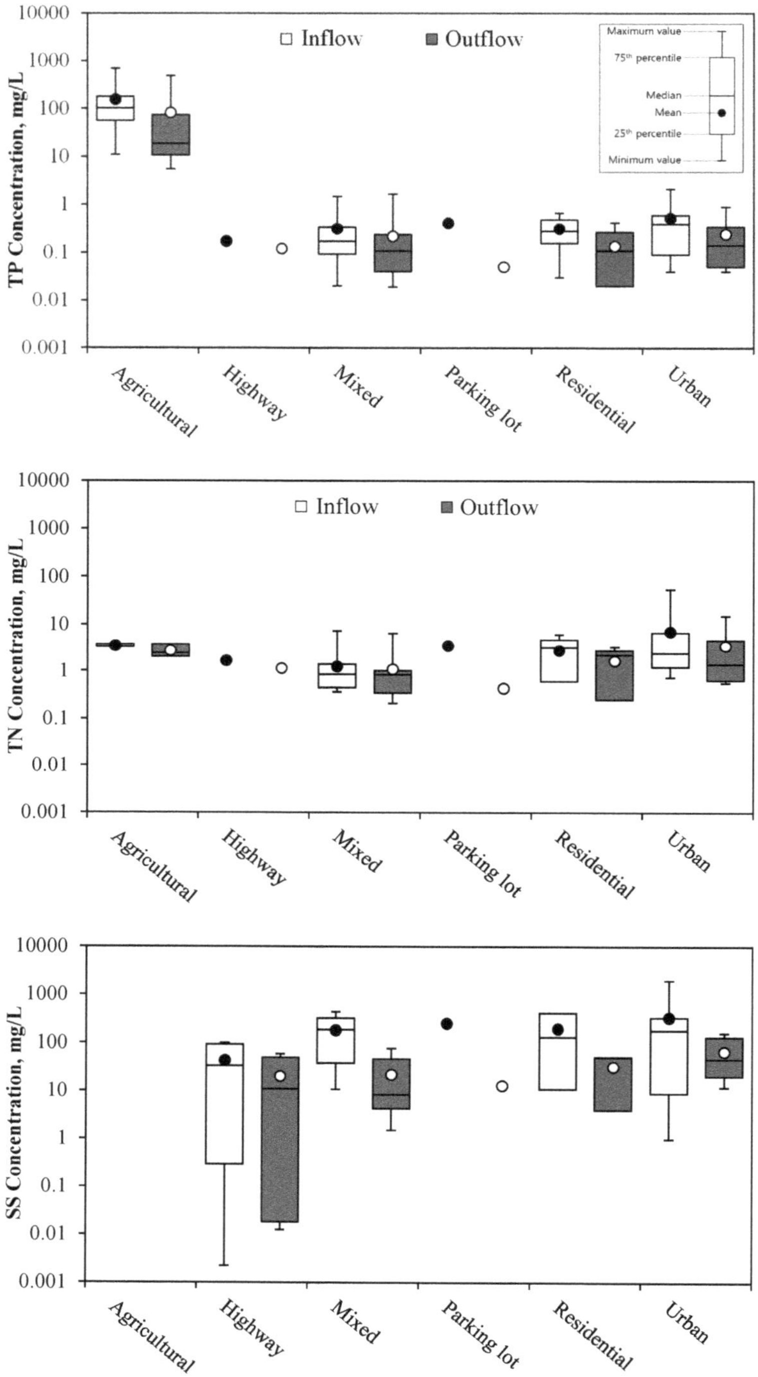

Figure 10. Synopsis of pollutant concentrations in the runoff and stormwater treatme wetlands effluent.

3.3. Knowledge Gaps and Future Research Directions

Stormwater contains various types of pollutants derived from natural and anthropogenic sources. The typical pollutants in stormwater include sediments, nutrients (nitrogen and phosphorus), organics, and heavy metals; however, the occurrence of micropollutants and other emerging pollutants in stormwater was also identified in recent inquiries. Piñon-Colin et al. (2020) detected the presence of plastic particles smaller than 5 mm, known as microplastics, in stormwater. Furthermore, it was found that stormwater is the primary mode of microplastic deposition in water bodies [90]. In the year-long survey conducted by Wicke et al. (2021), different organic micropollutants, such as plasticizers, flame retardants, and PAHs, were identified in stormwater. It was estimated that the stormwater collected from the catchment composed of different land uses contained 24 µg/L of an organic micropollutant mixture [91]. Recent advancements in analysis and instrumentation methods have also enabled the detection of compounds present in minute concentrations. The stormwater samples analyzed by Tran et al. (2019) were contaminated by compounds used in the synthesis of pharmaceuticals and personal care products [92]. Similarly, Fairbairn et al. (2018) reported the presence of 123 different compounds, classified as emerging pollutants, in stormwater. The analysis also revealed that emerging pollutant loads from stormwater may exceed the treated wastewater effluent loads to receiving water bodies [93]. The detection of emerging pollutants and other organic micropollutants in stormwater raised major environmental concerns; however, studies that explore the feasibility of using constructed wetlands as potential treatment systems for the new suites of pollutants are still limited. Based on the analysis of terms, most studies related to stormwater treatment wetlands only focused on the treatment of trace elements (i.e., heavy metals) and nutrients in stormwater. It is, therefore, necessary to conduct inquiries on the applicability of stormwater treatment wetlands in the treatment of emerging pollutants to maximize their water quality treatment benefits.

International collaborations open the platform for scientific productivity. The bibliographic information extracted from published literature revealed that only selected states (i.e., the USA and China) have a well-established network of authors that collaborate on scientific publications. Furthermore, European states were found to be more active in terms of publication as compared to states from other regions. It is recommended to promote knowledge sharing through international collaboration in order to increase scientific productivity and improve the functions, design, and benefits of stormwater treatment wetlands.

4. Conclusions

The number of publications related to stormwater treatment wetlands has considerably increased over the years, indicating that green stormwater infrastructures and NBS have become relevant approaches in stormwater management. USA and China were found to be the most productive states in terms of the number of scientific publications and research collaborations; however, further analyses revealed that Australia, Canada, and South Korea had the highest level of scientific productivity in terms of population-to-publication ratio. The typical design and components of stormwater treatment wetlands in different regions were identified through a comprehensive review of related scientific literature. FWS and FTW were the most common types of CWs used for stormwater treatment, and the size of facilities varied from 7 m^2 to more than 270 km^2. Sand and gravel were typically used as filter media for HSSF, VSSF, and hybrid treatment wetlands since these materials are abundant and have high pollutant removal performance. The most common plant genera used in CW systems include *Typha*, *Juncus*, *Carex*, *Phragmites*, and *Schoenoplectus*. These plants are considered hyperaccumulators of pollutants with high biomass yields and high resistance to toxic environments. Generally, studies on water quality parameters, including nutrients and heavy metals, were the most established field in stormwater treatment wetlands research; however, inquiries regarding the treatment of micropollutants and emerging pollutants by stormwater treatment wetlands are still lacking.

This study identified the need for future works that focus on addressing the aforementione research gaps that are necessary to optimize the benefits of stormwater treatment wetland Strengthening the collaboration among different states can also promote greater scientif productivity and new paradigms for the utilization of nature-based systems.

Supplementary Materials: The following supporting information can be downloaded at: http //www.mdpi.com/article/10.3390/su15032332/s1, Table S1: List of plants used in stormwate treatment wetlands.

Author Contributions: Conceptualization, N.J.D.G.R., F.K.F.G., and H.B.G.; methodology, N.J.D.G.R. software, N.J.D.G.R.; validation, N.J.D.G.R.; formal analysis, N.J.D.G.R.; investigation, N.J.D.G.R. F.K.F.G., and H.B.G.; resources, L.-H.K.; data curation, N.J.D.G.R.; writing—original draft preparatio N.J.D.G.R.; writing—review and editing, F.K.F.G., H.B.G., and L.-H.K.; visualization, N.J.D.G.R. supervision, F.K.F.G., H.B.G., and L.-H.K.; project administration, L.-H.K.; funding acquisition, L.-H. All authors have read and agreed to the published version of the manuscript.

Funding: This research was supported by the National University Development Project by th Ministry of Education in 2022.

Institutional Review Board Statement: Not applicable.

Informed Consent Statement: Not applicable.

Data Availability Statement: Not applicable.

Conflicts of Interest: The authors declare no conflict of interest.

References

1. Handayani, W.; Chigbu, U.E.; Rudiarto, I.; Surya Putri, I.H. Urbanization and Increasing Flood Risk in the Northern Coast c Central Java-Indonesia: An Assessment towards Better Land Use Policy and Flood Management. *Land* **2020**, *9*, 343. [CrossRef]
2. Mignot, E.; Li, X.; Dewals, B. Experimental Modelling of Urban Flooding: A Review. *J. Hydrol.* **2019**, *568*, 334–342. [CrossRef]
3. Reinhardt-Imjela, C.; Imjela, R.; Bölscher, J.; Schulte, A. The Impact of Late Medieval Deforestation and 20th Century Fores Decline on Extreme Flood Magnitudes in the Ore Mountains (Southeastern Germany). *Quat. Int.* **2018**, *475*, 42–53. [CrossRef]
4. Ma, Y.; Egodawatta, P.; McGree, J.; Liu, A.; Goonetilleke, A. Human Health Risk Assessment of Heavy Metals in Urban Stormwate *Sci. Total Environ.* **2016**, *557–558*, 764–772. [CrossRef] [PubMed]
5. Eriksson, E.; Baun, A.; Mikkelsen, P.S.; Ledin, A. Risk Assessment of Xenobiotics in Stormwater Discharged to Harrestrup Denmark. *Desalination* **2007**, *215*, 187–197. [CrossRef]
6. Weston, D.P.; Chen, D.; Lydy, M.J. Stormwater-Related Transport of the Insecticides Bifenthrin, Fipronil, Imidacloprid, an Chlorpyrifos into a Tidal Wetland, San Francisco Bay, California. *Sci. Total Environ.* **2015**, *527–528*, 18–25. [CrossRef]
7. Ma, Y.; Deilami, K.; Egodawatta, P.; Liu, A.; McGree, J.; Goonetilleke, A. Creating a Hierarchy of Hazard Control for Urba Stormwater Management. *Environ. Pollut.* **2019**, *255*, 113217. [CrossRef]
8. Yang, Y.Y.; Toor, G.S. Stormwater Runoff Driven Phosphorus Transport in an Urban Residential Catchment: Implications fc Protecting Water Quality in Urban Watersheds. *Sci. Rep.* **2018**, *8*, 11681. [CrossRef]
9. Lee, S.; Suits, M.; Wituszynski, D.; Winston, R.; Martin, J.; Lee, J. Residential Urban Stormwater Runoff: A Comprehensive Profil of Microbiome and Antibiotic Resistance. *Sci. Total Environ.* **2020**, *723*, 138033. [CrossRef]
10. Zgheib, S.; Moilleron, R.; Chebbo, G. Priority Pollutants in Urban Stormwater: Part 1—Case of Separate Storm Sewers. *Water Re* **2012**, *46*, 6683–6692. [CrossRef]
11. Pramanik, B.K.; Roychand, R.; Monira, S.; Bhuiyan, M.; Jegatheesan, V. Fate of Road-Dust Associated Microplastics and per- an Polyfluorinated Substances in Stormwater. *Process Saf. Environ. Prot.* **2020**, *144*, 236–241. [CrossRef]
12. Choi, H.; Reyes, N.J.D.; Jeon, M.; Kim, L.H. Constructed Wetlands in South Korea: Current Status and Performance Assessmer *Sustainability* **2021**, *13*, 10410. [CrossRef]
13. Stefanakis, A.I. The Role of ConstructedWetlands as Green Infrastructure for Sustainable Urban Water Management. *Sustainabili* **2019**, *11*, 6981. [CrossRef]
14. Kabenge, I.; Ouma, G.; Aboagye, D.; Banadda, N. Performance of a Constructed Wetland as an Upstream Intervention fc Stormwater Runoff Quality Management. *Environ. Sci. Pollut. Res.* **2018**, *25*, 36765–36774. [CrossRef] [PubMed]
15. Alihan, J.C.; Maniquiz-Redillas, M.; Choi, J.; Flores, P.E.; Kim, L.-H. Characteristics and Fate of Stormwater Runoff Pollutants i Constructed Wetlands. *J. Wetl. Res.* **2017**, *19*, 37–44. [CrossRef]
16. Grinberga, L.; Lagzdins, A. Nutrient Removal by Subsurface Flow Constructed Wetland in the Farm Mezaciruli. *Res. Rural De* **2017**, *1*, 160–165. [CrossRef]
17. McMaine, J.T.; Vogel, J.R.; Belden, J.B.; Schnelle, M.A.; Morrison, S.A.; Brown, G.O. Field Studies of Pollutant Removal fro Nursery and Greenhouse Runoff by Constructed Wetlands. *J. Environ. Qual.* **2020**, *49*, 106–118. [CrossRef]

Ockenden, M.C.; Deasy, C.; Quinton, J.N.; Surridge, B.; Stoate, C. Keeping Agricultural Soil out of Rivers: Evidence of Sediment and Nutrient Accumulation within Field Wetlands in the UK. *J. Environ. Manag.* **2014**, *135*, 54–62. [CrossRef]

Headley, T.; Tanner, C. *Application of Floating Wetlands for Enhanced Stormwater Treatment: A Review*; Auckland Regional Council: Hamilton, New Zealand, 2006. Available online: https://www.researchgate.net/profile/Tom-Headley/publication/266409739_Application_of_Floating_Wetlands_for_Enhanced_Stormwater_Treatment_A_Review/links/561c3d3a08ae044edbb3918c/Application-of-Floating-Wetlands-for-Enhanced-Stormwater-Treatment-A-Review.pdf (accessed on 1 December 2022).

Sharma, R.; Vymazal, J.; Malaviya, P. Application of Floating Treatment Wetlands for Stormwater Runoff: A Critical Review of the Recent Developments with Emphasis on Heavy Metals and Nutrient Removal. *Sci. Total Environ.* **2021**, *777*, 146044. [CrossRef]

Ingrao, C.; Failla, S.; Arcidiacono, C. A Comprehensive Review of Environmental and Operational Issues of Constructed Wetland Systems. *Curr. Opin. Environ. Sci. Heal.* **2020**, *13*, 35–45. [CrossRef]

Li, D.; Zheng, B.; Liu, Y.; Chu, Z.; He, Y.; Huang, M. Use of Multiple Water Surface Flow Constructed Wetlands for Non-Point Source Water Pollution Control. *Appl. Microbiol. Biotechnol.* **2018**, *102*, 5355–5368. [CrossRef] [PubMed]

Pranckutė, R. Web of Science (Wos) and Scopus: The Titans of Bibliographic Information in Today's Academic World. *Publications* **2021**, *9*, 12. [CrossRef]

Donthu, N.; Kumar, S.; Mukherjee, D.; Pandey, N.; Lim, W.M. How to Conduct a Bibliometric Analysis: An Overview and Guidelines. *J. Bus. Res.* **2021**, *133*, 285–296. [CrossRef]

Van Eck, N.J.; Waltman, L. Citation-Based Clustering of Publications Using CitNetExplorer and VOSviewer. *Scientometrics* **2017**, *111*, 1053–1070. [CrossRef] [PubMed]

Ubando, A.T.; Africa, A.D.M.; Maniquiz-Redillas, M.C.; Culaba, A.B.; Chen, W.H.; Chang, J.S. Microalgal Biosorption of Heavy Metals: A Comprehensive Bibliometric Review. *J. Hazard. Mater.* **2021**, *402*, 123431. [CrossRef] [PubMed]

Brito, R.; Rodríguez-Navarro, A. Evaluating Research and Researchers by the Journal Impact Factor: Is It Better than Coin Flipping? *J. Informetr.* **2019**, *13*, 314–324. [CrossRef]

McKiernan, E.C.; Schimanski, L.A.; Nieves, C.M.; Matthias, L.; Niles, M.T.; Alperin, J.P. Use of the Journal Impact Factor in Academic Review, Promotion, and Tenure Evaluations. *eLife* **2019**, *8*, e47338. [CrossRef]

Xu, B.; Wang, X.; Liu, J.; Wu, J.; Zhao, Y.; Cao, W. Improving Urban Stormwater Runoff Quality by Nutrient Removal through Floating Treatment Wetlands and Vegetation Harvest. *Sci. Rep.* **2017**, *7*, 7000. [CrossRef]

Winston, R.J.; Hunt, W.F.; Kennedy, S.G.; Merriman, L.S.; Chandler, J.; Brown, D. Evaluation of Floating Treatment Wetlands as Retrofits to Existing Stormwater Retention Ponds. *Ecol. Eng.* **2013**, *54*, 254–265. [CrossRef]

White, S.A.; Cousins, M.M. Floating Treatment Wetland Aided Remediation of Nitrogen and Phosphorus from Simulated Stormwater Runoff. *Ecol. Eng.* **2013**, *61*, 207–215. [CrossRef]

Borne, K.E.; Fassman, E.A.; Tanner, C.C. Floating Treatment Wetland Retrofit to Improve Stormwater Pond Performance for Suspended Solids, Copper and Zinc. *Ecol. Eng.* **2013**, *54*, 173–182. [CrossRef]

Payne, E.G.I.; Fletcher, T.D.; Russell, D.G.; Grace, M.R.; Cavagnaro, T.R.; Evrard, V.; Deletic, A.; Hatt, B.E.; Cook, P.L.M. Temporary Storage or Permanent Removal? The Division of Nitrogen between Biotic Assimilation and Denitrification in Stormwater Biofiltration Systems. *PLoS ONE* **2014**, *9*, e90890. [CrossRef] [PubMed]

Zhao, Y.; Zhang, C.; Yang, Z.; Yang, Y.; Huang, N.; Arku, J.E.; Mao, G.; Wang, Y. Global Trends and Prospects in the Removal of Pharmaceuticals and Personal Care Products: A Bibliometric Analysis. *J. Water Process Eng.* **2021**, *41*, 102004. [CrossRef]

Sharley, D.J.; Sharp, S.M.; Marshall, S.; Jeppe, K.; Pettigrove, V.J. Linking Urban Land Use to Pollutants in Constructed Wetlands: Implications for Stormwater and Urban Planning. *Landsc. Urban Plan.* **2017**, *162*, 80–91. [CrossRef]

Ioannidou, V.; Stefanakis, A.I. The Use of Constructed Wetlands to Mitigate Pollution from Agricultural Runoff. In *Contaminants in Agriculture: Sources, Impacts and Management*; Naeem, M., Ansari, A., Gill, S., Eds.; Springer: Cham, Switzerland, 2020; pp. 233–246. ISBN 9783030415525.

Senduran, C.; Gunes, K.; Topaloglu, D.; Dede, O.H.; Masi, F.; Kucukosmanoglu, O.A. Mitigation and Treatment of Pollutants from Railway and Highway Runoff by Pocket Wetland System; A Case Study. *Chemosphere* **2018**, *204*, 335–343. [CrossRef]

Horner, R. Constructed Wetlands for Urban Runoff Water Quality Control. In *EPA Seminar Publication*; Schultz, H., Ed.; EPA: Cincinnati, OH, USA, 1995; pp. 327–340.

García, J.; Solimeno, A.; Zhang, L.; Marois, D.; Mitsch, W.J. Constructed Wetlands to Solve Agricultural Drainage Pollution in South Florida: Development of an Advanced Simulation Tool for Design Optimization. *J. Clean. Prod.* **2020**, *258*, 120868. [CrossRef]

Zhao, H.; Piccone, T. Large Scale Constructed Wetlands for Phosphorus Removal, an Effective Nonpoint Source Pollution Treatment Technology. *Ecol. Eng.* **2020**, *145*, 105711. [CrossRef]

Zamorano, M.; Piccone, T.; Colon, S. *Baseline Soil Characterization for Compartment B in Stormwater Treatment Area 2 and Compartment C in Stormwater Treatment Area 5/6*; South Florida Water Management District: West Palm Beach,FL, USA, 2019.

Gorgoglione, A.; Torretta, V. Sustainable Management and Successful Application of Constructed Wetlands: A Critical Review. *Sustainability* **2018**, *10*, 3910. [CrossRef]

Molle, P.; Lombard Latune, R.; Riegel, C.; Lacombe, G.; Esser, D.; Mangeot, L. French Vertical-Flow Constructed Wetland Design: Adaptations for Tropical Climates. *Water Sci. Technol.* **2015**, *71*, 1516–1523. [CrossRef]

Kadlec, R.H. Comparison of Free Water and Horizontal Subsurface Treatment Wetlands. *Ecol. Eng.* **2009**, *35*, 159–174. [CrossRef]

45. Toudignant, E.; Frankhauser, O.; Hurd, S. *Guidance Manual for the Design, Construction and Operations of Constructed Wetlands Rural Applications in Ontario*; Agricultural Adaptation Council: Guelph, ON, Canada, 1999. Available online: https://atrium uoguelph.ca/xmlui/bitstream/handle/10214/15203/FDMR_wetlands_manual.pdf?sequence=1&isAllowed=y (accessed o December 2022).
46. Wadzuk, B.M.; Rea, M.; Woodruff, G.; Flynn, K.; Traver, R.G. Water-Quality Performance of a Constructed Stormwater Wetla for All Flow Conditions. *J. Am. Water Resour. Assoc.* **2010**, *46*, 385–394. [CrossRef]
47. Conn, R.M.; Fiedler, F.R. Increasing Hydraulic Residence Time in Constructed Stormwater Treatment Wetlands with Desigr Bottom Topography. *Water Environ. Res.* **2006**, *78*, 2514–2523. [CrossRef] [PubMed]
48. Tirpak, R.A.; Tondera, K.; Tharp, R.; Borne, K.E.; Schwammberger, P.; Ruppelt, J.; Winston, R.J. Optimizing Floating Treatm Wetland and Retention Pond Design through Random Forest: A Meta-Analysis of Influential Variables. *J. Environ. Manag.* **20** *312*, 114909. [CrossRef]
49. Hong, J.S.; Kim, L.-H. Assessment of Performances of Low Impact Development (LID) Facilities with Vegetation. *Ecol. Resili Infrastruct.* **2016**, *3*, 100–109. [CrossRef]
50. Choi, J.; Maniquiz-Redillas, M.C.; Hong, J.; Kim, L.H. Selection of Cost-Effective Green Stormwater Infrastructure (GSI) Applica in Highly Impervious Urban Catchments. *KSCE J. Civ. Eng.* **2018**, *22*, 24–30. [CrossRef]
51. Segismundo, E.Q.; Kim, L.H.; Jeong, S.M.; Lee, B.S. A Laboratory Study on the Filtration and Clogging of the Sand-Bottom A Mixture for Stormwater Infiltration Filter Media. *Water* **2017**, *9*, 32. [CrossRef]
52. Siriwardene, N.R.; Deletic, A.; Fletcher, T.D. Modeling of Sediment Transport through Stormwater Gravel Filters over Th Lifespan. *Environ. Sci. Technol.* **2007**, *41*, 8099–8103. [CrossRef]
53. Schwammberger, P.; Walker, C.; Lucke, T. Using Floating Wetland Treatment Systems to Reduce Stormwater Pollution fr Urban Developments. *Int. J. GEOMATE* **2017**, *12*, 45–50. [CrossRef]
54. Adyel, T.M.; Oldham, C.E.; Hipsey, M.R. Stormwater Nutrient Attenuation in a Constructed Wetland with Alternating Surf and Subsurface Flow Pathways: Event to Annual Dynamics. *Water Res.* **2016**, *107*, 66–82. [CrossRef]
55. Han, J.; Tao, W. Treatment Performance and Copper Removal Mechanisms of a Vegetated Submerged Bed Receiving Leach from ACQ-Treated Lumber. *Ecol. Eng.* **2014**, *70*, 162–168. [CrossRef]
56. Niu, S.; Wang, X.; Yu, J.; Kim, Y. Pollution Reduction by Recirculated Fill-and-Drain Mesocosm Wetlands Packed with Wo chip/Pumice Treating Impervious Road Stormwater. *Environ. Technol.* **2020**, *41*, 1627–1636. [CrossRef] [PubMed]
57. Zhang, X.; Wang, T.; Xu, Z.; Zhang, L.; Dai, Y.; Tang, X.; Tao, R.; Li, R.; Yang, Y.; Tai, Y. Effect of Heavy Metals in Mix Domestic-Industrial Wastewater on Performance of Recirculating Standing Hybrid Constructed Wetlands (RSHCWs) and Th Removal. *Chem. Eng. J.* **2020**, *379*, 122363. [CrossRef]
58. Kurniawan, S.B.; Ahmad, A.; Said, N.S.M.; Imron, M.F.; Abdullah, S.R.S.; Othman, A.R.; Purwanti, I.F.; Hasan, H.A. Macrophy as Wastewater Treatment Agents: Nutrient Uptake and Potential of Produced Biomass Utilization toward Circular Econo Initiatives. *Sci. Total Environ.* **2021**, *790*, 148219. [CrossRef]
59. Yan, X.; An, J.; Yin, Y.; Gao, C.; Wang, B.; Wei, S. Heavy Metals Uptake and Translocation of Typical Wetland Plants and Th Ecological Effects on the Coastal Soil of a Contaminated Bay in Northeast China. *Sci. Total Environ.* **2022**, *803*, 149871. [CrossR
60. Lolu, A.J.; Ahluwalia, A.S.; Sidhu, M.C.; Reshi, Z.A. Carbon Sequestration Potential of Macrophytes and Seasonal Carbon Inp Assessment into the Hokersar Wetland, Kashmir. *Wetlands* **2019**, *39*, 453–472. [CrossRef]
61. Sesin, V.; Davy, C.M.; Freeland, J.R. Review of Typha Spp. (Cattails) as Toxicity Test Species for the Risk Assessment Environmental Contaminants on Emergent Macrophytes. *Environ. Pollut.* **2021**, *284*, 117105. [CrossRef] [PubMed]
62. Chandra, R.; Yadav, S. Phytoremediation of Cd, Cr, Cu, Mn, Fe, Ni, Pb and Zn from Aqueous Solution Using Phragmi Cummunis, Typha Angustifolia and Cyperus Esculentus. *Int. J. Phytoremediat.* **2011**, *13*, 580–591. [CrossRef] [PubMed]
63. Nabuyanda, M.M.; Kelderman, P.; van Bruggen, J.; Irvine, K. Distribution of the Heavy Metals Co, Cu, and Pb in Sediments a Typha Spp. And Phragmites Mauritianus in Three Zambian Wetlands. *J. Environ. Manag.* **2022**, *304*, 114133. [CrossRef]
64. Müller, J.; Jantzen, C.; Wiedow, D. The Energy Potential of Soft Rush (Juncus Effusus L.) in Different Conversion Routes. *Ene Sustain. Soc.* **2020**, *10*, 26. [CrossRef]
65. Syranidou, E.; Christofilopoulos, S.; Kalogerakis, N. Juncus Spp.—The Helophyte for All (Phyto)Remediation Purposes? *Biotechnol.* **2017**, *38*, 43–55. [CrossRef]
66. Kao, J.T.; Titus, J.E.; Zhu, W.X. Differential Nitrogen and Phosphorus Retention by Five Wetland Plant Species. *Wetlands* **20** *23*, 979–987. [CrossRef]
67. Pappalardo, S.E.; Ibrahim, H.M.S.; Cerinato, S.; Borin, M. Assessing the Water-Purification Service in an Integrated Agricultu Wetland within the Venetian Lagoon Drainage System. *Mar. Freshw. Res.* **2017**, *68*, 2205–2215. [CrossRef]
68. Al-Homaidan, A.A.; Al-Otaibi, T.G.; El-Sheikh, M.A.; Al-Ghanayem, A.A.; Ameen, F. Accumulation of Heavy Metals in Macrophyte Phragmites Australis: Implications to Phytoremediation in the Arabian Peninsula Wadis. *Environ. Monit. Ass* **2020**, *192*, 202. [CrossRef] [PubMed]
69. Bonanno, G. Comparative Performance of Trace Element Bioaccumulation and Biomonitoring in the Plant Species Typ Domingensis, Phragmites Australis and Arundo Donax. *Ecotoxicol. Environ. Saf.* **2013**, *97*, 124–130. [CrossRef] [PubMed]
70. Rai, P.K. Heavy Metal Pollution in Aquatic Ecosystems and Its Phytoremediation Using Wetland Plants: An Ecosustainal Approach. *Int. J. Phytoremediat.* **2008**, *10*, 133–160. [CrossRef]

1. Uddin, M.N.; Robinson, R.W. Allelopathy and Resource Competition: The Efects of Phragmites Australis Invasion in Plant Communities. *Bot. Stud.* **2017**, *58*, 29. [CrossRef] [PubMed]

2. López, D.; Sepúlveda, M.; Vidal, G. Phragmites Australis and Schoenoplectus Californicus in Constructed Wetlands: Development and Nutrient Uptake. *J. Soil Sci. Plant Nutr.* **2016**, *16*, 763–777. [CrossRef]

3. Tanner, C.C. Plants for Constructed Wetland Treatment Systems—A Comparison of the Growth and Nutrient Uptake of Eight Emergent Species. *Ecol. Eng.* **1996**, *7*, 59–83. [CrossRef]

4. Li, Y.C.; Zhang, D.Q.; Wang, M. Performance Evaluation of a Full-Scale Constructed Wetland for Treating Stormwater Runoff. *Clean Soil Air Water* **2017**, *45*, 1600740. [CrossRef]

5. Maniquiz-Redillas, M.; Robles, M.E.; Cruz, G.; Reyes, N.J.; Kim, L.-H. First Flush Stormwater Runoff in Urban Catchments: A Bibliometric and Comprehensive Review. *Sustainability* **2022**, *9*, 63. [CrossRef]

6. Zhao, H.; Jiang, Q.; Ma, Y.; Xie, W.; Li, X.; Yin, C. Influence of Urban Surface Roughness on Build-up and Wash-off Dynamics of Road-Deposited Sediment. *Environ. Pollut.* **2018**, *243*, 1226–1234. [CrossRef] [PubMed]

7. Ziajahromi, S.; Drapper, D.; Hornbuckle, A.; Rintoul, L.; Leusch, F.D.L. Microplastic Pollution in a Stormwater Floating Treatment Wetland: Detection of Tyre Particles in Sediment. *Sci. Total Environ.* **2020**, *713*, 136356. [CrossRef] [PubMed]

8. Walaszek, M.; Bois, P.; Laurent, J.; Lenormand, E.; Wanko, A. Micropollutants Removal and Storage Efficiencies in Urban Stormwater Constructed Wetland. *Sci. Total Environ.* **2018**, *645*, 854–864. [CrossRef] [PubMed]

9. Page, D.; Miotliński, K.; Gonzalez, D.; Barry, K.; Dillon, P.; Gallen, C. Environmental Monitoring of Selected Pesticides and Organic Chemicals in Urban Stormwater Recycling Systems Using Passive Sampling Techniques. *J. Contam. Hydrol.* **2014**, *158*, 65–77. [CrossRef] [PubMed]

0. Reid, K.; Schneider, K.; McConkey, B. Components of Phosphorus Loss from Agricultural Landscapes, and How to Incorporate Them into Risk Assessment Tools. *Front. Earth Sci.* **2018**, *6*, 135. [CrossRef]

1. Behrouz, M.S.; Yazdi, M.N.; Sample, D.J.; Scott, D.; Owen, J.S. What Are the Relevant Sources and Factors Affecting Event Mean Concentrations (EMCs) of Nutrients and Sediment in Stormwater? *Sci. Total Environ.* **2022**, *828*, 154368. [CrossRef]

2. Baum, P.; Kuch, B.; Dittmer, U. Adsorption of Metals to Particles in Urban Stormwater Runoff—Does Size Really Matter? *Water* **2021**, *13*, 309. [CrossRef]

3. Smith, J.S.; Winston, R.J.; Tirpak, R.A.; Wituszynski, D.M.; Boening, K.M.; Martin, J.F. The Seasonality of Nutrients and Sediment in Residential Stormwater Runoff: Implications for Nutrient-Sensitive Waters. *J. Environ. Manag.* **2020**, *276*, 111248. [CrossRef]

4. Kong, Z.; Shao, Z.; Shen, Y.; Zhang, X.; Chen, M.; Yuan, Y.; Li, G.; Wei, Y.; Hu, X.; Huang, Y.; et al. Comprehensive Evaluation of Stormwater Pollutants Characteristics, Purification Process and Environmental Impact after Low Impact Development Practices. *J. Clean. Prod.* **2021**, *278*, 123509. [CrossRef]

5. Wijesiri, B.; Egodawatta, P.; McGree, J.; Goonetilleke, A. Understanding the Uncertainty Associated with Particle-Bound Pollutant Build-up and Wash-off: A Critical Review. *Water Res.* **2016**, *101*, 582–596. [CrossRef]

6. Byeon, C.W.; Nam, B.E. An Assessment of the Ecological Functions of a Sustainable Structured Wetland Biotope (SSB). *Ecol. Eng.* **2020**, *145*, 105723. [CrossRef]

7. Li, X.; Guo, Q.; Wang, Y.; Xu, J.; Wei, Q.; Chen, L.; Liao, L. Enhancing Nitrogen and Phosphorus Removal by Applying Effective Microorganisms to Constructed Wetlands. *Water* **2020**, *12*, 2443. [CrossRef]

8. Grinberga, L.; Lauva, D.; Lagzdins, A. Treatment of Storm Water from Agricultural Catchment in Pilot Scale Constructed Wetland. *Environ. Clim. Technol.* **2021**, *25*, 640–649. [CrossRef]

9. Howitt, J.A.; Mondon, J.; Mitchell, B.D.; Kidd, T.; Eshelman, B. Urban Stormwater Inputs to an Adapted Coastal Wetland: Role in Water Treatment and Impacts on Wetland Biota. *Sci. Total Environ.* **2014**, *485*, 534–544. [CrossRef] [PubMed]

0. Piñon-Colin, T.d.J.; Rodriguez-Jimenez, R.; Rogel-Hernandez, E.; Alvarez-Andrade, A.; Wakida, F.T. Microplastics in Stormwater Runoff in a Semiarid Region, Tijuana, Mexico. *Sci. Total Environ.* **2020**, *704*, 135411. [CrossRef] [PubMed]

1. Wicke, D.; Matzinger, A.; Sonnenberg, H.; Caradot, N.; Schubert, R.L.; Dick, R.; Heinzmann, B.; Dünnbier, U.; von Seggern, D.; Rouault, P. Micropollutants in Urban Stormwater Runoff of Different Land Uses. *Water* **2021**, *13*, 1312. [CrossRef]

2. Tran, N.H.; Reinhard, M.; Khan, E.; Chen, H.; Nguyen, V.T.; Li, Y.; Goh, S.G.; Nguyen, Q.B.; Saeidi, N.; Gin, K.Y.H. Emerging Contaminants in Wastewater, Stormwater Runoff, and Surface Water: Application as Chemical Markers for Diffuse Sources. *Sci. Total Environ.* **2019**, *676*, 252–267. [CrossRef]

3. Fairbairn, D.J.; Elliott, S.M.; Kiesling, R.L.; Schoenfuss, H.L.; Ferrey, M.L.; Westerhoff, B.M. Contaminants of Emerging Concern in Urban Stormwater: Spatiotemporal Patterns and Removal by Iron-Enhanced Sand Filters (IESFs). *Water Res.* **2018**, *145*, 332–345. [CrossRef]

forests

MDPI

iew

Review of Ecosystem Service Trade-Offs/Synergies: nlightenment for the Optimization of Forest Ecosystem unctions in Karst Desertification Control

ehua Deng [1], Kangning Xiong [1,2,*], Yanghua Yu [1,2], Shihao Zhang [1,2], Lingwei Kong [1,2] and Yu Zhang [3]

1 School of Karst Science, Guizhou Normal University, Guiyang 550001, China
2 State Engineering Technology Institute for Karst Desertification Control, Guiyang 550001, China
3 Department of Resource Management, Tangshan Normal University, Tangshan 063000, China
* Correspondence: xiongkn@gznu.edu.cn

Abstract: Ecosystem services provide regulation, provisioning, support, and cultural benefits for human survival, but it needs to be clarified how the trade-off/synergy relationships can be used to optimize function. Based on the Web of Science (WOS) and China National Knowledge Infrastructure (CNKI) databases, we collected 254 articles on the ecosystem trade-offs/synergies and functional optimization. Through a systematic review of the literature, this paper summarized the research progress and landmark achievements from three aspects: trade-offs/synergies, functional optimization, and evaluation methods. The results indicated the following: (1) In terms of the number of articles published, there were no reports before 2005; from 2006 to 2022, the annual number of published papers increased from 1 to 72, showing an overall growth trend year by year. This mainly includes three stages: initial (1970–2005), slow development (2005–2014), and rapid development (2014–2022). (2) In terms of research areas, focus was placed mainly on Asia, North America, and Europe, accounting for 40.47%, 25.55%, and 15.07% of all regions, respectively. (3) In the future, it is necessary to focus on scientific issues such as the improvement of forest ecosystem functions, the trade-off/synergy relationships between services, the scale of spatiotemporal research, and the driving factors and evaluation methods for the management of rocky karst desertification. The aim is to provide a theoretical basis to optimize the forest ecosystem service functions.

Keywords: desertification control; ecosystem services; forests; functions; trade-offs/synergies

tion: Deng, X.; Xiong, K.; Yu, Y.; ng, S.; Kong, L.; Zhang, Y. A iew of Ecosystem Service le-Offs/Synergies: Enlightenment he Optimization of Forest system Functions in Karst ertification Control. *Forests* **2023**, 8. https://doi.org/10.3390/ 10088

demic Editors: Panayiotis itrakopoulos, Mario A. Pagnotta, las Scholz and Arshiya Noorani

eived: 14 November 2022
ised: 29 December 2022
epted: 30 December 2022
lished: 3 January 2023

1. Introduction

Ecosystem services, as the term of a scientific period, began in 1970 with the publication of the UN University's report on "Human Impact on the Global Environment" [1]. Since then, Costanza [2] has proposed that ecosystem services are benefits derived directly or indirectly by human beings from the ecosystem, which are used to maintain the natural environmental conditions and utilities on which human beings depend for survival and development. Since the concept was proposed, it has attracted extensive attention from the academic community, and ecosystem services have been classified. People have widely accepted the classification of ecosystem services established by the Millennium Ecosystem Assessment, which is divided into provision, regulation, support, and cultural services [3]. Various factors regulate the services, and there are differences and overlaps in handling different services. This leads to complex interrelationships between other ecosystem services [4], manifesting as synergies and trade-offs between mutual gains and losses.

Trade-offs, as a fundamental concept, first appeared in economics and were defined as opportunity costs. Where resources are scarce, an individual or group must give up a certain amount of additional scarce resources to obtain more of the scarce resources [5]. Begong believes that a "trade-off" is a relationship between two life history traits. Due to the imbalance in the distribution of resources, an increase in the payoff from one feature is associated with a decrease in the payoff from the other feature [6]. Synergy was fi introduced by the German theoretical physicist Haken in 1971 [7]. He considered "syner as the joint action and collective behavior of the parts of a system that collaborate cooperate, leading to a new structure and characteristics of the whole system. In 20 the Millennium Ecosystem Assessment (MA) [3] defined ecosystem service trade-o as follows: within a particular spatial and temporal scale, when the supply level of ecosystem service is enhanced, it is at the cost of reducing the resilience of its ecosyst and the supply function of other ecosystem services. in contrast, a synergistic relations is manifested as a joint increase or decrease in various service supply capabilities [8]. I clear from this line of development that the study of ecosystem service trade-offs/synerg is still a hot topic for future research.

Karst desertification refers to the land degradation phenomenon caused by the c turbance and destruction of unreasonable human social and economic activities unc the fragile karst environment in the subtropical zone, which is manifested by soil e sion, gradual rock exposure, land productivity degradation, and a desert-like landsca on the surface [9]. Karst landscapes are widely distributed around the world, coveri 22 million km^2, accounting for 12–15% of the land area [10,11]. They are mainly distribut in the Mediterranean region, the east and central of North America, and southern Chi Due to the fragile ecological environment and unreasonable human activities in the areas, karst desertification is a major problem, which threatens the security of ecologi environment and restricts the development of the economy and society [9]. In order reduce land degradation and promote ecological and economic development, schol regard forest as the preferred goal of ecological restoration in this area [11,12]. The sults of management over the years show that forests, as the main provider of ecosyst services, have the functions of water conservation, soil and water conservation, carb sequestration, and climate change mitigation [13]. In particular, they play an irreplacea role in managing karst desertification. However, as time has passed, the problems of for ecosystems in karst rocky desertification areas have become more and more promine such as the competition between communities, leading to a reduction in biodiversity a the weakening of service functions. To promote the sustainable development of for ecosystems in karst desertification areas, it is necessary to review studies related to t trade-offs and synergies of forest ecosystem services.

In recent years, studies on ecosystem service trade-offs/synergies have mainly focus on watershed [14], wetland [15], and vulnerable functional areas [16], and have achiev good results. The karst rocky desertification forest ecosystem is an important part of t terrestrial ecosystem, but its trade-offs/synergies are rarely reported, which hinders t research on functional optimization strategies. In order to solve this problem, based a systematic review of the literature, this study summarizes the progress and landma achievements of global ecosystem service trade-offs/synergies. It summarizes the k scientific questions about the function of forest ecosystems in managing karst rocky dese fication, the trade-off/synergy relationships, the spatial research scale, driving factors, a evaluation methods. It aims to enlighten future research directions on optimizing for ecosystem service functions in stone desertification management and provide theoreti references to enhance its service supply capacity.

2. Materials and Methods

This study was based on platforms such as the CNKI (China National Knowled Infrastructure) and the Foreign Journal Resource Service System of Guizhou Normal Univ sity Library (Web of Science) (http://lib.gznu.edu.cn/data/weibu/waiwen/waiwen.h accessed on 22 October 2022). The search period was the maximum time range of bo databases, and the search time was up to 22 October 2022 (Figure 1). First, we search the WOS database by entering "ecosystem service" and "trade-offs/synergy", and ag by entering "forest ecosystems" and "function optimization." A total of 234 docume were retrieved. Second, the CNKI database was searched by entering "ecosystem servic "trade-offs/synergy", "forest ecosystems", and "function optimization", and a total of 399 Chinese documents were retrieved. After the manual screening, a total of 564 documents were obtained. Finally, the articles' titles, abstracts, and keywords were screened based on the research themes related to trade-offs/synergies and functional optimization of ecosystem services. Following this, we further combined the literature for ecosystem services, trade-offs/synergies among services, functional optimization, and assessment methods. After, the full text of the alternative literature was browsed, and a total of 254 articles were screened. Among them, 93 papers were obtained from the WOS database (92 articles and one conference paper), and 161 papers were obtained from the CNKI database (109 journal papers, 42 master's theses, five doctoral dissertations, and three conference papers).

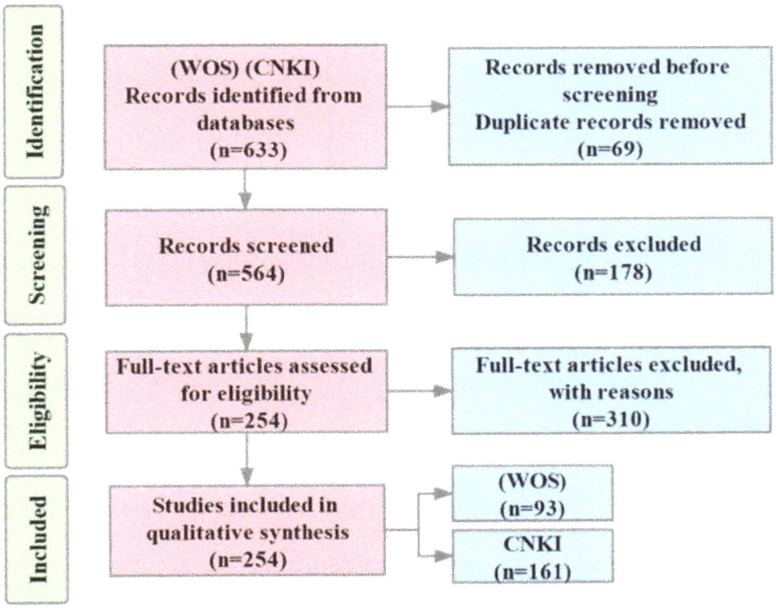

Figure 1. The process of the literature search.

3. Results

3.1. Annual Distribution of the Literature

From the annual distribution of the literature (Figure 2), the research related to ecosystem service trade-offs/synergies and functional optimization started in 2005 and reached its peak in 2022. This can be roughly divided into three stages. The first stage was from 2006 to 2014, when there were few studies on ecosystem service trade-offs/synergies and functional optimization, with only three literature articles in total; this was the starting stage. The second stage was from 2014 to 2018, and the literature showed a fluctuating growth trend, which was 16.3 times that of the first stage. In addition, there were already articles related to karst ecosystem research in this stage, but the number of types of literature was small; this is the development stage. The third phase was from 2017–2022, with a surge in the number of types of literature, 4.12 times more than in the second phase, but there were only 12 articles on the ecosystem service trade-offs/synergies and functional optimization of karst rocky desert ecosystem services. The overall trend shows that the number of articles on this topic will continue to grow.

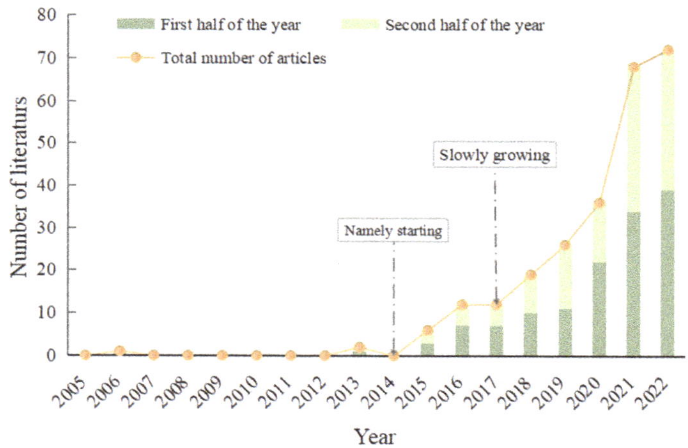

Figure 2. Annual distribution of the literature. The dark green is the first half of the year (January–June); the light green is the second half of the year (July–December).

3.2. *Content Distribution of Documents*

This paper classified and summarize all of the reviewed literature according to th classification method. It was also divided into studies related to ecosystem service trade offs/synergies, the optimization of ecosystem service functions, and assessment method according to the content of the studies. Ecosystem service trade-offs/synergies accoun for 61% of the literature, dominated by theoretical studies such as trade-offs/synergie spatiotemporal characteristics, and relationship identification. In addition, in this litera ture, studies related to wetlands, oceans, watersheds, and fragile functional areas wer dominant, and there were fewer studies on forest ecosystems in karst stone desert area The ecosystem service functional optimization literature accounts for 20%, which contain strategies for forest service functional optimization such as strategies for ecological corrido construction applied in northeast China, and ecological restoration and compensatio strategies used in karst stone desertification areas in southern China. Assessment method and other literature accounted for 11% and 8%, respectively, of which the assessmen methods included both ecosystem services and trade-offs/synergies. The proportion of thi literature indicates that the study of ecosystem service trade-offs/synergies is becomin more mature. However, the optimization of ecosystem service functions and assessmen methods is still in the exploration and development stage.

3.3. *Distribution of Research Institutions Publishing Documents*

The regional distribution of the literature shows that China, the United States, an Germany dominate the research on ecosystem service trade-offs/synergies and functiona optimization. The unique landscapes in China such as the northwest desert belt, fragil functional areas, and southern karst have attracted more Chinese scholars to study thes areas and publish their research results. Thus, the overall number of publications was th highest. Next, developed countries such as Canada, France, Spain, Finland, Sweden, an Australia also had more relevant studies. In addition, Brazil, South Africa, Singapore, an other countries such as Iran had few studies. In terms of the overall publication volume, Asi had the highest number of publications, followed by North American states, Europe, Oceani South American states, and Africa, accounting for 40.47%, 25.55%, 15.07%, 8.3%, 5.58%, an 5.03% of all regions, respectively. This correlated more significantly with each region's natura resources, social environment, policy support, and management experience.

The distribution of institutions publishing literature related to trade-offs/synergie and the functional optimization of ecosystem services was deciphered and analyzed. Th acquired institutions were divided into four categories: higher education institutions, research institutions (centers), business units, and grassroots organizations, with percentages of 77.1%, 13.50%, 5.7%, and 3.5%, respectively. The rates showed that higher education institutions and research institutes (centers) have paid more attention to research related to trade-offs/synergies and the functional optimization of ecosystem services. However, due to the limited space of the figure, only institutions with a higher number of publications are marked in Figure 3.

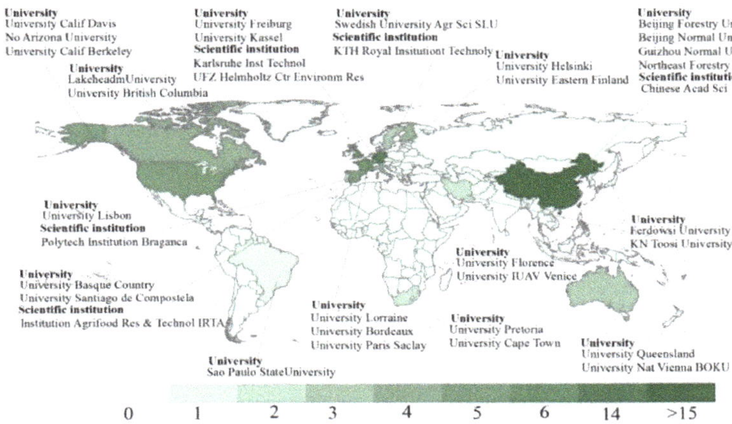

Figure 3. Distribution of countries and institutions studied in the literature. The different color bands and numbers at the bottom indicate the number of publications; the darker green the color, the higher the number of publications.

3.4. Research Stage Division

Ecosystem services appeared as a technical term in the Human Impact on the Global Environment report published by the United Nations University in 1970. Ecosystem service trade-offs/synergies emerged from the 2005 Millennium Ecosystem Assessment [3]. Ecosystem services are the basis, and an essential part, of the ecosystem service trade-off/synergy research. Thus, the first stage began in 1970 when ecosystem services were proposed. According to the research background of the changes in the theoretical studies, classification systems, and assessment methods, the development stages of ecosystem service trade-off/synergy and functional optimization were divided into the initial stage, the slow development stage, and the rapid development stage (Table 1).

Table 1. Division of the research stages.

Research Stages	Research Content	Background	Main Characteristics
Starting stage (1970–2005)	Conceptual and theoretical study of ecosystem services.	The relevant theoretical research was still at the beginning stage, and the depth of research was insufficient.	There were more qualitative studies, lacking systematic research and a single research method.
Slow growth stage (2005–2014)	The research methodology was diverse, using mainly statistical, spatial, model, and scenario analysis.	As the economy developed, attention begun to be paid to research related to trade-offs/synergies in ecosystem services and functional optimization.	The results were verified by the experiments, the classification system and evaluation method were gradually improved.

Table 1. *Cont.*

Research Stages	Research Content	Background	Main Characteristic
Rapid growth stage (2014–2022)	The mechanism of trade-offs/synergies and functional optimization of forest ecosystem services has been further clarified, and quantitative studies have gradually increased.	Ecosystem service trade-off/synergy research is developing rapidly, with an increasing number of cross-disciplinary studies and relatively mature technical approaches in China.	A variety of trade-off/synergy analysis methods. The research system presents a "pattern–process–service–sustainability" system.

3.5. Research Progress and Landmark Results

3.5.1. Ecosystem Service Trade-Off/Synergy

1. Spatial and temporal characteristics of ecosystem service trade-off/synergy

The trade-off/synergy among ecosystem services is spatial and temporal scale-depend and nonlinear, and the trade-off and synergistic relationships of ecosystem services m change at different spatial and temporal scales [17]. Current studies on the time scale of trade-offs and synergies of ecosystem services rarely use continuous time series and usua choose years with certain time intervals. For example, Niu et al. [18] analyzed the tra off/synergistic relationships of ecosystem services in the Songhua River basin in China duri 2000–2018. They concluded that there was a synergistic relationship between water harvesti soil conservation, biodiversity, and carbon sequestration, and a trade-off relationship betwe broadleaf forests, coniferous forests, and crops. Schroder et al. [19] found a short-term trade- and long-term synergistic relationship between national forest fire risk management (loggi and owl habitat protection and water quality regulation in the United States. Chen et al. [found an increasing trend in net primary productivity, water production, and soil retenti capacity from 2000 to 2018 during their study in a karst watershed. Still, the ecosyst synergy in the region was poor, and the trade-offs were particularly pronounced in karst sto desertification areas.

At the spatial scale, Zhang et al. analyzed the trade-off/synergistic effects of for ecosystem services in the Fuyu Mountain region. They found that the synergistic relati ship between services was better on the south slope than on the north slope, and that best synergistic relationship between services was found in the middle mountain deciduo broadleaf forest belt on the south slope. The worst synergistic relationship was found in low mountain deciduous broadleaf forest belt on the north slope [21]. Han et al. studi the trade-off and synergistic relationships between ecosystem services and land use char in the karst region response, finding synergistic relationships between soil water yield a soil retention, soil water yield and carbon storage, and soil retention and carbon storage. contrast, most of the relationships between other crops were trade-offs [22]. In summa the trade-off/synergistic relationships among services showed significant differences different temporal and spatial scales due to the natural recovery status and the selectiv of human use of the services.

2. Drivers of ecosystem service trade-off/synergy

The drivers of changes in the ecosystem service trade-offs and synergies are mair divided into anthropogenic and natural factors [23]. As one of the typical fragile ecos tems in the world, the generation of trade-off/synergistic relationships between servic in karst ecosystems mainly comes from the coupling effect of two significant facto anthropogenic and natural. Scholars have studied the causes of the changes in t synergistic relationships of service trade-offs in karst ecosystems. For example, H et al. suggested that climate is the leading natural factor for the transformation of serv trade-off/synergistic relationships, which changes the temperature and precipitation tensity in karst regions through climate change, thus affecting the distribution pattern of plants, causing competition for species' ecological niches, and indirectly changing the relationship between services [24]. Chen et al. found that lithology is also one of the factors influencing the trade-off/synergistic relationships, and that karst areas dominated by dolomite and limestone, where limestone is more susceptible to dissolution by running water, have a poorer soil retention capacity. Therefore, the trade-off relationship between soil conservation and services such as water production is more significant in this region [25]. In addition, some scholars also believe that factors such as soil erosion [10], vegetation degradation [9], and reduced biodiversity [26] have a more significant impact on the trade-off/synergistic relationship between services. With the rapid development of society, the economy, and urbanization, many karst ecosystems such as agricultural land, forests, and wetlands have been occupied by human activities. This approach has not only changed the land use pattern and disrupted the material and energy balance of the karst soil–vegetation system [27], but also induced the reverse evolution of the soil–vegetation system. In addition, the supply capacity of karst ecosystem services will also be reduced, directly affecting the relationship between the benefits of the karst ecosystem [28]. In addition, to meet the needs of human survival and development, a large amount of scrub was reclaimed for cultivation, which contributed to the loss of organic carbon from the surface layer of karst limestone soils and the intensification of surface erosion, resulting in the weakening of soil carbon and nitrogen sequestration capacity, which affected the relationship between the services to some extent [29]. Thus, it is clear that human activities are a key factor in the structural and functional changes of karst ecosystems.

3. Synergistic relationship of forest ecosystem service trade-offs

Forest ecosystems provide vital ecological services to the Earth's ecological environment, and the conservation of biodiversity and the development of forest industries are necessary ecological safeguards [12]. Karst forests are a special type of forest ecosystem developed on karst landscapes in the context of forest climate conditions [30]. However, long-term anthropogenic disturbances have exacerbated the problem of stone desertification in karst regions, reducing the stability of existing forest structures and leaving them in a state of fragmentation or secondary degradation succession. This significantly reduces the capacity of karst forest ecosystems in terms of water containment, soil conservation, biodiversity, and ecological product supply [31], creating trade-offs or synergistic relationships among services (Figure 4).

In recent years, some scholars have carried out studies related to the service relationships of karst forest ecosystems. For example, Wang et al. found that the relationship between the services in karst fallow forest areas (Grain for Green) was a trade-off between soil and water conservation and net primary productivity and water connotation, and a synergistic relationship between net primary productivity and water connotation [32]. Chen et al. found that the forest cover area in the karst area could enhance the forest's net primary productivity and soil erosion resistance. Still, expansion of the forest cover area will increase the uptake of water resources by vegetation, causing a decrease in surface water production service capacity so that the net primary productivity and water production form a trade-off relationship [33]. In turn, the trade-offs between services may impact the supply capacity of each service and hinder the ecological restoration process [9]. Therefore, based on the trade-offs between services, taking corresponding measures to optimize the service functions to maintain the structure and function of forest ecosystems is a critical way to achieve a "win–win" situation for economic development and ecological conservation [34].

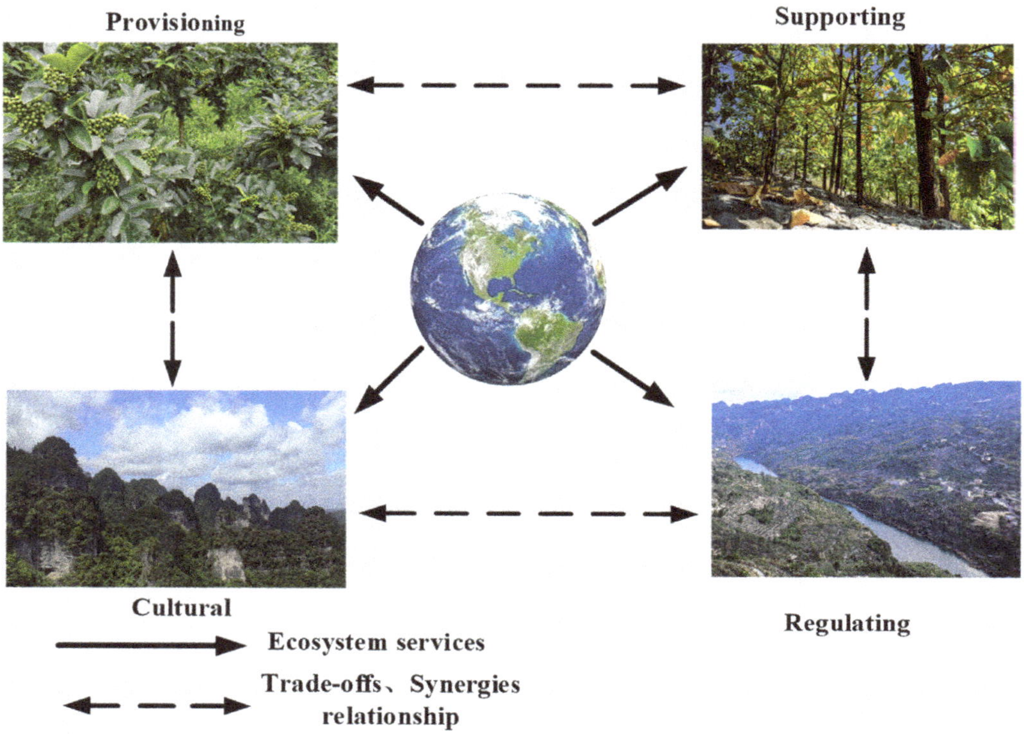

Figure 4. Trade-offs/synergies between forest ecosystem services.

3.5.2. Optimization of Ecosystem Service Functions

4. Ecological compensation

Ecological compensation and payment are differentiated regulation strategies for th interaction between ecosystem services and human welfare, which should be adapted t the current ecological environment and social and economic system. As a mechanism t stimulate ecological construction and environmental protection, ecological compensatio can realize the effect of internalizing the external ecological benefits [35]. Through th analysis of the spatial flow of supply, demand, and trade-offs, the key benefits and losse of ecosystem services are obtained, and the conditional payment transaction from servic users to service providers is facilitated [36]. In general, a combination of remote sensin data and location-based observations is used to measure the profit and loss of suppl regulation, support, and cultural services. Complementary compensation strategies hav been developed based on matching supply, demand, and trade-offs/synergies of ecosysten services. For example, Sun analyzed ecosystem service processes in crucial ecological fun tion areas in Xinjiang. According to the total value of ecosystem services, all counties in th autonomous region can be divided into three compensation levels: priority compensatio secondary compensation, and potential compensation areas [37]. Thus, the efficiency of th ecological compensation mechanism can be improved. It was more difficult to compensa for the particular habitat of the karst desert region, and further funds were used. Howeve certain shortcomings in financial compensation can result in poor compensation outcome As a result, alternative ways to precisely refine and structure economic compensation met ods exist such as improving the ecosystem structure stabilization and enhancing ecosysten service functions.

5. Enhance forest ecosystem service functions based on site conditions

A fragile ecological environment, fragmented surface morphology, and severe soil erosion and desertification characterize karst regions. The essence is the destruction of the ecosystem structure, which leads to a decline in and loss of ecosystem functions [38]. Given the uneven spatial distribution of forest ecosystem services and the selectivity of human use, the ecosystem service cascade framework was developed to increase the stability of the forest ecosystem structure, improve the function of forest ecosystem services, and promote regional synergies (Figure 5). According to the trade-off/synergy differentiation of the services, some scholars have proposed that plant communities can be adjusted and configured through functional groups to enhance the overall ecosystem service capacity [39]. Chen [40] offered optimal control strategies for the forest, irrigation, and grass communities in different rock desertification classes, considering the growth traits, adaptation strategies, ecological niches, and functional group types of tree species. Zhang [12] constructed a model for the ecological restoration of forests of different stone desertification classes with species adaptation strategies and ecological service functions. Zhang [41] proposed plant community optimization techniques by studying the plant species diversity, leaf functional traits of established species in plant communities, spatial patterns of plant communities, and interspecific correlations in the rocky desertification succession of different grades. Yu [39] took the upper reaches of the Chishui River as the study area, divided 32 water resource conservation forest tree species into seven water resource conservation function groups, and proposed optimization and adjustment strategies for the different function groups on this basis. Therefore, the conservation strategies for forest ecosystems in karst areas should be more targeted and specific.

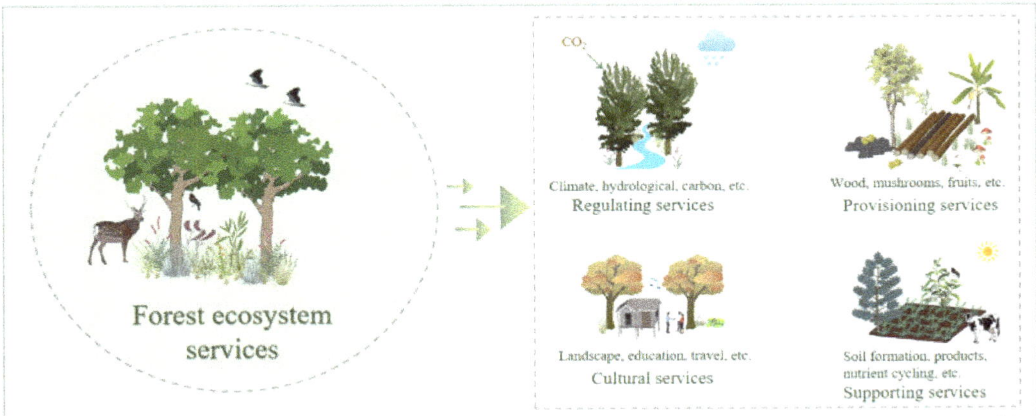

Figure 5. Forest ecosystem services.

3.5.3. Assessment Methods

Ecosystem service assessment is a complex process that needs to be refined and quantified. Currently, the main ecosystem service assessment methods include economical methods, plant functional traits, biophysical models, energy value conversion techniques [42], quality methods [43], modeling methods, and value quantity methods [44].

Several scholars have explored the spatial and temporal characteristics and the relationship between trade-offs and synergies of the forest ecosystem services. From the research perspective, the content mainly involves the expression form, dynamic change, scale effect, and scenario simulation of ecosystem service trade-off/synergy relationships [45]. The research scale includes country, region, and river basin [46]. The main research methods are statistical analysis [47], spatial analysis [48], scenario analysis [49], and service mobility analysis [50,51] (Table 2).

Table 2. Research methods of ecosystem service trade-offs/synergies.

	Research Method	Principles
Ecosystem service trade-off/ synergy research methods	Statistical methods	Mathematical models are used to study th quantitative relationships among variou forest ecosystem services.
	Spatial analysis methods	GIS, spatial statistics, and other methods are used to characterize the spatial patterr of forest ecosystem service trade-offs at specific spatial and temporal scales.
	Scenario analysis methods	Spatial mapping and regression analysis are used to analyze the dynamics of fores ecosystem service trade-offs and seek optimal scenarios.
	Service mobility analysis	"Source" and "sink" regions are identifie by analyzing single ecosystem service supply, transport, and utilization processe The impact of human activities on the supply and demand of forest ecosystem services is measured.

4. Discussion

4.1. Trade-Offs/Synergies and Functional Optimization of Annual Posting Volume Differences

The number of publications related to ecosystem service trade-offs/synergies a functional optimization is increasing rapidly with annual changes (Figure 2). For t forestry sector, this is both a challenge and an opportunity. Since the rise in ecosyste service trade-off/synergy research, the number of published articles has been increasi accounting for 74.4% of the total articles published, which will continue to grow in t future. Studies on the trade-off/synergy relationships are mainly related to river basi watersheds, and mountains. There needs to be more relevant studies on forests, as there currently a lack of articles on karst stone desertification forests. It is speculated that t complex environment of karst desertification zones, with significant differences in differe karst environments (canyons, depressions, rocky slopes, rocky forests, rocky troughs) [11] more difficult for scholars to study, resulting in a low number of publications in these are Regarding functional optimization, there are relatively few studies on forest ecosyste service functions, accounting for only 25.6%. The current studies on functional optimizati focus on the optimization of single services or the optimization of overall services, and the needs to be more studies on the optimization of services based on trade-off relationships is assumed that the concept and connotation of functional optimization need to be clea defined [52], leading scholars to ignore the trade-offs between services in studying t functional optimization strategies in forest ecosystems. The characteristics of trade- relationships and the driving mechanisms need to be better understood. The optimizati methods for transforming trade-off relationships into synergistic relationships cannot easily explored.

4.2. Differences in the Distribution of Study Areas

Regional differences in the natural conditions and social situations have led to t uneven development of research on ecosystem trade-offs/synergies and functional op mization (Figure 4). Regarding the number of publications, Asia has the highest numb with 40.47% of the total. Among the countries in Asia, China had the highest number publications, with a total share of 86.97%, which may be related to the attention of natio policies and research institutions [53], and possibly the use of the CNKI database. T emergence of a series of global problems such as resource and environmental degrad tion has made people deeply aware of the importance of the sustainable developme of ecosystems [54,55]. As a result, there is a growing interest in the benefits provided ecosystems, and the number of publications is increasing. South America had the second highest number of publications, with a total share of 25.55%. The region is advanced in education, science, and technology, but may have focused its research horizon on ecosystem function, biodiversity, and other related aspects, with less research on trade-offs/synergies and functional optimization. The lowest number of publications was found in Africa, accounting for 5.03% of the total number of publications. Compared with the Asian region, this region is poor in resources and lagging in education, science, and technology. More attention should be paid to the research on ecosystem trade-offs/synergies and functional optimization, and there were fewer relevant studies.

4.3. Key Scientific Questions to Be Addressed

1. Severe soil erosion, weak stability of vegetation communities, and deteriorating ecosystems in karst stone desertification areas. Karst systems have a unique, binary, three-dimensional spatial structure. However, due to their structural peculiarities, this exacerbates the water and soil leakage condition, making the soil layer in the area thin, causing challenges in retaining nutrients, along with a poor soil quality. In addition, to meet people's survival and development needs, the original vegetation in karst rocky desertification areas has been destroyed, so the existing forests are mostly planted and secondary forests. However, most planted and secondary forests have simple structures, poor biodiversity, low primary productivity, and poor soil and water retention functions. The vegetation shows a tendency to retreat, reducing the stability of the forest ecosystems [56] (Figure 6).

(a)

(b)

Figure 6. (**a**) Degradation of vegetation in karst desertification; (**b**) karst rocky desert landscape.

2. Karst stone desertification forest ecosystem services present trade-off/synergy relationships. This region has cultivated large areas of mature and overgrown forests using methods such as closed forestry and plantation, which have reached a certain high level of service capacity in terms of water content, net primary productivity, species diversity, and material production. However, the bias in human demand for forest ecosystem services and the tendency of forest communities to compete for ecological niche occupation [48] has led to trade-off/synergy relationships among services. Thus far, there are fewer studies on the trade-off/synergy of karst forest ecosystem services. During future research, understanding ecosystem services such as the water content and net primary productivity of karst forests should be enhanced to clarify the trade-off/synergistic relationships among services.

3. The spatial scale of forest ecosystem service trade-off/synergy studies is limited to large scales or single factors. Due to the successful development of 3S (remote sensing, geographic information systems, and global positioning systems) technologies, based on their ability to access spatial data quickly, the existing results of ecosystem service trade-offs/synergies are large-scale studies such as watersheds [57] and functional areas [58].

However, these studies typically use a single research scale, and the results cannot b verified. Especially in karst areas, the topographic conditions are complex, with numerou plateau mountains, peaks, and valleys, and the information identified by remote sensin needs to reflect the actual conditions. Therefore, in the future, we should strengthen th research on a small scale and obtain experimental data through the combination of fiel monitoring and 3S technology to clarify the process and mechanism of each ecosystem servic

4. The unclear trade-off/synergistic driving mechanisms of karst forest ecosysten services. The influences on ecosystem services arise from natural and socioeconomi systems, covering topography, soil, biology, climate, land use, socioeconomics, and huma activities [59]. Due to regional differences in development, the uneven regional distributio and diversification of socioeconomic factors such as population, education, social clas policies and regulations, religion and culture, urbanization, and economic level, loca people develop preferences for ecosystem services. Additionally, in the selection proces they indirectly or directly influence ecosystem services, creating trade-off/synergisti relationships. Therefore, in the future, it is essential to explore what factors play a dominan role at the spatial and temporal scales, the relevance of driving mechanisms at the spatia and temporal scales, and to clarify the coupling of various factors at different spatial an temporal scales [60].

5. Given the multitude of ecosystem service assessment methods, there needs to b more certainty in the results of trade-off/synergy studies among karst forest ecosysten services. Ecosystem service assessment involves many aspects including forestry, econc metrics, management, and ecology. There are more indicators of each service function an complex mechanisms of the action process. Additionally, there are significant difference among regions, research scales, and assessment methods. Especially in karst stone desertif cation areas, the ecological environment is special, and multiple processes can be selecte for comparative studies to enhance the scientific nature of the research methods.

4.4. Enlightenment on the Optimization of Forest Ecosystem Service Functions for Karst Desertification Control

Good social, economic, and ecological benefits have been achieved in participatin forest ecosystems in managing karst rock desertification [61]. However, the fragility of th karst rock desertification ecological environment [62] makes the forest ecosystem in thi area less able to resist external disturbances and less stable [63]. Once unreasonable huma activities occur and exacerbate the fragility of stone desertification habitats, they will chang the forest ecosystem's structural configuration and species composition [64] and affect th service provisioning capacity and trade-off/synergistic relationships. Therefore, the wor summarized and reviewed in this paper on the above ecosystem trade-off/synergy an functional optimization was synthesized. Insights into the optimization of forest ecosysten service functions in karst rock desertification management were provided regarding th forest ecosystem structure–function and trade-off relationships among services.

For the ecological structure–function, the karst stone desertification area has a shallov soil layer, discontinuous soil, high rock exposure rate, and rich soil calcium and magnesiur content [56]. The suitability for small habitats should be considered when selecting restor tion species, and drought-tolerant, calcium-loving, rocky, fast-growing, widely applicabl ecologically and economically valuable trees, bushes, vines, and grasses should be selecte for ecological restoration such as species of any bean, cedar, and lady's mantle [65]. Secon we should adjust the cultivation methods of natural forests and plantations, which ar sound ecological construction systems and natural gene pools with strong regulation an restoration abilities. The conservation of natural forests is mainly based on "no acces to protected forest areas" measures to preserve their species and genetic diversity an maintain the community structure and function [66]. For natural secondary forests in poo health, we adopted the measure of "managing the forest closed and adjusting the structur of the trees", and promoted better plant growth through artificial support measures such a replanting, nurturing, and inter-logging. For the planted plantations in natural forest area measures that support a "change of management approaches while protecting the forest in culturing" were adopted. Through the nurturing of mixed conifer and broadleaf forests, and the regular replanting of native species, we can induce succession in native forests and optimize the community structure and function [31]. Regarding the trade-offs between services, the trade-offs between forest ecosystem services in karst stone desertification areas resulted from a combination of natural and human factors.

Due to the poor soil in this area, plants compete with other species spatially to occupy more ecological niches during growth, leading to trade-off relationships between some services such as water connotation and soil conservation, nutrient netting, etc. [67]. Natural restoration or scientific fertilization can enhance soil quality in the stone desertification area and promote the balance of nutrient supply and demand. In addition, the appropriate stand density and planting depth need to be determined during planting to provide a wider space for plants to grow. The competition between plants for ecological niches should be reduced and the service capacity of water connotation and carbon sequestration and oxygen release should be improved while optimizing their trade-off relationship. The most important point is to reduce human interference in the forest ecosystem in stone desertification areas. The impact on the service function can be minimized by prohibiting deforestation, stepping on the protected forest area, reclaiming scrubland for cultivation, and planning the land for urban construction and highway construction in a reasonable way, in order to maintain the original state of the service function and reduce the formation of trade-off relationships.

5. Conclusions

This paper conducted a literature search for the trade-offs/synergies and functional optimization of forest ecosystem services through two databases, the Web of Science and CNKI. This study analyzed and reviewed 254 selected papers. The following conclusions were drawn. (1) The number of publications has shown a rapid growth trend since 2014, and the research process has gone through three stages: start (1970–2005), slow development (2005–2014), and rapid development (2014–2022). (2) Research regions are mainly concentrated in Asia, North America, and Europe, accounting for 40.47%, 25.55%, and 15.07% of all regions, respectively. (3) In the research, the content trade-off/synergy relationships were the most abundant, accounting for 61% of the total, followed by functional optimization and assessment methods, accounting for 20% and 11% of the total, respectively. (4) The following are future scientific questions to focus on and ideas for solutions: ① Given the severe soil erosion, weak stability of vegetation communities, and deteriorating ecosystems in karst stone desertification areas, a combination of forest sealing and artificial planting could enhance the vegetation cover, strengthen the ability to fix soil and retain water, and improve the system's stability. ② Given that karst stone desertification forest ecosystem services present trade-off/synergistic relationships, the reasons for the formation of inter-service trade-offs can be explored, and the trade-offs can be mitigated by enhancing the capacity of individual and multiple services. ③ Given that the spatial scale of forest ecosystem service trade-off/synergy studies is limited to large scales or single factors, the diversity of research dimensions should be increased, and a combination of remote sensing images and field monitoring should be used for research and validation. ④ Given the unclear trade-off/synergistic driving mechanisms of karst forest ecosystem services, the human-made and natural factors that cause trade-offs between services should be fully explored, and their specific drivers should be identified. ⑤ Given the multitude of ecosystem service assessment methods, there is a problem of uncertainty in the results of trade-off/synergy studies among karst forest ecosystem services. The research should be conducted using methods applicable to small watersheds such as habitat monitoring and analysis to enhance the scientific validity of the research results.

Author Contributions: Conceptualization, methodology, data curation, writing—original draft preparation, X.D.; Formal analysis, K.X.; Writing—review and editing, X.D., Y.Y. and S.Z.; Supervision, Y.Z.; Funding acquisition, K.X.; Software L.K. All authors have read and agreed to the published version of the manuscript.

Funding: This research was supported by the Key Science and Technology Program of Guizh Provence: Poverty alleviation model and technology demonstration for eco-industries derivated fr the karst desertification control (No. 5411 2017 Qiankehe Pingtai Rencai); the Science and Technolc Research Project of Higher Education Institutions in Hebei Province: Comparative study of mo and technology for characteristic high efficiency forestry from the karst desertification control North and South China Karst (No. QN2021412); the Scientific Foundation of Tangshan Norn University: Comparative analysis on vegetation restoration and characteristic high-efficiency fores in the desertification of South and North China Karst (No. 2021A07) and the project of Guizh Geographical Society: Research on characteristic high efficiency forestry from vegetation restorati of the karst desertification control (No. 2020HX05).

Data Availability Statement: The data presented in this study are available in the article.

Acknowledgments: The authors would like to thank Liyang Fu, Mingjun Feng, and Dong Chen their help with the grammar, translation, and revision of the drawings.

Conflicts of Interest: The authors declare no conflict of interest.

References

1. Daily, G.C. *Nature's Service: Societal Dependence on Natural Ecosystem*; Island Press: Washington, DC, USA, 1997; pp. 20–25.
2. Costanza, R.; d'Arge, R.; De Groot, R.; Farber, S.; Grasso, M.; Hannon, B.; Limburg, K.; Naeem, S.; O'neill, R.V.; Paruelo, J.; et The value of the world's ecosystem services and natural capital. *Nature* **1997**, *387*, 253–260. [CrossRef]
3. Millennium Ecosystem Assessment. *Ecosystems and Human Well-Being*; Island Press: Washington, DC, USA, 2005; Volum p. 563.
4. Lovrić, M.; Lovrić, N.; Mavsar, R. Mapping forest-based bioeconomy research in Europe. *For. Policy Econ.* **2020**, *110*, p1018 [CrossRef]
5. Groot, R.S.D.; Fisher, B.; Christie, M.; Aronson, J.; Braat, L.; Haines-Young, R.; Ring, I. *Integrating the Ecological and Econor Dimensions in Biodiversity and Ecosystem Service Valuation*; Earthscan: London, UK, 2010; pp. 9–40.
6. Begon, M.; Townsend, C.R. *Ecology: From Individuals to Ecosystems*; John Wiley & Sons: Hoboken, NJ, USA, 2020.
7. Haken, H. *Synergism-the Mystery of Nature's Composition*; Shanghai Translation Publishing House: Shanghai, China, 2005.
8. Baral, H.; Keenan, R.J.; Sharma, S.K. Economic evaluation of ecosystem goods and services under different landscape managem scenarios. *Land Use Policy* **2014**, *39*, 54–64. [CrossRef]
9. Xiong, K.N.; Zhu, D.Y.; Pang, T.; Yu, L.F.; Xue, J.H.; Li, P. Study on Ecological industry technology and demonstration for Ka rocky desertification control of the Karst Plateau-Gorge. *Acta Ecol.* **2016**, *36*, 7109–7113.
10. Yuan, D.X. World Comparison of Karst Ecosystems: Scientific Objectives and Implementation Plan. *Adv. Earth Sci.* **2001**, 461–466.
11. Yang, M.D. On the fragility of the Karst environment. *Yunnan Geogr. Environ. Res.* **1990**, *2*, 21–29.
12. Zhang, Y. *Tree, Shrub, Grass Restoration Mechanism and Characteristic High-efficiency Forestry Model in the Karst Rocky Desertificat Control*; Guizhou Normal University: Guiyang, China, 2020.
13. Pohjanmies, T.; Triviño, M.; Le, T.E.; Mazziotta, A.; Snäll, T.; Mönkkönen, M. Impacts of forestry on boreal forests: An ecosyst services perspective. *Ambio* **2017**, *46*, 743–755. [CrossRef] [PubMed]
14. Yuan, J.; Li, R.; Huang, K. Driving factors of the variation of ecosystem service and the trade-off and synergistic relationships typical karst basin. *Ecol. Indic.* **2022**, *142*, 109253. [CrossRef]
15. Yang, W.; Jin, Y.; Sun, T. Trade-offs among ecosystem services in coastal wetlands under the effects of reclamation activities. *E Indic.* **2018**, *92*, 354–366. [CrossRef]
16. Lang, Y.; Song, W. Trade-off analysis of ecosystem services in a mountainous karst area, China. *Water* **2018**, *10*, 300. [CrossRe
17. Haase, D.; Schwarz, N.; Strohbach, M.; Kroll, F.; Seppelt, R. Synergies, trade-offs, and losses of ecosystem services in urb regions: An integrated multiscale framework applied to the Leipzig-Halle Region, Germany. *Ecol. Soc.* **2012**, *17*, 22. [CrossRe
18. Niu, T.; Yu, J.; Yue, D. The Temporal and Spatial Evolution of Ecosystem Service Synergy/Trade-Offs Based on Ecolcgical Un *Forests* **2021**, *12*, 992. [CrossRef]
19. Schroder, S.A.; Tóth, S.F.; Deal, R.L.; Ettl, G.J. Multi-objective optimization to evaluate tradeoffs among forest ecosystem servi following fire hazard reduction in the Deschutes National Forest, USA. *Ecosyst. Serv.* **2016**, *22*, 328–347. [CrossRef]
20. Chen, J.; Pu, J.; Li, J.; Zhang, T. Trade-off and synergy analysis among ecosystem services in a karst watershed. *Geocarto Int.* **20** 1–17. [CrossRef]
21. Zhang, J.J.; Zhu, W.B.; Zhu, L.Q.; Li, Y.H. Multi-scale analysis of trade-off/synergy effects of forest ecosystem services in Funiu Mountain Region. *Acta Geogr. Sin.* **2020**, *75*, 975–988.
22. Han, H.; Yin, C.; Zhang, C.; Gao, H.; Bai, Y. Response of trade-offs and synergies between ecosystem services and land use char in the Karst area. *Trop. Ecol.* **2019**, *60*, 230–237. [CrossRef]
23. Feng, Y.; Cao, Y.G.; Li, S.P.; Wang, S.F.; Liu, S.H.; Bai, Z.K. Trade-offs and synergies of ecosystem services: Development histc and research characteristics. *J. Agric. Resour. Environ.* **2022**, *39*, 11–25.

4. Han, H.Y.; Su, Z.H. Research progress and prospect of karst ecosystem services. *Carssologica Sin.* **2017**, *36*, 352–358.
5. Chen, T.; Huang, Q.; Wang, Q. Differentiation characteristics and driving factors of ecosystem services relationships in karst mountainous area based on geographic detector modeling: A case study of Guizhou Province. *Acta Ecol. Sin.* **2022**, *42*, 6959–6972.
6. Chen, R.; Wang, S.J.; Bai, X.Y. Trade-offs and synergies of ecosystem services in southwestern China. *Environ. Eng. Sci.* **2020**, *37*, 669–678.
7. Liu, S.J.; Zhang, W.; Wang, K.L.; Pan, F.J.; Yang, S.; Shu, S.Y. Factors controlling accumulation of soil organic carbon along vegetation succession in a typical karst region in Southwest China. *Sci. Total Environ.* **2015**, *521*, 52–58. [CrossRef] [PubMed]
8. Feng, T.; Chen, H.S.; Polyakov, V.O.; Wang, K.L.; Zhang, X.B.; Zhang, W. Soil erosion rates in two karst peak-cluster depression basins of northwest Guangxi, China: Comparison of RUSLE model with 137 Cs measurements. *Geomorphology* **2016**, *253*, 217–224. [CrossRef]
9. Fu, Z.Y.; Chen, H.S.; Xu, Q.X.; Jia, J.T.; Wang, S.; Wang, K.L. Role of epikarst in near-surface hydrological processes in a soil mantled subtropical dolomite karst slope: Implications of field rainfall simulation experiments. *Hydrol. Process.* **2016**, *30*, 795–811. [CrossRef]
0. Wen, X.W.; Quan, J.L.; Wang, R.X.; Fu, L.T. Research Progress of Restoration of Degraded Karst Forest Bank. *Mod. Agric. Sci. Technol.* **2016**, *2*, 172–174+177.
1. Liu, S.R.; Dai, L.M.; Wen, Y.U.; Wang, H. A review on forest ecosystem management towards ecosystem services: Status, challenges, and future perspectives. *Acta Ecol.* **2015**, *35*, 1–9.
2. Wang, X.; Zhang, X.; Feng, X. Trade-offs and synergies of ecosystem services in karst area of China driven by grain-for-green Program. *Chin. Geogr. Sci.* **2020**, *30*, 101–114. [CrossRef]
3. Chen, T.T.; Wang, Y.X.; Zeng, X.L.; Wang, Q. Characteristics and the constraint relationship between ecosystem services and vegetation in the Southwest China. *Acta Ecol. Sin.* **2022**, *6*, 1–19.
4. Wang, N.; Chu, X.L.; Gou, M.M.; Li, L.; La, L.M.; Liu, C.F. Trade offs and Synergies Analysis on Forest Ecosystem Services in the Three Gorges Reservoir Area. *Ecol. Environ. Sci.* **2021**, *30*, 475–484.
5. Qiu, J.J.; Liu, Y.H.; Yu, A.L.; Chen, C.J.; Huang, Q.Y. Research progress and prospect of the interrelationship between ecosystem services and human well-being in the context of coupled human and natural system. *Prog. Geogr.* **2021**, *40*, 1060–1072. [CrossRef]
6. Wunder, S. Revisiting the concept of payments for environmental services. *Ecol. Econ.* **2015**, *117*, 234–243. [CrossRef]
7. Sun, Y.F.; Qin, W.; Fan, J.Y. Spatial selection of ecological compensation in key ecological functional areas of Xinjiang. *Arid Land Geogr.* **2021**, *44*, 565–573.
8. Wang, K.L.; Chen, H.S.; Yue, Y.M. Experiment and Demonstration on Degraded Mechanism and Its Adaptive Restoration of Karst Ecosystems in Northwest Guang xi. *Sci. Technol. Promot. Dev.* **2015**, *3*, 179–183.
9. Yu, Y.H.; Li, G.R.; Pi, F.J.; Yan, L.B.; Yu, L.F.; Huang, Z.S. Evaluation of water conservation function of main forests in the upper reaches of Chishui River. *J. Soil Water Conserv.* **2015**, *29*, 150–156.
0. Chen, W. Mechanism of Community Configuration and Optimized Regulation Technologies of Forest, Shrub and Grass in the Karst Rocky Desertification Control. Master's Thesis, Guizhou Normal University, Guiyang, China, 2019.
1. Zhang, S.H. Succession Law of Plant Communities and Structural Management Techniques in the Karst Rocky Desertification. Master's Thesis, Guizhou Normal University, Guiyang, China, 2020.
2. Yuan, Z.Y.Y.; Wan, R.R. A review on the methods of ecosystem service assessment. *Ecol. Sci.* **2019**, *38*, 210–219.
3. Huang, C.H.; Yang, J.; Zhang, W.J. Development of ecosystem services evaluation models: Research progress. *Chin. J. Ecol.* **2013**, *32*, 3360–3367.
4. Polasky, S. What's Nature Done for You Lately: Measuring the Value of Ecosystem Services. *Choices* **2008**, *23*, 469–476.
5. Cao, Q.W.; Wei, X.M.; Wu, J.S. A review on the trade offs and synergies among ecosystem services. *Chin. J. Ecol.* **2016**, *35*, 3012–3111.
6. Turner, K.G.; Odgaard, M.V.; Bocher, P.K.; Dalgaard, T.; Svenningb, C.J. Bundling ecosystem services in Denmark: Trade-offs and synergies in a cultural landscape. *Landsc. Urban Plan.* **2014**, *125*, 89–104. [CrossRef]
7. Guo, Z.Z. Trade-off and coordination of ecosystem service in Guanzhong-Tianshui Economic Zone. Master's Thesis, Shaanxi Normal University, Xi'an, China, 2015.
8. Li, S.C.; Zhang, C.Y.; Liu, J.L.; Zhu, W.B.; Ma, C.; Wang, J. The tradeoffs and synergies of ecosystem services: Research progress, development trend, and themes of geography. *Geogr. Res.* **2013**, *32*, 1379–1390.
9. Long, J.H. The Study on Evaluation and Trade-Off of Ecosystem Services in the He Gang Mining Area. Ph.D. Thesis, China University of Mining and Technology, Beijing, China, 2017.
0. Dai, E.F.; Wang, X.L.; Zhu, J.; Zhao, D. Methods, tools and research framework of ecosystem service trade-offs. *Geogr. Res.* **2016**, *35*, 1005–1016.
1. Bagstad, K.J.; Semmens, D.J.; Waage, S.; Winthropc, R. A comparative assessment of decision-support tools for ecosystem services quantification and valuation. *Ecosyst. Serv.* **2013**, *5*, 27–39. [CrossRef]
2. Jiang, S.; Xiong, K.; Xiao, J. Structure and Stability of Agroforestry Ecosystems: Insights into the Improvement of Service Supply Capacity of Agroforestry Ecosystems under the Karst Rocky Desertification Control. *Forests* **2022**, *13*, 878. [CrossRef]
3. Reidsma, P.; Feng, S.; van Loon, M.; Luo, X.; Kang, C.; Lubbers, M.; Kanellopoulos, A.; Wolf, J.; Van, I.M.K.; Qu, F. Integrated assessment of agricultural land use policies on nutrient pollution and sustainable development in Taihu Basin, China. *Environ. Sci. Policy* **2012**, *18*, 66–76. [CrossRef]

54. Kumar, B.M.; Singh, A.K.; Dhyani, S.K. South Asian Agroforestry: Traditions, Transformations, and Prospects. In *Agroforestry—Th Future of Global Land Use*; Springer: Dordrecht, The Netherlands, 2012; pp. 359–389.
55. Jordanka, S.; Sonja, B.; Krasimira, P.; Vladimir, P. Possibilities for agroforestry development in Bulgaria: Outlooks and limitation *Ecol. Eng.* **2007**, *29*, 382–387.
56. Sheng, M.Y.; Xiong, K.N.; Cui, G.Y.; Liu, Y. Plant diversity and soil physical-chemical properties in karst rocky desertificatio ecosystem of Guizhou, China. *Acta Ecol.* **2015**, *35*, 434–448.
57. Wang, B.; Zhao, J.; Hu, X.F. Analysison trade-offs and synergistic relationships among multiple ecosystem services in the Shiyan River Basin. *Acta Ecol.* **2018**, *38*, 7582–7595.
58. Pan, J.; Li, Z. Analysis on trade-offs and synergies of ecosystem services in arid inland river basin Transactions of the Chines *Soc. Agric. Eng.* **2017**, *33*, 280–289.
59. Zhao, W.W.; Liu, Y.; Feng, Q.; Wang, Y.P.; Yang, S.Q. Ecosystem services for coupled human and environment systems. *Pro Geogr.* **2018**, *37*, 139–151.
60. Ou, Z.R.; Sun, Y.Y.; Deng, Z.H.; Feng, D.F. Trade-offs in forest ecosystem services: Cognition, approach and driving. *Sci. So Water Conserv.* **2020**, *18*, 150–160.
61. Zou, Z.; Zeng, F.; Wang, K.; Zeng, Z.; Zhao, L.; Du, H.; Zhang, F.; Zhang, H. Emergy and Economic Evaluation of Seven Typica Agroforestry Planting Patterns in the Karst Region of Southwest China. *Forests* **2019**, *10*, 138. [CrossRef]
62. Wang, R.; Cai, Y.L. A model for remediation of degraded ecosystems in karst areas of southwest China. *J. Appl. Ecol.* **2010**, *2* 1070–1080.
63. Peng, W.X.; Wang, K.L.; Song, T.Q.; Zeng, F.P.; Wang, J.R. Controlling and restoration models of complex degradation of vulnerabl Karst ecosystem. *Acta Ecol. Sin.* **2008**, *28*, 811–820.
64. Yang, S.M.; Xiong, K.N.; Yu, Y.H.; Liu, X.Y.; Dong, X.C. Diagnosis and adjustment of forest and grass vegetation restoratio patterns in karst rocky desertifification areas in China. *World For. Res.* **2017**, *30*, 91–96.
65. Wang, K.L.; Yue, Y.M.; Chen, H.S.; Wu, X.B.; Xiao, J.; Qi, X.K.; Zhang, W.; Du, H. The comprehensive treatment of karst rock desertification and its regional restoration effects. *Acta Ecol. Sin.* **2019**, *39*, 7432–7440.
66. MacDicken, K.G.; Sola, P.; Hall, J.E.; Sabogal, C.; Tadoum, M.; de Wasseige, C. Global progress toward sustainable fores management. *For. Ecol. Manag.* **2015**, *352*, 47–56. [CrossRef]
67. Han, H.Q.; Yang, J.Q.; Chen, S.Y.; Xu, J.; Wang, T.G.; Hou, J.W. Spatial Grain Size Effect of Trade-off Synergy Relationship Betwee Fresh water Ecosystem Services in Karst Mountainous Areas. *J. Univ. Jinan Sci. Technol.* **2022**, *36*, 45–55.

land

ticle

Iomoge Internationally Important Wetland: Ecosystem ıtegrity Remote Assessment and Spatial Pattern Optimization Study

qi Han [1,2], Dongyan Wang [1,*] and Shuwen Zhang [2]

1 College of Earth Sciences, Jilin University, Changchun 130061, China
2 Northeast Institute of Geography and Agroecology, Chinese Academy of Sciences, Changchun 130012, China
* Correspondence: wang_dy@jlu.edu.cn

Abstract: Along the migration route between East Asia and Australia, numerous migratory birds use the Momoge Internationally Important Wetland as a habitat. Human activities and climate variability cause salinization and meadowization. We developed the "Quality-Pressure-Pattern-Service" remote assessment framework for ecosystem integrity, using a three level approach (TLA). The model was used to assess ecosystem integrity, identify improper wetland development, and provide spatial optimization strategies. The research region was dominated by wetlands, followed by dry fields. Wetlands continued to decrease between 1965 and 2019, as arable land and construction land continued to increase. Over the course of 54 years, ecosystem integrity declined. In 2019, around half of the areas had poor or extremely poor ecosystem integrity. Because the eastern study area contained many pristine inland beaches, the eastern study area displayed greater ecosystem integrity than the central and western areas. Priority should therefore be given to wetland restoration in the HJ core area (one of the three core areas of the reserve), where most of the herb marsh has been converted to arable land. This study revealed the integrity and authenticity of wetland ecosystems. Our results can aid in the protection of wetland habitats, encourage sustainable development, and help in the building of a national park in northeastern China.

Keywords: CORONA; environmental indicators; land use; national nature reserve; remote sensing

tion: Han, J.; Wang, D.; Zhang, S. noge Internationally Important land: Ecosystem Integrity Remote essment and Spatial Pattern imization Study. *Land* **2022**, *11*, 4. https://doi.org/10.3390/ 11081344

demic Editors: Panayiotis aitrakopoulos, Mario A. Pagnotta, las Scholz and Arshiya Noorani

eived: 7 July 2022
epted: 12 August 2022
lished: 18 August 2022

1. Introduction

The historical legacy of nature reserves is one of China's most significant environmental issues. In China's 474 national nature reserves, there are currently 29 urban built-up areas, 531 built-up areas consisting of formed townships, and 779 administrative villages [1]. Over the past three decades, scholars have created a variety of frameworks for assessing ecosystem integrity to characterize the natural and wilderness states of unaltered nature reserves. Initially, ecosystem integrity was recommended as a scientific evaluation criterion for ecosystem management in national parks in Canada [2]. Traditional frameworks for assessing ecosystem integrity and authenticity include the Index of Biotic Integrity (IBI) [3–5], the Ecosystem Integrity Assessment Framework (EIAF) [6], and the Ecosystem Integrity Assessment System Based on Essential Ecosystem Characteristics (EECs) [7]. In general, local and international scholars have centered their evaluations of ecosystem integrity on environmental stress, ecological processes, and biodiversity [8].

Wetlands exhibit some combined functions with those of terrestrial and aquatic ecosystems and they are among the most critical ecosystems [9–13]. However, the global wetland area has decreased by 35% during the past 50 years (https://www.global-wetland-outlook.ramsar.org/ (accessed on 17 December 2021)). Land use and development contribute significantly to the loss of wetland habitats. Various reports have discussed the loss of wetlands as a result of land use change, such as the extensive and rapid transformation of natural wetlands in Asia [14], the loss of 33% of the total domestic wetland in China between 1978 and 2008 [15], and the extensive conversion of marshland to construction and arable land China [16]. The conventional methodology used for assessing ecosystem integrity requi very high-altitude survey data and is challenging to apply to complex wetland habita The remote evaluation framework of the Three Level Approach (TLA) has establishe number of evaluation indicators that can be evaluated through remote sensing [17,18]. (this basis, numerous evaluations of ecosystem integrity have been performed through t adoption of a Geographic Information System (GIS) analysis [19,20].

There is an urgent need to develop ecosystem integrity assessment frameworks th can be applied more frequently and repeatedly to nature wetlands to evaluate, monitor, a report on the condition of the region's ecosystems. Natural wetland parcels are genera fragmented, and delineating multiple natural land use types at fine scales is a limit process. Wetland data from the Third National Land Survey of China (2019) and the Seco National Wetland Survey (2013) provide more fine-scale wetland vector data. However is difficult to obtain sufficient vector data for most wetland studies. Therefore, few previc studies have studied wetland changes over periods longer than 50 years and studies ha failed to explore the more primitive wetland landscapes in history.

Several research findings have emphasized the viability of employing CORON images to gain information on fine wetland types [21,22]. However, it has been not that CORONA images have certain deficiencies in terms of both feature identification a band calculation [23]. In this study, we gathered historical information on the resear area, such as county records, almanacs, and reports outlining the distribution of wetlan and hydrothermal conditions, to address this issue. In addition, we employed a hum visual interpretation technique to adjust the land use change data piece by piece, whi dramatically enhanced our ability to identify natural wetland types with precision.

Managing wetland areas with historical legacies is essential in order to safeguard co wetland areas and repair degraded wetlands. In response to numerous unknown fact concerning the degradation and development of wetland areas in the study area [24,2 an area of inappropriate development was identified, and the spatial pattern of the w land areas was modified to meet ecological requirements to the greatest extent possib In addition to the basic functions of natural wetland systems, wetlands with histori legacies are typically tightly linked to the production activities and living situations humans [26–28]. Consequently, resolving historical legacy issues in wetland nature serves plays a crucial role in fostering high-quality economic development and high-le ecological environmental protection through synergistic means.

On the basis of RS and GIS technologies, we developed a framework for assessi the integrity of ecosystems that is less expensive and requires less time. Such a framew will not replace assessment methods based on rigorous field surveys, but it will aid the constant monitoring of changes in ecosystem integrity and provide essential data addressing legacy concerns in nature reserves. In this study, we investigated the spatial a temporal aspects of a wetland from 1965 to 2019, utilizing the Momoge National Nat Reserve as a case study. On the basis of the "Quality-Pressure-Pattern-Service" ecosyst integrity remote assessment methodology, wetland ecosystem integrity was assessed. T optimization of the spatial pattern in the reserve was also performed to assist in providi fine control over the wetland's ecology.

2. Materials and Methods

2.1. Study Area

The Momoge Internationally Important Wetland is situated in Baicheng City, in t western portion of Jilin Province (Figure 1). The research area encompasses 144,000 and has a semi-humid, temperate continental monsoon climate. It sits at the weste boundary of the Sonnen Plain, which was produced by the second northern subsider zone of the Neocathasia tectonic structure. This structure is the Yingtai structure, whi faces north-northeast. In the reserve, swamps have evolved in lakefront depressio river floodplains, and low-lying places. The primary soil types of the reserve inclu meadow soil, chernozem soil, and alluvial soil. There are as many as 55 different varieties of vegetation in the Momoge wetland, which belongs to the Eurasian steppe region. The major plant communities consist of a *Phragmities australis* community, a *Deyeuxia angusrifolia* community, a *Carex spp* community, etc. It is one of the three greatest soda saline land distribution zones in the world and is a crucial stopover for 90 percent of the world's large waterfowls and wading birds, such as *Grus leucogeranus*, migrating between Siberia and Oceania. The area was Established as a natural reserve in 1981 and designated as a globally significant wetland in 2013.

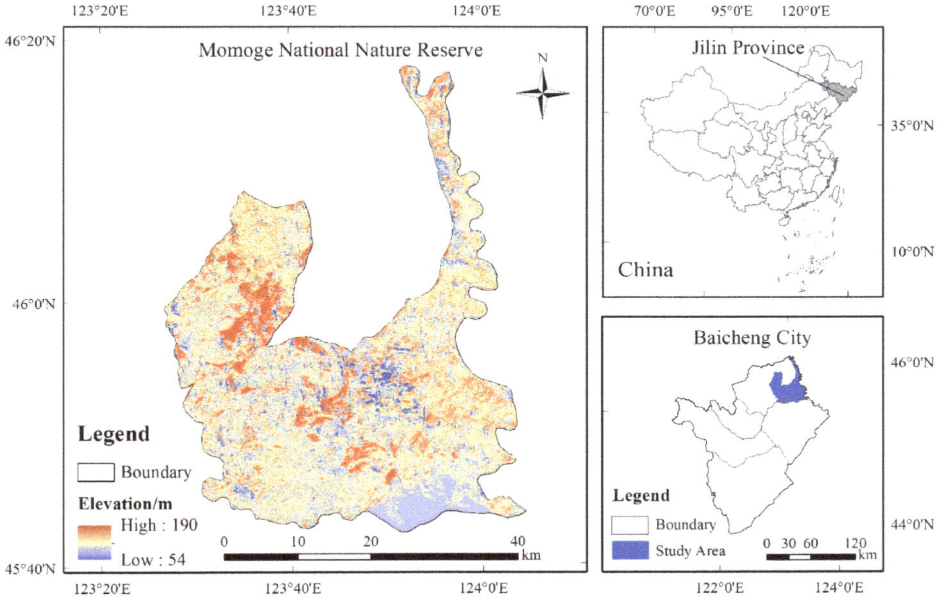

Figure 1. Location of Momoge Internationally Important Wetland.

2.2. Data Sources and Data Processing

2.2.1. Ecological Assessment Unit

To balance the time- consumption of land use modeling and the richness of picture information, a 1 km × 1 km grid was used as the assessment cell size. First, all variable data were transformed to single-precision TIF images with the same raster size, and then 1 km grid cells were counted.

2.2.2. Basic Geographic Data

In this study, we utilized meteorological data, DEM data, and land use data as the basic forms of geographic data. Meteorological data and some data on vegetation cover were collected from the National Earth System Science Data Center, National Science and Technology Infrastructure of China (http://www.geodata.cn/ (accessed on 9 December 2021)). For the years 1965, 2015, and 2019, the monthly precipitation and annual mean temperature of China were computed with a 1 km resolution. Since it was difficult to collect data on the vegetation cover in 1965, we used Landsat 3 images of the Momoge Nature Reserve from 1981 for our calculations. The geospatial data cloud (http://www.gscloud.cn/ (accessed on 10 May 2020)) provided the ASTERG DEMv2.0 data with a spatial resolution of 30 m. In this study, we made use of the 2013 wetland survey data from the forestry department, the 2013 land use change data from the Second National Land Survey, and the 2019 data from the Third National Land Survey.

2.2.3. Remote Sensing Images

The Third National Land Survey utilized satellite remote sensing images with resolution of over 1 m and classified wetlands in detail, whereas the Second National Lan Survey utilized SPOT-5 satellite images with a resolution of 2.5 m and did not classif wetland types. To differentiate wetland types using the 2013 land use change data, thi study should be complemented with by the use of additional remote sensing data with a r resolution. We chose the 1965 KH-4A satellite and the 2015 ZY-3 satellite as remote sensin information sources (Table 1).

Table 1. ZY-3 satellite and KH-4A satellite specification information.

Satellite	Year	Day/Month	Resolution (m)	Image Type	Source
ZY-3	2015	09/December	2.1	panchromatic	Satellite Environmental Application Center of the Ministry of Ecology and Environment (http://www.secmep.cn/ (accessed on 25 February 2020))
ZY-3	2015	2 6/October	2.1	panchromatic	
KH-4A	1965	23/September	2.7	panchromatic	the USGS (https://www.usgs.gov/ (accessed on 9 December 2019))

Coronal images have panoramic aberrations. We used corrected ZY-3 satellite data as reference for calibration. The root-mean-square error of the geometric correction results wa less than 1.0. The 1965 and 2015 land cover vector data in the research region were finall generated by changing the 2013 land use change data patches. To maintain consistenc in the classification accuracy of the data, we analyzed the 2019 land use vector data an combined road and ditch patches with widths of less than 2 m into adjacent land classe In addition, industrial and mining land patches were eliminated from the land use ma We evaluated the accuracy of the 2015 and 2019 classification results using high-resolutio Google Earth images. Due to the paucity of historical geographic data, we validated th correctness of the 1965 classification by creating a pseudo-color composite image of th Corona image using the density segmentation function. The results demonstrated that th overall accuracy of the secondary image classification reached 90%.

2.3. Method

The selection of 1965, 2015, and 2019 as the years for our analysis of land use type dat was based on three main considerations:

1. The accessibility of meter resolution remote sensing images;
2. The avoidance of years of extreme flooding and drought; and
3. The accessibility of the wetland vector database.

This paper is structured as the follows (Figure 2). First, we merge several mete resolution remote sensing photographs to obtain land use type data for 1965, 2015, an 2019 in the research area. Then, we model land use dynamics and discuss land use change We present the final version of the " Quality-Pressure-Pattern-Service " ecosystem integrit remote assessment framework based on the TLA remote assessment framework. Finall by evaluating the ecosystem integrity and authenticity of the study area, we identif the inappropriate development areas within the reserve and make recommendations fo optimizing its spatial pattern.

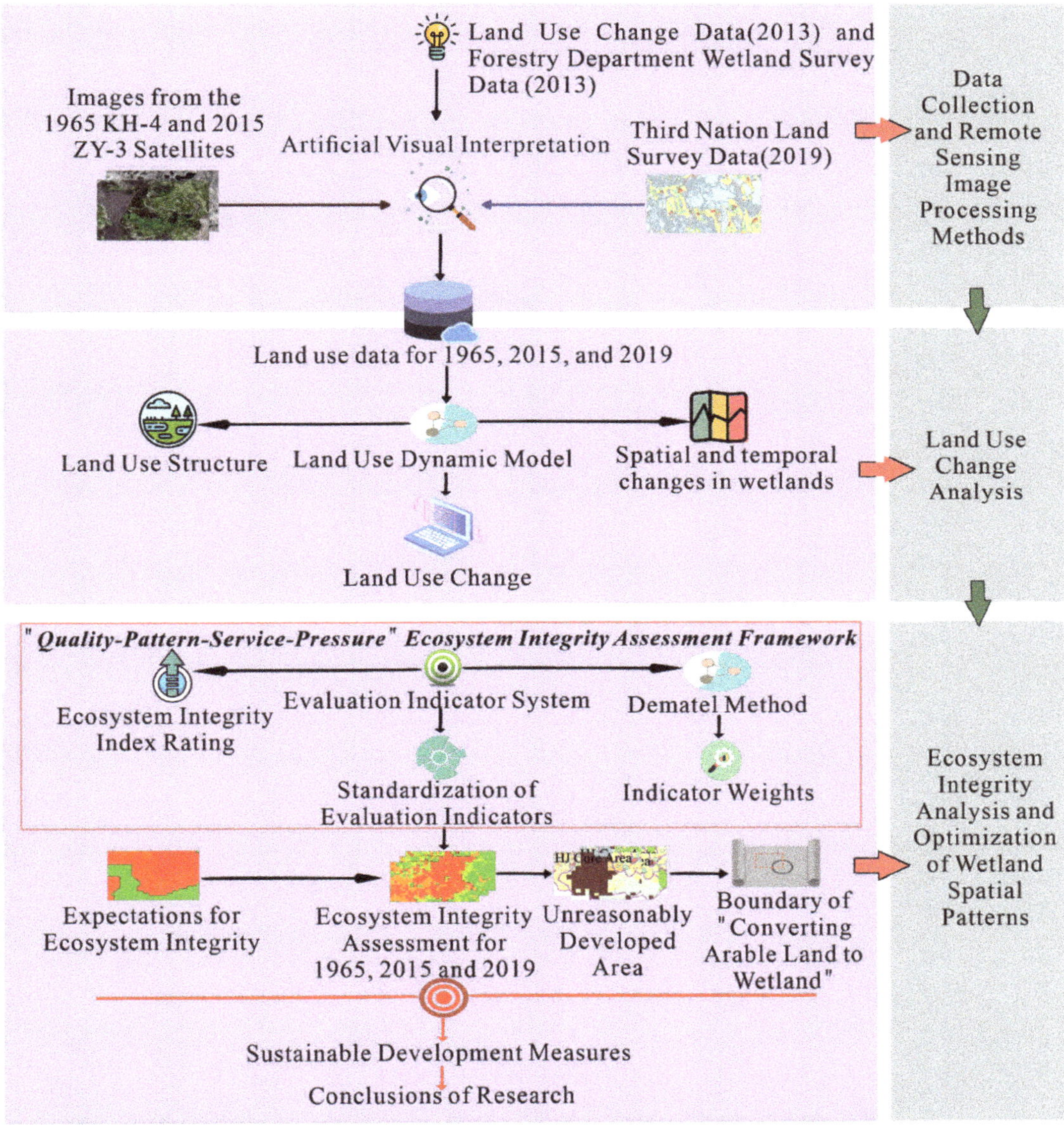

Figure 2. Flowchart of the study.

2.3.1. Land Classification System

We determined the land classification system employed in this work by referring to prior wetland classification studies, such as the Ramsar Convention, the Third National Land Survey Classification, and the National Forestry Administration wetland survey classification (Table 2). Our analysis focused on wetlands and we divided them into narrow wetlands and water.

Table 2. The land classification system used in this study.

Class 1	Class 2	Class 3
Wetland	Narrow wetland	Inland beach Salt marsh Herb marsh
	Water	—
Non-wetland	Arable land	Paddy field Dry field
	Woodland	—
	Grassland	—
	Construction land	—
	Saline land	—
	Other land	—

2.3.2. Land Use Dynamic Model

The single land use change rate is the rate of a specific land use or cover change in t research region, and it is computed as follows [29]:

$$V = \frac{B_1 - B_0}{B_0} \times \frac{1}{T} \times 100\%$$

In Equation (1), B_0 and B_1 are the area of a certain land use type at the beginning a conclusion of the study period, respectively, and T = 54 is the length of the study perio

2.3.3. Ecosystem Integrity Evaluation Index System

Ecosystem integrity can be understood in terms of the quality of the ecosystem, t stability of the landscape, the ecological services supplied by the ecosystem, and its capac to withstand less external pressure. Consequently, we developed a "Quality-Pressu Pattern-Service" ecosystem integrity remote assessment framework (Figure 3) by referri to the three-level approach (TLA).

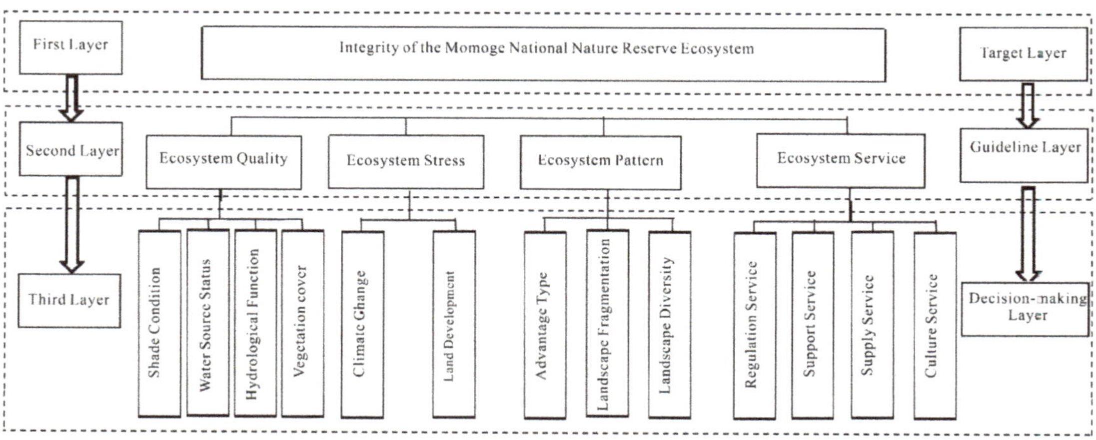

Figure 3. "Quality-Pressure-Pattern-Service" ecosystem integrity remote assessment framework.

The evaluation framework consists of four factors: ecosystem quality, ecological str sors, ecosystem patterns, and ecosystem services. The indicators were chosen based the characteristics of wetland ecosystems and the concepts of comprehensiveness, rep sentativeness, practicality, accessibility, and measurability. Table 3 displays the indica system used to evaluate the ecosystem integrity of the Momoge Internationally Importa Wetland. We selected 24 variables relevant to the integrity of wetland ecosystems, includi 7 natural attribute variables, 5 spatial density variables, 3 landscape pattern variables, a

9 functional value variables, based on previous research. We did not select socio-economic characteristics in consideration of the reserve's real population and economic growth.

Table 3. "Quality-Pressure-Pattern-Service" ecosystem integrity evaluation factor table.

Level 1	Level 2	Level 3	Calculation Methods	Indicator Sources
Ecosystem quality	Shade condition	DEM	-	Evaluation of waterfowl habitat suitability [30,31].
		Slop	Slop tool in ArcGIS 10.8	
	Water source status	River density	Kernel density analysis in ArcGIS 10.8	The PCFAEI and EECs evaluation frameworks both contain indicators of hydrology and plant life [7,32].
		Lake density		
	Hydrological function	TWI	[33,34]	
	Vegetation cover	NDVI	[35]	
Ecosystem stress	Climate change	WI	[36]	Climate change indicators under the PCFAEI and EECs assessment frameworks [7,32].
		HI		
	Land development	Ecosystem comprehensive anthropogenic disturbance index (ECADI)	[37]	Principal stressors in the evaluation frameworks of the PCFAEI, EIAF, and EECs [6,7,32].
		Road density	Kernel density analysis in ArcGIS 10.8	
		Ditch and dyke density		
		Density of residential land		
Ecosystem pattern	Advantage type	LPI	The index was computed in Fragstats 4.2 employing the moving window method with a 1000 m window size	Indicators of landscape pattern in the PCFAEI assessment framework [19,32].
	Landscape diversity	SHDI		Indicators of ecosystem quality within the EECs assessment system [7]
	Landscape fragmentation	CONTAG		
Ecosystem service	Regulation service	Gas regulation	Details of the method are provided in Section 2.3.4.	Xie et al. improved China's ecological valuation process for terrestrial ecosystem services [38,39].
		Climate regulation		
		Waste disposal		
		Water conservation		
	Support service	Soil conservation		
		Biodiversity conservation		
	Supply service	Food production		
		Raw material production		
	Culture service	Aesthetic landscape		

2.3.4. Ecosystem Service Indicators

Using the global ecosystem service function evaluation model [40], Gao et al. determined the ecosystem service values per unit area in China [41]. Their findings established one standard equivalence factor for the net profit of food production per unit area of a farmland ecosystem. For research and comparison purposes, the 2007 grain price (449.1 CNY/hm^2) was used as 1 ecological service value equivalent factor.

To make the evaluation model more regionally applicable, we revised the ecological service value per unit area of Chinese ecosystems (2007) [41]. The biomass factor in Jilin Province was 0.96 [3], compared to a national average biomass factor of 1. First, we used this factor to adjust the value of ecological services per unit area of the ecosystem and obtained an initial value factor appropriate for the Momoge Nature Reserve. The ecological value coefficients of farmland were then adjusted using the equivalence factors of paddy fields and drylands (Table 4) [38]. Finally, the ecological service value per unit area suited to the actual conditions of the research region was determined (Table 5).

Table 4. Equivalent value per unit area of cropland ecosystem services in China (2011).

Ecological Service Types	Paddy Field	Dry Field
Gas regulation	1.11	0.67
Climate regulation	0.57	0.36
Headwater conservation	2.72	0.27
Waste treatment	0.17	0.1
Soil formation and protection	0.01	1.03
Biodiversity protection	0.21	0.13
Food production	1.36	0.85
Raw materials	0.09	0.4
Cultural and recreation	0.09	0.06
Total	6.33	3.87

Table 5. Nature Reserve ecological service values per unit area (CNY/haa^{-1}).

Ecological Service Types	Paddy Field	Dry Field	Woodland	Grassland	Water	Wetland	Saline Land
Gas regulation	344.56	207.98	1862.51	646.70	219.88	1039.04	25.87
Climate regulation	238.38	150.55	1754.73	672.58	888.14	5841.90	55.04
Headwater conservation	902.98	89.63	1763.35	655.32	8092.43	5794.46	30.18
Waste treatment	101.88	59.93	741.55	569.10	6402.37	6208.36	112.10
Soil formation and protection	6.34	652.79	1733.16	965.74	176.76	857.96	73.30
Biodiversity protection	92.35	57.17	1944.42	806.23	1478.79	1590.89	172.45
Food production	586.34	366.47	142.27	185.39	228.50	155.21	8.62
Raw materials	35.31	67.26	1284.79	155.21	150.90	103.47	17.24
Cultural and recreation	6.60	4.40	896.76	375.09	1914.24	2022.03	103.47
Total	2314.73	1656.17	12,123.54	5031.36	19,552.02	23,613.32	599.28

2.3.5. Weighting and Standardization of Evaluation Indicators

The Decision-making Trial and Evaluation Laboratory (Dematel) was used to calculat the degree of influence of each element on the other elements and the degree of influence t determine the causal relationship between the indicators and the weight of each indicatio in the system [42]. Table 6 displays the relative importance of the ecosystem integrity ind cators used in this study. The min–max normalization approach was used to standardiz the evaluation indexes to make them comparable [43].

2.3.6. The Integrity Index and Grading Standards for Ecosystems

The ecosystem integrity index is a composite score that evaluates the significanc and influence of each evaluation index on ecosystem integrity [20,44]. In this study, w used the comprehensive index method to calculate the Momoge Internationally Importar Wetland's ecosystem integrity index. The ecosystem integrity index is determined by th following formula:

$$EI = \sum_{i=1}^{n} \omega_i A_i \quad (2$$

In Equation (2), EI is the ecosystem integrity index, n is the number of indicators, ω is the weight value of the ith indicator, and A_i is the normalized value of the ith indicato The higher the EI, the more intact the ecosystem is.

To examine the relative levels of ecosystem integrity in various regions, in this stud we constructed the Momoge Internationally Important Wetland Ecosystem Integrity Evalu ation Scale based on the natural breakpoint method (Table 7).

Table 6. "Quality-Pressure-Pattern-Service" ecosystem integrity evaluation factor weighting table.

Level 1	Weight Value	Level 2	Weight Value	Level 3	Weight Value	Indicator Orientation
Ecosystem quality	0.306	Shade condition	0.040	DEM	0.014	−
				Slop	0.026	−
		Water source status	0.137	River density	0.084	+
				Lake density	0.053	+
		Hydrological function	0.027	TWI	0.027	+
		Vegetation cover	0.102	NDVI	0.102	+
Ecosystem stress	0.371	Climate change	0.052	WI	0.025	−
				HI	0.027	+
		Land development	0.319	ECADI	0.068	−
				Road density	0.092	−
				Ditch and dyke density	0.057	−
				Density of residential land	0.102	−
Ecosystem pattern	0.109	Advantage type	0.026	LPI	0.026	+
		Landscape diversity	0.039	SHDI	0.039	−
		Landscape fragmentation	0.044	CONTAG	0.044	+
Ecosystem service	0.214	Regulation service	0.117	Gas regulation	0.025	+
				Climate regulation	0.023	+
				Waste disposal	0.035	+
				Water conservation	0.034	+
		Support service	0.049	Soil conservation	0.022	+
				Biodiversity conservation	0.027	+
		Supply service	0.028	Food production	0.019	+
				Raw material production	0.009	+
		Culture service	0.020	Aesthetic landscape	0.020	+

Table 7. Evaluation and grading of Ecosystem Integrity results.

Level	Index	Ecosystem Integrity Status
Excellent	>0.8	Ecosystem structure, composition, and function all shift within the range of natural disturbance. There are no or few ecological issues, and the anthropogenic disturbance pressure is minimal.
Good	0.6–0.8	The structure, composition, and function of ecosystems fluctuate within the range of natural disturbances. There are mild ecological concerns and low anthropogenic disturbance pressure.
Medium	0.4–0.6	The structure, composition, and function of the ecosystem vary within the range of natural disturbances, with certain ecological concerns and human disturbance pressure.
Poor	0.2–0.4	Changes in ecosystem structure, composition, and function of the ecosystem are beyond natural disturbance. There are serious ecological problems and high pressures from human interference.
Extremely poor	0–0.2	Changes in the structure, composition, and function of ecosystems beyond the scope of natural disturbances. There are serious ecological problems, high pressure from anthropogenic disturbance, and ecological processes that are difficult to reverse.

2.3.7. Spatial Pattern Optimization Method

Based on the above ecosystem integrity remote assessment framework and the m surement method presented by Wei et al. [45], the integrated ecological function assessm value was subdivided into four degrees to illustrate the strength of the ecological functi in different regions.

For the classification of the levels, we took into account the probability distribution the values under all grids and established a, b, and c as the cut-off values for the integrat ecological functioning of the district, ranging from the lowest to the highest. Then, w reference to the functional area border of the Nature Reserve designated by the St Environmental Protection Administration (SEPA) in 2007, the principle of establishi the important ecological protection area as the area with the highest expected value ecosystem integrity was chosen. According to the system integrity index distribution of units assessed, the expected value of the core area was designated a a, and the expect value of the pilot area was designated a d (the minimum value of ecosystem integrity the pilot area). The expected value for the buffer zone decreased with increasing distar from the core zone (divided into 25 classes with a value range of d–a). Finally, the actu value of the comprehensive evaluation of wetland ecological functioning was subtract from the expected value, and if the result was less than a predetermined threshold val (we set this threshold value to −0.04), this indicated that the proposed development of t wetland was disproportionate.

The ultimate purpose of assessing ecosystem integrity is to provide scientific evider for the transformation of future land use in a sensible manner. We assumed the followi conditions for the future restoration of wetland areas:

- Built-up construction land would be maintained;
- The built-up wetland types in the pilot area and buffer zone of the reserve would r go back to their original state; and
- Any occasional changes to the land were ignored.

3. Results

3.1. Land Use Change Analysis of the Momoge Internationally Important Wetland

3.1.1. Land Use Structure Investigation

The real spatial distribution of Momoge Internationally Important Wetland in 19 2015, and 2019 is depicted in Figure 4. The wetland maps for different years had compara spatial distributions. The main land cover type in the research region was always wetla followed by dry field.

The dominant species of wetland was inland beaches, followed by water. Inla beaches and water were primarily distributed along the Nengjiang River in the east a the Tao'er River in the south; salt marshes were primarily distributed in the northwe where the terrain was slightly elevated; and herb marshes were primarily distributed in t low-lying regions of the south-central portion of the study area. The middle portion of t research region was dominated by arable land. The majority of saline land was found the depressions surrounding the western lakes and marshes.

3.1.2. Analysis of Wetland Change Spatial Patterns

We focused on two degraded regions, which showed wetlands changed in protect areas, in order to investigate the spatial features of wetland at the regional scale duri different years. The first area shown in Figure 5, was one of the places where wetland lo is the worst, and inland beaches were the most common type of wetland there.

This typical region was considered agricultural terrain. The government carried c agricultural reclamation in the reserve between 1965 and 2015 for the purpose of increasi food production. For instance, the southern herb marsh was almost entirely converted t mixture of arable land and grassland, whereas the western salt marsh was also reclaim and even degraded to saline land. From 2015 to 2019, the government converted d fields to paddy fields in an effort to increase the productivity of arable land use. Th measure slowed the encroachment of agricultural space on wetland ecosystems. However, in general, degradation continued to dominate the spatial change of narrow wetlands.

The second location was chosen to illustrate the evolution of water in the western portion of the research area (Figure 6). A sufficient water supply is required to prevent the salt marsh from degrading into saline land. From 1965 to 2019, the amount of water in this region decreased before increasing. From 1965 to 2015, the conversion of huge dry areas to paddy fields in the study area led to a dramatic increase in water demands for agricultural irrigation, resulting in a substantial decrease in the area of water. Between 2015 and 2019, the area of water in the research region increased by 6344.70 ha, or 31.21 percent. Different hydraulic engineering methods were employed to make sure that there was enough water to keep the wetland ecosystem healthy and to stop the wetland from becoming more salty and growing meadows.

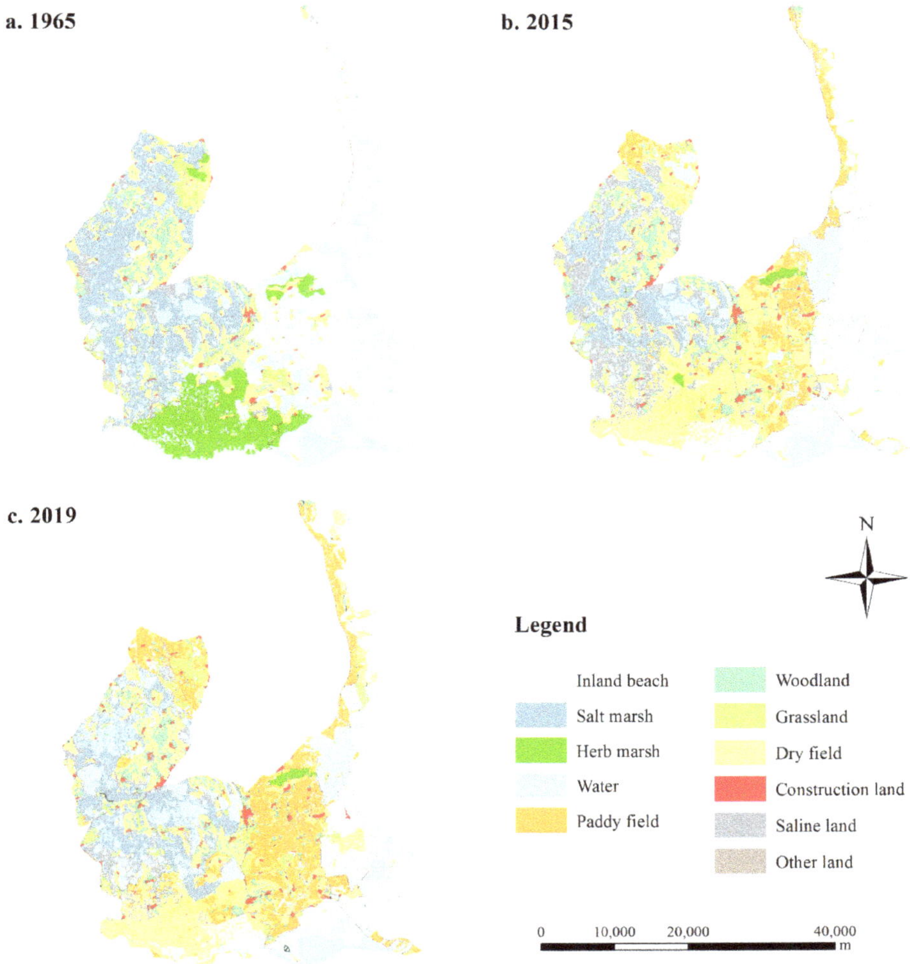

Figure 4. Map of land use types in the Momoge Internationally Important Wetland.

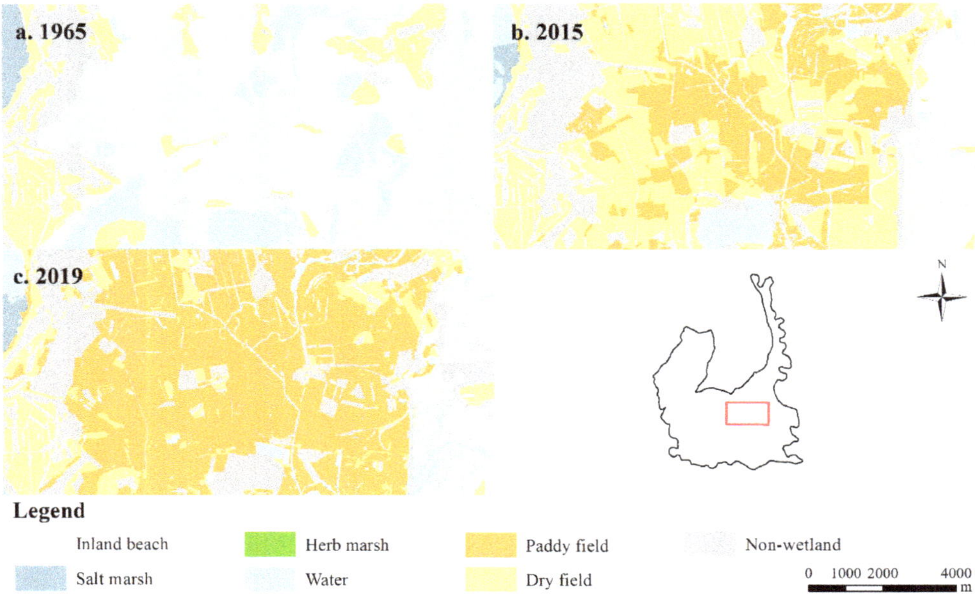

Figure 5. Maps of typical water area changes from 1965 to 2019.

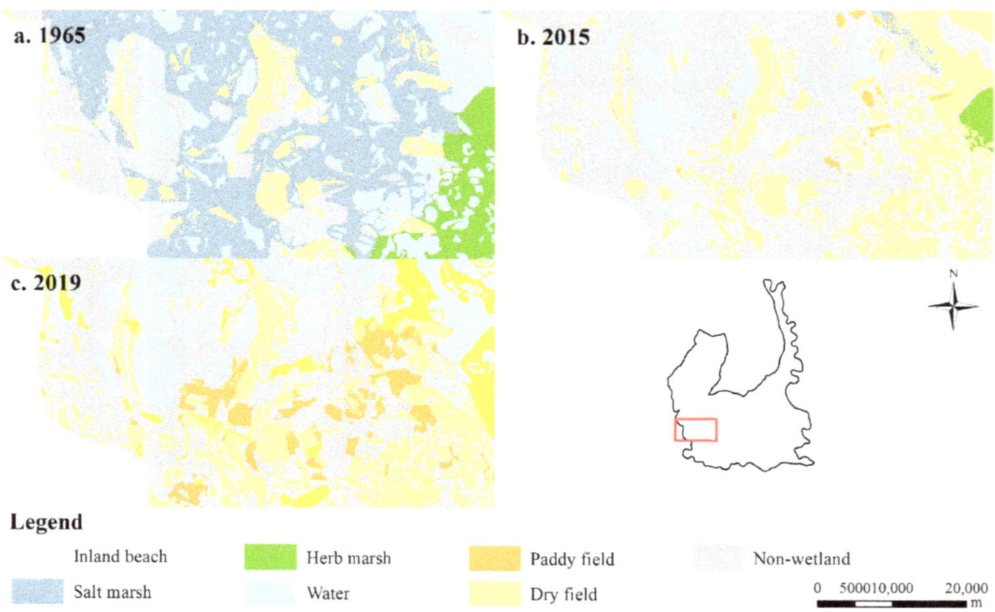

Figure 6. Maps of typical narrow wetland changes from 1965 to 2019.

3.1.3. Analysis of Wetland Changes throughout Time

To investigate the temporal aspects of wetland changes, we computed the change in wetland area from 1965 to 2019 (Figure 7). Over the two time periods, the areas c both arable land and construction land increased continuously. Wetlands were the onl land type of which proportion of the area continued to decline, falling from 75.06 percent to 45.62 percent. The area of every subcategory of wetland except water and salt marsh dropped.

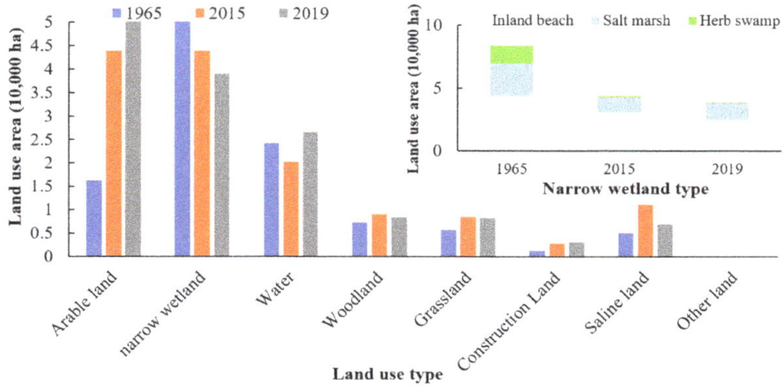

Figure 7. Momoge Internationally Important Wetland dynamic land use change map.

From 1965 to 2019, land use in the Momoge Internationally Important Wetland changed greatly (Figure 8), and the rates of change for different types of land were very different.

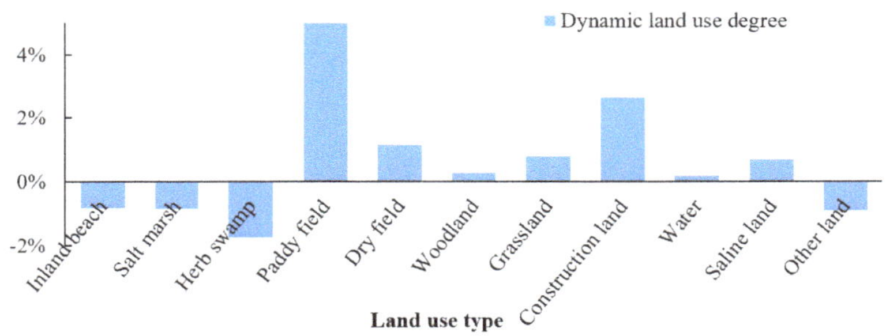

Figure 8. Momoge Internationally Important Wetland dynamic land use degree map.

From 1965 to 2019, the area of paddy fields expanded by 24,936.17 ha at the highest yearly growth rate. This was followed by construction land, which increased by 142.67 percent with an annual growth rate of 2.64 percent. With an annual growth rate of 0.69 percent, the amount of saline land grew by 37.49 percent. The area of a narrow wetland was reduced by 53.43 percent, at a rate of 0.99 percent per year. It is likely that non-wetland areas such as arable land, construction land, grassland, and saline land grew at the expense of wetland in the research area.

3.2. Analysis of the Ecosystem Integrity Index

The Momoge Nature Reserve can be separated into three types of areas. The pilot area, buffer area, and core area are listed in descending order of the degree of human activity allowed in each area. The core area consists of the HJ core area, the HR core area, and the HL core area.

In terms of ecosystem integrity rating (Figure 9), the majority of the Momoge Internationally Important Wetland was in a relatively pristine and untouched state in 1965, with 31.64 percent, 38.83 percent, and 7.81 percent of the territory receiving excellent, good, and medium ratings, respectively. In 2015, 17.95 percent, 23.62 percent, and 13.86 percent of the research area received excellent, good, and medium ratings for ecosystem integrity.

Ecosystem integrity was broken in some places, as 36.79% of the areas were rated as havi low ecosystem integrity and 7.78% as having extremely poor ecosystem integrity. The 20 assessment revealed that over half of the study area had an ecosystem integrity rating poor or extremely poor. Notably, the number of places with excellent and good ratings ecosystem integrity increased by 0.42 percent and 1.45 percent, respectively.

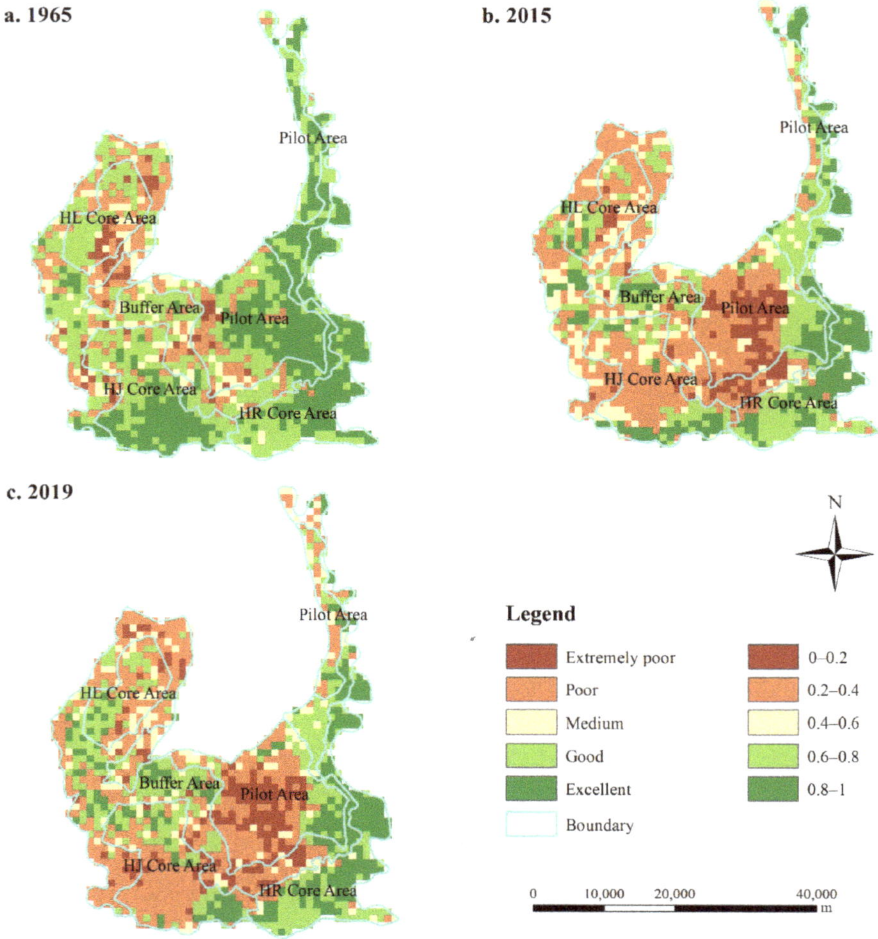

Figure 9. Momoge Internationally Important Wetland ecosystem integrity index map.

Figure 9 shows that the ecosystem integrity of the eastern part of the Momoge Inter tionally Important Wetland was much higher than those of the middle and western pa Affected by the zoning of the nature reserve, the pilot area in the center of the research a contained the majority of the regions with low ecosystem integrity in 2015 and 2019. T was mostly due to the extensive conversion of herb marsh into paddy field. Secondly, t northwest region of the research area had poor ecosystem integrity. This region was m susceptible to wetland salinization. The regions with a moderate level of environmen integrity were primarily located near dry field and salt marsh. Most of the places w exceptional ecosystem integrity were located near the Nengjiang and Tao'er Rivers, whi are very important for conservation.

The analysis of stacked single indicators revealed that elevation and hydrological conditions influence ecosystem quality components. Rare ducks were able to find better places to breed and grow in the HR core area near the water source. The overall distribution of ecological pressure in the research area exhibited a pattern of high pressure in the center, low pressure in the east and west, high pressure in the north, and low pressure in the south. The areas of high ecological pressure were centered on densely populated rural settlements and arable land, and were distributed in points and clusters. We determined that the central and western regions had a high intensity of human disturbance, a high degree of landscape fragmentation, and poor connectivity between wetland patches. Human activities are were weaker in the eastern zone, and wetland patches exhibited positive clustering and extension trends.

3.3. Identification of Unreasonably Developed Areas in the Momoge Internationally Important Wetland

The irresponsible overexploitation of the historical Momoge region ultimately resulted in the rapid degeneration of wetland ecological processes. The results of this study can be used to effectively diagnose historically overdeveloped areas (Figure 10).

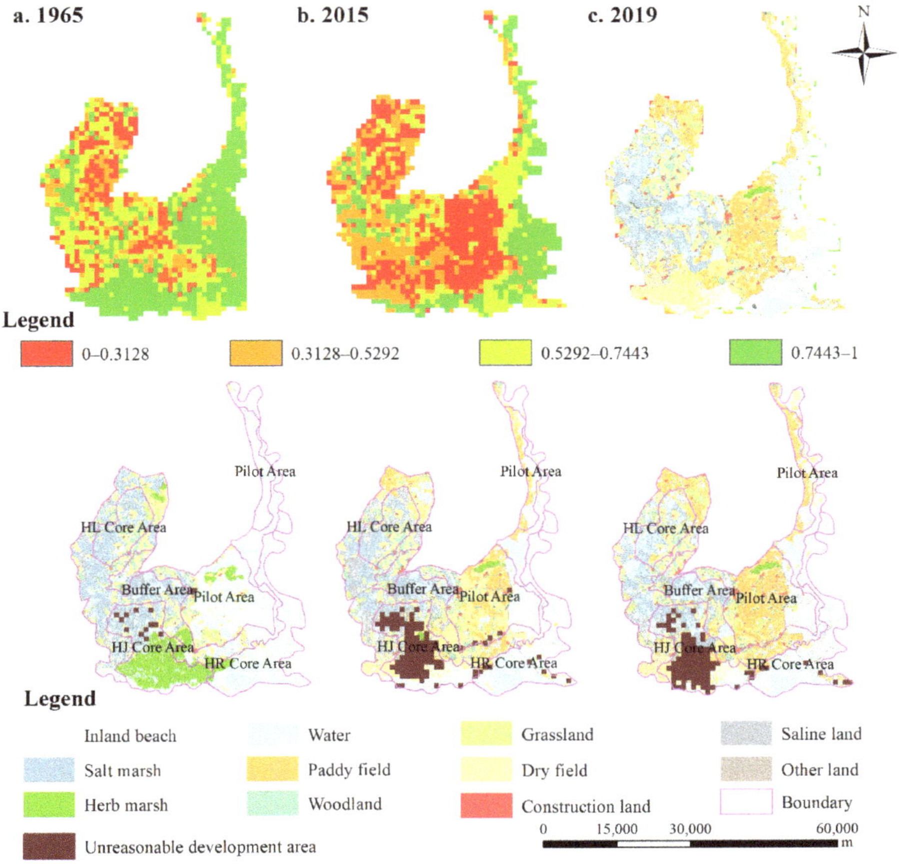

Figure 10. Ecosystem integrity assessment and identification of unreasonable development areas for the Momoge Internationally Important Wetland in 1965, 2015, and 2019.

First, we separated the integrated ecological function evaluation value into four grade for the three years to demonstrate the strength of ecological function in different portions c the Momoge Internationally Important Wetland, with the ecological function being stronge the higher the ecological function value. We found that the inappropriate developmer area in 2015 and 2019 was primarily concentrated in the HJ core area and the adjoinin paddy fields within the HR core region that were closer to the buffer zone.

3.4. Delineation for Projects Converting Arable Land to Wetland in the Momoge Internationally Important Wetland

To determine the future boundaries for converting arable land to wetland in th Momoge Internationally Important Wetland, we focused primarily on the identification c the historically over-exploited wetland regions in 2019 and the distribution of wetlands i the area's original state in 1965. Figure 11 depicts the precise spatial pattern distribution.

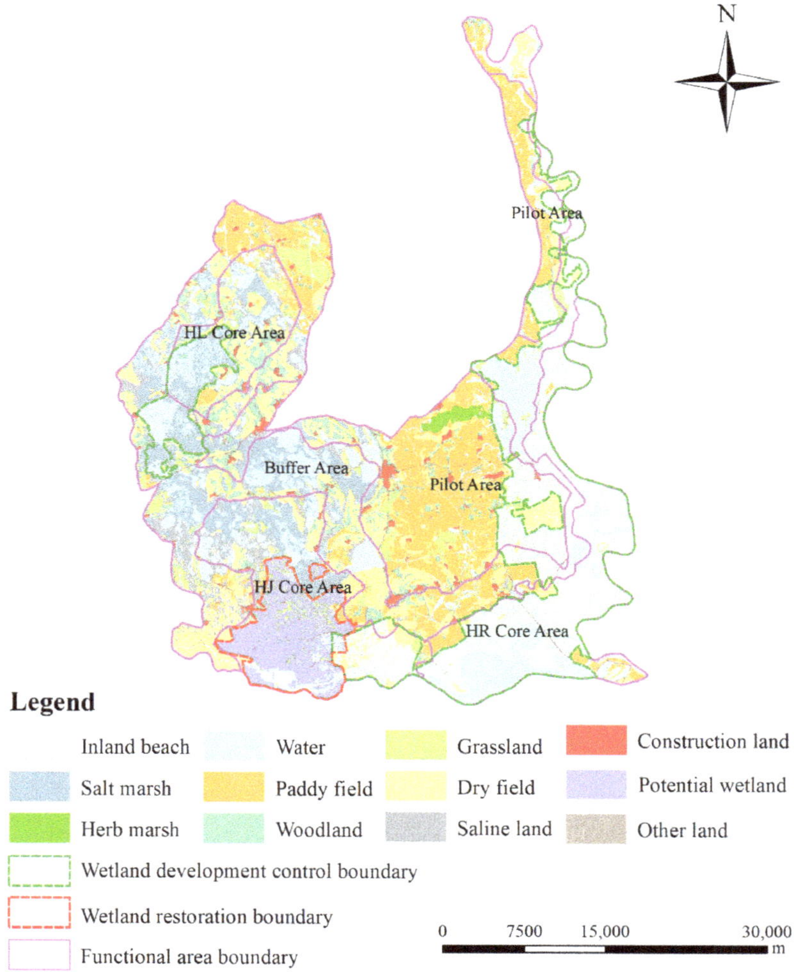

Figure 11. Momoge Internationally Important Wetland conservation spatial pattern optimizatio for 2019.

Since agricultural land in the pilot area of the Momoge Internationally Important Wetland has already been acquired on a large scale, it would not be possible to convert the arable land to inland beach in the foreseeable future. Therefore, we proposed to maintain the area of arable land in the pilot region and further improve ecological function by converting dry fields to paddy fields (refer to 2015, where areas of dry field have changed to areas of paddy field).

The Momoge Internationally Important Wetland, arable land, woodland, grassland, construction land, and saline land will be adjusted to 72,565.84 hectares, 8332.43 hectares, 4564.55 hectares, 3115.80 hectares, 8389.77 hectares, and 6979.55 hectares, respectively. Based on our assessment of the ecosystem's integrity and the nature reserve's future development trends, we drew the wetland control development boundaries of the research area to conserve wildlife-habitat wetlands. Both the eastern and western halves of the nature reserve contain sufficient wetland habitats. In addition, in the pilot area, moderately developed for agricultural purposes could be permitted. For the inappropriate development area in the southern HJ core region, the project of converting arable land to wetland was given priority.

4. Discussion

4.1. Comparative Analysis of the Results

Taking the Momoge National Nature Reserve as an example, the main objective of this study was to evaluate the ecosystem integrity of the study area in 1965, 2015, and 2019. On this basis, unreasonable developed regions were identified, and the restoration of wetland habitats was prioritized. First, we presented a framework for the remote assessment of the integrity and authenticity of ecosystems. Land use data were used to assess the dynamic changes of wetlands. Using the framework provided in this article, we investigated the spatial and temporal trends of ecosystem integrity. Based on the ecosystem integrity index, we identified inappropriately developed regions and determined wetland restoration boundaries.

Most previous studies have concentrated on a single aspect of a wetland's historical evolution or ecological function evaluation [46,47]. However, few studies have been conducted on wetland restoration through the evaluation of ecosystem integrity and authenticity. Ecosystem authenticity and integrity are mutually integrated and holistic concepts, and integrity is more widely discussed in ecology than authenticity.

Land use change is the primary cause of wetland degradation [48–50]. Agriculture, the most common cause of land use change, has destroyed more than 50 percent of the internationally significant wetland habitats worldwide [51,52]. Dong, Z. et al. and Wang, Z. et al. researched the processes and causes underlying wetland fragmentation and contraction, and they discovered that agricultural growth in the context of climate change was the primary cause of the massive loss of wetland in the western Sonnen Plain [53,54].

The simultaneous stress of climate change and human disturbance has destabilized the wetland's basic ecological structure and constitutes a grave threat to the ecosystem [55]. The lack of water in wetlands is the primary cause of the considerable decline in ecosystem integrity in Momoge National Nature Reserve. Changes in land use type were spatially concentrated in the HJ core and the central pilot area, where wetland development has historically been more intense. Jilin Province's 14th Five-Year Plan for Ecological Protection projected a 60 percent effective rate of wetland protection by 2035.

In environmental integrity and authenticity evaluations, one may use an undisturbed or slightly disturbed ecosystem as a reference standard. In the current context of human growth, few ecosystems remain untouched by human activity. As a result, the natural state (e.g., the state closest to the natural habitat in a regional ecosystem) is typically used as the benchmark in studies. In 1965, the study region had relatively few human disturbances. Therefore, the spatial pattern of the study area was optimized using the level of ecosystem integrity in 1965 as a standard. In this study, the restoration boundary and protection boundary of the wetland was not limited by those of the three original types of functio areas of protected areas.

4.2. Sustainable Development Measures

The following suggestions are provided for the future development of the Momo Internationally Important Wetland.

(1) Promoting the project of converting arable land to wetland and increase the quality arable land. From 1965 to 2019, the population density of the study region increas from 16 to 29 people per square kilometer, resulting in an increase in the local popu tion's demand for food. In this setting, if managers continue to increase the amou of arable land by reclaiming wetland, woodland, and grassland to fulfill increas agricultural production needs, the study area's ecology will be affected by even mo dire problems. Several studies have suggested that measures such as "rewetting" a water penetration can be used to mitigate the adverse effects of overexploitation wetland ecosystems [56,57]. Wetland managers must give the ecological environme the highest priority. The project of converting arable land to wetland is execut gradually in historically overexploited areas, with farmers receiving compensati for the loss of land, in compliance with national legislation. Farmers are also urged enhance the quality of arable land through reclamation or preparation.

(2) Enhancing the hydrological connectivity of wetlands and restoring wildlife habit In the past 50 years, climate-related droughts and human activities have contribut to a diminishing supply of water in the Momoge wetland. To restore the qual of waterfowl habitats, we can, on the one hand, execute water system penetrati projects and remove unneeded ditches to restore the hydrological system in the stu area to its original state. Alternatively, we should conserve the natural reed b surrounding the habitats of waterfowls, such as whooping cranes, to separate hum activity from the birds' breeding grounds. These techniques will aid in maintaining t stability of the wetland ecosystem to restore these habitats for birds and other anim

4.3. Uncertainties and Prospects

There are numerous unknown factors relating to the deterioration and developme of wetlands in the research area, including policy shifts, population expansion, and clim change. The objective of wetland spatial pattern optimization is not to predict the futu exactly, as there are innumerable uncertainties involved, but to explore the potential the restoration of wetlands in the future in a way that meets biological needs as mu as possible.

The remote evaluation approach provides quick assessments of ecosystem integr although the framework has some shortcomings. First, the data processing metho used for certain indicators must be enhanced. Due to the limited availability of da we did not consider many variables that are strongly related to biodiversity. Seconc the accounting technique for ecological service values must be enhanced. Thirdly, t assessment framework did not effectively take into account the indicators that are used judge how pristine an ecosystem is. In future research, we will thus consider introduci the spatial and temporal distributions of rare species such as *Grus leucogeranus*, utilizi the InVEST model to address the effects of climate change or anthropogenic disturban on ESV [57], and introducing remote metrics aimed at characterizing the authenticity ecosystems.

5. Conclusions

In this study, we combined the EIAF, TLA, and PCFAEI models to develop a "Quali Pressure-Pattern-Service" remote assessment framework, investigating the ecosystem tegrity pattern of the study area, and identifying wetland restoration boundaries a wetland development control boundaries. This framework is an efficient tool for assessi the environmental impact of complex ecosystems. Our research was less costly and ap cable on a broader regional scale than conventional methods. In general, the framework can be easily updated and repeated regularly, providing additional data for continued adaptive-management.

The "Quality-Pressure-Pattern-Service" ecosystem integrity remote assessment framework can be broken down into four indices: the ecosystem qualityindex, the ecosystem stress index, the ecosystem pattern index, and the ecosystem service value index. These indices can be used as a set of indicators for natural resource agencies and organizations to monitor the condition of ecosystems. The information gathered to create these indexes can be used to create databases and atlases depicting the condition and trends of important nature reserves. The framework can also be used to identify areas of unreasonable development that need to be prioritized for ecological restoration, while highlighting areas that need to be protected. It can help managers to improve spatial patterns and create ecological restoration targets for entire ecosystems.

In future studies, this remote evaluation system will be integrated with a spatial pattern simulation model. The consequences of land use changes on ecosystem integrity will be investigated under several scenarios based on future simulations.

Author Contributions: Conceptualization, D.W. and S.Z.; methodology, J.H.; software, J.H.; validation, J.H.; formal analysis, J.H.; investigation, J.H.; resources, D.W., S.Z. and J.H.; data curation, J.H.; writing—original draft preparation, J.H.; writing—review and editing, D.W.; visualization, J.H.; supervision, D.W.; project administration, S.Z.; funding acquisition, D.W. All authors have read and agreed to the published version of the manuscript.

Funding: This research was funded by the National Natural Science Foundation of China, grant number 42071255.

Institutional Review Board Statement: Not applicable.

Informed Consent Statement: Not applicable.

Data Availability Statement: All data, models, and codes generated or used during the study appear in the submitted article.

Acknowledgments: Acknowledgement for the data support from "National Earth System Science Data Center, National Science & Technology Infrastructure of China. (http://www.geodata.cn (accessed on 1 July 2022))".

Conflicts of Interest: The authors declare no conflict of interest.

eferences

Xu, J.; Zhang, Z.; Liu, W.; McGowan, P.J.K. A review and assessment of nature reserve policy in China: Advances, challenges and opportunities. *Oryx* **2012**, *46*, 554–562. [CrossRef]

Dearden, P.; Berg, L.D. Canada national-parks—A model of administrative penetration. *Can. Geogr.-Geogr. Can.* **1993**, *37*, 194–211. [CrossRef]

Huang, X.; Xu, J.; Liu, B.; Guan, X.; Li, J. Assessment of Aquatic Ecosystem Health with Indices of Biotic Integrity (IBIs) in the Ganjiang River System, China. *Water* **2022**, *14*, 278. [CrossRef]

Liao, J.-Q.; Huang, Y. Research progress on using index of biological integrity to assess aquatic ecosystem health. *Ying Yong Sheng Tai Xue Bao = J. Appl. Ecol.* **2013**, *24*, 295–302.

Karr, J.R. Biological Integrity: A Long-Neglected Aspect of Water Resource Management. *Ecol. Appl. Publ. Ecol. Soc. Am.* **1991**, *1*, 66–84. [CrossRef]

Tierney, G.L.; Faber-Langendoen, D.; Mitchell, B.R.; Shriver, W.G.; Gibbs, J.P. Monitoring and evaluating the ecological integrity of forest ecosystems. *Front. Ecol. Environ.* **2009**, *7*, 308–316. [CrossRef]

Staszak, L.A.; Armitage, A.R. Evaluating Salt Marsh Restoration Success with an Index of Ecosystem Integrity. *J. Coast. Res.* **2013**, *29*, 410–418. [CrossRef]

Fraser, R.H.; Olthof, I.; Pouliot, D. Monitoring land cover change and ecological integrity in Canada's national parks. *Remote Sens. Environ.* **2009**, *113*, 1397–1409. [CrossRef]

Moreno-Mateos, D.; Power, M.E.; Comin, F.A.; Yockteng, R. Structural and Functional Loss in Restored Wetland Ecosystems. *Plos Biol.* **2012**, *10*, e1001247. [CrossRef]

Menendez, P.; Losada, I.J.; Torres-Ortega, S.; Narayan, S.; Beck, M.W. The Global Flood Protection Benefits of Mangroves. *Sci. Rep.* **2020**, *10*, 4404. [CrossRef]

11. Colvin, S.A.R.; Sullivan, S.M.P.; Shirey, P.D.; Colvin, R.W.; Winemiller, K.O.; Hughes, R.M.; Fausch, K.D.; Infante, D.M.; Olde J.D.; Bestgen, K.R.; et al. Headwater Streams and Wetlands are Critical for Sustaining Fish, Fisheries, and Ecosystem Service *Fisheries* **2019**, *44*, 73–91. [CrossRef]
12. Cavanaugh, K.C.; Kellner, J.R.; Forde, A.J.; Gruner, D.S.; Parker, J.D.; Rodriguez, W.; Feller, I.C. Poleward expansion of mangrove is a threshold response to decreased frequency of extreme cold events. *Proc. Nat. Acad. Sci. USA* **2014**, *111*, 723–727. [CrossRe [PubMed]
13. Sheaves, M.; Baker, R.; Nagelkerken, I.; Connolly, R.M. True Value of Estuarine and Coastal Nurseries for Fish: Incorporatin Complexity and Dynamics. *Estuar. Coasts* **2015**, *38*, 401–414. [CrossRef]
14. Davidson, N.C. How much wetland has the world lost? Long-term and recent trends in global wetland area. *Mar. Freshw. Re* **2014**, *65*, 934–941. [CrossRef]
15. Niu, Z.; Zhang, H.; Wang, X.; Yao, W.; Zhou, D.; Zhao, K.; Li, N.; Huang, H.; Li, C.; Yang, J.; et al. Mapping wetland changes i China between 1978 and 2008. *Chin. Sci. Bull.* **2012**, *57*, 2813–2823. [CrossRef]
16. Shen, X.; Liu, B.; Jiang, M.; Lu, X. Marshland Loss Warms Local Land Surface Temperature in China. *Geophys. Res. Lett.* **2020**, *4* e2020GL087648. [CrossRef]
17. Mack, J.J. Landscape as a predictor of wetland condition: An evaluation of the Landscape Development Index (LDI) with a larg reference wetland dataset from Ohio. *Environ. Monit. Assess.* **2006**, *120*, 221–241. [CrossRef]
18. Hansen, A.J.; Noble, B.P.; Veneros, J.; East, A.; Goetz, S.J.; Supples, C.; Watson, J.E.M.; Jantz, P.A.; Pillay, R.; Jetz, W.; et al. Towar monitoring forest ecosystem integrity within the post-2020 Global Biodiversity Framework. *Conserv. Lett.* **2021**, *14*, e1282 [CrossRef]
19. Brooks, R.P.; Wardrop, D.H.; Bishop, J.A. Assessing wetland condition on a watershed basis in the Mid-Atlantic region usin synoptic land-cover maps. *Environ. Monit. Assess.* **2004**, *94*, 9–22. [CrossRef]
20. Zeleny, J.; Mercado-Bettin, D.; Mueller, F. Towards the evaluation of regional ecosystem integrity using NDVI, brightnes temperature and surface heterogeneity. *Sci. Total Environ.* **2021**, *796*, 148994. [CrossRef]
21. Akin, A.; Berberoglu, S.; Erdogan, M.A.; Donmez, C. Modelling Land-Use Change Dynamics In A Mediterranean Coastal Wetlan Using Ca-Markov Chain Analysis. *Fresenius Environ. Bull.* **2012**, *21*, 386–396.
22. Yan, F. Large-Scale Marsh Loss Reconstructed from Satellite Data in the Small Sanjiang Plain since 1965: Process, Pattern an Driving Force. *Sensors* **2020**, *20*, 1036. [CrossRef] [PubMed]
23. Brinkmann, K.; Hoffmann, E.; Buerkert, A. Spatial and Temporal Dynamics of Urban Wetlands in an Indian Megacity over th Past 50 Years. *Remote Sens.* **2020**, *12*, 662. [CrossRef]
24. Shi, S.; Chang, Y.; Li, Y.; Hu, Y.; Liu, M.; Ma, J.; Xiong, Z.; Wen, D.; Li, B.; Zhang, T. Using Time Series Optical and SAR Data t Assess the Impact of Historical Wetland Change on Current Wetland in Zhenlai County, Jilin Province, China. *Remote Sens.* **202** *13*, 4514. [CrossRef]
25. Wang, Y.; Feng, J.; Lin, Q.; Lyu, X.; Wang, X.; Wang, G. Effects of Crude Oil Contamination on Soil Physical and Chemic Properties in Momoge Wetland of China. *Chin. Geogr. Sci.* **2013**, *23*, 708–715. [CrossRef]
26. Dong, J.; Xiao, X.; Kou, W.; Qin, Y.; Zhang, G.; Li, L.; Jin, C.; Zhou, Y.; Wang, J.; Biradar, C.; et al. Tracking the dynamics of padd rice planting area in 1986–2010 through time series Landsat images and phenology-based algorithms. *Remote Sens. Environ.* **201** *160*, 99–113. [CrossRef]
27. Vymazal, J.; Bfezinova, T. The use of constructed wetlands for removal of pesticides from agricultural runoff and drainage: review. *Environ. Int.* **2015**, *75*, 11–20. [CrossRef]
28. Wang, X.; Xiao, X.; Zou, Z.; Hou, L.; Qin, Y.; Dong, J.; Doughty, R.B.; Chen, B.; Zhang, X.; Cheng, Y.; et al. Mapping coast wetlands of China using time series Landsat images in 2018 and Google Earth Engine. *Isprs J. Photogramm. Remote Sens.* **2020**, *16* 312–326. [CrossRef]
29. Wang, X.; Bao, Y. Exploration of methods for studying dynamic land use change. *Prog. Geogr.* **1999**, 83–89.
30. Ikhumhen, H.O.; Li, T.; Lu, S.; Matomela, N. Assessment of a novel data driven habitat suitability ranking approach for Laru relictus specie using remote sensing and GIS. *Ecol. Model.* **2020**, *432*, 109221. [CrossRef]
31. Zhu, Y.; Wang, H.; Guo, W. The impacts of water level fluctuations of East Dongting Lake on habitat suitability of migratory bird *Ecol. Indic.* **2021**, *132*, 108277. [CrossRef]
32. Liu, X.; Liu, C.; Zhang, J.; Wei, Y.; Huang, B. Ecosystem integrity and authenticity assessment framework in Qinghai-Tibet Platea National Park Cluster. *J. Ecol.* **2021**, *41*, 833–846.
33. Beven, K.J.; Kirkby, M.J. A physically based, variable contributing area model of basin hydrology/Un modèle à base physique c zone d'appel variable de l'hydrologie du bassin versant. *Hydrol. Sci. J.* **1979**, *24*, 43–69. [CrossRef]
34. Beven, K.J.; Kirkby, M.; Schofield, N.; Tagg, A. Testing a physically-based flood forecasting model (TOPMODEL) for three U catchments. *J. Hydrol.* **1984**, *69*, 119–143. [CrossRef]
35. Kriegler, F.; Malila, W.; Nalepka, R.; Richardson, W. Preprocessing transformations and their effects on multispectral recognitio *Remote Sens. Environ. VI* **1969**, 97.
36. Xu, W. Kira's heat index and its application to vegetation in China. *J. Ecol.* **1985**, *3*, 35–39.
37. Zhao, G.; Liu, J.; Kuang, W.; Ouyang, Z.; Xie, Z. Disturbance impacts of land use change on biodiversity conservation priori areas across China:1990–2010. *J. Geogr. Sci.* **2015**, *25*, 515–529. [CrossRef]

Xie, G.; Zhang, C.; Zhang, L.; Chen, W.; Li, S. Improvement of the evaluation method for ecosystem service value based on perunit area. *J. Nat. Resour.* **2015**, *30*, 1243–1254.
Xie, G.; Zhang, C.; Zhang, C.; Xiao, Y.; Lu, C. The value of ecosystem services in China. *Resour. Sci.* **2015**, *37*, 1740–1746.
Costanza, R.; dArge, R.; de Groot, R.; Farber, S.; Grasso, M.; Hannon, B.; Limburg, K.; Naeem, S.; Oneill, R.V.; Paruelo, J.; et al. The value of the world's ecosystem services and natural capital. *Nature* **1997**, *387*, 253–260. [CrossRef]
Xie, G.; Zhen, L.; Lu, C.; Xiao, Y.; Chen, C. Expert knowledge based valuation method of ecosystem services in China. *J. Nat. Resour.* **2008**, *23*, 911–919.
Si, S.-L.; You, X.-Y.; Liu, H.-C.; Zhang, P. DEMATEL Technique: A Systematic Review of the State-of-the-Art Literature on Methodologies and Applications. *Math. Probl. Eng.* **2018**, *2018*, 3696457. [CrossRef]
Singh, D.; Singh, B. Investigating the impact of data normalization on classification performance. *Appl. Soft Comput.* **2020**, *97*, 105524. [CrossRef]
Zhao, C.; Shao, N.; Yang, S.; Ren, H.; Ge, Y.; Zhang, Z.; Zhao, Y.; Yin, X. Integrated assessment of ecosystem health using multiple indicator species. *Ecol. Eng.* **2019**, *130*, 157–168. [CrossRef]
Ou, W.; Xiao, J.; Li, W. Spatial pattern optimization simulation of coastal wetland use based on BP neural network and cellular automata—A case of Dafeng coastal wetland. *J. Nat. Resour.* **2014**, *29*, 744–756.
Dar, S.A.; Rashid, I.; Bhat, S.U. Linking land system changes (1980–2017) with the trophic status of an urban wetland: Implications for wetland management. *Environ. Monit. Assess.* **2021**, *193*. [CrossRef]
Hamandawana, H.; Eckardt, F.; Chanda, R. Linking archival and remotely sensed data for long-term environmental monitoring. *Int. J. Appl. Earth Obs. Geoinf.* **2005**, *7*, 284–298. [CrossRef]
Herbert, E.R.; Boon, P.; Burgin, A.J.; Neubauer, S.C.; Franklin, R.B.; Ardon, M.; Hopfensperger, K.N.; Lamers, L.P.M.; Gell, P. A global perspective on wetland salinization: Ecological consequences of a growing threat to freshwater wetlands. *Ecosphere* **2015**, *6*, 1–43. [CrossRef]
Hou, M.; Ge, J.; Gao, J.; Meng, B.; Li, Y.; Yin, J.; Liu, J.; Feng, Q.; Liang, T. Ecological Risk Assessment and Impact Factor Analysis of Alpine Wetland Ecosystem Based on LUCC and Boosted Regression Tree on the Zoige Plateau, China. *Remote Sens.* **2020**, *12*, 368. [CrossRef]
Hoang Huu, N.; Dargusch, P.; Moss, P.; Aziz, A.A. Land-use change and socio-ecological drivers of wetland conversion in Ha Tien Plain, Mekong Delta, Vietnam. *Land Use Policy* **2017**, *64*, 101–113. [CrossRef]
Verones, F.; Pfister, S.; Hellweg, S. Quantifying Area Changes of Internationally Important Wetlands Due to Water Consumption in LCA. *Environ. Sci. Technol.* **2013**, *47*, 9799–9807. [CrossRef] [PubMed]
Zedler, J.B.; Kercher, S. Wetland resources: Status, trends, ecosystem services, and restorability. *Annu. Rev. Environ. Resour.* **2005**, *30*, 39–74. [CrossRef]
Wang, Z.; Huang, N.; Luo, L.; Li, X.; Ren, C.; Song, K.; Chen, J.M. Shrinkage and fragmentation of marshes in the West Songnen Plain, China, from 1954 to 2008 and its possible causes. *Int. J. Appl. Earth Obs. Geoinf.* **2011**, *13*, 477–486. [CrossRef]
Dong, Z.; Wang, Z.; Liu, D.; Song, K.; Li, L.; Jia, M.; Ding, Z. Mapping Wetland Areas Using Landsat-Derived NDVI and LSWI: A Case Study of West Songnen Plain, Northeast China. *J. Indian Soc. Remote Sens.* **2014**, *42*, 569–576. [CrossRef]
Roe, S.; Streck, C.; Obersteiner, M.; Frank, S.; Griscom, B.; Drouet, L.; Fricko, O.; Gusti, M.; Harris, N.; Hasegawa, T.; et al. Contribution of the land sector to a 1.5 degrees C world. *Nat. Clim. Chang.* **2019**, *9*, 817. [CrossRef]
Zhang, D.; Sun, J.; Cui, Q.; Jia, X.; Qi, Q.; Wang, X.; Tong, S. Plant growth and diversity performance after restoration in Carex schmidtii tussock wetlands, Northeast China. *Community Ecol.* **2021**, *22*, 391–401. [CrossRef]
Xiang, H.; Wang, Z.; Mao, D.; Zhang, J.; Xi, Y.; Du, B.; Zhang, B. What did China's National Wetland Conservation Program Achieve? Observations of changes in land cover and ecosystem services in the Sanjiang Plain. *J. Environ. Manag.* **2020**, *267*, 110623. [CrossRef]

plants MDPI

rticle

ree Species Composition and Forest Community Types along :nvironmental Gradients in Htamanthi Wildlife Sanctuary, Ayanmar: Implications for Action Prioritization n Conservation

Iyo Min Latt [1,2] and Byung Bae Park [1,*]

1 Department of Environment and Forest Resources, Chungnam National University, Daejeon 34134, Korea
2 Department of Natural Resources Management, University of Forestry and Environmental Science, Yezin 15013, Myanmar
* Correspondence: bbpark@cnu.ac.kr

Abstract: The identification of forest community types is essential for prioritizing choices and targets in species and community conservation purposes amid climate change impacts on forest community dynamics. Here, we determined the tree species composition, species diversity, and the forest community types across contrasting topographic and edaphic conditions in Htamanthi Wildlife Sanctuary (HWS), Myanmar. All tree species with diameter at breast height (DBH) ≥10 cm were recorded in 66 plots (625 m^2), from which the species diversity, density, frequency, dominance, and importance value (IV) of each tree species were measured. The soil hardness (Hd), bulk density (BD), moisture content (MC), organic matter content (OM), texture, pH, total N, and available P, K, Ca, Na, and Mg concentrations were also analyzed. The elevation (ELV) and slope (SLP) were also measured as the topographic factors. Cluster analysis resulted in five distinct forest communities and the soil Ca, Mg, clay proportion, soil hardness, and elevation were the major influencing factors. The species diversity in HWS ranged from low to very high relative values, with 209 tree species belonging to 119 genera and 55 families. Identification of these community types and understanding the diversity levels and major factors influencing the community structure may play a key role in the planning, prioritization, and implementation of species and community conservation strategies amid the unpredictable impacts of climate change on forest community dynamics.

Keywords: edaphic factors; forest communities; Htamanthi; Myanmar; soil hardness; species diversity; topographic factors

tation: Latt, M.M.; Park, B.B. Tree ecies Composition and Forest ommunity Types along ivironmental Gradients in tamanthi Wildlife Sanctuary, yanmar: Implications for Action ioritization in Conservation. *Plants* **22**, *11*, 2180. https://doi.org/ .3390/plants11162180

cademic Editors: Panayiotis mitrakopoulos, Mario A. Pagnotta, iklas Scholz and Arshiya Noorani

eceived: 22 May 2022
ccepted: 18 August 2022
ublished: 22 August 2022

1. Introduction

Climate change has been altering the structure and functions of forest ecosystems worldwide. The unprecedented rise in global temperature may lead to the modification of various environmental variables and, thus, may shift in species composition and alterations in forest community structure at the local scale. Even minor changes in microclimatic conditions with elevation may also result in unusual changes in local diversity [1,2]. Environmental gradients, which refer to variations in site characteristics (i.e., edaphic, climatic, and physiographic variables), influence the patterns of tree species composition and distribution of forest communities [3]. Many interacting biotic and abiotic factors, including changes in elevation, slope, soil properties, and site index, affect species richness [4]. Climatic factors vary with elevation and exert a strong influence on plant distribution across ecosystems [5]. Thus, understanding the pattern of tree species composition and the types of forest communities across contrasting environmental conditions may play a key role in the planning, prioritization, and implementation of species and community conservation strategies amid the unpredictable impacts of climate change on forest community dynamics.

The pattern in tree species composition and forest community may vary amon groups of plants and from one area to another due to differences in resource availabilit plant life-history traits, and adaptive strategies of plants, as influenced by the prevailin environmental conditions [6,7]. Soil nutrient and moisture availability, for instance, ca influence natural regeneration, seedling establishment, and species dominance dependin on elevation [8–10]. Climatic factors, such as air temperature and precipitation, have a interactive role in the effects of soil nutrients on species dominance [6]. A study reporte that a high soil nutrient supply, particularly nitrogen and phosphorus, can lead to a low species richness and high soil moisture can result in a higher species richness [11]. It ha also been suggested that soil pH exerts a strong influence on species composition at th local scale [12] by limiting the number of species that can adapt to the extreme ends of th pH gradient [13]. While numerous studies have focused on large geographical areas, the are fewer studies about tree species composition and distribution of forest communitie along environmental gradients conducted in smaller landscapes or mountains. Becaus global changes in environmental conditions have a direct impact on local ecological system localized studies may contribute to global biodiversity conservation programs throug prioritization of habitats and conservation strategies.

Htamanthi Wildlife Sanctuary (HWS), an ASEAN heritage park in Myanmar an probably the largest contiguous forest landscape in Asia, lies in a transition zone of thre biodiversity hotspots [14]. The sanctuary is globally important because of its proximit to the Northern Forest Complex (NFC), one of the largest remaining contiguous fores in Southeast Asia. Despite the ecological importance of HWS, most earlier studies dea with faunal resources and less attention was given to the floral resources, particular trees. By knowing the patterns of tree species composition and types of forest comm nities along environmental gradients in this biodiversity hotspot, we could advance ou understanding of the local forest communities and how the environmental variables affe these communities.

The identification of forest community types can play a key role in prioritizing co servation choices in biodiversity hotspots since nature-based solutions are emerging a an integrated approach to conserving biodiversity and ecosystem services [15]. One o the practical strategies in biodiversity conservation is the identification of priority area featuring exceptional need for immediate conservation, particularly biodiverse areas expe encing high degree of habitat loss and risks [16]. This is because one of the major challeng commonly encountered by conservation practitioners is how to efficiently allocate limite resources to many focal ecosystems or forest plant communities needing conservation [17 Thus, information on forest community types in Htamanthi Wildlife Sanctuary can su sequently be useful for the development of more efficient ecosystem-based conservatio approaches amid limited resources.

Consequently, we determined the tree species composition, species diversity, and th forest community types along environmental gradients in HWS. The role of the environme tal factors in shaping forest communities in HWS remains poorly studied. Identificatio of species and community structure is very relevant to understanding the status of tre populations, natural regeneration for species, and community conservation purposes ami climate change impacts on forest community dynamics [18].

2. Results

2.1. Forest Community Assemblages

In this study, the Bray–Curtis dissimilarity cluster analysis grouped the evergree forest in Htamanthi Wildlife Sanctuary into five distinct forest communities (Figure 1). Th biggest group is *Vatica maingayi* forest (VMF) with 36 plots, *Quercus glauca* forest (QGF *Nothaphoebe condensa* forest (NCF), and *Diospyros toposia* forest (DTF) are intermediate, wit 5–13 plots, and the smallest group is the bamboo forest (BF), with only two plots.

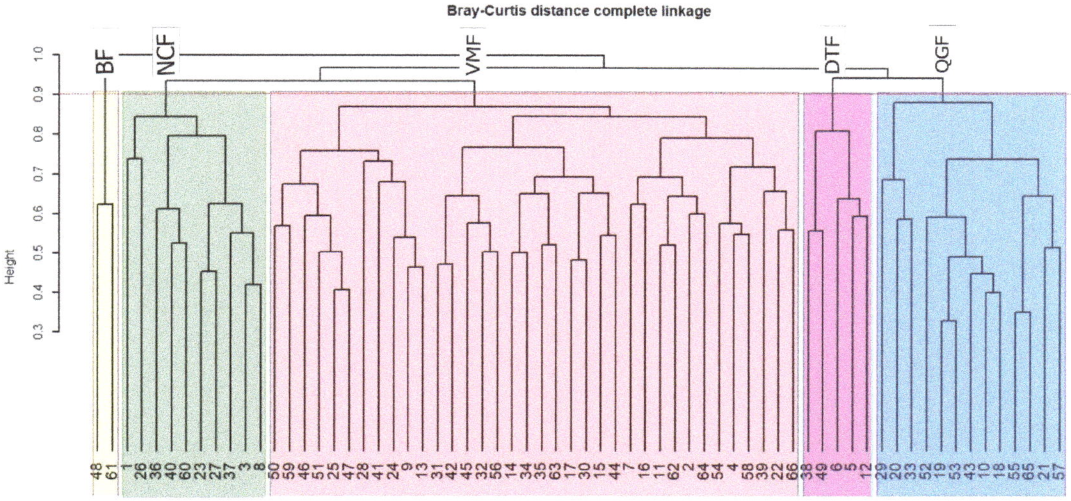

Figure 1. Dendrogram of all plots based on Bray–Curtis Distance complete linkage showing the five distinct forest communities of evergreen forest ecosystem in Htamanthi Wildlife Sanctuary in Myanmar. Forest communities: Bamboo forest (BF), *Vatica maingayi* forest (VMF), *Quercus glauca* forest (QGF), *Nothaphoebe condensa* forest (NCF), and *Diospyros toposia* forest (DTF).

2.2. Species Diversity and Importance Values among Forest Communities

In this study, we identified 209 tree species belonging to 119 genera and 55 families across the five forest communities (Table S1). The VMF and BF forest communities have very high (3.965) and very low (0.086) Shannon (H) diversity values, respectively (Table 1). The same pattern was observed in Evenness (J′) values for these two forest communities. The QGF has a high diversity value of H = 3.25. The other forest communities (NCF and DTF) have moderate H values (2.659–2.807). The VMF is a major forest community composed of a higher number of species, genera and family, followed by NCF, QGF, DTF, and BF (Table S1). In terms of IV, the most dominant tree species is *Vatica maingayi* in VMF, *Nothaphoebe condensa* in NCF, *Quercus glauca* in QGF, and *Diospyros toposia* in DTF (Table 1 and Table S1).

Table 1. Species diversity values among forest community types in Htamanthi Wildlife Sanctuary in Myanmar. Values in parenthesis are the importance values of the dominant tree species.

Forest Communities	Richness (S)	Shannon-Wiener Diversity Index (H)	Evenness (J′)	Dominant Species and Importance Values (IV)
BF	10	0.086	0.037	*Aglaia perviridis* (17.3%)
DTF	51	2.659	0.676	*Diospyros toposia* (25.2%)
NCF	95	2.807	0.616	*Nothaphoebe condensa* (26.2%)
QGF	65	3.254	0.780	*Quercus glauca* (30.9%)
VMF	174	3.965	0.769	*Vatica maingayi* (35.9%)

2.3. Variations in Topography and Soil Characteristics among Forest Communities

There was a significant difference in the elevation and slope across the four forest communities (Table 2). The highest elevation was in DFT, intermediate in NCF, and the lowest value was found in both QGF and VMF. In terms of slope, DFT is the steepest among the forest communities.

Table 2. Topographic factors in the identified forest community types in Htamanthi Wildlife Sanctu in Myanmar. Values in parenthesis are the standard deviations from the mean. Different lowerc letters indicate statistical significance between community types at $\alpha = 0.05$.

FC	n	Elevation (masl)	Slope (°)
DTF	5	274 (57) ^{a}	29 (19) ^{a}
NCF	10	188 (32) ^{ab}	8 (6) ^{b}
QGF	13	179 (24) ^{b}	16 (12) ^{ab}
VMF	36	177 (31) ^{b}	16 (14) ^{ab}

At 0–15 cm soil depth, the MC, sand, and clay proportions varied significantly acr forest communities (Table 3 and Table S2). The soil moisture (MC) in NCF and VM were higher than DTF and QGF. Further, DTF and QGF have similarly high sand and cl proportions compared the other forest communities. At 15–30 cm soil depth, the Hd, M and clay proportion were significantly different across sites. Both DTF and NCF hav higher Hd than QGF or VMF. The DTF and QG have similarly higher MC than the otl forest community types and a similar pattern was observed in clay proportions.

Table 3. Soil physical properties in the *Diospyros toposia* forest (DTF), *Nothaphoebe condensa* for (NCF), *Quercus glauca* forest (QGF), and *Vatica maingayi* forest (VMF). Values in parenthesis the standard deviations from the mean. Different lowercase letters indicate statistical significa between forest community types at $\alpha = 0.05$. Values in parenthesis are the standard deviations fr the mean.

Soil Depth (cm)	Forest Community	n	Soil Hardness (kg/cm^2)	Moisture Content (%)	Bulk Density (%)	Organic Matter (%)	Sand (%)	Silt (%)	Clay (%)
0–15	DTF	5	2.50 (0.55) ^{a}	10.30 (3.19) ^{a}	1.18 (0.06) ^{a}	6.00 (1.87) ^{a}	54.2 (4.2) ^{c}	29.0 (4.2) ^{a}	16.2 (3.5)
	NCF	10	2.79 (0.88) ^{a}	20.10 (4.32) ^{b}	1.18 (0.14) ^{a}	7.30 (2.11) ^{a}	39.4 (13.1) ^{a}	30.3 (7.6) ^{a}	29.2 (7.3)
	QGF	13	3.31 (0.72) ^{a}	17.89 (8.04) ^{ab}	1.10 (0.09) ^{a}	6.69 (1.25) ^{a}	50.0 (5.3) ^{bc}	29.8 (3.7) ^{a}	19.1 (3.8)
	VMF	36	3.00 (1.11) ^{a}	20.10 (9.80) ^{b}	1.19 (0.14) ^{a}	6.00 (1.87) ^{a}	46.6 (6.7) ^{ab}	27.6 (5.0) ^{a}	24.5 (6.1)
15–30	DTF	5	3.06 (1.18) ^{a}	12.00 (2.95) ^{a}	1.13 (0.09) ^{a}	6.20 (1.48) ^{a}	46.8 (4.3) ^{a}	27.0 (3.2) ^{a}	22.6 (3.0)
	NCF	10	2.89 (0.87) ^{a}	21.80 (3.50) ^{b}	1.16 (0.11) ^{a}	7.70 (1.34) ^{a}	42.1 (2.5) ^{a}	28.9 (8.2) ^{a}	32.6 (5.9)
	QGF	13	3.94 (0.48) ^{b}	16.30 (2.52) ^{a}	1.05 (0.10) ^{a}	7.15 (1.14) ^{a}	46.2 (2.2) ^{a}	28.8 (3.7) ^{a}	23.1 (4.3)
	VMF	36	3.05 (1.02) ^{ab}	19.80 (4.63) ^{b}	1.13 (0.09) ^{a}	6.97 (1.13) ^{a}	42.3 (1.3) ^{a}	26.6 (4.8) ^{a}	27.9 (6.8) ^{a}

In terms of soil chemical characteristics, the available phosphorus (AP) was higher DDT than those in the other forest communities at 0–15 cm soil depth (Table 4 and Table S An almost similar pattern was observed in Ca and Mg concentrations for both DDT a NCF. At 0–15 cm soil depth, the AP, Ca, and Mg were also significantly higher in D and/or NCF than the other forest community types (Table 4).

2.4. Redundancy Analysis (RDA) Biplot and Important Topographic and Edaphic Variables

To provide an additional quantification of the proportion of the variance in the da set and confirm that environmental variables influence the structure and classification forest communities, we performed an RDA. At 0–15 cm soil depths, the relationship amo the variables is significant ($p = 0.001$, Table S4). The first RDA axis and second axis expla 34.38% and 22.81% of the total variance, respectively (Figure 2). RDA1 is strongly a positively associated with ELV, Ca, and Mg, which are well represented in DTF commun or *D. toposia* (DioTop)-dominated forests (Figure 2b). RDA1 is also strongly and negativ associated with clay proportion in VMF. RDA2 was highly associated with Hd in Q community or *Q. glauca* (QurcGl)-dominated forests (Figure 2b).

Table 4. Soil chemical properties in the *Diospyros toposia* forest (DTF), *Nothaphoebe condensa* forest (NCF), *Quercus glauca* forest (QGF), and *Vatica maingayi* forest (VMF). Values in parenthesis are the standard deviations from the mean. Different lowercase letters indicate statistical significance between forest community types at α = 0.05. Values in parenthesis are the standard deviations from the mean. Abbreviations: TN—total nitrogen, AP—available phosphorus, K—extractable potassium, Ca—extractable calcium, Na—extractable sodium, Mg—extractable magnesium.

Soil Depth (cm)	Forest Community	n	pH	TN (g/kg)	AP (mg/kg)	K (mg/100 g)	Ca (mg/100 g)	Na (mg/100 g)	Mg (mg/100 g)
0–15	DTF	5	5.12 (0.19) a	0.590 (0.136) a	80.0 (7.10) a	8.40 (3.78) a	63.80 (6.44) a	0.30 (0.07) a	154.40 (144.0) a
	NCF	10	4.94 (0.09) a	0.657 (0.285) a	47.0 (35.0) b	5.20 (3.52) a	10.20 (6.25) b	0.32 (0.09) a	21.60 (28.7) ab
	QGF	13	4.85 (0.23) a	0.611 (0.189) a	55.4 (13.9) b	4.15 (1.57) a	5.15 (3.63) b	0.22 (0.15) a	18.77 (18.2) ab
	VMF	36	4.89 (0.11) a	0.663 (0.141) a	50.0 (20.6) b	5.14 (3.09) a	7.36 (8.59) b	0.28 (0.14) a	12.19 (18.0) b
15–30	DTF	5	4.98 (0.12) a	0.622 (0.955) a	80.0 (21.5) a	8.80 (4.32) a	13.60 (12.60) a	0.24 (0.11) a	72.40 (49.0) a
	NCF	10	4.97 (0.09) a	0.572 (0.174) a	44.0 (32.4) b	5.00 (3.56) a	7.60 (3.92) ab	0.34 (0.18) a	19.80 (29.8) b
	QGF	13	5.08 (0.84) a	0.600 (0.129) a	45.4 (32.4) b	4.08 (1.93) a	3.15 (2.61) b	0.19 (0.13) a	19.46 (21.6) b
	VMF	36	4.94 (0.13) a	0.618 (0.172) a	42.8 (16.1) b	5.28 (3.09) a	6.39 (7.53) ab	0.26 (0.16) a	19.19 (54.7) b

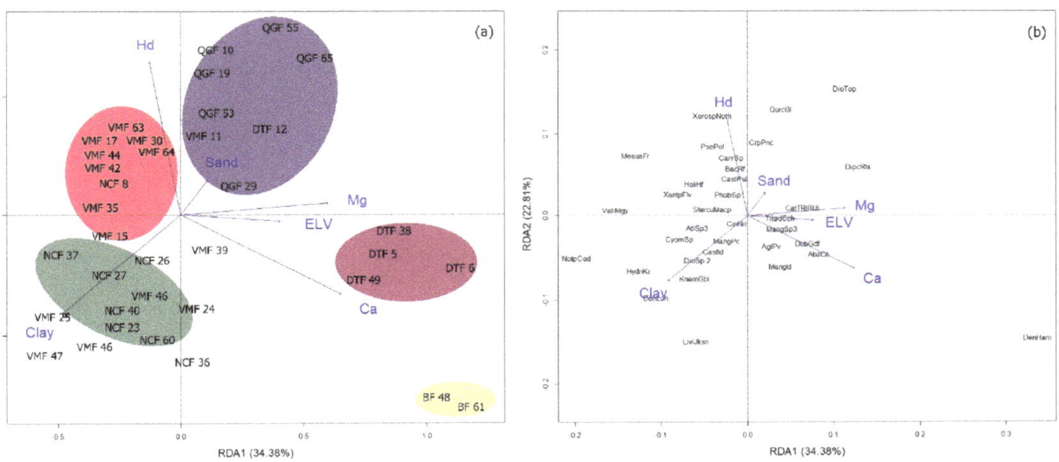

Figure 2. Redundancy analysis (RDA) biplot showing the (**a**) relationship among environmental variables and forest communities and (**b**) relationship among environmental variables and dominant tree species at 0–15 cm soil depth in Htamanthi Wildlife Sanctuary. Abbreviations: ELV—elevation, Hd—soil hardness, Mg—extractable magnesium, Ca—extractable calcium; forest communities: *Vatica maingayi* forest (VMF), *Quercus glauca* forest (QGF), *Nothaphoebe condensa* forest (NCF), and *Diospyros toposia* forest (DTF). The number after the abbreviated name of the forest community in panel (**a**) indicates the plot number. In panel (**b**), the black lowercase letters indicate the abbreviated scientific names of dominant tree species (Table S1).

At 15–30 cm soil depth, the relationship among topographic factors, the properties of soil at 15–30 cm soil depths, and ecologically important tree species and five forest communities is significant (p = 0.001, Table S4). The first two RDA axes accounted for 68.14% of the variations in the data set (Figure 3). Specifically, RDA1 accounted for 37.49% of the variation and was highly and positively related to Ca and ELV in DTF community. RDA2 accounted for 30.65% of the variation and was positively related to Hd in QGF and negatively to BD in VMF community.

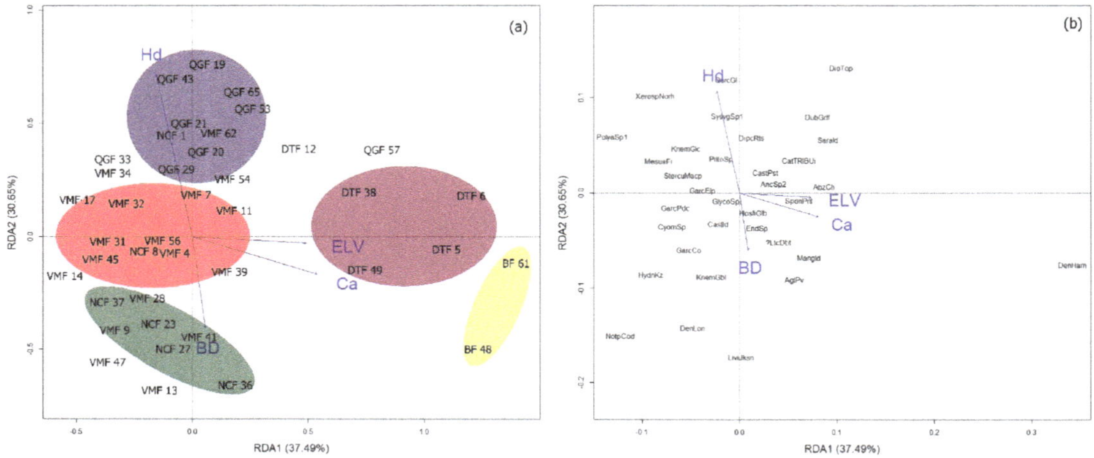

Figure 3. Redundancy analysis (RDA) biplot showing the (**a**) relationship among environment variables and forest communities and (**b**) relationship among environmental variables and dominan tree species at 15–30 cm soil depth in Htamanthi Wildlife Sanctuary in Myanmar. Abbreviation ELV—elevation, Hd—soil hardness, Mg—extractable magnesium, Ca—extractable calcium; fore communities: *Vatica maingayi* forest (VMF), *Quercus glauca* forest (QGF), *Nothaphoebe condensa* fore (NCF), and *Diospyros toposia* forest (DTF). The number after the abbreviated name of the fore community in panel (**a**) indicates the plot number. In panel (**b**), the black lowercase letters indica the abbreviated scientific names of dominant tree species (Table S1).

Further, the forward selection approach revealed Ca (p = 0.001), clay proportio (p = 0.01), and Hd (p = 0.05) as the most contributive factors affecting the composition c tree species and classification of forest communities in the study area at 0–15 cm soil dept (Table S5). At 15–30 cm soil depth, the ELV, Hd, and Ca were the major factors affecting th classification of forest communities in Htamanthi Wildlife Sanctuary (Table S6).

3. Discussion

In this study, the identified forest community types showed a significant variatio in their tree species richness, diversity, and evenness. Results revealed that the VMF an QGF have high to very high Shannon (H) diversity and Evenness (J’) values compare with the other community types. This variation could be ascribed to the differences in th communities in terms of topographic and edaphic factors through their influence on th dispersion behavior of tree species [19]. From a community ecology perspective, result suggest that VMF and QGF community types are more stable and resilient to natura disturbances and accelerating rates of environmental change than the other types of fores communities. Under a range of environmental disruptions that could occur in the futur amid climate change, VMF and QGF may be able to provide essential forest ecosyster functions for the region. However, the impacts of environmental disruptions on ecosyster functions depend on the characteristics of forest communities that are related to resilienc including species diversity, relative abundance, and ability of the dominant species t resist regime shifts and recover functions following disturbance [20]. Here, the RDA biplc shows that the clay proportion and soil bulk density (BD) are highly associated with VM suggesting that any significant disturbance in these factors could potentially influence th community structure and functions. This is supported by the presence of the dominar species, *V. maingayi*, which is a dipterocarp tree species, typical of clay soil in lowlan dipterocarp forests [21]. Two of the important factors influencing clay formation are th effects of soil moisture and temperature on weathering processes, which are the importar aspects of regional climate change impacts on soil [22,23]. Similarly, the elevated globa temperature may increase soil BD via climate change stresses (e.g., drought) and forest management activities [24]. Thus, our results suggests that, as climate change worsens, monitoring the changes in clay formation or weathering processes and soil bulk density is essential for the maintenance of VMF community structure and functions. Dealing with these two major factors influencing VMF community may represent a crucial step for determining priorities in plant community conservation.

The RDA biplot shows that the QGF community is highly associated with soil hardness (Hd), which is a good indicator of soil compaction and strength. This result suggests that HD may be an important factor controlling the plant community in QGF through either detrimental effects on subsequent natural regeneration in the area or enhancement of plant growth and accumulation of soil nutrients. The growth and survival of tree saplings can significantly decrease in compacted soil [25] through a reduction in root cell size, root penetration and, thus, acquisition of essential elements [26]. This may explain the observed negative correlation between Hd and Ca availability in soil. Contrarily, a study found that moderate-level soil compaction improved the plant uptake of P, K, Mg, Ca, and other elements, suggesting that soil compaction effects may vary depending on the severity [26]. Moreover, because soil hardness or soil compaction increases with decreasing soil moisture, the QGF may be more vulnerable to drought than the other forest communities, especially in years with little rainfall. The abundance of the slow-growing and drought-tolerant species, *Q. glauca,* can further explain the high association between Hd and the QGF community. As a drought-tolerant species, *Q. glauca* can thrive and flourish in low-resource and harsh environments, including those which are poorly drained and have tight soil spaces [27]. It has been projected that climate change will increase the frequency of drought and, thus, the effects on the plant community structure [28]. Results suggest that it is necessary to manage and prevent the causes of either surface or subsurface hardness constraints in the QGF community.

The DTF and NCF have moderate diversity values. This can be attributed to the possible effects of elevation (ELV) and soil nutrient availability, particularly Ca and Mg. The RDA biplot shows that the DTF community is highly associated with ELV, Ca, and Mg. The DTF has the highest ELV among the identified forest communities and the value is nearly similar to that of NCF. Results support the findings of Ohdo and Takahashi [29], who reported that the number of tree species decreased at high elevations and the pattern was attributed to soil nutrient availability. In the Himalaya Mountains, it was observed that soil nutrients (N, P, K, and Mg) decreased significantly with elevation [30]. In this study, the concentrations of Ca and Mg were higher in both DTF and NCF communities compared with the other community types. Higher elevation in the two communities may have influenced the concentration of exchangeable cations present in O and/or A horizons, i.e., Ca and Mg may have increased as elevation increased. Such an increase may be due to the decline in tree species composition as elevation increases. Fewer plants may decrease the demand for Ca and Mg ions, resulting in a greater number of ions left in the soil. This can be supported by the presence of the dominant species, *D. toposia,* which is typical of undisturbed forests with high amounts of nutrients [31]. *D. toposia* is also an evergreen tree species and evergreen plants usually exhibit a more conservative strategy, resulting in greater resource conservation.

4. Materials and Methods

4.1. Study Site and Sampling Method

The study was conducted in Htamanthi Wildlife Sanctuary (HWS, 25°45′52″ N to 25°45′15″ N, 95°16′55″ E to 95° 56′ 55″ E), which is a part of Himalaya biodiversity hotspot area in the northern part of Myanmar (Figure 4a). It was declared a protected area in 1974 with the extent of 215,073 hectares and, recently, it was declared the newest ASEAN heritage park. It is a dense forest comprising different habitats, including evergreen and semi-evergreen forests, swamp forests, and upper mixed dry deciduous forests. Some unique and ecologically important flora can also be found in the area, including *Tectona grandis*, *Xylia xylocarpa*, and *Shorea robusta*. The heritage park is generally pristine becau of the minimal disturbance and absence of tourism-related activities. The annual temperature and annual rainfall in the study site ranged from 17.7 to 34.3 °C and 343 m respectively, based on the 10-year climatic data from the local meteorological station ne the HWS. The soil types in the study area are generally yellowish red, derived from acri parent material.

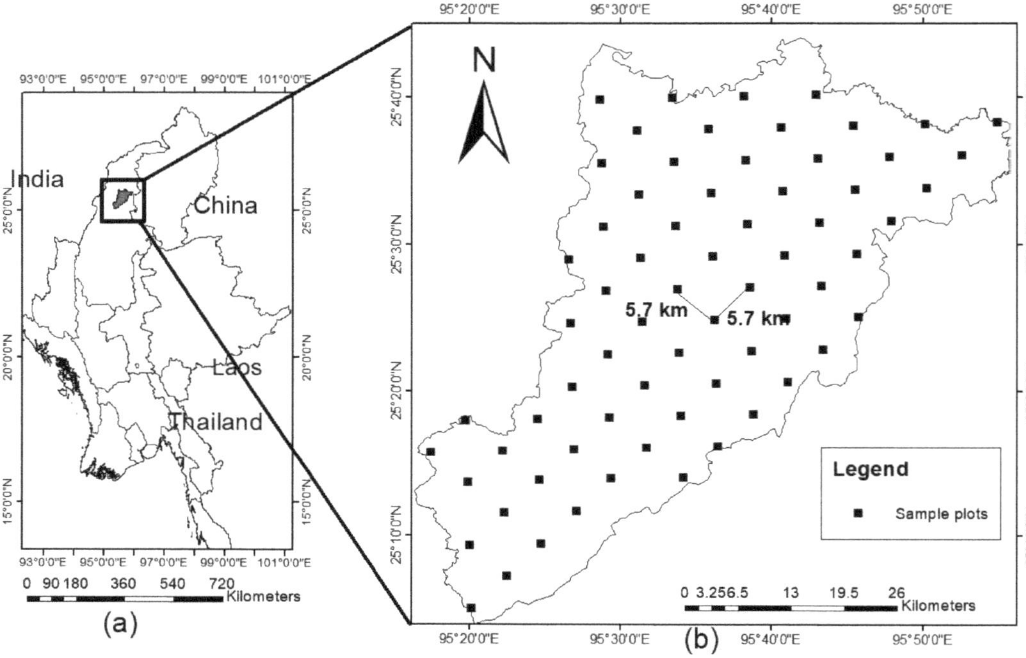

Figure 4. (**a**) Location of study site (Htamanthi Wildlife Sanctuary) and (**b**) distribution of sa ple plots.

Sixty-six plots having a size of 25 m × 25 m were established in the study site by empl ing a systematic sampling (Figure 4b). The distance between plots was 5.7 km × 5.7 km. adult trees with diameter at breast height (DBH) ≥ 10 cm were recorded in 25 m × 25 pl The vegetative inventory was conducted from November to February in 2019 and 2021

4.2. Determination of Topographic and Edaphic Variables

The physiographical variables, namely, geographic coordinates, elevation (ELV), slo (SLP), and aspect (ASP) were recorded for each plot using a GPS (Garmin GPSMAP 6 device. The ALOS PALSAR RTC DEM (12.5 m) images were obtained from the Alas Satellite Facility (UAF). Thereafter, ELV, SLP, and ASP of the sample plots were extract from images and processed using ArcMap software (version 10.8, Esri, CA, USA).

Soil samples were collected from the three pre-determined points (10 m apart) in t 25 × 25 m plot using soil core sampler (5.0 cm in diameter). Three samples (c.a. 100 g) we collected from each soil depth (0–15 m and 15–30 cm) in each point and then composit into one bag. There was a total of 132 bags of soil (i.e., 1 bag × 2 soil depths × 66 plots = 1 The soil hardness (Hd) was measured at each soil depth and plot using a penetrometer. T soil bulk density (BD), the content of moisture content (MC), organic matter compositi (OM), texture (sand, silt, clay), and seven chemical properties of soil; pH, total N, Availa P, exchangeable K, Ca, Na, and Mg, were also analyzed using c.a. 100 g of composited s sample at the Forest Soil Laboratory, Forest Research Institute (FRI), Yezin, Myanmar.

4.3. Data and Statistical Analyses

In this study, the species density, dominance, and frequency were calculated for all trees identified in each plot using the formulae in Table 5. The importance value (IV in %) of tree species encountered in the sampled plots were computed by obtaining the summation of the relative values of stem density, dominance, and frequency. The IV measures a given species' dominance in a forest area based on species and stand structure. In this study, the IV was determined to describe the ecological significance of the species in HWS.

Table 5. Ecological/biotic parameters measured in Htamanthi Wildlife Sanctuary. Abbreviations: **SDen**—species density; **RDen**—relative density; **SFre**—species frequency; **RFre**—relative frequency; **SDom**—species dominance; **RDom**—relative dominance; **IV**—importance values; **BA**—basal area; **AS**—area sampled; DBH—diameter at breast height.

Parameters	Description	Formula
Species density	The actual size or number of individuals of one species per unit area.	Den = No. of individuals of each species/AS
Relative density	The density of one species as a percent of the total density of all species.	RDen = (Den for a species/total density for all identified species) × 100
Species frequency	The number of times a plant species is present in a given number of plots or quadrats.	Fre = No. of plots in which species occur/ total number of plots sampled
Relative frequency	The frequency of one species as a percent of the total frequency of all species.	RFre = (Fre for a species/total frequency for all identified species) × 100
Dominance	A species that is most commonly found or dominant based on basal area or percent coverage.	Dom = BA of one species/AS
Relative dominance	The dominance of one species as a percent of the total dominance of all species.	RDom = (Dom for a species/total dominance for all identified species) × 100
Importance values	The IV is a measure of how dominant a species is in a given forest area.	IV = RDen + RFre + RDom

The Bray–Curtis cluster analysis was performed in RStudio software (version 4.2.0) using the vegan package to classify the vegetation into distinct forest community types. The similarity and dissimilarity among forest community types were determined using the Jaccard similarity index based on the species abundance data. The Jaccard's similarity index formula is: $J\ (i, j) = a/(a + b + c)$, where a = number of species in common between the communities; b = number of species unique to the first community; c = number of species unique to the second community.

The species richness (S), evenness (J), and Shannon–Wiener diversity index (H') were computed using the R software. The Shannon diversity index (H') was computed from the equation: $H = -\Sigma pi \times \ln(pi)$, where H is Shannon–Wiener diversity index, pi is the proportion of individual tree species.

Kruskal–Wallis test was applied to assess the significant differences in topographic and physico-chemical characteristics across the different forest communities. Pairwise comparisons among the different forest communities were conducted by Dunn–Bonferroni post hoc test method.

Here, redundancy analysis (RDA) ordination was used in describing the relationship between forest communities, topographic factors, and physico-chemical properties of soil. The global model test with permutations = 4999 was conducted to test for significance.

The forward selection of variables was conducted to select the driving variables of a semblage of forest communities as well as to explain the variation in species with highl correlated variables. Monte Carlo test was applied to prove the significant correlation wit 4999 permutations. Plot × species matrices and the estimation of IV were conducted i Microsoft Office Excel and all the statistical analyses were processed in RStudio softwar (version 4.2.0) at a 95% confidence level.

5. Conclusions

The present study revealed that there are four major evergreen forest communities, i.e *Vatica maingayi* forest (VMF), *Quercus glauca* forest (QGF), *Nothaphoebe condensa* forest (NCF and *Diospyros toposia* forest (DTF), and one minor forest community, i.e., bamboo forest (BF in Htamanthi Wildlife Sanctuary (HWS). Based on Shannon diversity index estimatio three of the five communities have high to very high species diversity, while the other tw communities have low to moderate species diversity. Here, the soil Ca, Mg, clay proportio soil hardness, and elevation were identified as the major factors in classifying the fore community types. Identification of these community types and understanding of the majo factors influencing their structure is an important nature-based solution in biodiversit conservation for a biodiversity hotspot like HWS. The present work will enhance ou understanding on how to efficiently implement an ecosystem-based conservation approac through conservation prioritization.

Supplementary Materials: The following supporting information can be downloaded at: http //www.mdpi.com/article/10.3390/plants11162180/s1, Table S1: List of tree species surveyed i Htamanthi Wildlife Sanctuary; Table S2: *p* values estimated by Kruskal–Wallis test for soil physic properties across soil depths and forest communities; Table S3: *p* values estimated by Kruskal–Wall test for soil chemical properties across soil depths and forest communities; Table S4: Significance test of the global model with all explanatory variables of topographic factors and soil properties; Tab S5: Selection of important variables of topographic factors and soil properties at 0–15 cm depth soil; Table S6: Selection of important variables of topographic factors and soil properties at 15–30 c depth of soil.

Author Contributions: Conceptualization, B.B.P. and M.M.L.; methodology, B.B.P. and M.M.L software and data analysis, data collection, writing—original draft preparation, M.M.L.; writing- review and editing, B.B.P. and M.M.L.; supervision, B.B.P. All authors have read and agreed to th published version of the manuscript.

Funding: This study was carried out with the support of 'R&D Program for Forest Science Technolog (Project No. 2021379B10-2223-BD02 and 2020184C10-2222-AA02)' provided by Korea Forest Servic (Korea Forestry Promotion Institute).

Data Availability Statement: The data used are primarily reflected in the article. Other relevant dat are available from the authors upon request.

Acknowledgments: This study was carried out according to the project of laboratory of forest ecolog and silviculture, Chungnam National University (CNU), Republic of Korea (ROK) to analyze th diversity of flora and wildlife of Htamanthi Wildlife Sanctuary. The authors are grateful to th Forest Department of Myanmar for the permit and the support as well as to the local people an graduate students from the University of Forestry and Environmental Science, Yezin, for the effo and participation during our forest inventory in the protected areas. The authors acknowledg Wildlife Conservation Society (WCS) Myanmar program for supporting logistics and accommodatio during the survey. Last but not the least, we deeply appreciate Jonathan O. Harnandez for kindl sharing his knowledge and discussion during this research.

Conflicts of Interest: The authors declare no conflict of interest.

ferences

Chen, J.; Saunders, S.C.; Crow, T.R.; Naiman, R.J.; Brosofske, K.D.; Mroz, G.D.; Brookshire, B.L.; Franklin, J.F. Microclimate in forest ecosystem and landscape ecology. *BioScience* **1999**, *49*, 288–297. [CrossRef]
Duchicela, S.A.; Cuesta, F.; Tovar, C.; Muriel, P.; Jaramillo, R.; Salazar, E.; Pinto, E. Microclimatic warming leads to a decrease in species and growth form diversity: Insights from a tropical alpine grassland. *Front. Ecol. Evol.* **2021**, *9*, 673655. [CrossRef]
Rahman, I.U.; Afzal, A.; Iqbal, Z.; Bussmann, R.W.; Alsamadany, H.; Calixto, E.S.; Shah, G.M.; Kausar, R.; Shah, M.; Ali, N.; et al. Ecological gradients hosting plant communities in Himalayan subalpine pastures: Application of multivariate approaches to identify indicator species. *Ecol. Inform.* **2020**, *60*, 101162. [CrossRef]
Hawkins, B.A.; Field, R.; Cornell, H.V.; Currie, D.J.; Guégan, J.-F.; Kaufman, D.M.; Kerr, J.T.; Mittelbach, G.G.; Oberdorff, T.; O'Brien, E.M.; et al. Energy, water, and broad-scale geographic patterns of species richness. *Ecology* **2003**, *84*, 3105–3117. [CrossRef]
Miao, L.; Jianmeng, F. Biogeographical interpretation of elevational patterns of genus diversity of seed plants in Nepal. *PLoS ONE* **2015**, *10*, e0140992. [CrossRef]
Nepali, B.R.; Skartveit, J.; Baniya, C.B. Impacts of slope aspects on altitudinal species richness and species composition of Narapani-Masina Landscape, Arghakhanchi, West Nepal. *J. Asia-Pac. Biodivers.* **2021**, *14*, 415–424. [CrossRef]
Cornwell, W.K.; Grubb, P.J. Regional and local patterns in plant species richness with respect to resource availability. *Oikos* **2003**, *100*, 417–428. [CrossRef]
Toro Manríquez, M.D.R.; Cellini, J.M.; Lencinas, M.V.; Peri, P.L.; Peña Rojas, K.A.; Martínez Pastur, G.J. Suitable conditions for natural regeneration in variable retention harvesting of southern Patagonian nothofagus pumilio forests. *Ecol. Process.* **2019**, *8*, 1–12. [CrossRef]
Davis, E.L.; Hager, H.A.; Gedalof, Z. Soil properties as constraints to seedling regeneration beyond alpine treelines in the Canadian Rocky Mountains. *Arct. Antarct. Alp. Res.* **2018**, *50*, e1415625-1-15. [CrossRef]
Wenk, E.H.; Dawson, T.E. Interspecific differences in seed germination, establishment, and early growth in relation to preferred soil type in an Alpine community. *Arct. Antarct. Alp. Res.* **2007**, *39*, 165–176. [CrossRef]
Palpurina, S.; Wagner, V.; Wehrden, H.V.; Hájek, M.; Horsák, M.; Brinkert, A.; Hölzel, N.; Wesche, K.; Kamp, J.; Hájková, P.; et al. The relationship between plant species richness and soil pH vanishes with increasing aridity across Eurasian dry grasslands. *Glob. Ecol. Biogeogr.* **2016**, *26*, 425–434. [CrossRef]
Tyler, G. Some ecophysiological and historical approaches to species richness and calcicole/calcifuge behaviour—Contribution to a debate. *Folia Geobot.* **2003**, *38*, 419–428. [CrossRef]
Currie, D.J.; Mittelbach, G.G.; Cornell, H.V.; Field, R.; Guegan, J.F.; Hawkins, B.A.; Kaufman, D.M.; Kerr, J.T.; Oberdorff, T.; O'Brien, E.; et al. Predictions and tests of climate-based hypotheses of broad-scale variation in taxonomic richness. *Ecol. Lett.* **2004**, *7*, 1121–1134. [CrossRef]
Naing, H.; Ross, J.; Burnham, D.; Htun, S.; Macdonald, D.W. Population density estimates and conservation concern for clouded leopards neofelis nebulosa, marbled cats pardofelis marmorata and tigers panthera tigris in htamanthi wildlife sanctuary, Sagaing, Myanmar. *Oryx* **2017**, *53*, 654–662. [CrossRef]
Seddon, N.; Chausson, A.; Berry, P.; Girardin, C.A.; Smith, A.; Turner, B. Understanding the value and limits of nature-based solutions to climate change and other global challenges. *Philos. Trans. R. Soc. B Biol. Sci.* **2020**, *375*, 20190120. [CrossRef] [PubMed]
Myers, N.; Mittermeier, R.A.; Mittermeier, C.G.; da Fonseca, G.A.; Kent, J. Biodiversity hotspots for conservation priorities. *Nature* **2000**, *403*, 853–858. [CrossRef] [PubMed]
Rodríguez-Rodríguez, E.J.; Beltrán, J.F.; El Mouden, E.H.; Slimani, T.; Márquez, R.; Donaire-Barroso, D. Climate change challenges IUCN conservation priorities: A test with Western Mediterranean amphibians. *SN Appl. Sci.* **2020**, *2*, 216. [CrossRef]
Hernandez, J.O.; Maldia, L.S.J.; Park, B.B. Research trends and methodological approaches of the impacts of windstorms on forests in tropical, subtropical, and temperate zones: Where are we now and how should research move forward? *Plants* **2020**, *9*, 1709. [CrossRef]
Dibaba, A.; Soromessa, T.; Warkineh, B. Plant community analysis along environmental gradients in moist afromontane forest of Gerba Dima, south-western Ethiopia. *BMC Ecol. Evol.* **2021**, *22*, 1–17. [CrossRef]
Oliver, T.H.; Heard, M.S.; Isaac, N.J.B.; Roy, D.B.; Procter, D.; Eigenbrod, F.; Freckleton, R.; Hector, A.; Orme, C.D.; Petchey, O.L.; et al. Biodiversity and resilience of ecosystem functions. *Trends Ecol. Evol.* **2015**, *30*, 673–684. [CrossRef]
Davies, S.J.; Tan, S.; LaFrankie, J.V.; Potts, M.D. Soil-related floristic variation in the hyperdiverse dipterocarp forest in Lambir Hills, Sarawak. In *Pollination Ecology and Rain Forest Diversity, Sarawak Studies*; Roubik, D.W., Sakai, S., Hamid, A., Eds.; Springer: New York, NY, USA, 2005; pp. 22–34.
Huszár, T.; Mika, J.; Lóczy, D.; Molnár, K.; Kertész, Á. Climate change and soil moisture: A case study. *Phys. Chem. Earth Part A Solid Earth Geod.* **1999**, *24*, 905–912. [CrossRef]
Deepthy, R.; Balakrishnan, S. Climatic control on clay mineral formation: Evidence from weathering profiles developed on either side of the Western Ghats. *J. Earth Syst. Sci.* **2005**, *114*, 545–556. [CrossRef]
Birkás, M. Tillage, impacts on soil and environment. In *Encyclopedia of Agrophysics. Encyclopedia of Earth Sciences Series*; Gliński, J., Horabik, J., Lipiec, J., Eds.; Springer: Dordrecht, The Netherlands, 2011; pp. 903–906. [CrossRef]
Ampoorter, E.; de Frenne, P.; Hermy, M.; Verheyen, K. Effects of soil compaction on growth and survival of tree saplings: A meta-analysis. *Basic Appl. Ecol.* **2011**, *12*, 394–402. [CrossRef]

26. Wang, M.; He, D.; Shen, F.; Huang, J.; Zhang, R.; Liu, W.; Zhu, M.; Zhou, L.; Wang, L.; Zhou, Q. Effects of soil compaction plant growth, nutrient absorption, and root respiration in soybean seedlings. *Environ. Sci. Pollut. Res.* **2019**, *26*, 22835–228 [CrossRef] [PubMed]
27. Hernandez, J.O.; An, J.Y.; Combalicer, M.S.; Chun, J.P.; Oh, S.K.; Park, B.B. Morpho-anatomical traits and soluble su concentration largely explain the responses of three deciduous tree species to progressive water stress. *Front. Plant Sci.* **2021**, 738301. [CrossRef]
28. Walter, J.; Hein, R.; Auge, H.; Beierkuhnlein, C.; Löffler, S.; Reifenrath, K.; Schädler, M.; Weber, M.; Jentsch, J. How do extre drought and plant community composition affect host plant metabolites and herbivore performance? *Arthropod-Plant Inter* **2012**, *6*, 15–25. [CrossRef]
29. Ohdo, T.; Takahashi, K. Plant species richness and community assembly along gradients of elevation and soil nitrogen availabi *AoB PLANTS* **2020**, *12*, plaa014. [CrossRef]
30. Drollinger, S.; Müller, M.; Kobl, T.; Schwab, N.; Böhner, J.; Schickhoff, U.; Scholten, T. Decreasing nutrient concentrations in so and trees with increasing elevation across a treeline ecotone in Rolwaling Himal, Nepal. *J. Mt. Sci.* **2017**, *14*, 843–858. [CrossR
31. Lalfakawma; Sahoo, U.K.; Roy, S.; Vanlalhriatpuia, K.; Vanalalhluna, P.C. Community composition and tree population struct in undisturbed and disturbed tropical semi-evergreen forest stands of north-east India. *Appl. Ecol. Environ. Res.* **2009** 303–318. [CrossRef]

forests

MDPI

rticle

Knowledge about Plant Coexistence during Vegetation Succession for Forest Management on the Loess Plateau, China

ilong Tian [1,2,3], Xiaoping Zhang [1,2,3,4,*], Xiaoming Xu [4], Haijie Yi [1,2,3], Jie He [4], Liang He [4] and Weinan Sun [4]

1 The Research Center of Soil and Water Conservation and Ecological Environment, Chinese Academy of Sciences and Ministry of Education, Xianyang 712100, China
2 Institute of Soil and Water Conservation, Chinese Academy of Sciences and Ministry of Water Resources, Xianyang 712100, China
3 University of Chinese Academy of Sciences, Beijing 100049, China
4 Institute of Soil and Water Conservation, Northwest A & F University, Xianyang 712100, China
* Correspondence: zhangxp@ms.iswc.ac.cn; Tel.: +86-13709124604

Abstract: Coexistence between species within plant communities is a key issue in the practice of revegetation, forest management, and biodiversity conservation. Vegetation restoration is critical to control soil erosion and improve the ecological environment on the Loess Plateau. Here, we investigate the interspecific relationships of dominant plants during natural vegetation succession on the Loess Plateau. The results suggest that, under the ecological process of environmental filtering, species within communities can reduce interspecific competition and promote species coexistence via spatial heterogeneity and temporal asynchronous differences. The ecological niche overlap index (O_{ik}) significantly and positively correlated with the strength of interspecific associations. Most species pairs had weak competition and more stable interspecific relationships. The results of the χ^2 test showed that 317 species pairs were positively associated and 118 were negatively associated. The community is in a positive succession process, and the interaction relationship between species tends to be neutral. We should enhance the protection of positively associated species and pay attention to negatively associated species during forest management. Results revealed that *Carex lanceolata* Boott and *Lespedeza bicolor* Turcz coexisted easily with other species for mutual benefit, which could help build artificial forestland of native species to improve the ecological function.

Keywords: ecological niche; overlap; plant communities; interspecific relationship; biodiversity conservation; spatial heterogeneity

tation: Tian, Q.; Zhang, X.; Xu, X.; H.; He, J.; He, L.; Sun, W. owledge about Plant Coexistence ring Vegetation Succession for rest Management on the Loess ateau, China. *Forests* **2022**, *13*, 1456. tps://doi.org/10.3390/f13091456

ademic Editors: Panayiotis mitrakopoulos, Mario A. Pagnotta, klas Scholz and Arshiya Noorani

ceived: 12 August 2022
cepted: 8 September 2022
blished: 10 September 2022

1. Introduction

Coexistence between species has long been a central question in community ecology research [1,2]. Investigating coexistence relationships can objectively reflect species distribution in a community and the degree of species adaptation to the environment, reveal community structure, type, and plant substitution mechanisms, and predict population extinction [3–5]. Modern coexistence theory suggests that the causes of species coexistence in communities depend on the balance of niche overlap and competitive asymmetry [6]. Therefore, such interspecific relationships are often quantified by species' ecological niches and interspecific association characteristics [7]. Ecological niches are the sum of all resources available to various organisms in a community and their functional relationships with related populations, describing the role played by organisms in the community [8]. Under the pressure of ecological processes, species change their characteristics through trait evolution, and this may lead to changes in species interactions and their "ecological niche" in the community. This affects species coexistence and changes in community stability [9]. Interspecific associations represent the interconnectedness of the spatial distribution of different species, reflecting the differences in adaptation to different habitats among species in the community [10]. Analysis of interspecific relationships can provide insight into the competitive exclusionary effects between species. In general, positive interspecif correlations indicate similarity in resource use by species; negative correlations indicate di advantages for one or both parties, such as interspecific competition and disturbance [1 This is essential to reveal the formation and evolution of communities [12]. Ultimatel this may help us control interspecies relationships and maintain the stability of ecosyste biodiversity [13,14].

The Loess Plateau of China is a global hotspot of land degradation with a fragil ecological environment and extremely serious soil erosion [15,16]. To change this situatio the Chinese government implemented the Grain for Green (GFG) Project in 1999, in whic artificial planting was the main practice [17]. Nowadays, the GFG project area has reache 3.33×10^5 km^2, and the vegetation cover of the Loess Plateau has increased from abou 49% in 1998 to 63% in 2018 [18,19]. The ecological environment has been greatly improve Soil erosion decreased to 595 t/km^2/year during 2011–2015, and the amount of sedime entering the sea from the Loess Plateau between 2005–2015 was only 8.9% of that in th 1950s [20,21]. Vegetation restoration is regarded as a fundamental measure to combat so erosion [22]. However, in the processes, irrational plantation and monoculture plantin presented negative impacts, such as vegetation degradation, soil dryness, groundwate drought, and more than a 50% reduction in runoff, posing a serious challenge to th sustainability of vegetation growth and the stability of ecological functions [23–27]. Thus, is important to figure out how native species can live together and work together, becaus mixed forests of native species can make a region more resistant to drought, help plant survive, and keep ecosystem functions stable [28,29].

Ziwuling is located in the Loess Plateau's hinterland after a 150-year natural restor tion period, making it an ideal site for researching the Loess Plateau's vegetation successio law [30–32]. Vegetation succession refers to the process by which a plant communit evolves when one type of vegetation is replaced by another, and qualitative change occur [33,34]. It is demonstrated through the interdependence and competition of c existing species within the community, as well as through the interaction and influenc of habitats [12]. This provides an excellent natural template to reveal plant coexistenc relationships; however, reports related to the coexistence of species in the local area are rar The research objectives of this study were to (1) identify the dominant species of herbs an woody plants in the process; (2) analyze the ecological niche and interspecific associatio characteristics of dominant species; (3) determine reliable species coexistence relationshi and mechanisms through holistic analysis. Answers to these questions might help buil local plantation forests, protect biodiversity, manage forests, conservation of rare specie the control of biological invasions, the forecasting of climate change impacts, and improv the ecosystem [14].

2. Materials and Methods

2.1. Study Area

Ziwuling Nature Reserve is located in the temperate zone of the Loess Plateau, Chin (34°50′–36°50′ N, 107°30′–109°40′ E) (Figure 1). The study area is approximately 200 k long from north to south and 95 km wide from east to west, covering a total area c approximately 3.79×10^4 km^2. Its altitude range is 553–1856 m above sea level, with relative height difference of 400 m. It has an East Asian monsoon climate with an averag annual precipitation of 588 mm, which is unevenly distributed throughout the year, an 70% of precipitation occurs from July to September. The average annual temperature c the area is 7.4–8.5 °C, with a minimum recorded temperature of −27.7 °C and a maximu recorded temperature of 36.7 °C [35]. The soil layer in the mountains is shallow, bedrock exposed in some areas, and zonal soil is grayish brown with a pH value of 7.5–8.2. The so is primary or secondary loess, loose and easy to wet, and sinks because of its poor erosic resistance [36]. The ground zone vegetation is a warm temperate deciduous broad-leave forest with *Quercus mongolica* Fisch. Ex Ledeb as the dominant species and a temperat coniferous forest with *Pinus tabuliformis* Carrière as the dominant species [37].

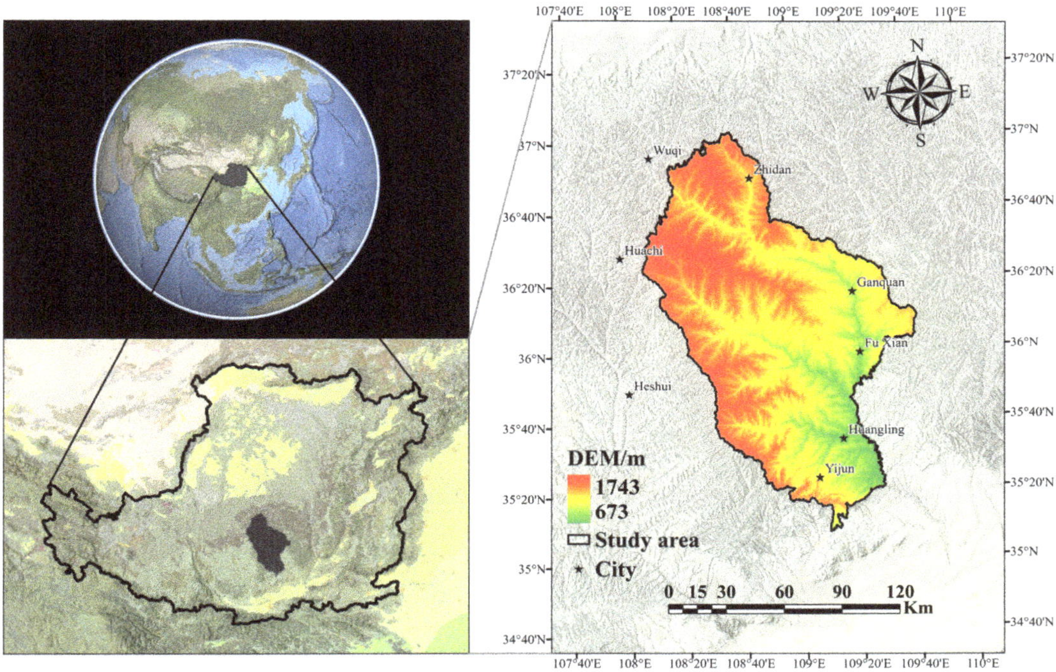

Figure 1. Ziwuling study area on the Chinese Loess Plateau.

2.2. Experimental Design and Investigation

Since 1899, ecologists have commonly used a "space-for-time" substitution method for vegetative succession studies to predict vegetation chronosequence change [38,39]. This study adopted a "space for time" approach to select 48 sites. Then, based on the results of the relevant forestry department and a comprehensive analysis based on related reports by Zou et al. [35], Fan et al. [36], Deng et al. [40], Zhao et al. [41], and Liu et al. [42], standard sample plots of typical vegetation types in eight different restoration periods, with six sites for each period, were set up from 2020 to 2021 with essentially identical environmental conditions and similar soil texture. Standard sample plots of 20 × 20 m (trees), 10 × 10 m (shrubs), and 1 × 1 m (grasses) were established, and the names and numbers of species occurring in each sample plot were recorded. The Latin names of the species were cross-checked using The Plant List (http://www.theplantlist.org, accessed on: 20 December 2021) with reference to the Flora of China [43]. A total of 128 species of seed plants, 79 herbs, and 49 woody species were found during this process. The specific information is shown in Table 1.

Table 1. Vegetation survey sample site information at the Ziwuling study area on the Chinese Loess Plateau.

Restoration Age (Year)	Succession Stage	Numbers of Sites	Representative Plants
0	Agricultural land	6	*Viola collina* Besser; *Agropyron cristatum* (L.) Gaertn.
10	Natural restoration of 10 years of grass	6	*Bothriochloa ischaemum* (L.) Keng; *Artemisia chamaemelifolia* Vill; *Lespedeza bicolor* Turcz.

Table 1. *Cont.*

Restoration Age (Year)	Succession Stage	Numbers of Sites	Representative Plants
20	Natural restoration of 20 years of grass	6	*Artemisia codonocephala* Diels; *Agrimonia pilosa* Ledeb; *Lespedeza bicolor* Turcz.
40	Shrub stage	6	*Sophora davidii* (Franch.) Pavol; *Lespedeza bicolor* Turcz; *Agropyron cristatum* (L.) Gaertn; *Anemone chinensis* Bunge.
70	Pioneer arbor stage	6	*Betula platyphylla* Sukaczev; *Spiraea salicifolia* L.; *Lespedeza bicolor* Turcz; *Carex lanceolata* Boott.
120	Sub-top stage	6	*Pinus tabuliformis* Carrière; *Lespedeza bicolor* Turcz, *Carex lanceolata* Boott.
135	Sub-top to top transition stage	6	*Pinus tabuliformis* Carrière; *Quercus mongolica* Fisch. Ex Ledeb; *Lespedeza bicolor* Turcz; *Carex lanceolata* Boott.
150	Top Stage	6	*Quercus mongolica* Fisch. Ex Ledeb; *Lespedeza bicolor* Turcz; *Carex lanceolata* Boott.

2.3. Statistical Analyses

Competition theory suggests that interspecific associations can change depending external conditions and that different environments may exhibit different associations [To ensure the reliability of the results, this research combined data from eight major succ sional stages during 150 years of natural vegetation restoration for the overall analysis.

2.3.1. Species Selection

The plants identified in the vegetation survey were classified as herbaceous or woo plants. The top 15 plants of each category were ranked according to importance value (and identified as the dominant plants [45].

$$IV = \frac{\left(\frac{a_i}{\sum_{i=1}^{S} a_i} + \frac{f_i}{\sum_{i=1}^{S} f_i}\right)}{2}$$

where a_i is the number of individuals in population i, f_i is the number of quadrats in wh the population i appears, and S is the total number of species.

2.3.2. Ecological Niche Features

To calculate ecological niche breadth (B_i), the weighted modified Levins index meth described by Colwell and Futuyma [46] was used

$$B_i = \frac{1}{r \sum_{j=1}^{r} P_{ij}^2}$$

where P_{ij} is the ratio of the number of individuals of species i at resource site j to the to number of individuals of species i at all resource sites; r is the total number of samp squares, and each sample square represents one resource site.

To determine the ecological niche overlap index (O_{ik}), the Pianka index was used [

$$O_{ik} = \frac{\sum_{j=1}^{N}\left(P_{ij} \cdot P_{kj}\right)}{\sqrt{\sum_{j=1}^{N} P_{ij}^2 \cdot \sum_{j=1}^{N} P_{kj}^2}}$$

where O_{ik} is the ecological niche overlap value of species i and species k, where P_{ij} a P_{kj} are the importance value shares of species i and species k, respectively, in resou niche j, and the value range is [0, 1]; the larger the value, the higher the ecological nic overlap degree.

2.3.3. Overall Association

The variance ratio method (*VR*) was used, and the statistic *W* was calculated to test the significance level of the overall association [48]

$$VR = \frac{s_T^2}{\delta_T^2} = \frac{\frac{1}{N}\sum_{j=1}^{N}(T_j - t)^2}{\sum_{l=1}^{S} P_i(1 - P_i)}. \tag{4}$$

$$W = VR \times N \tag{5}$$

where P_i is the frequency of species *i*, *N* is the total number of samples, *S* is the total number of species, T_i is the total number of species occurring in sample *j*, and *t* is the average number of species in the sample. When $VR > 1$, the overall species are positively associated, and when $VR < 1$, the overall species are negatively associated. The interspecific association is significant if $\chi^2_{(0.95,\ N)} < W < \chi^2_{(0.05,\ N)}$; otherwise, the association is not significant.

2.3.4. Interspecific Association

Based on a 2 × 2 column table [49], the χ^2 statistic was used for qualitative analysis. The continuous correction factor of Yates was used to correct for the sampling discontinuities, which can cause biased underestimation, with the formula [50]

$$\chi^2 = \frac{N[|ad - bc| - 0.5N]^2}{(a+b)(a+c)(c+d)(b+d)} \tag{6}$$

$$V = \frac{[(a+d) - (b+c)]}{a+b+c+d} \tag{7}$$

where *a* indicates the number of samples in which both species occur, *b* indicates the number of samples in which species *B* occurs, but species *A* does not, *c* indicates the number of samples in which species *A* occurs but species *B* does not, and *d* indicates the number of samples in which neither species occurs. Generally, interspecific associations were considered highly significant when $\chi^2 > 6.635$, i.e., $p < 0.01$; significant when $3.841 \leq \chi^2 < 6.635$, i.e., $0.01 < p < 0.05$; and insignificant when $\chi^2 < 3.841$, i.e., $p > 0.05$. The indicator of *V* was constructed to show the interspecific association. $V < 0$ indicated negative associations, and $V > 0$ indicated positive associations [51].

2.3.5. Strength of Associations

The χ^2 statistic can only qualitatively describe whether the association between species is significant, but it cannot determine the strength of interspecific associations. Therefore, the association coefficient (*AC*) and percentage co-occurrence (*PC*) are commonly used to express the strength of association [52]. *PC* is more accurate than *AC* for reflecting the positive association strength and avoiding the effects of high *AC* due to high d values or low *AC* due to low *a* values. Therefore, *PC* and *AC* are usually analyzed together to improve the accuracy of the results.

The *AC* calculation formula is as follows

$$AC = \frac{ad - bc}{(a+b)(b+d)}(ad \geq bc) \tag{8}$$

$$AC = \frac{ad - bc}{(a+b)(a+c)}(ad < bc, d \geq a) \tag{9}$$

$$AC = \frac{ad - bc}{(b+d)(d+c)}(ad < bc, d < a) \tag{10}$$

The value of *AC* is in the range of $[-1, 1]$: the closer *AC* is to 1, the stronger the positive association of interspecific association; the closer *AC* is to -1, the stronger the negative association of interspecific association; when $AC = 0$, each species is completely independent.

The *PC* calculation formula is as follows

$$PC = \frac{a}{a+b+c} \quad (1$$

The *PC* value is [0, 1]. The closer the *PC* value is to 1, the stronger the degree o interspecies association, and a *PC* value of 0 indicates that interspecies association doe not exist.

This research used Excel 2019 for data preparation (Microsoft Corp., Redmond, WA USA) and R 4.0.4 (R Foundation for Statistical Computing, Vienna, Austria; program packagesspaa and corrplot) for statistical analysis [53,54]. Mapping was done by ArcGI 10.5 (Esri, Redlands, CA, USA) and Origin 2021b (OriginLab, Northampton, MA, USA).

3. Results and Analysis

3.1. Importance Values and Ecological Niche Breadth of Dominant Plants

A total of 128 species in 99 genera and 39 families of seed plants, 79 species of herb ceous plants, and 49 species of woody plants were identified during the vegetation su cession in our survey. The sum of the importance values of the top 15 herbaceous specie accounted for 65.85% of all herbaceous plants, and the sum of the importance values o the top 15 woody species accounted for 79.56% of all woody plants (Table 2). Therefor these 30 species were suitable as the main plants for researching species interrelationship in this process. The ecological niche breadth ranged from [1.07, 12.8] for herbaceous plan to [1.43, 9.42] for woody plants. Among herbaceous plants, *C. lanceolata* had the highes importance value at 14.43%, while among the woody plants, *L. bicolor* had the highes importance value at 20.98%. These two plants also had the largest ecological niche breadt values (Table 1). The order of the dominant species importance values and the sizes of the ecological niches were not always the same.

Table 2. Importance value and ecological niche breadth of selected 30 species at the Ziwulin study area on the Chinese Loess Plateau. Abbreviations: IV—importance value; B_i—ecologic niche breadth.

No.	Herb	IV/%	B_i	No.	Woody	IV/%	B_i
1	*Carex lanceolata* Boott	14.43	12.80	16	*Lespedeza bicolor* Turcz	20.98	9.42
2	*Artemisia argyi* H.Lév. & Vaniot	8.7	1.07	17	*Quercus mongolica* Fisch. Ex Ledeb	11.41	5.25
3	*Agrimonia pilosa* Ledeb	7.53	1.07	18	*Sophora davidii* (Franch.) Pavol	5.67	4.36
4	*Miscanthus sinensis* Andersson	5.77	3.01	19	*Spiraea salicifolia* L.	5.55	2.11
5	*Artemisia chamaemelifolia* Vill.	5.17	3.49	20	*Pinus tabuliformis* Carrière	5.37	7.61
6	*Viola philippica* Cav	4.41	1.70	21	*Acer tataricum subsp. ginnala* (Maxim.) Wesm	4.34	3.68
7	*Potentilla chinensis* Ser	3.1	3.48	22	*Ostryopsis davidiana* Decne	4.32	2.67
8	*Bothriochloa ischaemum* (L.) Keng	2.67	2.27	23	*Betula platyphylla* Sukaczev	4	5.96
9	*Aster hispidus* Thunb	2.54	2.69	24	*Cotoneaster multiflorus* Bunge	3.63	3.34
10	*Artemisia lancea* Vaniot	2.21	3.81	25	*Lonicera japonica* Thunb	3.23	6.21
11	*Viola collina* Besser	2.18	3.38	26	*Campylotropis macrocarpa* (Bunge) Rehder	2.97	5.48
12	*Cyperus compressus* L.	1.84	1.45	27	*Koelreuteria paniculata* Laxm	2.75	1.43
13	*Sophora flavescens* Aiton	1.83	2.45	28	*Periploca sepium* Bunge	1.98	3.73
14	*Anemone chinensis* Bunge	1.76	2.23	29	*Rubus parvifolius* L.	1.84	2.29
15	*Agropyron cristatum* (L.) Gaertn	1.71	3.77	30	*Rhamnus utilis* Decne	1.52	1.54

3.2. Niche Overlap

The O_{ik} ranged from 0 to 0.9996. In general, 56 pairs (12.87% of the total) had $O_{ik} \geq 0.$ and these pairs were subject to intense competition when environmental resources wer limited. Another 127 pairs (29.2% of the total) had O_{ik} between 0.1 and 0.5, and these pair were subject to a certain degree of competition but could still coexist when resources wer sufficient. The remaining 252 pairs (57.93% of the total) had $O_{ik} < 0.1$, and competition among these pairs was weak (Figure 2).

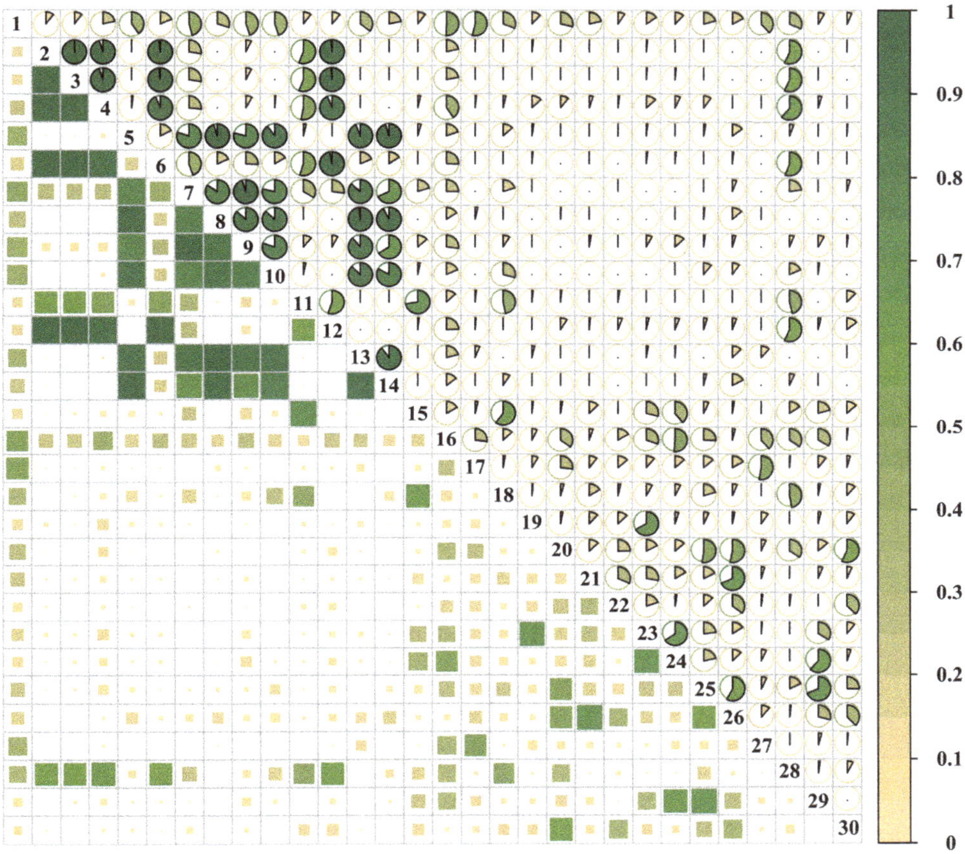

Figure 2. The ecological niche overlap index. The numbers represent the plants in Table 1. The numbers of 1 to 15 represent herbaceous plants, and 16 to 30 indicate woody plants. The larger the square or sector, the darker the color, the larger the O_{ik}.

Sorted by life form, 37 pairs of herbs had $O_{ik} \geq 0.5$, accounting for 35.24% of the total number of pairs of herbs; 29 pairs of herbs had O_{ik} between 0.1 and 0.5, accounting for 27.62% of the total; and 39 pairs of herbs had $O_{ik} < 0.1$, accounting for 37.14% of the total. *A. pilosa* and *A. argyi* had the largest O_{ik} value at 0.9996. Among the woody plants, 11 species pairs had $O_{ik} \geq 0.5$, accounting for 10.47% of the total; 49 pairs had O_{ik} between 0.1 and 0.5, accounting for 46.67% of the total; and 45 pairs had $O_{ik} < 0.1$, accounting for 42.86% of the total. *L. japonica* and *R. parvifolius* had the largest O_{ik} value at 0.6875 (Figure 2).

3.3. Overall Association Analysis

As shown in Table 3, the *VR* of herbaceous and woody plants as a whole, herbaceous and woody plants during vegetation succession were 1.20, 2.40, and 2.49, respectively. The *VR*s of all three scenarios vegetation is greater than 1, indicating a positive association among the dominant species in general. The χ^2 table was checked according to the degrees of freedom, and the test statistics (*W*) did not fall under χ^2 critical values, indicating that the interspecies association was significant ($p < 0.05$). Therefore, the overall associati between the dominant species in this succession was significantly positive.

Table 3. Overall association of the 30 dominant species at the Ziwuling study area on the Chin Loess Plateau.

Life Form	Variance Ratio (*VR*)	Test Statistics (*W*)	$\chi^2_{(0.95,N)}$, $\chi^2_{(0.05,N)}$	Test Results
Herb	2.40	115.20	33.10, 65.17	Significant association
Woody	2.49	119.52	33.10, 65.17	Significant association
Herb + Woody	1.20	57.82	33.10, 65.17	Not a significant association

Note: Herb + Woody denotes all associations formed by herbaceous and woody plants in this study.

3.4. Species Association Analysis

3.4.1. Test of Species Associations

The results of the χ^2 tests (Figure 3, Supplementary Materials) showed that, the statistic was less than 3.841 for more than 90% of the species pairs in all three scenari There were 25 significantly associated species pairs. Meanwhile, Table 4 shows that t strongest positive linkage among herbaceous species pairs, followed by woody plants, a the weakest herb + woody in the three scenarios. There were 16 pairs of significantly pc tively associated species pairs and 9 pairs of significantly negatively associated species pa

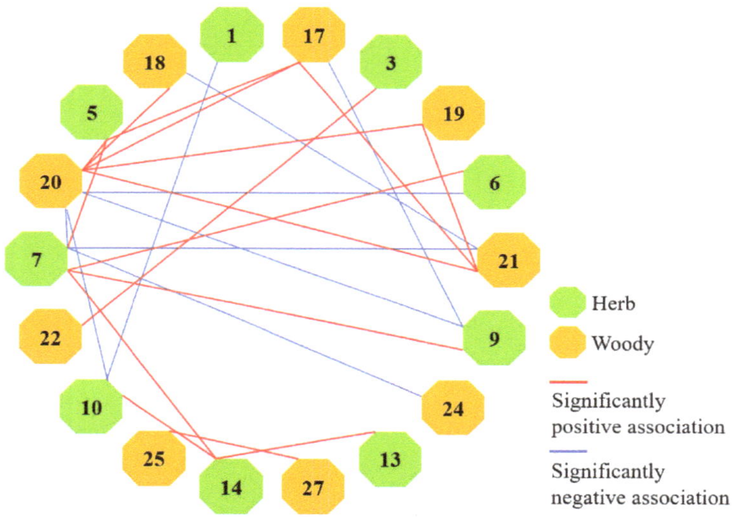

Figure 3. Results of χ^2 test of significantly association. The numbers represent the plants in Table

Table 4. *V* values of interspecific association among the 30 species at the Ziwuling study area on Chinese Loess Plateau.

Life Form	Positive Association (*p*)	Negative Association (*N*)	*p*/*N*
Herb	85	20	4.25
Woody	78	27	2.89
Herb + Woody	317	118	2.69

3.4.2. Strength of Species Associations

The results of the association coefficient (*AC*) and percentage co-occurrence (*PC*) dominant species demonstrate the strength of association between species (Figure 4 a Table 5). The number of positive association species pairs was 61, 63, and 211, w corresponding positive and negative species pair association ratio values of 1.39, 1.54, and 0.95 in the Herb, Woody, and Herb + Woody Scenarios, respectively, based on the *AC* results (Table 5).

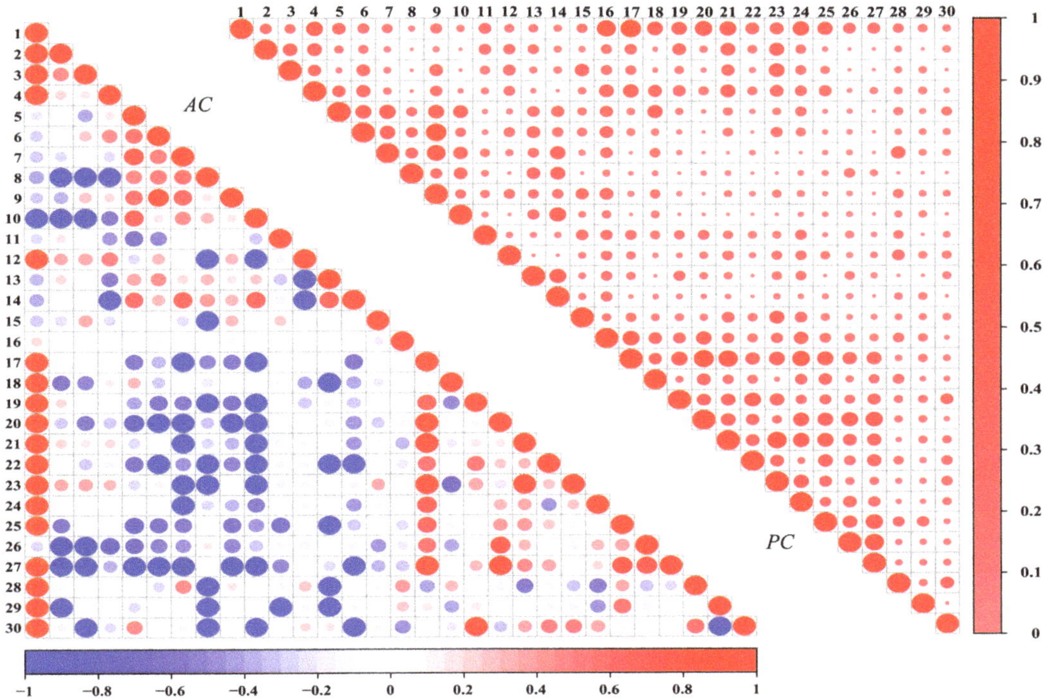

Figure 4. Interspecific associations strength. Lower is *AC* values, upper is *PC* values. The numbers represent the plants in Table 1. The larger the circle, the redder the color and the larger the value; the smaller the circle, the bluer the color and the smaller the value.

Table 5. Interspecific associations strength analysis among the 30 species at the Ziwuling study area on the Chinese Loess Plateau.

Index	Type	Strength of Association	Herb		Woody		Herb + Woody	
			Species Pair Number	%	Species Pair Number	%	Species Pair Number	%
AC	Positive association	$AC \geq 0.6$	12	11.43	13	12.38	38	8.74
		$0.2 \leq AC < 0.6$	31	29.52	30	28.57	74	17.01
		$0 < AC < 0.2$	18	17.14	20	19.05	99	22.76
	No association	$AC = 0$	0	0.00	1	0.95	3	0.69
	Negative association	$-0.2 \leq AC < 0$	8	7.62	18	17.14	59	13.56
		$-0.6 \leq AC < -0.2$	23	21.91	21	20.00	93	21.38
		$AC < -0.6$	13	12.38	2	1.91	69	15.86
PC		$0.5 \leq PC < 1$	6	5.71	11	10.48	22	5.06
		$0 < PC < 0.5$	88	83.81	93	88.57	360	82.76
		$PC = 0$	11	10.48	1	0.95	53	12.18

Among the three scenarios, the woody scenario had the largest proportion of species pairs with a high degree of positive interspecific association ($AC \geq 0.6$) at 12.38%; the Herb + Woody scenario had the largest proportion of species pairs with a high degree of negative interspecific association ($AC \geq 0.6$) at 15.86%; the degree of association coefficient was average ($0.2 \leq AC < 0.6$ and $-0.6 \leq AC < -0.2$) and tended to be independent ($0 < AC < 0.2$ and $-0.2 \leq AC < 0$) ranged from about 30% to 50% and 25% to 35%, respectively; there were three and one pair of species pairs with Herb + Woody and Woody scenarios with complete independence ($AC = 0$). The number of species pairs with wea connectivity ($0 < PC < 0.5$) for all three scenarios exceeded 80%. The largest percentag of species pairs with a high degree of linkage ($0.5 \leq PC < 1$) was 10.48% for the Wood scenario. The number of species pairs with no association ($PC = 0$) between Herb, Wood and Herb + Woody scenarios was 11, 1 and 53, respectively (Table 5).

3.5. Regression Analysis between Association Strength and Niche Overlap

The regression analyses of interspecific association coefficients and ecological nich overlap index during vegetation succession in this research are shown in Figure 5. Th p-values were less than 0.05 for interspecific association coefficients and O_{ik} of the dominan species in the three scenarios. Also, the slopes of the plots were all greater than 0, indicatin that a significant linear positive correlation existed between them. Thus, the stronger th positive interspecific association is, the more likely it is for interspecific pairs to live togethe and the larger the O_{ik}. Conversely, the stronger the negative interspecific association is, th more independent the pairs are and the smaller the O_{ik}.

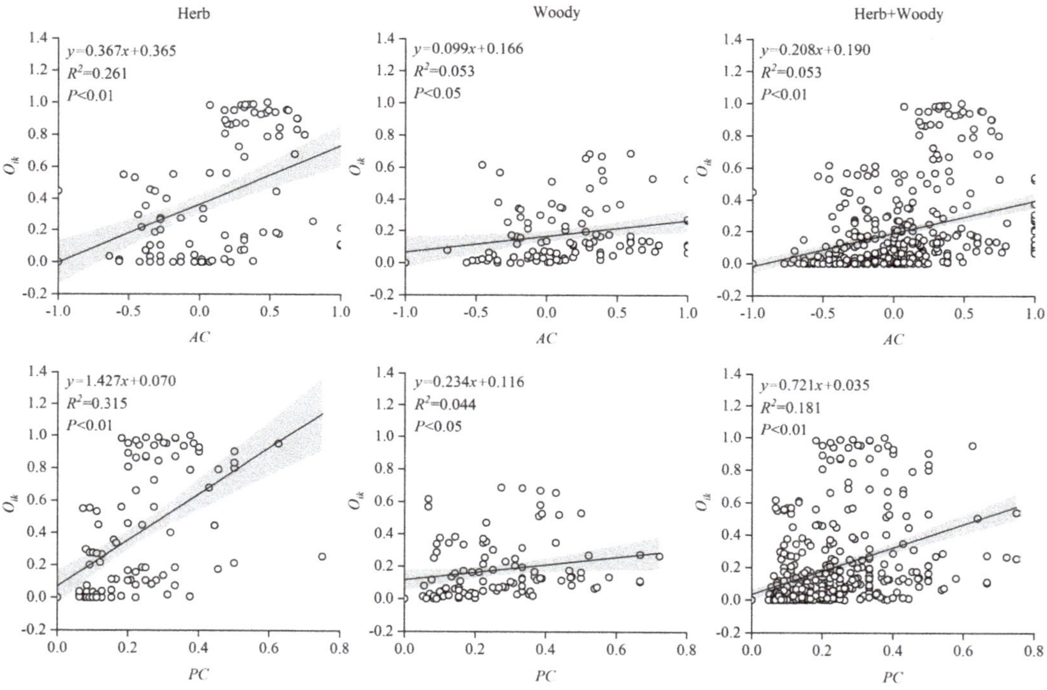

Figure 5. Regression analyses between the interspecific association coefficients (AC and PC) and O for herbs, woody plants and, herbs + woody plants. The dark color lines indicate the fits of the linea model, while the ribbon are the 95% confidence intervals of these linear models.

4. Discussion

4.1. Importance Values and Ecological Niches of Dominant Plants

C. lanceolata and *L. bicolor* were the most important species in the process and had highly functional position in the community. They are strongly competitive, use resource efficiently, adapt to a wide ecological range, and have a strong storage effect on generalize species [55]. This may be because both species are temperate and fit the local climati characteristics [43]. In addition, they are positively associated with each other, are engage in symbiosis and synergistic evolution, and serve as a basis for community constructio playing important roles in promoting positive succession and maintaining communit stability [56]. However, other plants with high importance values and relatively narrow ecological niche breadth indicate that they are sensitive to environmental changes, have a poor adaptive capacity, play important roles in only a few stages of succession, are more likely to be specialized (for a particular habitat), and are vulnerable to elimination and replacement during the development of the community [57].

Ecological niche overlap is an important indicator of differences in the ability of species to use environmental resources and engage in competitive relationships. A relatively large overlap of ecological niches indicates that two species have similar life forms and ecological needs for environmental resources and that there may be fierce competition between them. At the same time, species can also adjust their resource use and competitive ability to cope with different ecological and evolutionary pressures through adaptive changes in traits [9,58,59]. Species exhibit feedback effects between ecological processes and evolution to drive community succession development [9]. The degree of overlap of ecological niches among dominant species was found to be generally low in this research, which indicates that there is strong competition among a few species pairs and weak competition among most species pairs and that interspecific relationships are stable. The degree of ecological niche overlap for herbaceous plants was greater than that for woody plants, indicating a greater similarity in environmental resource requirements among dominant herbaceous plant species. Meanwhile, *A. Pilosa—A. argyi*, and *L. japonica—R. parvifolius* interspecies pairs showed non-significant positive associations. These species have the potential to co-occur in the community as companion species, thus enhancing the utilization of resources.

4.2. Interspecific Association Strength and Niche Overlap Relationships

Interspecific associations reflect species interactions and community dynamics. A positive correlation means that they have the same or similar needs for environmental resources, strong complementarity between species, fuller use of resources, and an ecological compensation effect [12,60]. Contrary of that, negative correlations indicate an adaptation of species pairs to environmental heterogeneity due to large differences in biological characteristics, leading to exclusion and ecological niche separation [9]. The results of the overall association analysis and the results of χ^2 test in this research were consistent, showing that most species pairs were positively associated in all three scenarios. This suggests that dominant species in the community weaken interspecific competition and help each other during this process. The community structure stabilizes and adapts to the environment, with the community moving toward top succession [61,62].

However, the overall non-significant positive associations between herbaceous and woody plants suggest that species pairing between different life forms may still enhance. This effect is in line with the forest secondary succession pattern, developing from a single herbaceous community with simple structural functions to an advanced multifunctional community with multiple life forms that coexist to maximize resource environment utilization [63,64]. This finding suggests that vertical spatial complementation of plants with different functional traits is an important mechanism for species coexistence [65].

χ^2 tests showed that the dominant species pairs were mostly non-significantly associated in all three scenarios, with weak interspecific associations and relatively strong independence. This is mainly because the environment experienced by plants during the 150-year natural restoration period was highly heterogeneous, and habitat filtering may still be an important ecological process for species to achieve coexistence patterns at each stage. This finding supports those of Tilman [66], Pedersen et al. [67], and Wu et al. [68] that, when vegetation is restored, competition tends to decrease, and interactions between species tend to be neutral.

In this research, the significant positive correlations were found between the interspecific association coefficients (*AC* and *PC*), and O_{ik} among dominant species, which showed that the degree of O_{ik} is greater among positively associated species pairs. This reflects the consistency of habitat requirements for these species. By comparing Tables 2 and 4, we also found that the species pairs are negatively associated, but their O_{ik} values are necessarily small (e.g., *C. lanceolata—A. lancea*). The causes of negative associations complex. The associativity between species pairs correlates with the species' ecologi niches and with the frequency of distribution, habitat, and resource utilization capac of the species pairs, which may explain these findings. This conclusion has also be confirmed in many studies on ecological niches and interspecific associations [69,70].

4.3. Inspirations and Prospects

Past two decades, the introduced species black locust (*Robinia pseudoacacia* L.) is wid monocultured on the Loess Plateau due to its rapid growth and high tolerance to drou and poor soils [71,72]. Nevertheless, its high-water consumption makes it difficult for nat plantations to survive, with an unbalanced community structure and lower plant divers than natural secondary forests [73–76]. As future droughts intensify [77,78], the rising r of decline and mortality of *R. pseudoacacia* stands could severely impact sustainability ecological functions [72,76,79].

As shown in Table 1 and Supplementary Materials, our findings provide a go template for plantation forest construction and theoretical support for local ecosyst function restoration. This research demonstrates that, under the ecological process environmental filtering, species within communities can ultimately reduce interspeci competition and promote species coexistence through storage effects via the use of spa heterogeneity and temporal asynchronous differences. This is consistent with the findir of Usinowicz et al. [80] and Levine et al. [2]. We also find weak associations amo dominant species during the 150 years of natural vegetation restoration succession. T communities were all in a positive and healthy development period. So, we should contin to protect the native environment, prevent anthropogenic disturbance, and enhance t protection of not significantly positively associated species, such as *A. chamaemelifc* and *A. tataricum* subsp. *ginnala*. Among herbaceous plants, *P. chinensis, A. hispidus*, a *A. lancea* formed the most significant negative associations with woody plant species pairs, suggesting that they are susceptible to replacement by woody plant influences on microenvironment, such as light resources, during community formation. The status these negatively associated species should be given attention when the plantation for is constructed.

Meanwhile, we can consider adding *C. lanceolata* and *L. bicolor*, two plants that ger ally coexist easily with other species for mutual benefit, to enhance the community functi and stability of the plantation forest. Thus, in the reconstruction and restoration of a co munity, it is important to fully understand the ecological and biological traits of each t species, take into account how different habitats affect the relationships between species different times, and choose tree species that can adapt well to their environment and coex well with other species for collocation planting, which will prevent extreme interspeci competitive exclusion.

5. Conclusions

During 150 years of natural vegetation restoration succession, species within t community reduced interspecific competition and promoted coexistence through spa heterogeneity and temporal asynchronous differences with storage effects. The ranges ecological niche breadth variation in this research are [1.07, 12.80] for herbaceous pla and [1.43, 9.42] for woody plants. *C. lanceolata* and *L. bicolor* are the most import species in this process. Local plantation forest construction can consider adding these tw plants to enhance ecological functions, maintain community stability, and promote t healthy development of the community. Overall, 379 pairs of ecological niche over index O_{ik} < 0.5 were identified, accounting for 87.13% of the total, and most of the pa had weak interspecific competition and stable interspecific relationships. The similar in environmental resource requirements among dominant species of herbaceous plants greater than that of woody plants. The overall positive association among the 30 domin plants indicates that the community is in a positive succession process. Plant coexistence relationships in Supplementary Materials can provide a reference for plantation forest construction. The results of the *AC*, *PC*, and χ^2 tests were essentially consistent, all showing that the dominant species pairs were mostly insignificantly associated. The degree of association between species pairs was weak, while independence was relatively strong. Meanwhile, a significant positive correlation was found between *AC*, *PC*, and O_{ik}.

Habitat filtering is an important ecological process for species at each stage to achieve coexistence patterns. With vegetation restoration, competition tends to lessen in intensity, and interactions between species tend to be neutral. In the future, we should enhance the protection of positively associated species and pay attention to negatively associated species during forest management. Then, to get more complete analysis results, the soil, topography, climate, and vertical structure of the forest should be considered.

Supplementary Materials: The following supporting information can be downloaded at: https://www.mdpi.com/article/10.3390/f13091456/s1, Figure S1. Half matrix graph of the interspecific association χ^2 test among the 30 plant species at the Ziwuling study area on the Chinese Loess Plateau.

Author Contributions: Conceptualization, Q.T. and X.Z.; methodology, Q.T. and X.Z.; software, Q.T.; formal analysis, Q.T.; investigation, Q.T., X.Z., X.X., H.Y., J.H., L.H. and W.S.; data curation, Q.T. and X.X.; writing—original draft preparation, Q.T.; writing—review and editing, Q.T. and X.Z.; visualization, Q.T.; project administration and funding acquisition, X.Z. All authors have read and agreed to the published version of the manuscript.

Funding: This research was funded by the National Key R & D Program of China (2022YFE0115300) and National Natural Science Foundation of China, grant number 41877083, 41440012 and 41230852.

Data Availability Statement: Data is contained within the article or supplementary material.

Acknowledgments: We would like to thank Haojia Wang, Miaoqian Wang, Yujie Zhang, Fan Xue, Yadong Zou for help with data processing and discussion for this article.

Conflicts of Interest: The authors declare no conflict of interest.

eferences

Tilman, D. Resource Competition and Community Structure. (MPB-17), Volume 17. In *Resource Competition and Community Structure. (MPB-17)*; Princeton University Press: Princeton, NJ, USA, 2020; Volume 17. [CrossRef]

Levine, J.M.; Bascompte, J.; Adler, P.B.; Allesina, S. Beyond pairwise mechanisms of species coexistence in complex communities. *Nature* **2017**, *546*, 56–64. [CrossRef] [PubMed]

Callaway, R.M.; Brooker, R.W.; Choler, P.; Kikvidze, Z.; Lortie, C.; Michalet, R.; Paolini, L.; Pugnaire, F.I.; Newingham, B.; Aschehoug, E.T.; et al. Positive interactions among alpine plants increase with stress. *Nature* **2002**, *417*, 844–848. [CrossRef] [PubMed]

Jia, S.; Wang, X.; Yuan, Z.; Lin, F.; Ye, J.; Lin, G.; Hao, Z.; Bagchi, R. Tree species traits affect which natural enemies drive the Janzen-Connell effect in a temperate forest. *Nat. Commun.* **2020**, *11*, 286–289. [CrossRef] [PubMed]

Zhang, H.; Bearup, D.; Nijs, I.; Wang, S.; Barabás, G.; Tao, Y.; Liao, J. Dispersal network heterogeneity promotes species coexistence in hierarchical competitive communities. *Ecol. Lett.* **2021**, *24*, 50–59. [CrossRef] [PubMed]

Yang, Y.; Hui, C. How competitive intransitivity and niche overlap affect spatial coexistence. *Oikos* **2021**, *130*, 260–273. [CrossRef]

Brooker, R.W.; Maestre, F.T.; Callaway, R.M.; Lortie, C.L.; Cavieres, L.A.; Kunstler, G.; Liancourt, P.; Tielbörger, K.; Travis, J.; Anthelme, F.; et al. Facilitation in plant communities: The past, the present, and the future. *J. Ecol.* **2008**, *96*, 18–34. [CrossRef]

Pocheville, A. The ecological niche: History and recent controversies. In *Handbook of Evolutionary Thinking in the Sciences*; Springer: Dordrecht, The Netherlands, 2015; pp. 547–586. [CrossRef]

Pastore, A.I.; Barabás, G.; Bimler, M.D.; Mayfield, M.M.; Miller, T.E. The evolution of niche overlap and competitive differences. *Nat. Ecol. Evol.* **2021**, *5*, 330–337. [CrossRef]

). Liu, J.H.; Wang, Z.W.; Han, G.D. Effects of heavy grazing on the interspecific relationship of main plant species and community stability in a desert steppe. *Chin. J. Ecol.* **2019**, *38*, 2595–2602. (In Chinese)

. Gao, W.; Huang, Y.; Lin, J.; Huang, M.; Wu, X.; Lin, W.; Huang, S. Interspecific Correlations among Dominant Populations of Natural Forest of Endangered Species *Betula fujianensis*. *Sci. Silvae Sin.* **2021**, *57*, 1–14. [CrossRef]

?. Jin, S.S.; Zhang, Y.Y.; Zhou, M.L.; Dong, X.M.; Chang, C.H.; Wang, T.; Yan, D.F. Interspecific Association and Community Stability of Tree Species in Natural Secondary Forests at Different Altitude Gradients in the Southern Taihang Mountains. *Forests* **2022**, *13*, 373. [CrossRef]

3. Tokeshi, M. *Species Coexistence: Ecological and Evolutionary Perspectives*; John Wiley & Sons: Hoboken, NJ, USA, 2009.

14. Adler, P.B.; Smull, D.; Beard, K.; Choi, R.T.; Furniss, T.; Kulmatiski, A.; Meiners, J.M.; Tredennick, A.T.; Veblen, K.E. Competitic and coexistence in plant communities: Intraspecific competition is stronger than interspecific competition. *Ecol. Lett.* **201** *21*, 1319–1329. [CrossRef] [PubMed]
15. Ren, Y.; Lü, Y.; Fu, B.; Comber, A.; Li, T.; Hu, J. Driving factors of land change in china's loess plateau: Quantification usir geographically weighted regression and management implications. *Remote Sens.* **2020**, *12*, 453. [CrossRef]
16. Xie, H.; Zhang, Y.; Wu, Z.; Lv, T. A bibliometric analysis on land degradation: Current status, development, and future directio *Land* **2020**, *9*, 28. [CrossRef]
17. Hua, F.; Wang, X.; Zheng, X.; Fisher, B.; Wang, L.; Zhu, J.; Tang, Y.; Yu, D.W.; Wilcove, D.S. Opportunities for biodiversity gai under the world's largest reforestation programme. *Nat. Commun.* **2016**, *7*, 12717. [CrossRef] [PubMed]
18. Shan, L.; Xu, B.C. Discussion on some issues about returning Farmland to Forest or Grassland on Loess Plateau in new era. *Bu Soil Water Conserv.* **2019**, *39*, 3. (In Chinese)
19. Kou, P.; Xu, Q.; Jin, Z.; Yunus, A.P.; Luo, X.; Liu, M. Complex anthropogenic interaction on vegetation greening in the Chine Loess Plateau. *Sci. Total Environ.* **2021**, *778*, 146065. [CrossRef] [PubMed]
20. Li, J.; Liu, Q.; Feng, X.; Shi, W.; Fu, B.; Lü, Y.; Liu, Y. The synergistic effects of afforestation and the construction of check-dams o sediment trapping: Four decades of evolution on the Loess Plateau, China. *Land Degrad. Dev.* **2019**, *30*, 622–635. [CrossRef]
21. Jin, F.; Yang, W.; Fu, J.; Li, Z. Effects of vegetation and climate on the changes of soil erosion in the Loess Plateau of China. *S Total Environ.* **2021**, *773*, 145514. [CrossRef]
22. Wang, Y.H.; Bonell, M.; Feger, K.H.; Yu, P.; Xiong, W.; Xu, L. Changing forestry policy by integrating water aspects in forest/vegetation restoration in dryland areas in China. *Bull. Chin. Acad. Sci.* **2012**, *26*, 59–67. [CrossRef]
23. Sun, G.; Zhou, G.; Zhang, Z.; Wei, X.; McNulty, S.G.; Vose, J.M. Potential water yield reduction due to forestation across Chin *J. Hydrol.* **2006**, *328*, 548–558. [CrossRef]
24. Wang, X.; Chen, F.; Hasi, E.; Li, J. Desertification in China: An assessment. *Earth-Sci. Rev.* **2008**, *88*, 188–206. [CrossRef]
25. Li, S.; Liang, W.; Fu, B.; Lü, Y.; Fu, S.; Wang, S.; Su, H. Vegetation changes in recent large-scale ecological restoration projects an subsequent impact on water resources in China's Loess Plateau. *Sci. Total Environ.* **2016**, *569*, 1032–1039. [CrossRef] [PubMed]
26. Shao, M.A.; Wang, Y.; Xia, Y.; Jia, X. Soil drought and water carrying capacity for vegetation in the critical zone of the Loes Plateau: A review. *Vadose Zone J.* **2018**, *17*, 170077. [CrossRef]
27. Wang, C.; Wang, S.; Fu, B.; Lü, Y.; Liu, Y.; Wu, X. Integrating vegetation suitability in sustainable revegetation for the Loes Plateau, China. *Sci. Total Environ.* **2021**, *759*, 143572. [CrossRef]
28. Huang, Y.; Chen, Y.; Castro-Izaguirre, N.; Baruffol, M.; Brezzi, M.; Lang, A.; Li, Y.; Härdtle, W.; Von Oheimb, G.; Yang, X.; et a Impacts of species richness on productivity in a large-scale subtropical forest experiment. *Science* **2018**, *362*, 80–83. [CrossRef]
29. Gong, C.; Tan, Q.; Xu, M.; Liu, G. Mixed-species plantations can alleviate water stress on the Loess Plateau. *For. Ecol.* **202** *458*, 117767. [CrossRef]
30. Wang, K.B.; Ren, Z.P.; Deng, L.; Zhou, Z.C.; Shangguan, Z.P.; Shi, W.Y.; Chen, Y.P. Profile distributions and controls of so inorganic carbon along a 150-year natural vegetation restoration chronosequence. *Soil Sci. Soc. Am. J.* **2016**, *80*, 193–202. [CrossRe
31. Deng, L.; Wang, K.; Zhu, G.; Liu, Y.; Chen, L.; Shangguan, Z.P. Changes of soil carbon in five land use stages following 10 years vegetation succession on the Loess Plateau, China. *Catena* **2018**, *171*, 185–192. [CrossRef]
32. Deng, L.; Kim, D.G.; Peng, C.; Shangguan, Z.P. Controls of soil and aggregate-associated organic carbon variations followir natural vegetation restoration on the Loess Plateau in China. *Land Degrad. Dev.* **2018**, *29*, 3974–3984. [CrossRef]
33. Grime, J.P. *Plant Strategies, Vegetation Processes, and Ecosystem Properties*; John Wiley and Sons: Hoboken, NJ, USA, 2006.
34. West, D.C.; Shugart, H.H.; Botkin, D.F. (Eds.) *Forest Succession: Concepts and Application*; Springer Science and Business Medi Berlin, Germany, 2012.
35. Zou, H.Y.; Liu, G.B.; Wang, H.S. The vegetation development in North Ziwuling forest region in last fifty years. *Acta Bota Boreali-Occiden. Sin.* **2002**, *22*, 1–8. (In Chinese)
36. Fan, W.Y.; Wang, X.A.; Guo, H. Analysis of plant community successional series in the Ziwuling area on the Loess Plateau. *Ac Ecol. Sin.* **2006**, *26*, 706–714. (In Chinese)
37. Cheng, J.; Cheng, J.; Shao, H.; Zhao, L.; Yang, X. Soil seed banks and forest succession direction reflect soil quality in Ziwulin Mountain, Loess Plateau, China. *Clean–Soil Air Water* **2012**, *40*, 140–147. [CrossRef]
38. Cowles, H.C. The Ecological Relations of the Vegetation on the Sand Dunes of Lake Michigan. Part I.-Geographical Relations the Dune Floras. *Bot. Gaz.* **1899**, *27*, 95–117. [CrossRef]
39. Blois, J.L.; Williams, J.W.; Fitzpatrick, M.C.; Jackson, S.T.; Ferrier, S. Space can substitute for time in predicting climate-chang effects on biodiversity. *Proc. Natl. Acad. Sci. USA* **2013**, *110*, 9374–9379. [CrossRef] [PubMed]
40. Deng, L.; Wang, K.B.; Chen, M.L.; Shangguan, Z.P.; Sweeney, S. Soil organic carbon storage capacity positively related to fore succession on the Loess Plateau, China. *Catena* **2013**, *110*, 1–7. [CrossRef]
41. Zhao, Y.G.; Liu, X.F.; Wang, Z.L.; Zhao, S.W. Soil organic carbon fractions and sequestration across a 150-yr secondary fore chronosequence on the Loess Plateau, China. *Catena* **2015**, *133*, 303–308. [CrossRef]
42. Liu, Y.; Zhu, G.; Hai, X.; Li, J.; Shangguan, Z.P.; Peng, C.; Deng, L. Long-term forest succession improves plant diversity an soil quality but not significantly increase soil microbial diversity: Evidence from the Loess Plateau. *Ecol. Eng.* **2020**, *142*, 10563 [CrossRef]
43. Li, J. Flora of China. *Harv. Pap. Bot.* **2007**, *13*, 301–302. [CrossRef]

Schoener, T.W. Resource Partitioning in Ecological Communities: Research on how similar species divide resources helps reveal the natural regulation of species diversity. *Science* **1974**, *185*, 27–39. [CrossRef]
Song, Y.C. *Vegetation Ecology*; East China Normal University Press: Shanghai, China, 2001. (In Chinese)
Colwell, R.K.; Futuyma, D.J. On the measurement of niche breadth and overlap. *Ecology* **1971**, *52*, 567–576. [CrossRef]
Pianka, E.R. The structure of lizard communities. *Annu. Rev. Ecol. Syst.* **1973**, *4*, 53–74. [CrossRef]
Schluter, D. A variance test for detecting species associations, with some example applications. *Ecology* **1984**, *65*, 998–1005. [CrossRef]
Dice, L.R. Measures of the amount of ecologic association between species. *Ecology* **1945**, *26*, 297–302. [CrossRef]
Dai, J.; Liu, H.; Xu, C.; Qi, Y.; Zhu, X.; Zhou, M.; Liu, B.; Wu, Y. Divergent hydraulic strategies explain the interspecific associations of co-occurring trees in forest–steppe ecotone. *Forests* **2020**, *11*, 942. [CrossRef]
Grieg-Smith, P. *Quantitative Plant Ecology (Studies in Ecology)*; Blackwell Science Ltd: Hoboken, NJ, USA, 1983.
Zhang, J.T. *Quantitative Ecology*; Science Press: Beijing, China, 2004. (In Chinese)
Zhang, J.L. *Spaa: Species Association Analysis, R Package Version 0.2.2*; R Foundation for Statistical Computing: Vienna, Austria, 2016.
Wei, T.; Simko, V.R. Package "Corrplot": Visualization of a Correlation Matrix (Version 0.84). 2021. Available online: https://github.com/taiyun/corrplot (accessed on 20 August 2022).
Chesson, P. Mechanisms of maintenance of species diversity. *Annu. Rev. Ecol. Syst.* **2000**, *31*, 343–366. [CrossRef]
Álvarez-Yépiz, J.C.; Búrquez, A.; Dovčiak, M. Ontogenetic shifts in plant–plant interactions in a rare cycad within angiosperm communities. *Oecologia* **2014**, *175*, 725–735. [CrossRef] [PubMed]
Gu, L.; O'Hara, K.L.; Li, W.Z.; Gong, Z.W. Spatial patterns and interspecific associations among trees at different stand development stages in the natural secondary forests on the Loess Plateau, China. *Ecol. Evol.* **2019**, *9*, 6410–6421. [CrossRef]
Fielding, A.P.; Pantel, J.H. Eco-evolutionary feedbacks and the maintenance of metacommunity diversity in a changing environment. *Genes* **2020**, *11*, 1433. [CrossRef]
Tsafack, N.; Wang, X.; Xie, Y.; Fattorini, S. Niche overlap and species co-occurrence patterns in carabid communities of the northern Chinese steppes. *ZooKeys* **2021**, *1044*, 929–949. [CrossRef]
Yuan, Z.; Wei, B.; Chen, Y.; Jia, H.; Wei, Q.; Ye, Y. How do similarities in spatial distributions and interspecific associations affect the coexistence of *Quercus* species in the Baotianman National Nature Reserve, Henan, China. *Ecol. Evol.* **2018**, *8*, 2580–2593. [CrossRef]
Stubbs, W.J.; Bastow, W.J. Evidence for limiting similarity in a sand dune community. *J. Ecol.* **2004**, *92*, 557–567. [CrossRef]
Nathan, R. Long-distance dispersal of plants. *Science* **2006**, *313*, 786–788. [CrossRef] [PubMed]
Petritan, I.C.; Marzano, R.; Petritan, A.M.; Lingua, E. Overstory succession in a mixed *Quercus petraea–Fagus sylvatica* old growth forest revealed through the spatial pattern of competition and mortality. *For. Ecol. Manag.* **2014**, *326*, 9–17. [CrossRef]
Ma, F.; Pan, G.; Li, X.; Han, Y. Interspecific relationship and canonical correspondence analysis within woody plant communities in the karst mountains of Southwest Guangxi, southern China. *J. Beijing For. Univ.* **2017**, *39*, 32–44. [CrossRef]
Kikuzawa, K.; Onoda, Y.; Wright, I.J.; Reich, P.B. Mechanisms underlying global temperature-related patterns in leaf longevity. *Glob. Ecol. Biogeogr.* **2013**, *22*, 982–993. [CrossRef]
Tilman, D. Competition and biodiversity in spatially structured habitats. *Ecology* **1994**, *75*, 2–16. [CrossRef]
Pedersen, R.Ø.; Bollandsås, O.M.; Gobakken, T.; Næsset, E. Deriving individual tree competition indices from airborne laser scanning. *For. Ecol. Manag.* **2012**, *280*, 150–165. [CrossRef]
Wu, S.; Wen, L.; Dong, S.; Gao, X.; Xu, Y.; Li, S.; Dong, Q.; Wessell, K. The Plant Interspecific Association in the Revegetated Alpine Grasslands Determines the Productivity Stability of Plant Community Across Restoration Time on Qinghai-Tibetan Plateau. *Front. Plant Sci.* **2022**, *13*, 850854. [CrossRef]
Shao, L.Y.; Zhang, G.F. Niche and interspecific association of dominant tree populations of *Zelkova schneideriana* communities in eastern China. *Bot. Sci.* **2021**, *99*, 823–833. [CrossRef]
Gu, L.; Gong, Z.W.; Li, W.Z. Niches and interspecific associations of dominant populations in three changed stages of natural secondary forests on Loess Plateau, PR China. *Sci. Rep.* **2017**, *7*, 6604. [CrossRef]
Fu, B.; Wang, S.; Liu, Y.; Liu, J.; Liang, W.; Miao, C. Hydrogeomorphic ecosystem responses to natural and anthropogenic changes in the Loess Plateau of China. *Annu. Rev. Earth Planet. Sci.* **2017**, *45*, 223–243. [CrossRef]
Li, G.; Zhang, X.; Huang, J.; Wen, Z.; Du, S. Afforestation and climatic niche dynamics of black locust (*Robinia pseu-doacacia*). *For. Ecol. Manag.* **2018**, *407*, 184–190. [CrossRef]
Du, S.; Wang, Y.L.; Kume, T.; Zhang, J.G.; Otsuki, K.; Yamanaka, N.; Liu, G.B. Sapflow characteristics and climatic responses in three forest species in the semiarid Loess Plateau region of China. *Agric. For. Meteorol.* **2011**, *151*, 1–10. [CrossRef]
Cao, S.; Sun, G.; Zhang, Z.; Chen, L.; Feng, Q.; Fu, B.; McNulty, S.; Shankman, D.; Tang, J.; Wang, Y.; et al. Greening china naturally. *Ambio* **2011**, *40*, 828–831. [CrossRef] [PubMed]
Jiao, L.; Lu, N.; Fu, B.; Wang, J.; Li, Z.; Fang, W.; Liu, J.; Wang, C.; Zhang, L. Evapotranspiration partitioning and its implications for plant water use strategy: Evidence from a black locust plantation in the semi-arid Loess Plateau, China. *For. Ecol. Manag.* **2018**, *424*, 428–438. [CrossRef]
Wei, J.S.; Li, Z.S.; Feng, X.Y.; Zhang, Y.; Chen, W.L.; Xing, W.U.; Jiao, L.; Wang, X.C. Ecological and physiological mecha-nisms of growth decline of *Robinia pseudoacacia* plantations in the Loess Plateau of China: A review. *Chin. J. Appl. Ecology* **2018**, *29*, 2433–2444. [CrossRef]

77. Yan, W.; Zhong, Y.; Shangguan, Z. Responses of different physiological parameter thresholds to soil water availability in f plant species during prolonged drought. *Agric. For. Meteorol.* **2017**, *247*, 311–319. [CrossRef]
78. Peng, S.; Ding, Y.; Wen, Z.; Chen, Y.; Cao, Y.; Ren, J. Spatiotemporal change and trend analysis of potential evapo-transpirat over the Loess Plateau of China during 2011–2100. *Agric. For. Meteorol.* **2017**, *233*, 183–194. [CrossRef]
79. Zhang, Z.; Huang, M.; Yang, Y.; Zhao, X. Evaluating drought-induced mortality risk for *Robinia pseudoacacia* planta-tions alc the precipitation gradient on the Chinese Loess Plateau. *Agric. For. Meteorol.* **2020**, *284*, 107897. [CrossRef]
80. Usinowicz, J.; Chang-Yang, C.-H.; Chen, Y.-Y.; Clark, J.S.; Fletcher, C.; Garwood, N.C.; Chia-Hao, C.-Y.; Johnstone, J.; Lin, Y.; M M.; et al. Temporal coexistence mechanisms contribute to the latitudinal gradient in forest diversity. *Nature* **2017**, *550*, 105– [CrossRef]

sustainability MDPI

ata Descriptor

Vave Height Attenuation over a Nature-Based Breakwater of 'loating Emergent Vegetation

anhong Li [1,*], Dongliang Zhao [2], Guoliang Yu [1,*] and Liquan Xie [3]

[1] KLMIES, MOE, State Key Laboratory of Ocean Engineering, School of Naval Architecture, Ocean and Civil Engineering, Shanghai Jiaotong University, Shanghai 200240, China
[2] CCCC Second Harbor Engineering Company Ltd., Wuhan 430040, China
[3] Department of Hydraulic Engineering, Tongji University, Shanghai 200092, China; xie_liquan@tongji.edu.cn
* Correspondence: yyhli@sjtu.edu.cn (Y.L.); yugl@sjtu.edu.cn (G.Y.)

Abstract: The nature-based breakwater of floating emergent vegetation (BFEV) provides protection for water banks and various engineering structures from wave erosion. Compared with the convenient hard breakwater, the BFEV is beneficial to the resilient and sustainable development of rivers, lakes, coasts, and marine areas because it is free of new pollution. As a new breakwater, the unrevealed effect and efficiency of the BFEV on wave attenuation are to be investigated through a set of 312 physical tests in a rectangular indoor water flume in the present study. Results show that the wave height attenuates by 38–62%. Based on statistical methods, the main influencing factors of the wave transmitted coefficient (C_t) are found to be closely dependent on three conventional and newly proposed dimensionless parameters (λ_1, λ_2, λ_3, λ_4). Three conventional parameters include the wave orbital velocity, wave period, and the BFEV-width and stem spacing-based parameter (λ_1, λ_2), and the ratio of stem spacing to wave height (λ_3). The newly proposed parameter (λ_4) is the ratio of gravity to wave orbital acceleration, which is significantly positively related to the wave height attenuation. A multiple linear regression formula for C_t based on these four parameters is obtained with a high correlation coefficient of 0.958. This study is expected to supplement the wave attenuation data of this new breakwater and provide fundamental theory for the design and construction of the BFEV.

Keywords: breakwater of floating emergent vegetation; wave attenuation; wave transmitted coefficient; multiple linear regression

itation: Li, Y.; Zhao, D.; Yu, G.; Xie, Wave Height Attenuation over a ature-Based Breakwater of Floating nergent Vegetation. *Sustainability* **23**, *15*, 10749. https://doi.org/ .3390/su151410749

ademic Editors: Panayiotis mitrakopoulos, Mario A. Pagnotta, iklas Scholz and Arshiya Noorani

ceived: 16 May 2023
vised: 26 June 2023
cepted: 30 June 2023
blished: 8 July 2023

1. Introduction

Wave erosion has endangered the stability of the banks of rivers, lakes, coasts, and marine areas for a long time [1,2]. Further, the possible wave resonance would increase wave load and induce rapid growth of the motion response of coastal and offshore engineering structures [3–6], which would cause the failure of these structures. The breakwater of floating emergent vegetation (BFEV) has been used as a new nature-based breakwater to prevent water banks from wave erosion, and its further application potential to the protection of engineering structures is huge due to its predominant ecological benefit. Compared with various hard breakwaters of concrete, stone, or metal, the BFEV is more habitat-adaptive for rare birds, more scenery-pleasant, more economy-saving, more construction-convenient, and less polluting. Further, compared with ecological macrophytes including the mangrove forest, the salt marsh, and the water-bed-growing emergent vegetation, the BFEV is more site-adaptive and more exclusive of the limitation of water depth. Therefore, it is beneficial to the resilient and sustainable development of rivers, lakes, coasts, and marine areas and is thus expected to be a good alternative to hard breakwaters.

The vegetative strip, including kelp farms, salt marsh, mangrove forests, reed or reed-like emergent macrophyte aquatic vegetation, and even the novel Bragg breakwater, has been found to significantly affect the dissipation of wave energy and thus the attenuatio of wave height at certain wave and vegetation conditions [7–11]. Zhu et al. [12] found i a physical experimental case that over a kelp farm, the wave energy is dissipated up 29–43% at the incident wave height of 1 m and the wave period of 6 s. Van Rooijen et al. [1 reported in experiments that the submerged vegetation canopy attenuated the wave heigl by up to 40% when the wave orbital excursion length was greater than the characterist length of canopy drags. Huang et al. [14] found in a series of indoor experiments th the rigid emergent vegetation of various lengths and porosities caused the solidary wav height to attenuate by 18–62%. Cassidy and Tomiczek [15] observed in a 1:10 geometr scale physical model that the wave heights were attenuated by up to 65% in the mangrov forest. Cao et al. [16] investigated the mangrove field on the south China sea coast an found that wave attenuation occurred over the 100 m width of the mangrove. Massel [1 computed that over 50 m wide in a dense mangrove forest, the wave height was attenuate even by 99%. Zhou et al. [18] observed that the mangrove field caused the attenuation wave height by nearly 58.33% over a transport distance of 275 m and approximately 80 over a transport distance of 1000 m.

Wave attenuation has been observed to be closely dependent on multiple vegetatio characteristics. Cassidy and Tomiczek [15] observed that the attenuation increased with ve etation distribution density and the row width of the mangrove forest. Augustin et al. [1 conducted 46 experimental case tests involving various stem densities and wave properti under emergent and near-emergent vegetation conditions. They found the wave heigh attenuated by 10.1–24.9% over the first 3 m wide of vegetation with higher density an then further attenuated by 5.7–17.4% over the last 3 m wide with lower density. The also found that emergent vegetation led to higher wave attenuation than near-emerge vegetation. Lee et al. [20] experimentally observed the wave attenuation during a stor event and found the density and width of mangroves were positively correlated to the wav height attenuation. They compared roots, trunks, and canopies and found that mangrov roots contributed more to wave height attenuation. Their observation found no statistic differences between mangrove types, incident wave heights, or water levels.

Simultaneously, wave properties are also reported as important factors influencin wave attenuation. Jadhav et al. [21] found that wave attenuation increased with wav height due to increased orbital wave particle velocities. Cao et al. [16] discovered by spectr analysis that wave energy dissipation depended on wave frequency. Sun et al. [22] modele vegetation with rigid cylinders physically and numerically under 15 wave conditions various wave heights, wave periods, and wavelengths, establishing that wave energ dissipation decreases with the increase in wave period. Gao et al. [11] found the Brag breakwater can utilize the Bragg reflection to significantly alleviate harbor resonance an thus protect the coast for the first time. They investigated the influence of incident wav height on floating box moments. They found the ratios of the second-order components t the corresponding first-order ones around the resonant frequency are normally larger tha those at the frequencies far from the resonant frequency, and the larger the incident wav height is, the larger the ratios around the resonant frequency become [4]. Gao et al. [further found that the transient gap resonance affected the reflection and energy loss waves significantly.

The multiple characteristics of vegetation and waves are usually integral to two simp factors, drag force and inertia force, in various analytical and computational models c wave attenuation. Drag force caused by the viscous effect and form drag around the ste is positively related to square velocity and is therefore usually more significant in valu compared with inertia force, which is caused by the acceleration of the surrounding flui and is positively related to a relatively minor accelerated velocity. Consequently, the inerti effect is usually considered a small constant or neglected. Wu and Marsooli [23] numericall simulated long waves over rigid vegetation with various drag coefficients and a consta inertia coefficient of 2.0, finding that the wave height attenuation was positively related t vegetation density. Ozeren et al. [24] experimentally observed the linear and regular wave

They derived the drag coefficient by attributing the wave-damping effect of vegetation only to drag force and thus neglecting inertia force.

Vuik et al. [25] (cited by Veelen et al. [26]) found that in salt marsh flows, the drag coefficient ranged from 0.13 to 5.75, differing by a factor of 44. This suggested the influencing factors of drag force were complex. Zhou et al. [18] calculated the wave attenuation by integrating the effects of vegetation into a drag coefficient, finding that young trees of nearly 0.55 m high were more effective in attenuating wave energy than the stem part of grown trees of nearly 1.2 m high. This suggested that the variations in the inundation of trees induced by water level fluctuations might affect wave damping. Van Veelen et al. [26] found that the best fit for the drag coefficient is a function of the *KC* number, and the velocity attenuation inside the vegetation is a function of the ratio of wave excursion to stem spacing and the ratio of stem spacing to stem diameter. Different exponents and constants under rigid and flexible vegetation conditions were successfully fitted for the analytical expression of the drag coefficient, including the *KC* number proposed by Kobayashi et al. [27]. Jadhav et al. [28] confirmed an inverse relationship between the *KC* number and the drag coefficient based on measurements in a Spartina alterniflora marsh. Ozeren et al. [24] found the drag coefficient was dependent on the *KC* number when $KC < 10$, increasing with the decrease in *KC* value.

The complexity of vegetated waves causes uncertainty in estimating drag effects and, thus, in the prediction of energy dissipation and wave height attenuation. Work has been conducted to include more parameters of vegetation and waves to obtain more accurate drag estimation and wave dissipation prediction. Tanino and Nepf [29] found in their experiments that the cross-sectionally averaged drag coefficient decreased with the increase in cylinder Reynolds number and increased with the increase in solid volume fraction, and viscous drag per unit cylinder length is independent of solid volume fraction in the range of 0.15–0.35. Stone and Shen [30] replaced the apparent velocity with a cross-sectional velocity dependent on the stem diameter and the lateral spacing of vegetation stems and obtained a more accurate drag coefficient. He et al. [31] estimated the drag coefficients proposed by Dalrymple et al. [32] and Kobayashi et al. [26] based on 112 sets of indoor experiments and further proposed a semi-empirical analytical equation for wave attenuation. They predicted a transmitted wave coefficient ranging from 0.21 to 0.83 due to vertically non-uniform vegetation properties in the root, stem, and canopy. However, differences in the prediction of wave height attenuation are still present. The dimensionless parameters involving multi-characteristics of vegetation, flow, and wave, which include the stem Reynolds number, Froude number, and vegetation volume fraction, have been introduced and applied successfully. While comparison of numerical models using different drag coefficient formulae based on these three dimensionless parameters showed inconsistent wave height attenuation.

It has been found that some new factors might also be important in the prediction of wave attenuation. Jadhav et al. [28] found that regardless of whether *KC* or *Re* is used, there is a reduction in drag coefficient for increased orbital wave particle velocities associated with higher waves. Gao et al. [4] found that both the vertical and horizontal wave forces on the floating body were closely related to the incident wave height and the resonant wave frequency. Augustin et al. [19] numerically simulated the wave attenuation by integrating the effect of vegetation into a bulk drag coefficient. Interestingly, the ratio of numerical to observed transmittance was lower than 1, at 0.91–0.97 over the first 3 m width of denser-distributed vegetation, while it was higher than 1, reaching 1.01–1.04 over the last 3 m width of sparser vegetation. They explained that this error and bias could result from the inherent error in the empirical equations for friction factor, missing two-horizontal-dimension terms in these equations, or some neglected turbulent or vertically varying physics not captured by the Boussinesq simulation.

The previous study discovered the critical influence of various parameters on wave dissipation. However, a more accurate prediction of wave height attenuation still needs more attempts with new parameters. As a new vegetated breakwater, the BFEV's effect and efficiency on wave attenuation still need to be revealed. The influencing paramet applied for the previous breakwaters need to be tested, and new parameters are wor of attempting. The limitation is mainly due to the availability of data. Therefore, aimi at supplementing data, selecting appropriate parameters for the BFEV, and attempti new parameters that might influence wave attenuation significantly, the present stu will conduct a series of indoor physical experiments under various BFEV and wa conditions. Based on these experiments, the dependence of wave height attenuati on multiple parameters, including the previous and newly proposed ones, will analyzed. The results are expected to provide some new parameters for wave heig attenuation and provide a fundamental guide for the design and construction of BFEV in application.

2. Theoretical Background and Methodology

Wave dissipation over the vegetation strip and the BFEV is affected by the physical a geometric properties of the vegetation strip, including strip width, vegetation density, a stem spacing, as well as the wave and water features, including incident wave height, wa period, and water depth. Other parameters, including submergence ratio, stem diame biomass, and flexibility of the vegetation, remained constant in all the experimental te and thus they were found to have no significant relationship with wave height attenuati

The drag force and inertia force, and thus the wave height attenuation, are under combined effects of the complex features of vegetation and waves. Based on the previc study and the observed data of this study, four combined dimensionless parameters found to be monotonically related to the transmitted coefficient of wave and thus are us to predict the wave height attenuation in regressive analysis.

The Keulegan–Carpenter (KC) number represents the contrast effect of drag fo and inertia force. In this study, the width of the BFEV (B) is used in the KC number the characteristic length of a solid since it is the largest length scale of the BFEV, which defined as

$$\lambda_1 = KC = \frac{VT}{B}$$

where λ_1 is the first dimensionless parameter that will be used in the regressive analysis this study; V is the horizontal orbital velocity of the incident wave; T is the wave perio

As the second largest length scale of the BFEV, the stem spacing also affects the dr force and the inertia force, which composes the second parameter

$$\lambda_2 = \frac{VT}{D}$$

where λ_2 is the second dimensionless parameter chosen by the authors, which is related the stem density of the BFEV; D is the stem spacing of the vegetation.

The third dimensionless parameter, the ratio of wave height and the stem spacing vegetation, is also closely related to wave height attenuation, which is expressed as

$$\lambda_3 = \frac{H}{D}$$

where H is wave height.

And a new dimensionless parameter was found closely related to wave height atte ation, which is defined as

$$\lambda_4 = \frac{g}{(V/T)} = \frac{1}{F_r^2}\frac{VT}{h}$$

where F_r is the wave Froude number (Zhu et al. [12]); h is the water depth in a still state

The transmission coefficient is expressed as the ratio of transmitted wave height incident wave height,

$$C_t = \frac{H_t}{H}$$

where H_t is the transmitted wave height.

In this study, the dependent relationships of C_t and the four dimensionless parameters (λ_1, λ_2, λ_3, λ_4) are represented through physical experiments. Further, a prediction formula for C_t is to be regressed based on these relationships.

3. Experimental Setup

Experiments were conducted in a rectangular cross-sectional water flume with two glass side walls in the laboratory of harbor, coastal, and hydraulic engineering at Shanghai Jiaotong University. The flume is 20 m long, 1.0 m wide, and 0.8 m high. A swaying wave generator was installed at the upstream end of the flume to generate two-dimensional regular waves (Figure 1). Three rectangular steel plates with uniformly punched holes were set at a slope of 1:4 at the downstream end of the flume, which were parallel to each other to dissipate wave energy and avoid the reflection effect of waves (Figure 1).

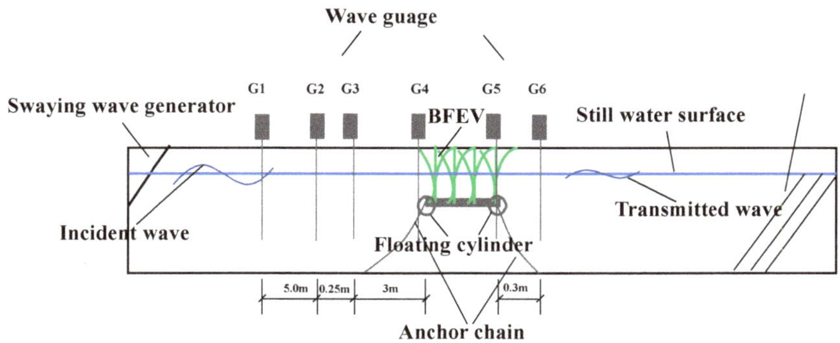

Figure 1. Side view of the water flume.

The BFEV model was installed in the middle part, 10 m from the upstream end of the water flume. Two hollow cylinders with a diameter of 5 cm were installed under the BFEV as floaters. Two mooring chains with a stiffness of 1.8 KN/m were designed to anchor the BFEV to meet the requirements of the real application as well (Figure 1). The BFEV emerged from the water surface, with its submerged part deeper than the experimental wave height. Corresponding to the two-dimensional waves in the water flume, the BFEV was installed, occupying the full cross width of the flume (Figure 2). The individual vegetation model used the real foliage branch of a plant with a uniform stem diameter (d) of 0.006 m. The scale of the model vegetation diameter to the prototype was selected at 1:10. The height of BFEV was 35 cm, with 20 cm emerging under and 15 cm emerging above the still water surface. The baseboard of the BFEV was 18 cm below the water surface. The modulus of elasticity in the bending of a single vegetation stem was 0.98×10^{10} Pa and thus was considered rigid. *Morinda citrifolia* L. was used to model the coastal bush vegetation. Vegetation models were in alignment in both lines and rows (Figure 3). The vegetation models of foliage branches were fixed on a bottom board made of Perspex, and the bottom board was moored on the bottom of the water flume (Figure 4). Both the width of the BFEV (B) and the stem spacing (D) vary in different experimental cases.

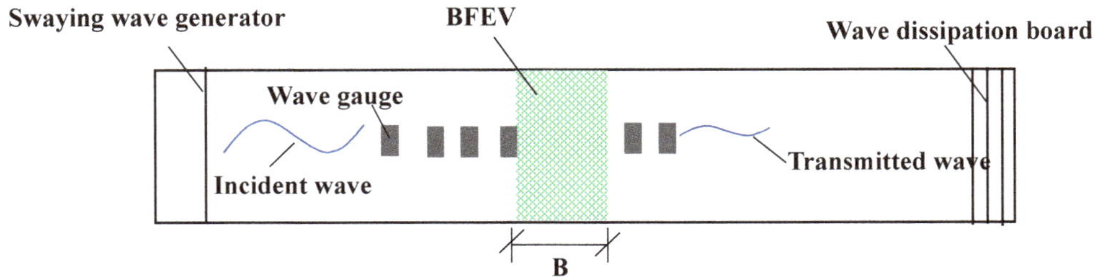

Figure 2. Horizontal planer view of the water flume.

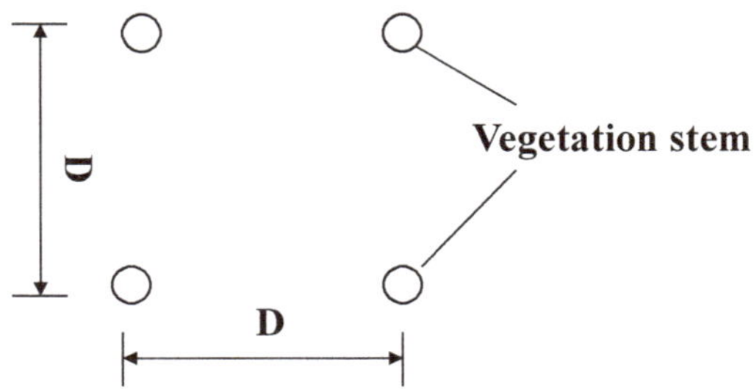

Figure 3. Arrangement of vegetation stem.

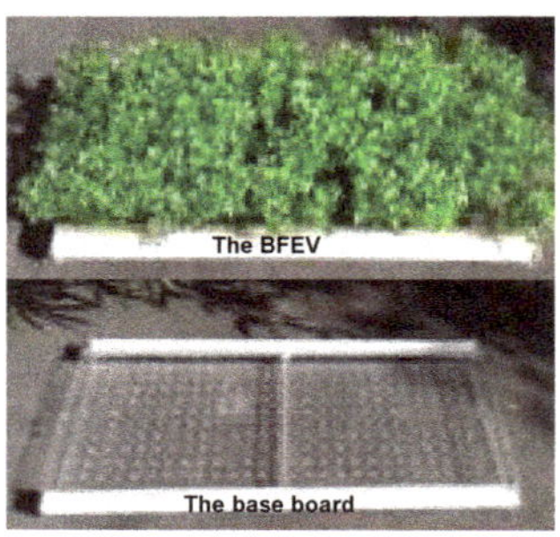

Figure 4. Camera photo of the BFEV and the baseboard.

The incident wave height (*H*), wave period (*T*) and horizontal wave orbital velocit (*V*) varied in the experiments, but the water depth was constant at 0.35 m for all th experiments. A total of 312 sets of experimental cases were designed by cross-combinatio of different BFEV and wave conditions, including five BFEV widths, eight stem space two incident wave heights, four wave periods, and four wave propagating velocities. Th parameters of different experimental conditions are summarized in Table 1.

Table 1. Parameters of the FBEF and wave.

B (m)	*D* (m)	*d* (m)	*H* (m)	*T* (s)	*V* (m/s)
0.45, 0.35, 0.25, 0.20, 0.15	0.0214, 0.0237, 0.0268, 0.0306, 0.0355, 0.0428, 0.0538, 0.0713	0.006	0.06, 0.08	0.881	1.192
				0.784	1.084
				0.693	1.082
				0.64	1.016

The SDA100 system of wave sensors and data collection was used to observe the wa parameters. Six wave gauges were set along the wave direction to obtain the wave heig

Guages 1 to 4 were set 8.35 m, 3.25 m, 3 m, and 0 m upstream of the BFEV, and gauges 5 and 6 were set 0 m and 0.3 m downstream of the BFEV. Gauges 2 and 3 were used to obtain the incident wave height by the two-point separation method of Goda and Suzuki [33]. Gauge 6 was used to obtain the transmitted wave height. Gauges 1, 4, and 5 were used to monitor the variation of wave height along the wave direction.

A control test was conducted to estimate the effects of flume walls, the baseboard, and the installation devices of the BFEV on the wave attenuation. It shows that these effects account for less than 1% of the total attenuation of the wave.

4. Results

A control experiment without the floating emergent vegetation was conducted to test the influence of the framework of the BEFV and the water flume, and the results show that their influence on wave attenuation is less than 0.15%, so they are not considered in the analysis.

Among the total experimental set of 312 tests, the wave height attenuation ranges from 22.6% to 71.5%, with the transmitted coefficient C_t ranging from 0.285 to 0.774 (Figures 5–8), which validates the significant effect of the BFEV on wave attenuation. To further understand the influencing mechanism of BFEV on wave attenuation, the dependences of C_t on λ_1, λ_2, λ_3, and λ_4 are analyzed, respectively. Then, to predict the wave height attenuation over the BFEV, the regression function of C_t is analyzed based on these four dimensionless parameters.

(a)

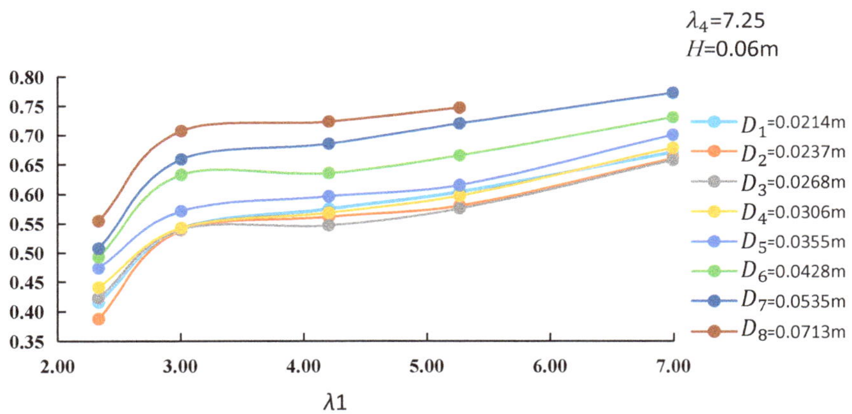

(b)

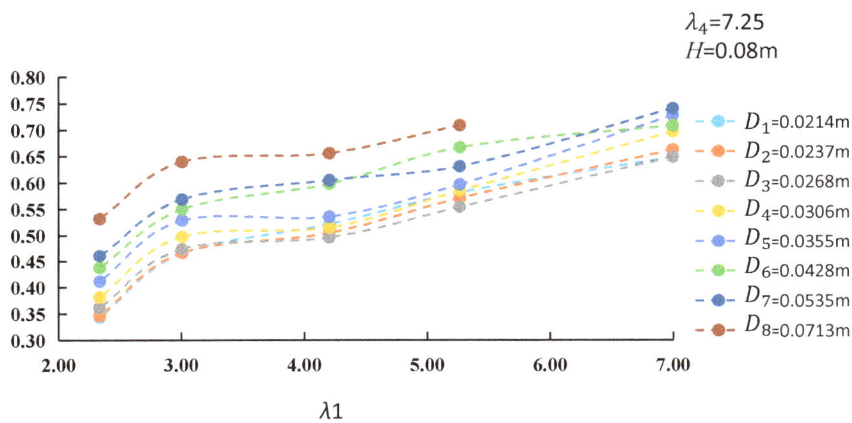

Figure 5. *Cont.*

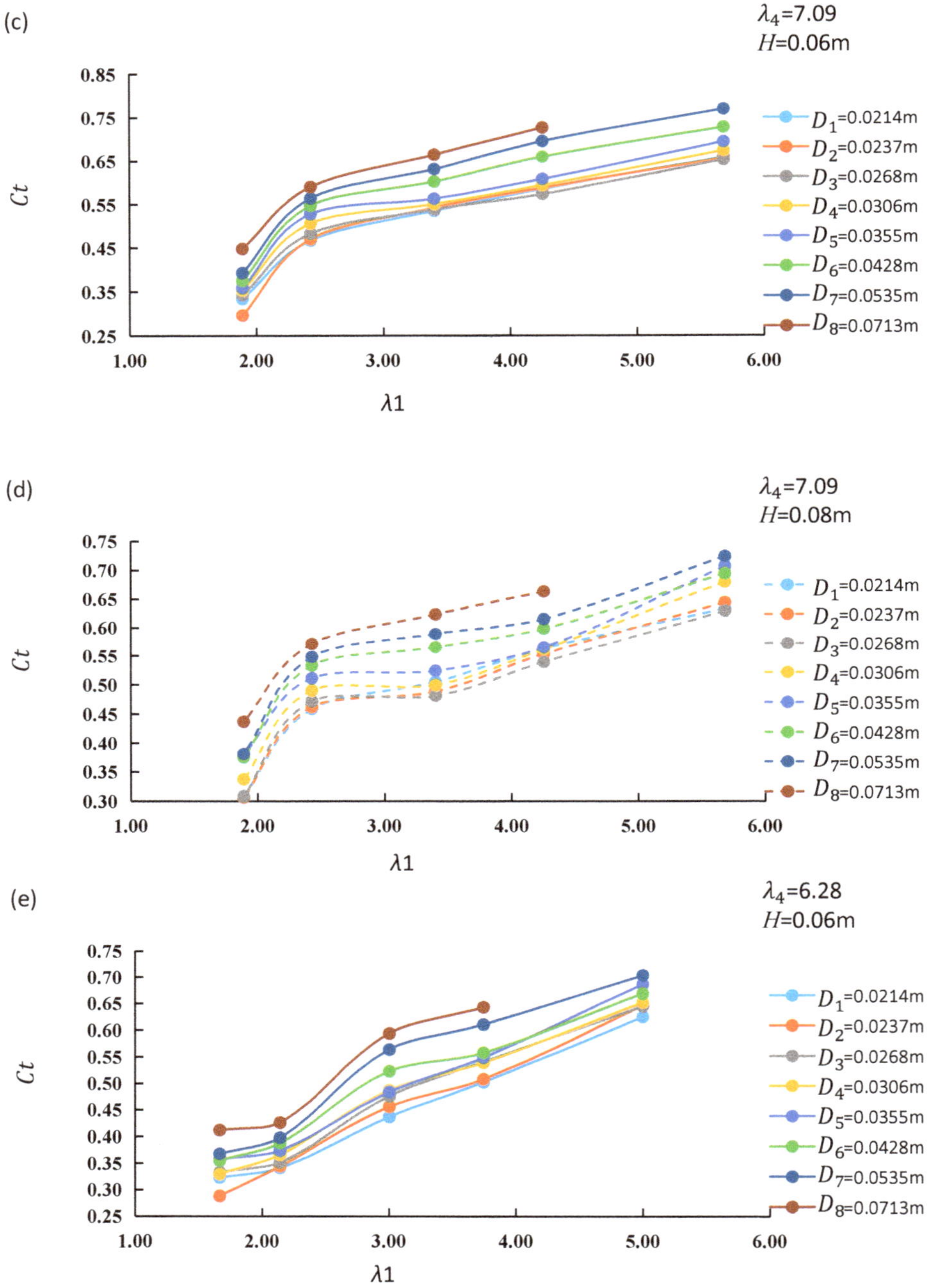

Figure 5. *Cont.*

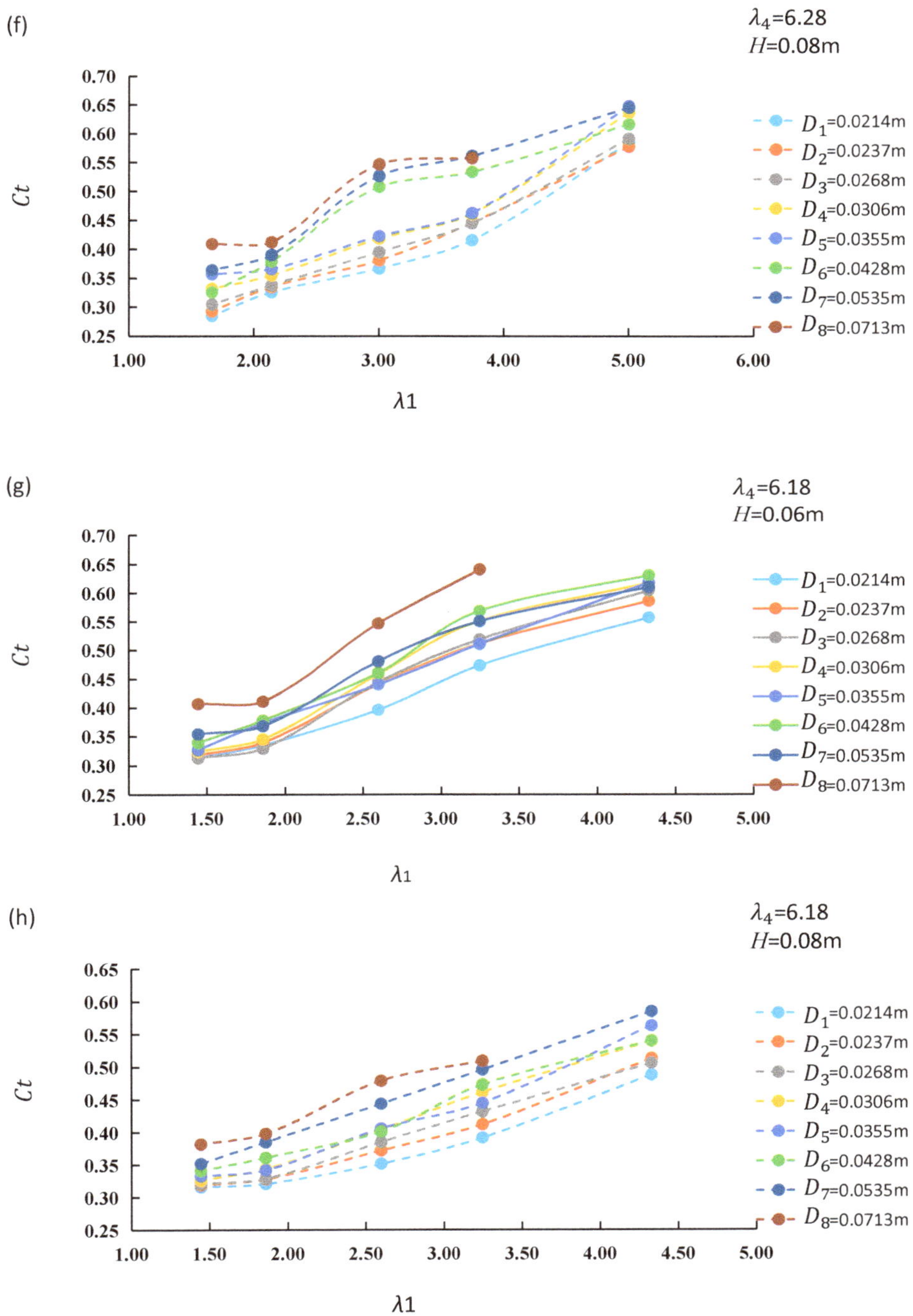

Figure 5. Dependence of the wave transmitted coefficient (C_t) on the first conventional parameter ($\lambda_1 = KC = \frac{VT}{B}$) under eight stem spacing conditions (D = 0.0214 m, 0.0237 m, 0.0268 m, 0.0306 m, 0.0355 m, 0.0428 m, 0.0535 m and 0.0713 m). Four subfigure pairs (**a**) and (**b**), (**c**) and (**d**), (**e**) and (**f**), and (**g**) and (**h**) are at $\lambda_4 = \frac{g}{(V/T)}$ = 7.25, 7.09, 6.28 and 6.18, respectively, with each subfigure pair corresponding to the wave height H = 0.06 m and 0.08 m.

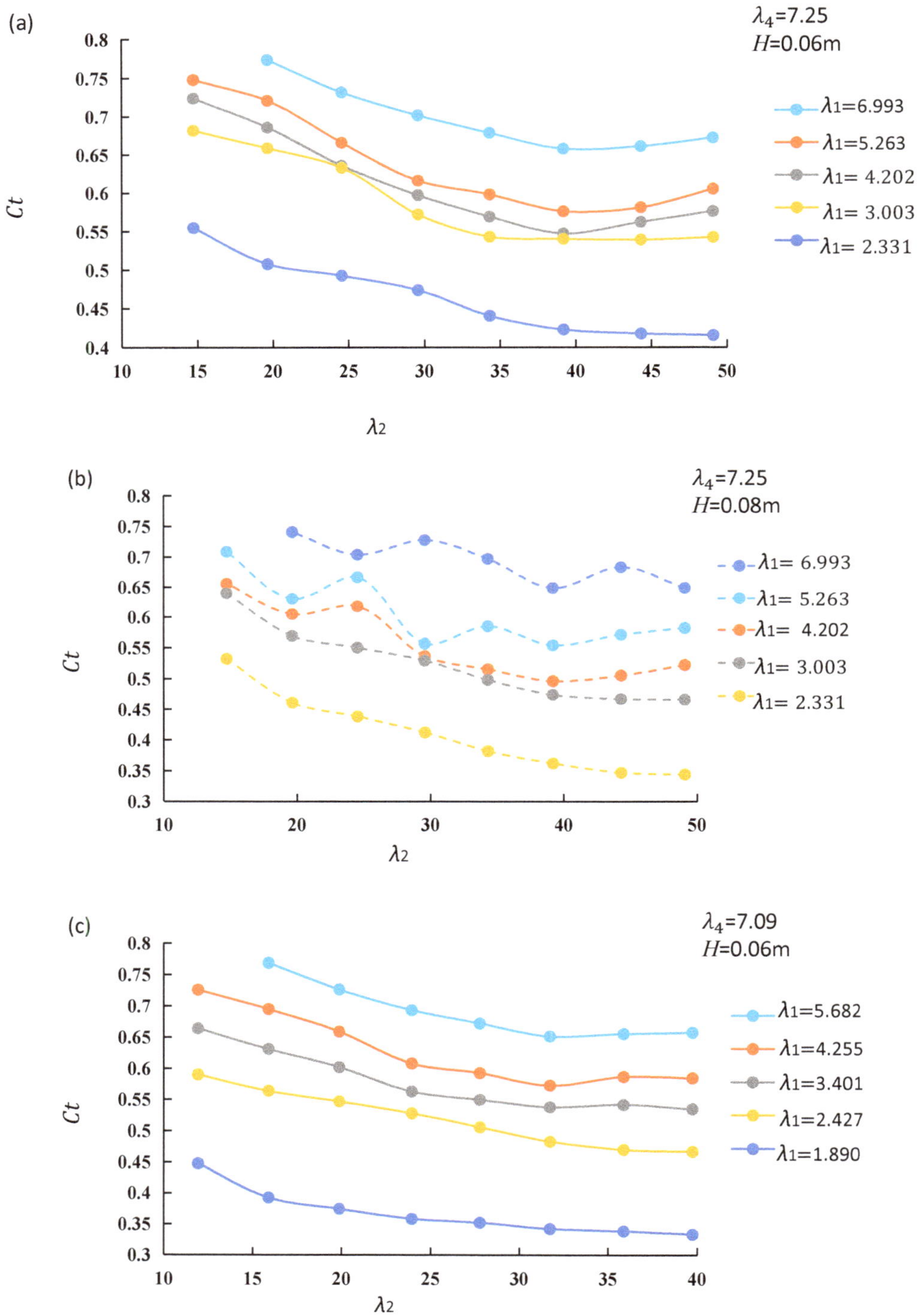

Figure 6. *Cont.*

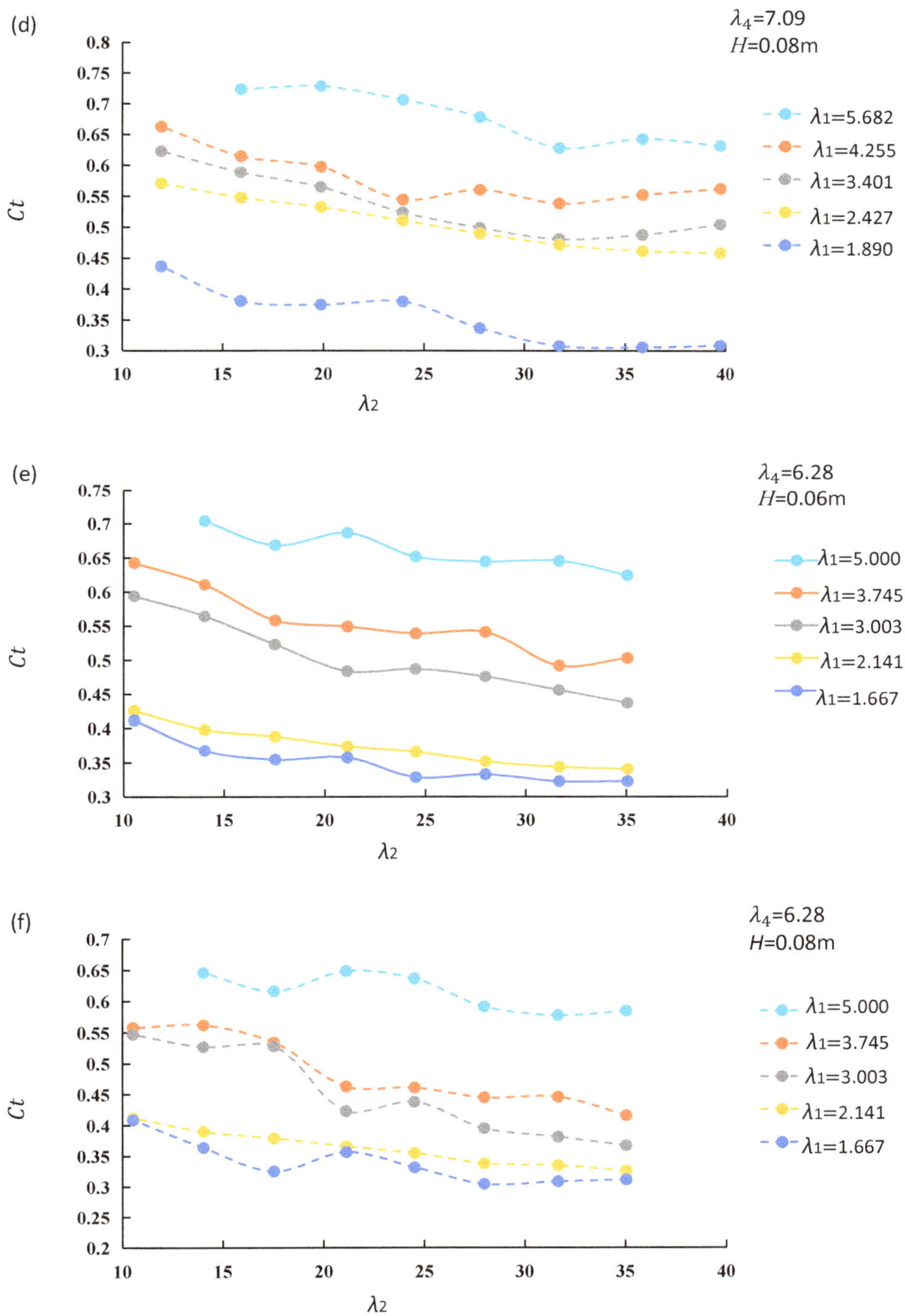

Figure 6. *Cont.*

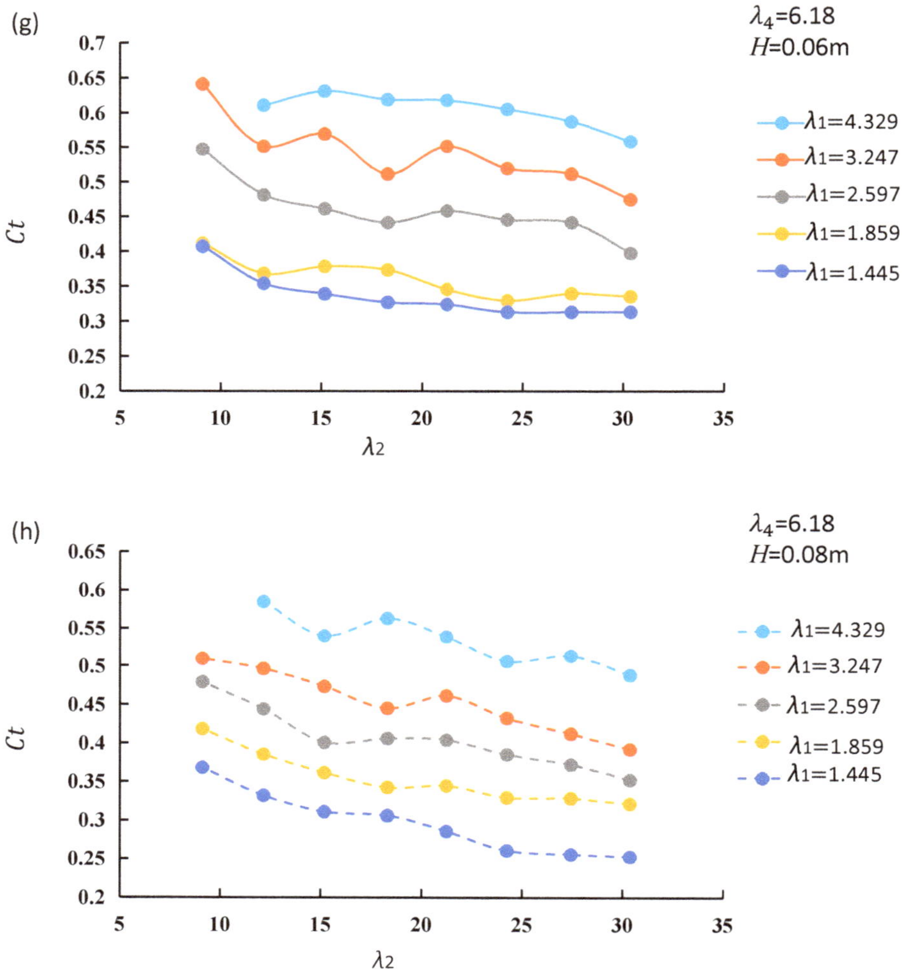

Figure 6. Dependence of the wave transmitted coefficient (C_t) on the second conventional para eter ($\lambda_2 = \frac{VT}{D}$). Four subfigure pairs (**a**) and (**b**), (**c**) and (**d**), (**e**) and (**f**), and (**g**) and (**h**) are $\lambda_4 = \frac{g}{(V/T)}$ = 7.25, 7.09, 6.28 and 6.18, respectively, with each subfigure pair corresponding to t wave height H = 0.06 m and 0.08 m. Subfigure pair (**a**) and (**b**) are at $\lambda_1 = \frac{VT}{B}$ = 6.993, 5.263, 4.2 3.003 and 2.331; Subfigure pair (**c**) and (**d**) are at λ_1 = 5.682, 4.255, 3.401, 2.427 and 1.890; Subfigure p (**e**) and (**f**) are at λ_1 = 5.000, 3.745, 3.003, 2.141 and 1.667; Subfigure pair (**g**) and (**h**) are at λ_1 = 4.3 3.247, 2.597, 1.859 and 1.445.

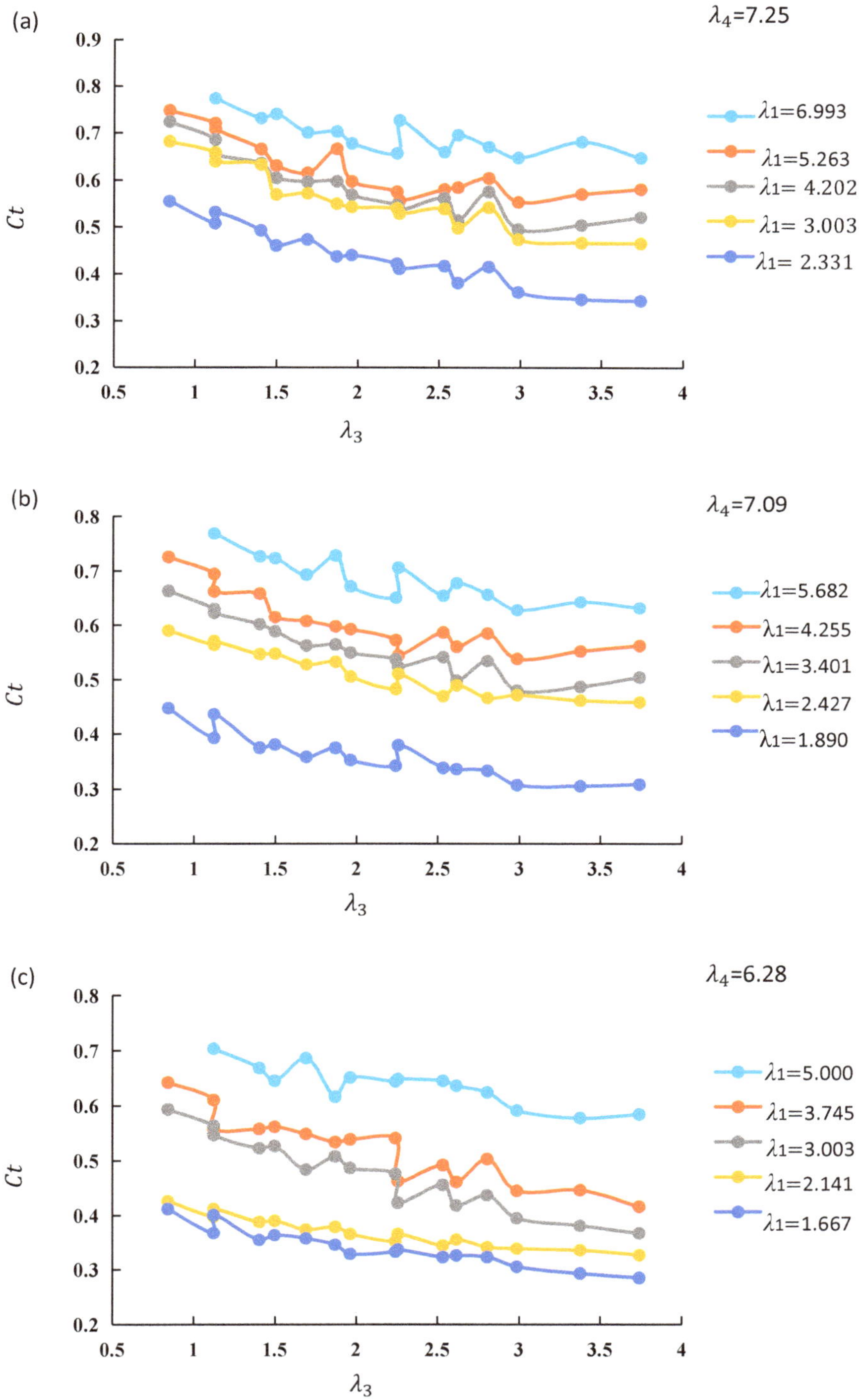

Figure 7. *Cont.*

(d)

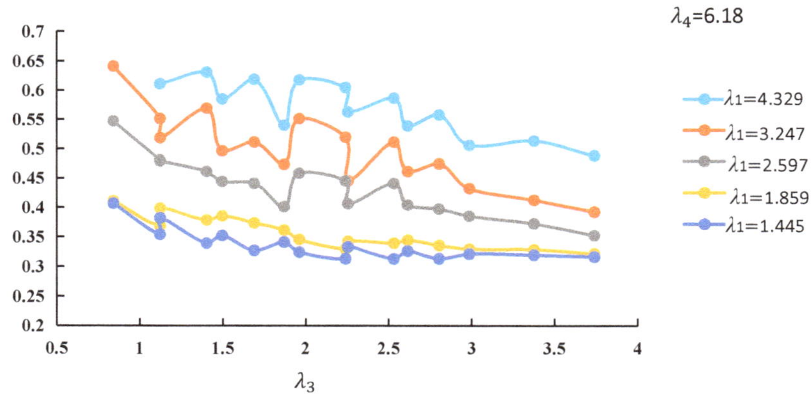

Figure 7. Dependence of the wave transmitted coefficient (C_t) on the third conventional paramete ($\lambda_3 = \frac{H}{D}$). Subfigure (**a**) is at $\lambda_4 = \frac{g}{(V/T)}$ = 7.25, $\lambda_1 = \frac{VT}{B}$ = 6.993, 5.263, 4.202, 3.003 and 2.33 Subfigure (**b**) is at λ_4 = 7.09, λ_1 = 5.682, 4.255, 3.401, 2.427 and 1.890; Subfigure (**c**) is at λ_4 = 6.2 λ_1 = 5.000, 3.745, 3.003, 2.141 and 1.667; Subfigure (**d**) is at λ_4 = 6.18, λ_1 = 4.329, 3.247, 2.597, 1.85 and 1.445.

(a)

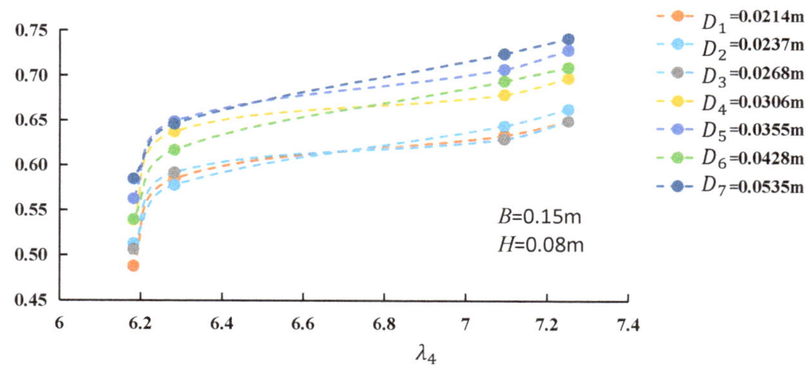

(b)

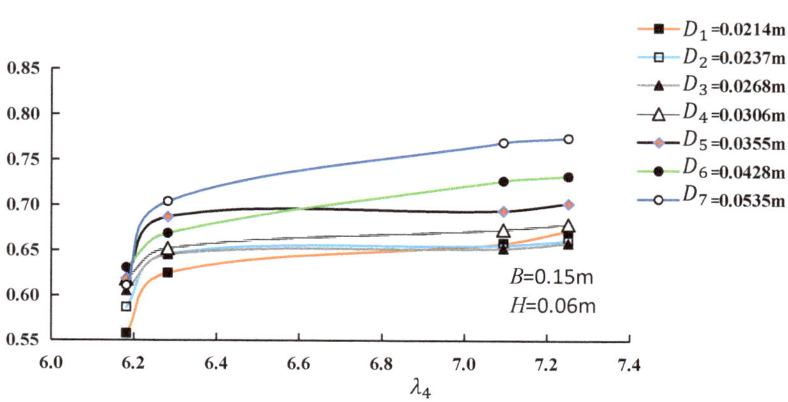

Figure 8. *Cont.*

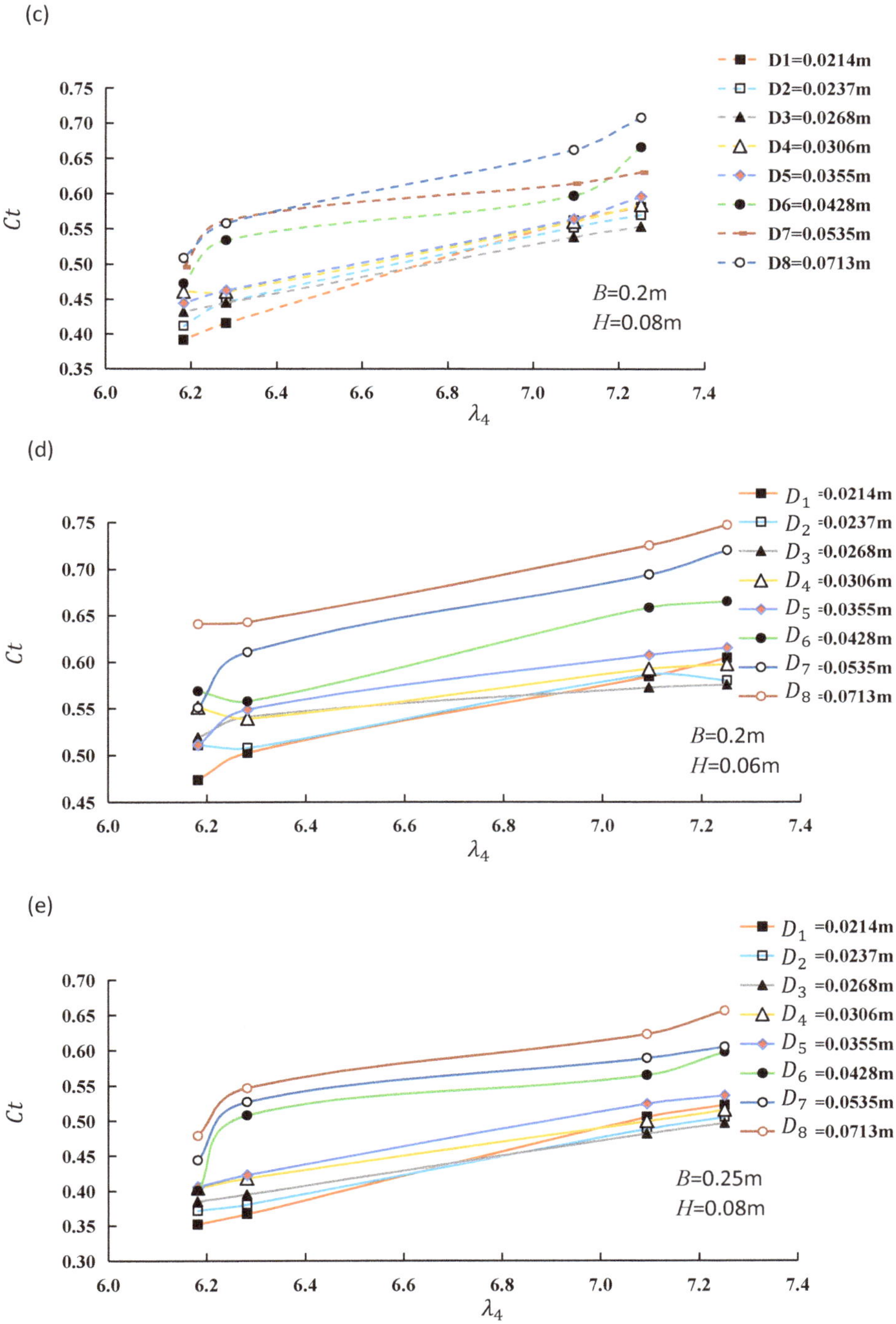

Figure 8. *Cont.*

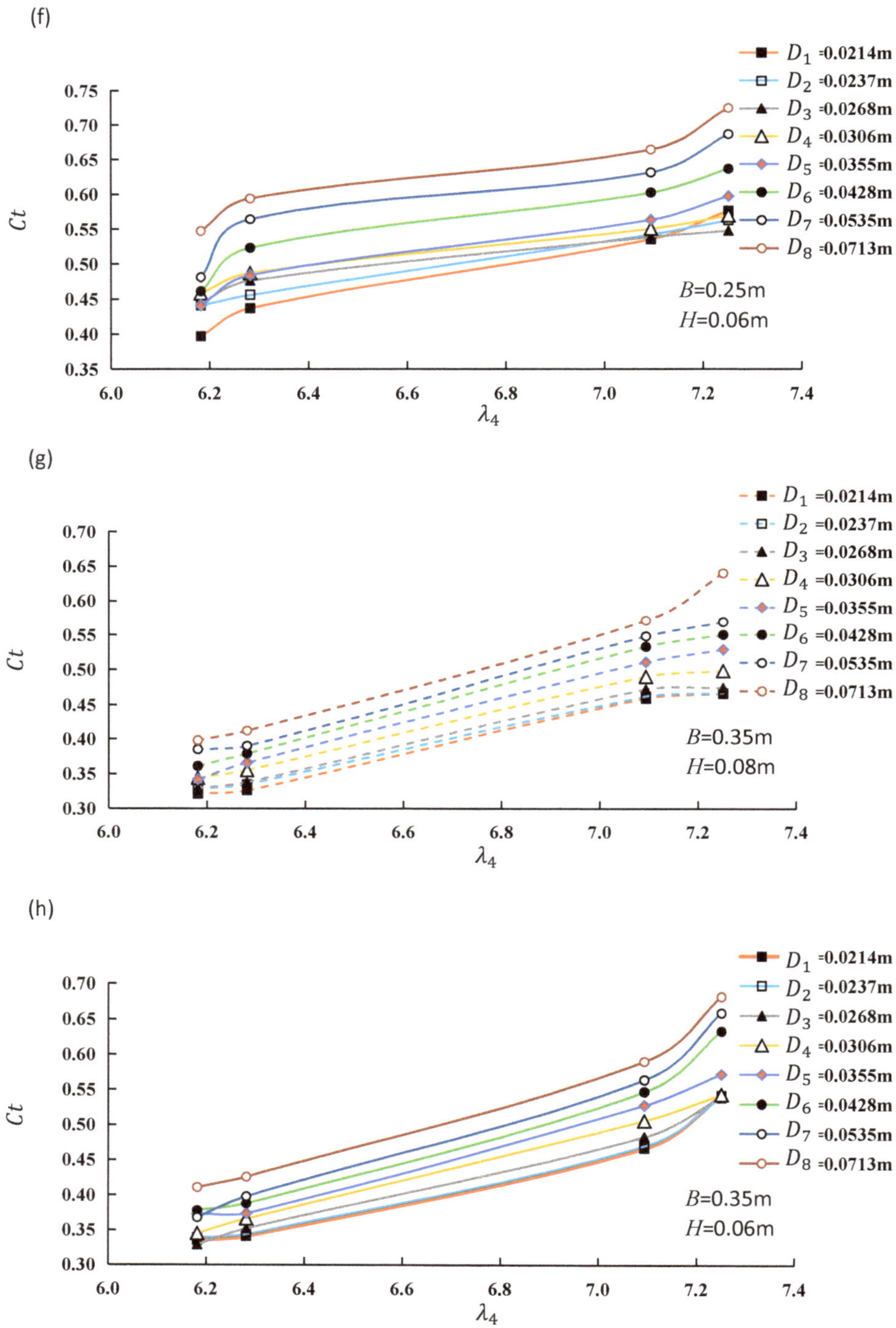

Figure 8. *Cont.*

(i)

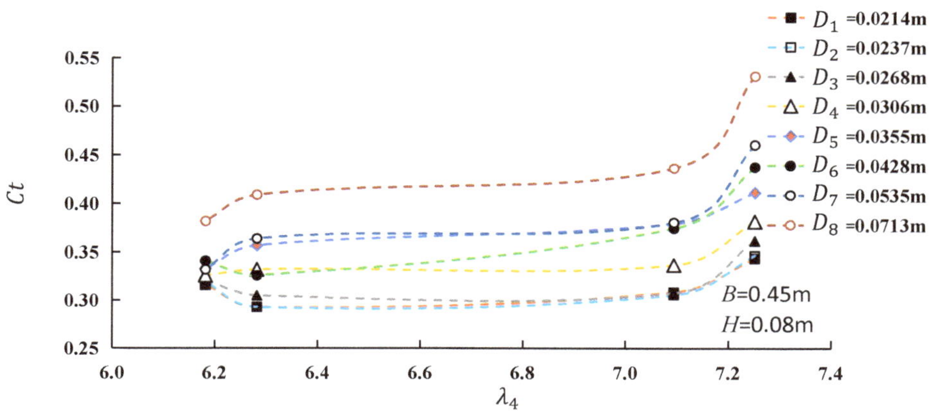

(j)

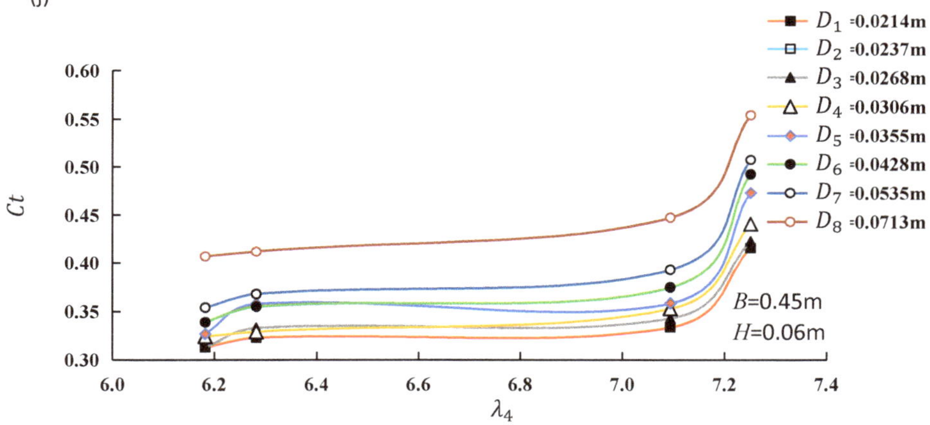

Figure 8. Dependence of the wave transmitted coefficient (C_t) on the newly proposed parameter ($\lambda_4 = \frac{g}{(V/T)} = \frac{1}{F_r^2}\frac{VT}{h}$) under seven or eight stem spacing D conditions (D = 0.0214 m, 0.0237 m, 0.0268 m, 0.0306 m, 0.0355 m, 0.0428 m, 0.0535 m and 0.0713 m), with Subfigure (**a**) showing seven D conditions and sugbfigures (**b–j**) showing eight D conditions. Five subfigure pairs (**a**) and (**b**), (**c**) and (**d**), (**e**) and (**f**), (**g**) and (**h**), and (**i**) and (**j**) are at the BFEV's length $B = 0.15$ m, 0.2 m, 0.25 m, 0.35 m and 0.45 m, respectively, with each subfigure pair corresponding to the wave height H = 0.06 m and 0.08 m.

4.1. Dependence of C_t on λ_1

The relationship of C_t and λ_1 at four λ_4, two H, and eight D conditions is shown in Figure 5a–h with each subfigure under a given combination of λ_4 and H conditions. Figure 5 shows that, in general, C_t increases with the increase of λ_1, that is, the wave attenuation decreases with the increase of λ_1. It is reasonable since both the increase in wavelength and the decrease in the width of BFEV, which increase the value of λ_1 can lead to a decline in wave attenuation and thus an increase in C_t. A comparison of eight lines in each subfigure shows that a smaller D corresponds to a smaller value of C_t, indicating denser stem spacing causes more significant wave attenuation.

Comparisons of Figure 5a and b, c and d, e and f, and g and h show that at given λ_4 and D, the value of C_t is smaller at the higher wave height (H = 0.08 m) than at the lower wave height (H = 0.06 m). The difference between these two wave height conditions is more significant at sparser stem spacing. The largest difference of 0.132 occurs at $\lambda_4 = 6.1$ and D = 7.13, with the value of C_t at H = 0.08 m being smaller than that of H = 0.06 r by 20.6%. This indicates the efficiency of wave attenuation over the BFEV is improved a higher wave heights, and this is more significant under sparser stem spacing conditions.

4.2. Dependence of C_t on λ_2

The relationship of C_t and λ_2 at four λ_4, two H and five λ_1 conditions is show in Figure 6a–h with each subfigure under a given combination of λ_4 and H condition Figure 6 shows that, in general, C_t decreases with the increase of λ_2, that is, the wav attenuation increases with the increase of λ_2. The main reason is that the decrease i stem spacing, which increases the value of λ_2 can lead to improved wave attenuation an thus a decrease in C_t. This suggests that the effect of stem spacing might be relativel important compared with that of the wavelength under the experimental conditions. comparison of five lines in each subfigure shows that the smaller value of λ_2 correspond to a smaller value of C_t and thus leads to a more significant attenuation of the wave, whic is in agreement with that of Figure 5.

Comparisons of Figure 6a and b, c and d, e and f, and g and h show that at given λ and λ_1, the value of C_t is smaller at the higher wave height (H = 0.08 m) than that of th lower wave height (H = 0.06 m). The difference ranges from 0.011 to 0.072. The large difference of 0.072 occurs at $\lambda_4 = 7.25$, λ_1 = 2.33 and λ_2 = 49.06, with the value of C_t a H = 0.08 m being smaller than that of H = 0.06 m by 17.3% (Figure 6a,b). This indicate the efficiency of wave attenuation over the BFEV is improved at the higher wave heigh condition, which is consistent with that of Figure 5.

4.3. Dependence of C_t on λ_3

The relationship of C_t and λ_3 at four λ_4, two H and five λ_1 conditions is show in Figure 7a–d, with each subfigure under a given combination of λ_4 and H condition Figure 7 shows that, in general, C_t decreases with the increase of λ_3, indicating the wav attenuation is improved with the increase of λ_3. The main reason is that the increase i wave height and the decrease in stem spacing, which increase the value of λ_3 can lead t an improvement in wave attenuation and thus a decrease in C_t. A comparison of five line in each subfigure shows that the smaller value of λ_1 corresponds to the smaller value of C and thus leads to a more significant attenuation of the wave, which is in agreement wi that of Figures 5 and 6.

4.4. Dependence of C_t on λ_4

The relationship of C_t and λ_4 at five B, two H, and eight D conditions is show in Figure 8a–j with each subfigure under a given combination of B and H condition Figure 8 shows that, in general, C_t increases with the increase of λ_4, which indicates th wave attenuation is improved with the increase of λ_4. This suggests that the increase i horizontal inertia of waves might cause an improvement in wave attenuation over th BFEV. Comparison of the seven (Figure 8a,b) or eight lines (Figure 8c–j) under different conditions in each subfigure shows that the value of C_t decreases with the decrease of This is consistent with that of Figure 5, indicating the wave attenuation over the BFEV improved under denser vegetation conditions.

In consistency with that of Figures 5–7, comparisons of Figure 8a and b, c and d, and f, and g and h also show that the incident wave height affects the wave attenuatio significantly. The wave attenuating efficiency over the BFEV is improved at the high wave height (H = 0.08 m) than at the lower wave height (H = 0.06 m).

4.5. Regression Analysis of C_t on λ_1, λ_2, λ_3, and λ_4

Figures 5–8 indicate that the wave height attenuation is closely dependent on the fou dimensionless parameters of λ_1, λ_2, λ_3, and λ_4 (Equations (1)–(4)). Therefore, a multip linear regression analysis of C_t on these four predominant parameters is conducted.

Let

$$C_t = f(\lambda_1, \lambda_2, \lambda_3, \lambda_4) \tag{6}$$

where f is a function.

Choose a nonlinear multiplication model for the function f, obtaining:

$$C_t = \beta(\lambda_1)^{x_1}(\lambda_2)^{x_2}(\lambda_3)^{x_3}(\lambda_4)^{x_4} \tag{7}$$

where β, x_1, x_2, x_3 and x_4 are regression coefficients of the function f.

Then, based on the total set of 312 experimental cases, the regressed formula of C_t is obtained.

$$C_t = 0.133(\lambda_1)^{0.508}(\lambda_2)^{-0.016}(\lambda_3)^{-0.207}(\lambda_4)^{0.488} \tag{8}$$

with the correlation coefficient $R = 0.958$.

The calculated and observed C_t shown in Figure 9 indicate they are in good agreement. The agreement is better at the smaller C_t, and the deviation is relative larger at the higher C_t. The possible reason might be that the wave attenuation was more affected by the bulk swaying motion of the frame of BFEV when the effect of the four dimensionless parameters was relatively smaller at the higher C_t.

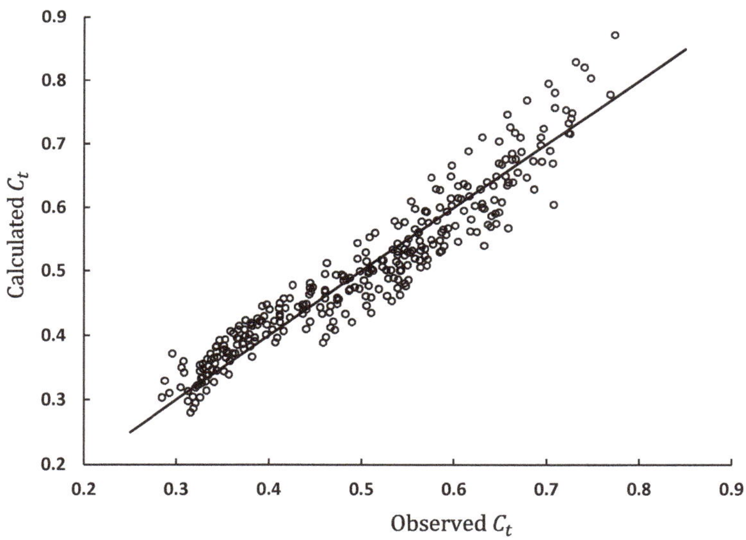

Figure 9. The calculated wave transmitted coefficient (C_t) versus the observed ones.

5. Discussion

Although the regressed Formula (8) agrees with the observed value C_t, there is still one point that needs to be noted. The two parameters λ_1 and λ_2 have the same form, but they have an inverse effect on wave attenuation and the transmitted coefficient C_t. The underlined mechanism is that the width of the BFEV (B) is a characteristic length positively related to the resistant force of the BFEV, but the stem spacing (D) is negatively related to the resistant force.

For a group of cylinder-conceptualized rigid vegetation stems, the forces, including the drag and inertia effects acting on vegetation, can be estimated by the Morison equation [23,34] ($Fi = \frac{1}{2}\rho C_D N_v A_v U_{vi}\sqrt{U_{vi}U_{vi}} + \rho C_M N_v V_v \frac{\partial U_{vi}}{\partial t}$, where U_{vi} is the apparent velocity acting on the vegetation elements in the ith direction, C_D is the drag coefficient, C_M is the inertia coefficient, N_v is the vegetation density defined as the number of vegetation elements per unit horizontal (bed) area, A_v is the projected area defined as the frontal area of a vegetation element projected to the plane normal to the stream-wise flow direction, and V_v is the volume of a vegetation element). Stone and Shen [30] proposed an alternati drag coefficient C_{Dm} based on the constricted cross-sectional apparent flow velocity U with $U_v = U_{vm}(1 - d/D)$. The modified C_{Dm} is more accurate because it is closer to drag coefficient of a single cylinder and has less variation for a wide range of values vegetation density, stem size, and cylinder Reynolds number in comparison with C_D. T Morison equation and C_{Dm} have been validated by many researchers [15,19,20,23]. In t present study, the width B is positively related to the total projected area of vegetati A_v, and the stem spacing D is negatively related to U_v and N_v, therefore λ_1 and λ_2 ha inverse effects on the bulk resistance of the BFEV.

6. Conclusions

A series of 312 experimental tests were conducted in an indoor water flume to vestigate the effect of the BFEV on wave attenuation, which supplements the data this new nature-based type of breakwater. Three conventional and one newly propos dimensionless parameters λ_1, λ_2, λ_3, and λ_4 are found to have a significant effect on wa height attenuation, with the transmitted coefficient C_t being positively related to λ_1 and while negatively related to λ_2 and λ_3.

A regressed formula of the transmitted coefficient C_t on the four parameters is obtain based on the 312 experimental tests, with a correlation coefficient reaching up to 0.9 The calculated and observed C_t are in good agreement. The relationship between C_t a the four predominant parameters, as well as the regression formula of C_t, obtained in t study, are expected to provide fundamental support for the design and construction the BFEV for bank and structure protection from wave erosion in rivers, lakes, coasts a marine environments.

Author Contributions: Conceptualization, Y.L. and G.Y.; methodology, Y.L.; validation, L.X. and D formal analysis, Y.L. and D.Z.; investigation, L.X.; resources, G.Y.; data curation, G.Y.; writing—origi draft preparation, Y.L.; writing—review and editing, Y.L. and D.Z.; supervision, G.Y.; project admi tration, G.Y.; funding acquisition, L.X. All authors have read and agreed to the published versior the manuscript.

Funding: This research was funded by the National Foundation of the People's Republic of Chi grant numbers 51479109 and 51479137.

Institutional Review Board Statement: Not applicable.

Informed Consent Statement: Not applicable.

Data Availability Statement: The data are available at https://doi.org/10.6084/m9.figshare.228 57 (accessed on 29 June 2023).

Acknowledgments: We thank the anonymous reviewers, editors and assistant editors for th constructive comments and suggestions!

Conflicts of Interest: The authors declare no conflict of interest.

References

1. Luom, T.T.; Phong, N.T.; Anh, N.T.; Tung, N.T.; Tu, L.X.; Duong, T.A. Using fine-grained sediment and wave attenuation as a n measure for evaluating the efficacy of offshore breakwaters in stabilizing an eroded muddy coast: Insights from Ca Mau, Mekong Delta of Vietnam. *Sustainability* **2021**, *13*, 4798. [CrossRef]
2. Park, Y.H.; Edge, B.L. Beach erosion along the Northeast Texas Coast. *J. Coast. Res.* **2011**, *27*, 502–514. [CrossRef]
3. Gao, J.; Ma, X.; Dong, G.; Chen, H.; Liu, Q.; Zang, J. Investigation on the effects of Bragg reflection on harbor oscillations. *Co Eng.* **2021**, *170*, 103977. [CrossRef]
4. Gao, J.; Zang, J.; Chen, L.; Chen, Q.; Ding, H.; Liu, Y. On hydrodynamic characteristics of gap resonance between two fixed bod in close proximity. *Ocean Eng.* **2019**, *173*, 28–44. [CrossRef]
5. Gao, J.; Lyu, J.; Wang, J.; Zhang, J.; Liu, Q.; Zang, J.; Zou, T. Study on transient gap resonance with consideration of the motior floating body. *China Ocean Eng.* **2022**, *36*, 994–1006. [CrossRef]
6. He, Z.; Gao, J.; Shi, H.; Zang, J.; Chen, H.; Liu, Q. Investigation on effects of vertical degree of freedom on gap resonance betw two side-by-side boxes under wave actions. *China Ocean Eng.* **2022**, *36*, 403–412. [CrossRef]

Li, C.; Zhang, H.H.; Sun, B.; Shaollin, Y. Wave-attenuation and hydrodynamic properties of twin pontoon floating breakwater with kelp. *Appl. Ocean Res.* **2022**, *124*, 103213. [CrossRef]

Mancheño, A.G.; Peter, M.J.H.; Jonkman, S.N.; Urrutia, S.; Kazi, I.; van Ledden, M. Mapping mangrove opportunities with open access data: A case study for bangladesh. *Sustainability* **2021**, *13*, 8212. [CrossRef]

Rosenberger, D.; Marsooli, R. Benefits of vegetation for mitigating wave impacts on vertical seawalls. *Ocean Eng.* **2022**, *250*, 110974. [CrossRef]

Yin, K.; Xu, S.; Gong, S.; Zhou, R.; Wang, Y. Effects of wave nonlinearity on submerged flexible vegetation dynamics and wave attenuation. *Ocean Eng.* **2021**, *241*, 110103. [CrossRef]

Gao, J.; Shi, H.; Zang, J.; Liu, Y. Mechanism analysis on the mitigation of harbor resonance by periodic undulating topography. *Ocean Eng.* **2023**, *281*, 114923. [CrossRef]

Zhu, L.; Huguenard, K.; Fredriksson, D.W.; Lei, J. Wave attenuation by flexible vegetation (and suspended kelp) with blade motion: Analytical solutions. *Advances Water Resour.* **2022**, *162*, 104148. [CrossRef]

van Rooijen, A.; Lowe, R.; Ghisalberti, M.; McCall, R.; Hansen, J. Modelling wave attenuation through submerged vegetation canopies using a subgrid canopy flow model. *Coast. Eng.* **2022**, *176*, 104153. [CrossRef]

Huang, Z.; Yao, Y.; Sim, S.Y.; Yao, Y. Interaction of solitary waves with emergent, rigid vegetation. *Ocean Eng.* **2011**, *38*, 1080–1088. [CrossRef]

Cassidy, C.; Tomiczek, T. Exploring the effects of density and configuration on wave attenuation through mangrove living shorelines. In Proceedings of the Oceans, San Diego-Porto, San Diego, CA, USA, 20–23 September 2021; pp. 20–23.

Cao, H.; Chen, Y.; Tian, Y.; Feng, W. Field investigation into wave attenuation in the mangrove environment of the South China Sea coast. *J. Coast. Res.* **2016**, *32*, 1417–1427. [CrossRef]

Massel, S.R.; Furukawa, K.; Brinkman, R.M. Surface wave propagation in mangrove forests. *Fluid Dynamics Res.* **1999**, *24*, 219–249. [CrossRef]

Zhou, X.; Dai, Z.; Pang, W.; Wang, J.; Long, C. Wave attenuation over mangroves in the Nanliu Delta, China. *Front. Mar. Sci.* **2022**, *9*, 874818. [CrossRef]

Augustin, L.N.; Irish, J.L.; Lynett, P.L. Laboratory and numerical studies of wave damping by emergent and near-emergent wetland vegetation. *Coast. Eng.* **2009**, *56*, 332–340. [CrossRef]

Lee, W.K.; Tay, S.H.X.; Ooi, S.K.; Friess, D.A. Potential short wave attenuation function of disturbed mangroves. *Estuar. Coast. Shelf Sci.* **2020**, *248*, 106747. [CrossRef]

Jadhav, R.S.; Chen, Q. Probability distribution of wave heights attenuated by salt marsh vegetation during tropical cyclone. *Coast. Eng.* **2013**, *82*, 47–55. [CrossRef]

Sun, B.; Zhang, H.; Li, C.; Li, Z. Wave-attenuation performance and hydrodynamic characteristics of a plant ecological floating breakwater. *Ships Offshore Struct.* **2022**, 1–16. [CrossRef]

Wu, W.; Marsooli, R. A depth-averaged 2D shallow water model for breaking and non-breaking long waves affected by rigid vegetation. *J. Hydraul. Res.* **2012**, *50*, 558–575. [CrossRef]

Ozeren, Y.; Wren, D.G.; Wu, W. Experimental investigation of wave attenuation through model and live vegetation. J.; Waterway, Port, Coast. *Ocean Eng.* **2014**, *140*, 04014019.

Vuik, V.; Jonkman, S.N.; Borsje, B.W.; Suzuki, T. Nature-based flood protection: The efficiency of vegetated foreshores for reducing wave loads on coastal dikes. *Coast. Eng.* **2016**, *116*, 42–56. [CrossRef]

van Veelen, T.J.; Fairchild, T.P.; Reeve, D.E.; Karunarathna, H. Experimental study on vegetation flexibility as control parameter for wave damping and velocity structure. *Coast. Eng.* **2020**, *157*, 103648. [CrossRef]

Kobayashi, N.; Raichle, A.W.; Asano, T. Wave attenuation by vegetation. J.; Waterway, Port, Coast. *Ocean Eng.* **1993**, *119*, 30–48.

Jadhav, R.S.; Chen, Q.; Smith, J.M. Spectral distribution of wave energy dissipation by salt marsh vegetation. *Coast. Eng.* **2013**, *77*, 99–107. [CrossRef]

Tanino, Y.; Nepf, H.M. Laboratory investigation of mean drag in a random array of rigid, emergent cylinders. *J. Hydraul. Eng.* **2008**, *134*, 34–41. [CrossRef]

Stone, B.M.; Shen, H.T. Hydraulic resistance of flow in channels with cylindrical roughness. *J. Hydraul. Eng.* **2002**, *128*, 500–506. [CrossRef]

He, F.; Chen, J.; Jiang, C. Surface wave attenuation by vegetation with the stem, root and canopy. *Coast. Eng.* **2019**, *152*, 103509. [CrossRef]

Dalrymple, R.A.; Kirby, J.T.; Hwang, P.A. Wave diffraction due to areas of energy dissipation. *J. Waterw. Port Coast. Ocean Eng.* **1984**, *110*, 67–79. [CrossRef]

Goda, Y.; Suzuki, Y. Estimation of incident and reflected waves in random wave experiments. In Proceedings of the 15th Conference on Coastal Engineering, ASCE, Honolulu, HI, USA, 11–17 July 1976; pp. 828–845.

Morison, J.R.; O'Brien, M.P.; Johnson, J.W.; Schaaf, S.A. The force exerted by surface waves on piles. *J. Petroleum Tech.* **1950**, *2*, 149–154. [CrossRef]

plants

icle

ew Polyfunctional Biorationals Use to Achieve Competitive ïeld of Organic Potatoes in the North-West Russian Ecosystem

ıa Novikova [1], Vladislav Minin [2,*], Julia Titova [1], Anton Zakharov [2], Irina Krasnobaeva [1], Irina Boikova [1]
d Evgeniy Murzaev [2]

[1] All Russia Research Institute for Plant Protection Federal State Budget Scientific Institution (FSBSI VIZR), 196608 Saint-Petersburg, Russia; irina_novikova@inbox.ru (I.N.); juli1958@yandex.ru (J.T.); krasnobaeva08@mail.ru (I.K.); irina_boikova@mail.ru (I.B.)

[2] Branch of Federal State Budgetary Scientific Institution Institute for Engineering and Environmental Problems in Agricultural Production (FSBSI FSAC VIM), 196625 Saint-Petersburg, Russia; bauermw@mail.ru (A.Z.); murzaev.e.a@mail.ru (E.M.)

* Correspondence: minin.iamfe@mail.ru; Tel.: +7-9218728003

Abstract: To increase the organic potato yield, it is necessary to provide the crop with sufficient nutrients and effective means of biocontrol the diseases. The research goal was to characterize the biorationals' efficacy to achieve competitive organic potatoes' yield under various weather conditions. A 4-year trial was carried out in the Leningrad region using Udacha variety potatoes. The tests used liquid forms of new polyfunctional biologicals Kartofin based on highly active *Bacillus subtilis* I-5-12/23 and organic fertilizer BIAGUM obtained from poultry manure by aerobic fermentation in a closed biofermenter. Significant stimulation in plant growth and development to the flowering phase regardless of the hydrothermal conditions of the growing season was noted. The stimulating effect was determined by the combined use of biorationals pro rata to BIAGUM dose. Kartofin biologicals and BIAGUM almost doubled the potato tubers' yield compared to the control, regardless of the growing season conditions. At the flowering phase, the biological efficacy in potato fungal diseases incidence and development was near 90% under optimal and 50–75% under drought hydrothermal conditions. At the end of vegetation, the efficiency in fungal diseases incidence and development made up 45–65% under optimal and 45–70% under dry conditions. BIAGUM effectiveness in reducing disease development reached 45–50% regardless of growing season conditions.

Keywords: polyfunctional biologics; biorationals; compost; potato diseases management; organic potatoes; weather conditions; stimulating biologic effect

.tion: Novikova, I.; Minin, V.; va, J.; Zakharov, A.; Krasnobaeva, oikova, I.; Murzaev, E. New functional Biorationals Use to ieve Competitive Yield of anic Potatoes in the North-West sian Ecosystem. *Plants* **2022**, *11*, https://doi.org/10.3390/ ıts11070962

demic Editors: Panayiotis Dimitrakopoulos, Mario Pagnotta, Miklas Scholz and ıiya Noorani

ɔived: 21 February 2022
ɔpted: 28 March 2022
lished: 1 April 2022

lisher's Note: MDPI stays neutral regard to jurisdictional claims in ished maps and institutional affil- ns.

1. Introduction

The potato (*Solanum tuberosum* L.) is the fourth most important food crops for human nutrition, after maize, wheat and rice. In addition, a significant amount of potato is used for livestock feed and for seed purposes [1]. Russia, producing about 20 million tons of potatoes annually, is among the top 5 countries with the largest potato production. Given the great diversity of natural and climatic conditions for growing potatoes in Russia, as well as the large number cultivated varieties of Russian and foreign breeding, in recent years, various technologies for growing potatoes have been developed [2]. In the Northwestern region of Russia, extensive research has been carried out into potato cultivation. The result of these studies has been to develop zonal technologies [3]. In the near future, the world's growing population will require more and more food resources, which inevitably leads to the intensification of agriculture. However, intensive agriculture consumes more and more natural resources, causing significant and often irreparable damage to the environment [1]. Data compiled by a team of authors [4] on the global supply of food, feed and other industrial uses of agricultural products suggest that potato-based agri-food systems provide significant opportunities for food security and income generation in the face of anticipated trends in population growth, climate change, conflict, migration, inequality and the rec effects of the COVID-19 crisis.

Currently, organic farming is quite actively developing in Russia [5–7]. If the requ ment for the production of organic crop products is observed, environmental conditic are significantly improved and natural biodiversity in agroecosystems is preserved The same requirements pose new challenges in the production of organic potatoes, wh must be solved in order to obtain a high yield that competes with intensive production Organic potato growers face two major challenges—disease control and providing pla with sufficient nutrients [10,11].

It should be noted that biological control is one of the most promising approaches organic and sustainable agriculture to protect crop plants from disease, increase yields a improve food quality [12–14]. Currently, organic production for plant protection agai disease is increasingly focused on biologicals [15]. As biologicals in a broad sense, b living cultures of microorganisms and biologically active compounds synthesized by th can be used as biocontrol agents [13,15].

Biorationals as natural, environmentally non-toxic products obtained from non-microb sources (activators, adjuvants, elicitors, organic materials, soil improvers, phytohormor extracts from biomass and plants, waste, special composts, etc.) have direct and indir effect on plants, accelerating their growth and development, thereby increasing prod tivity [16–18]. Such biorationals form a large heterogeneous group of biotechnologi products called biostimulants [19–22]. Biological products produced using living micro ganisms' cells have both direct and indirect multifactorial effects on plants, leading t stable increase in productivity and yield [23–25]. Such biologicals have a positive effect the yield [26,27]. Beneficial microorganism strains that form the basis of biologicals are part of the soil microbiome, rhizosphere, rhizoplane, as well as the phyllosphere, includi the phylloplane, of cultivated crops [28–30]. The introduction of microbial strains into agroecosystem of a cultivating crop can affect the biocenotic interactions between plar phytopathogenic species and the environment and lead to significant changes in agr cosystems. This requires a comprehensive assessment of the possible effects of biocont agents. The authors [31] point out that the philosophy of plant disease control needs shift from a concept that deals only with crop productivity to a multifaceted paradig that includes an environmental impact study, social acceptability and the availability biocontrol technologies.

Microorganisms of biologicals secrete hormones and also change the concentration phytohormones secreted by the plant, affecting root and root hair growth, shoot devel ment, flowering and seed growth, aging, various physiological processes, including c division, gene expression and response to abiotic and biotic stresses [32–36].

The indirect effect of biopesticides, whether they are biorationals and/or biologic based on microbes' living cells is to control plant pathogens, which reduces crop losses fr their development [37,38]. Microbe-antagonist strains possess the ability to synthesize a secrete as secondary metabolites a lot of biologically active substances (BAS) of various ture (antibiotics, siderophores, hydrolytic enzymes, volatile organic compounds, hydrog cyanide, etc.). This ability has been called polyfunctionality, which determines the varic target activity of biologicals' producer strains [13,14,39]. Producer strains' BAS suppr the phytopathogens development on plants, reducing their virulence and aggressivene thereby saving the crop [40–43]. The producer strains polyfunctionality provides anta nism to the phytopathogens together with inducing the plant systemic resistance [44– Due to the producer strain polyfunctionality modern biologicals combine the properties biofertilizers, biostimulants and biopesticides in their formulations, providing a sustaina increase in crop yields when used [48–52]. In addition, polyfunctional biologics applicati improves the crop quality contributing to carbohydrates, proteins, vitamins, macro- a microelements accumulation in agricultural products [53–55]. In particular, this applies the quantitative and qualitative indicators of the potato harvest [56–59].

Biologicals based on microorganisms' living cells that have a direct stimulating effect on the plants development and provide an increase in their productivity are combined into a group of biofertilizers [60–63]. Bacterial inoculants directly increase the availability of N, P, K, Zn, S and other nutrients for plants, which leads to the accelerated growth of roots and shoots and increased yields [48,64–66].

Aerobic composting technologies developed in recent years in closed-type biofermenters can significantly reduce the processing time, improve the quality of the resulting compost and reduce the negative impact on the environment [67]. Regulating the aeration regime has a positive effect on the viability of mesophilic and thermophilic microorganisms, which decontaminate and change the physical and chemical conditions of manure into forms accessible to plants [68,69]. Under Russian geographical conditions, drum biofermenters are of the greatest preference. They provide a short processing time (up to 3–4 days) and high preservation of nutrients, which compensates for the higher capital costs of implementation. The ability to automate the recycling process relatively easily leads to lower operating costs, which ultimately increases the profitability of compost production [70,71].

Our research goal is to characterize the efficacy of new polyfunctional biologicals and special organic compost to provide maximum possible yields in organic potatoes under the preponderant weather conditions.

To achieve this goal, the following objectives were established:

- To develop the experimental batches of polyfunctional biologics liquid formulations (Kartofin, SC; Kartofin MR, SC) and organic fertilizer BIAGUM (special organic compost);
- To evaluate the effect of polyfunctional biologics liquid formulations (Kartofin, SC; Kartofin MR, SC) and organic fertilizer BIAGUM application on biometric indicators of the organic potato plants development during optimal and dry growing seasons (by hydrothermal conditions);
- To estimate the biological efficacy of polyfunctional biologics and special organic compost BIAGUM application in organic potatoes' (plants and tubers) protection from diseases;
- To estimate the effect of new polyfunctional biologics and special organic compost BIAGUM in organic potatoes' healthy yield formation and its productivity during various vegetation periods.

2. Materials and Methods

The research work was carried out in the Microbiocontrol Laboratory (ML) FSBSI VIZR, its edible mushrooms' pilot farm and in the production potato plantings at the IEEP branch Experimental Station of the FSBSI FSAC VIM. The IEEP branch Experimental Station of the FSBSI FSAC VIM is the long-term site for biological farming development and improvement. Its cultivable land has not been treated with chemicals (fungicides, insecticides, herbicides, etc.) for about 20 years. This is a complex system of multi-field organic crops rotation. For phytosanitary optimization and increasing the yield on these lands, only biorationals are used. Such a long absence of chemical pesticides' in the topsoil makes it possible to obtain completely organic agricultural products of all crops cultivated on the site. The crop rotations at this long-term site are used in field trials of the biorationals efficacy for their use in organic crop production.

2.1. Materials

The research materials were semi-synthetic and natural nutrient media such as dried nutrient medium (DNM): pancreatic sprat hydrolysate—15 g L^{-1}; NaCl—4.59 g L^{-1}; microbiological agar—20 g L^{-1}; H_2O—1 L; pH = 7.2 (Microgen Co. Ltd., Moscow, Russia); corn-molasses nutrient medium (CMNM): molasses—15 g L^{-1}; corn extract—30 g L^{-1}; H_2O—1 L; pH = 7.8 (Research and Production Association "ALTERNATIVE", Moscow, Russia).

Another research material was natural nutrient medium, based on multirecycle (MR) spent mushroom substrates' (SMS) aqueous extracts. Such nutrient medium i obtained by multirecycling of industrial and agricultural waste from commercial substrate (CS) for cultivating *Lentinula edodes* (Berk.) Pegler (CSLe)—shiitake mushroom. CSLe i a noninvolved state has the following percentage composition per weight of substrat having 70% moisture content: oak sawdust—88.9%; wheat bran—10%; $CaCO_3$—0.1% $CaSO_4 \times 2H_2O$—1%. For industrial edible mushrooms cultivating the commercial spawr *L. edodes* 4080 and *Pleurotus ostreatus* (Jacq.) P. Kumm. HK-35 (Sylvan, Inc.; Moscow, Russia were used. Shiitake mushrooms were grown on sterilized CSLe by the industrial metho of low-volume solid-phase fermentation in the FSBSI VIZR pilot farm for 3 months a 18–23 °C and 85–95% air humidity. Shiitake basidiomata and spent shiitake mushroor substrate (SMSLe) were obtained. SMSLe: Once recycled/spent CSLe containing fung protein as *L. edodes* 4080 mycelium derivate. There were used multirecycled (MR) SM left after double edible macromycetes cultivation on the same substrate. When preparin SMSLe for double recycling, the substrates components were crushed to 0.5–2.5 cm piece and soaked in water for 20–24 h to complete moisturizing. The prepared SMSLe havin 70–80% moisture content and stabilized acidity pH = 7.0–7.5 were packed in 1500 g eac in 1 l polypropylene bags. The bags were sealed up and sent for 1 h steam sterilizatio at 133 °C (202.7 kPa) in autoclave 5075ELVPV D (Tuttnauer Europe BV, Netherlands) fc subsequent inoculating by *P. ostreatus* HK-35—oyster mushroom. SMSLePo was obtaine by 2-fold biorecycling of CSLe. Firstly, basidiomata as food product was obtained b *L. edodes* 4080 for consumers and spent substrate (SMSLe), and then *P. ostreatus* HK-35 wa sequentially obtained in turn from the fruit bodies and spent substrate: double fermente SMSLe containing *L. edodes* 4080 and *P. ostreatus* HK-35 mycelium's derivates of thes 2 fungi [14,72]. There was a nutrient medium developed for biological producer strain cultivated based on double spent SMS (MR SMS)–SMSLePo aqueous extracts. The SMSLeP was preliminarily milled and boiled in the amount of 200 g l^{-1} for 1 h, filtered and restore to the previous 1 l volume, thus obtaining MR SMS-extract: double spent SMS—200 g L^{-} H_2O—1 L; pH = 6.5–7.5 (FSBSI VIZR, St. Petersburg, Russia).

Sterilization modes for the nutrient media were 30–60 min at 50.7–81.1 kPa [14,73].

Organic material for the special compost BIAGUM was bedding poultry manur supplied by Leningrad Region poultry farm [74].

2.2. Objects

The research objects were the experimental batches of polyfunctional biologicals liqui formulations and special organic compost BIAGUM.

Experimental batches of polyfunctional biologicals liquid formulations (suspensio concentrates–SC) under the names Kartofin, SC; Kartofin MR, SC were developed usin highly active producer strain *Bacillus subtilis* (Ehren.) Cohn I-5 12/23. The producer strai is certified, deposited and maintained at FSBSI VIZR State Collection of Microorganism Pathogenic for Plants and Their Pests registered on 28.01.1998 No. 760 in the Worl Federation for Culture Collections, World Data Center for Microorganisms (WFCC WDCM Japan). Biologicals from Kartofin-series (titer not less than $\times 10^{10}$ colony forming unit (CFU) mL^{-1}) are intended to protect agricultural crops from fungal and bacterial disease during vegetation and harvest storage. The active biologicals' ingredients are the cell (spores) and the complex of *B. subtilis* I-5 12/23 metabolites.

The Kartofin, SC and Kartofin MR, SC experimental batches were developed in ac cordance to FSBSI VIZR-approved regulations and specifications. The biologicals' exper imental batches were prepared in submerged cultivating using liquid inoculums in th ML FSBSI VIZR. The producer strain inoculum was preliminarily stored and develope in test tubes on DNM. The submerged cultivations were carried out in shaking flask with 750 mL volume, containing 100 mL of appropriate nutrient medium (CMNM, MR SMS-extract). Liquid-phase producer strain inoculum was grown at 27–28 °C for 3 day with aeration (180 rpm, New Brunswick™ Innova® 44 Shaker Incubator, Eppendorf, Han burg, Germany). To control the bacterial growth stage and the contaminating microbiota presence the samples were taken away and microscoped every day. The fermentation process was stopped when 85–90% spores had been produced in the culture liquid (CL). The CLs were concentrated for 10 min at 3000 rpm in the centrifuge OS-6MC (Dastan Inc., Bishkek, Kyrgyzstan), and then 0.2% potassium sorbate was added to the spore suspension concentrates which were put into 1 L wide-mouth bottles, PE-LD (VITLAB GmbH, Großostheim, Germany). The initial CL titers were not less than$\times$ 10^{10} CFU mL^{-1}, the finished SC Kartofin-series titers were near $\times 10^{11}$ CFU mL^{-1}. Experimental batches of new polyfunctional biologicals Kartofin, SC and Kartofin MR, SC based on *B. subtilis* strain I-5 12/23 were developed for field trials of their biological efficacy.

Organic compost BIAGUM was produced in IEEP's (branch of FSAC VIM) Laboratory of organic waste bioconversion by the aerobic fermentation of bedding poultry manure [74]. The aerobic solid-state fermentation was carried out in the special closed installatios (biofermenter). The compost was characterized by dry matter (nearly 40%). The carbon content and essential nutrients were as follows: C—21.3%; N—2.1%; P—1.5%; K—0.9%; and Ca—1.4%. Special compost BIAGUM has a maximum shelf life of 2 years from the production date under −20–+30 °C air temperature and 60–75% air humidity. The BIAGUM doses 80 and 160 kg N ha^{-1} were used providing potato productivity in medium and high levels. These doses corresponded to 4.3 t ha^{-1} and 8.6 t ha^{-1} of compost application by weight.

2.3. Methods

The following research methods were used: inoculum and stock cultures preparation, SMS preparation for multirecycling, development of biologicals experimental batches, quality assessment of titers and biologicals experimental batches, conducting field tests, field testing results and potato yield accounting, statistical analysis and visualization [14,72–80].

2.3.1. Inoculum and Stock Cultures Preparation

Preparation and steam sterilization of agarized and liquid nutrient media under autoclaving conditions 30–60 min at 50.7–81.1 kPa followed by inoculation with *B. subtilis* I-5 12/23 pure culture and/or liquid-phase fermentation in 750 mL shaker flasks with 100 mL of nutrient medium on an orbital shaker at 180 rpm and t = 28 °C for 3 days.

2.3.2. SMS for Multirecycling Preparation and MR SMS-Extract Production

SMS components were crushed into 0.5–2.5 cm pieces and soaked in water for 20–24 h to complete moisturizing. Acidity was stabilized at pH = 7.0–7.5, then components were packed into 1500 g batches, each packed in 1 L polypropylene bags. Packed products were steam sterilized at 133 °C for 1 h (202.7 kPa), Followed by MR SMS milling, boiling in amount of 200 g L^{-1} for 1 h in 1 L H_2O, filtering and restoring to 1 L volume.

2.3.3. Biologicals Experimental Batches Production

The products were submerged cultivating in 750 mL shaker flasks with 100 mL of CMNM/MR SMS-extract on an orbital shaker at 180 rpm and 27–28 °C for 3 days with aeration. The fermentation was stopped when 85–90% of spores were produced in CL. The CL was concentrated for 10 min at 3000 rpm and then 0.2% potassium sorbate was added and put into 1 L wide-mouth bottles.

2.3.4. Titers and Biologicals Experimental Batches, Quality Assessment

Serial dilutions method was used for the inoculum and experimental batches titers and quality assessment (on DNM), and we conducted 10-fold replications.

2.3.5. Conducting Field Trials

The field trials were carried out in 2018 to 2020 at the Experimental Station of the Institute for Engineering and Environmental Problems in Agricultural Production (IEEP)—branch of FSAC VIM. The experimental area was located at 59°65 N and 30°38 E near Pavlov town (Leningrad region, Russia). According to Russian classification, the soil of the exp mental plots is sod-podzolic light loamy gleyic soil on residual carbonate moraine lo It has a weak acidic reaction (pH = 6.5–6.6), high organic matter content (5.6%) and medium to high levels of available P and K.

The compost BIAGUM was applied before ridging with its subsequent embedding disking. In order to eliminate the compaction zones in the row-spacing after planting, following strategies were employed:

- Deep loosening of the soil directly in the potato rooting zone;
- Hilling and destroying weeds;
- Using an experimental multipurpose row-crop chisel cultivator, designed and dev oped by the IEEP—BRANCH OF FSAC VIM.

The space between the ridges was loosened by a chisel cultivator up to a depth 30 cm from the bottom of the furrow. The use of deep loosening between rows of orga potatoes as an inter-row treatment resulted in a decrease in soil compaction, both direc in the aisle and in the zone of root formation.

In the field trials, the new potato variety Udacha, super elite and elite classes, obtai from the seed farm was used. This variety has an average yield of 30–50 t ha^{-1} and a sta content of 12–15%; it matures early and is adapted to various types of soil and clima zones. It is midresistant to late blight, rhizoctonia disease, wrinkled mosaic, black leg, rot and common scab (State register of selection achievements approved for use in Russi Federation from FSBI "State Breeding Commission", 2021). It is characterized by a go taste and a smooth tuber surface [74].

Potatoes were treated with the biologicals at the time of planting and then by fol spray after 10 days and 20 days at the biologics' consumption rate 3 and 6 L ha^{-1} un optimal and arid hydrothermal growing conditions, respectively. The total number treatments with biological products was three treatments. A specially designed sprayer w installed on the planter and cultivator for this purpose. Inter-row cultivation was carri out regularly, starting from the second week after planting, using an experimental specim of a row-crop cultivator of an original design that provides deep loosing of inter-ro Weed vegetation was removed mechanically using small rotary harrow BRU-0.7 harro mounted on the cultivator.

Measurement of biomass growth rate (by the plant development phases) and s properties were carried out regularly. Soil samples were taken from the arable horiz (0–250 mm) 4 times per season. Analytical studies were performed at FSBSI VIZR and IE (branch of FSAC VIM) Chemical Analytical Laboratory in accordance with Standards of Interstate Council for Standardization, Metrology and Certification (ISC) and Internatio Organization for Standardization (ISO) Standards of the Russian Federation: 26951- (Soils. Determination of nitrates by ionometric method) Group C09.

The field trials were conducted by mutual orthogonal organization with continu placement of organized repetitions for standard arrangement of test options: 4 replicates a complete randomized design, accounting plot size was 20–60 m^2, according to growi season conditions. Biologicals were not applied in the control. Since the IEEP bran Experimental Station of the FSBSI FSAC VIM is a long-term site for the organic farmi development, the application of non-biological treatments as a control was not used. T biologicals' efficacy was compared with special compost BIAGUM's effective impact the organic potatoes yield.

2.3.6. Field Testing Results and Potato Yields Accounting

Phytoregulatory activity and biological efficacy of biologicals experimental batc were evaluated in 4 series of field trials using standard methods [75–78]. Potato plant hei (sm) and productive/flowering stems number per bush (pieces) parameters and the yi of healthy tubers were used to evaluate phytoregulatory activity. To assess the dama of fungal disease complex to plants and potato tubers, the standard phytopathologi indicators were used (the disease incidence and disease development on potato plants and on tubers, crop losses, biological effectiveness). The dominant fungal diseases identified on the organic Udacha variety potato during research period included leaf spots, late blight, fusarium wilt and rhizoctoniosis. On tubers, the dominate diseases were reticular form of rhizoctoniosis, common scab, silver scab, anthracnose and late blight.

The field trials results were recorded in 5 stages:

- Two biometric accountings with onset of disease symptoms fixation were carried out on 3–5-week-old seedlings in the 1–3^{d} leaf tiers phase and on 6–7-week-old potato plants in the 9–10th leaf tiers phase.
- Two phytopathological accountings of diseases incidence and development were carried out at the flowering phase beginning and ending.
- One accounting was carried out when harvesting tubers on accounting plots (tuber analysis).

Every third potato plant in the trial plot size 20–60 m^2 (255–285 plants) was examined. Replication was 4-fold. In one account, commonly 1000–1150 potato plants were examined per test option. [75–77].

2.3.7. Statistical Analyses

All the results obtained in the experiments were arranged in a database [80], which allows for the statistically obtained results to be processes using the Microsoft Excel 2010 and Statistica 10.0 software packages (StatSoft, Inc., Tulsa, OK, USA), including checking the analyzed data's normal distribution with the help of Shapiro–Wilk's *W*-test, analysis of variance (ANOVA), calculating the mean values (M) as well as standard errors (±SEM) calculation. The data's mean comparison was made using the least significant difference (LSD) test at a 5% error probability. The statistical differences significance in options pairwise comparison was assessed by Student's *t*-test [79,80].

3. Results

3.1. Weather Conditions

The weather conditions during the summer period in the research years differed significantly from each other (Table 1). The month of May seemed to be the warmest in 2018, and the most precipitation fell during this month in 2021. The weather conditions in 2019 and 2020 were characterized by fairly comfortable temperatures and good rainfall during the period of active potato development. The maximum amount of precipitation for the entire growing season fell in 2020.

Table 1. The weather conditions during summer period in the research years.

Month	Weather Indicator, Hydrothermal Coefficient (HTC)	Yearly Average				Average for Cumulative Years
		2018	2019	2020	2021	-
May	Temperature °C	15.1	12.1	10.0	11.6	11.3
	Precipitation, mm	14.0	79.3	53.0	172.0	46.0
	HTC	0.6	2.1	0.6	0.3	-
June	Temperature °C	16.2	18.7	19.2	20.9	15.7
	Precipitation, mm	35.2	79.3	129.4	16.6	71.0
	HTC	1.0	1.4	2.3	0.3	-
July	Temperature °C	20.8	16.5	17.6	22.0	18.8
	Precipitation, mm	152.0	179.8	186.2	16.8	79.0
	HTC	2.9	3.5	3.4	0.3	-
August	Temperature °C	15.7	17.0	17.2	15.8	16.9
	Precipitation, mm	60.1	94.6	195.9	109.2	83.0
	HTC	2.0	1.8	3.8	2.2	-

Selyaninov's Hydrothermal Coefficient (HTC) allows us to calculate the aridity cond tions for a certain period in a given area. It helped in defining the driest conditions whic occurred during the active development of potatoes in June and July 2021. Only 33.4 mr precipitation fell during these two months. The soil temperature in the tuber formatio zone rose above 20 degrees in late June to July, which led to some delay in potato tuber development. At the same time, the soil moisture was reduced to 15% (Figure 1).

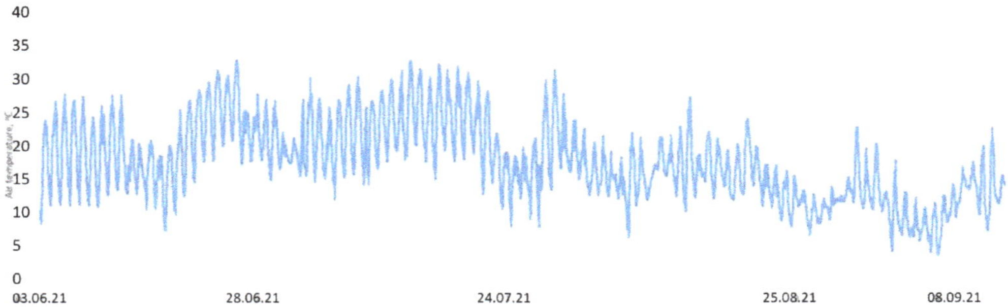

(**A**). Air temperature

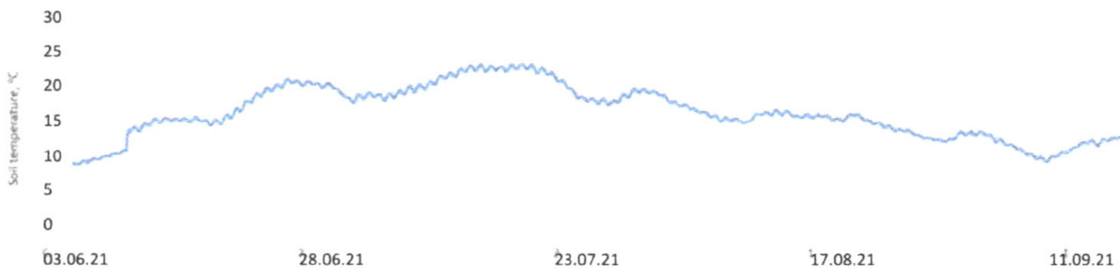

(**B**). Soil temperature in the tuber formating zone

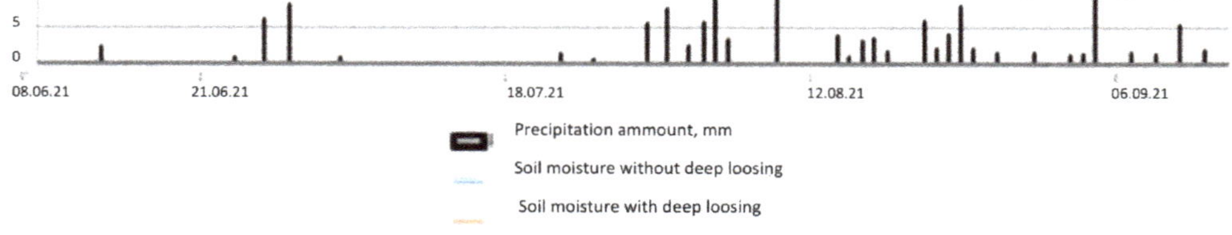

(**C**). Soil moisture (%) in the tuber formating zone

Figure 1. Air and soil temperatures (**A**,**B**) and soil moisture (**C**) in the tuber forming zone during th period of active potato development in 2021.

3.2. Field Experiments

When assessing the phytoregulatory activity of biologicals Kartofin, SC and Kartofin MR, SC and against the background of organic fertilizer BIAGUM in doses of 80 and 160 kg N ha^{-1} in field trials, the significant stimulating effect of the tested biologicals on the growth of potato cultivar Udacha and its development up to the flowering phase were observed as compared to the control (Figures 2 and 3). Under optimal hydrothermal conditions of the development (June to July 2019–2020), the plants' heights were significantly ($p \leq 0.05$) increased by Kartofin, SC together with BIAGUM in a dose 160 kg N ha^{-1} (Figure 2). Arid development conditions (June to July 2021) affected the potato plants' heights only in the budding phase, decreasing this indicator value by 1.5 times (Figure 3). In the flowering phase, the differences were largely leveled, and the potato plants' height in optimal and dry growing seasons (by hydrothermal conditions) was 370–400 and 303–398 mm, respectively (Figures 2 and 3). Under arid conditions in the budding phase, the treatments with BIAGUM in a dose of 160 kg N ha^{-1} and Kartofin, SC together with BIAGUM in a dose of 160 kg N ha^{-1} differed significantly from the control in this indicator. In other test options, significant differences were not noted (Figure 3).

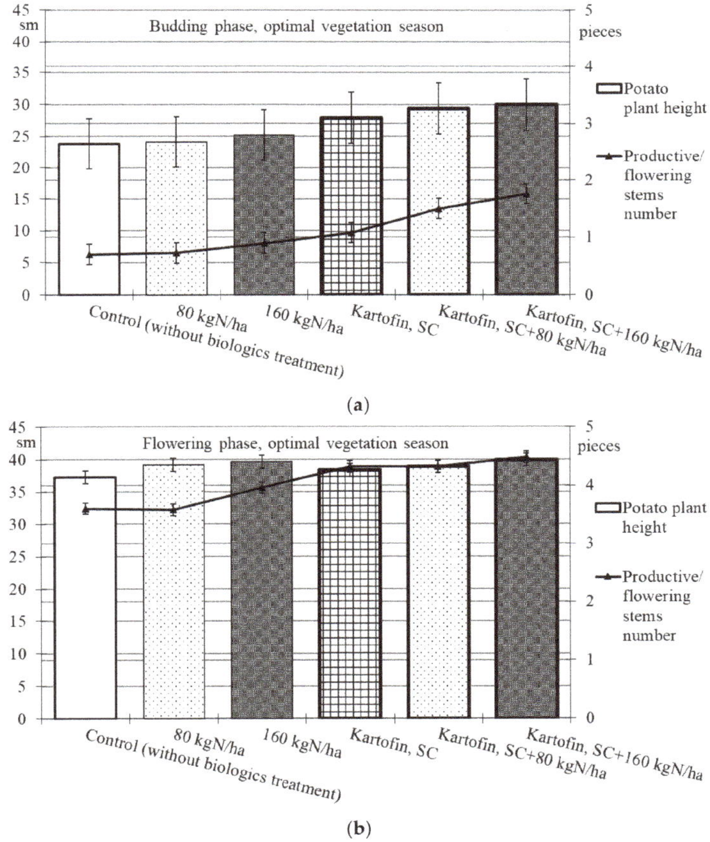

Figure 2. Effect of application polyfunctional biologicals (Kartofin, SC; Kartofin MR, SC) and organic fertilizer BIAGUM in doses of 80 and 160 kg N ha^{-1} on biometric indicators (plant height—sm; productive stems number—pieces) of the potato plants development in organic growing during optimal vegetative season: (**a**)—in budding phase; (**b**)—in flowering phase.

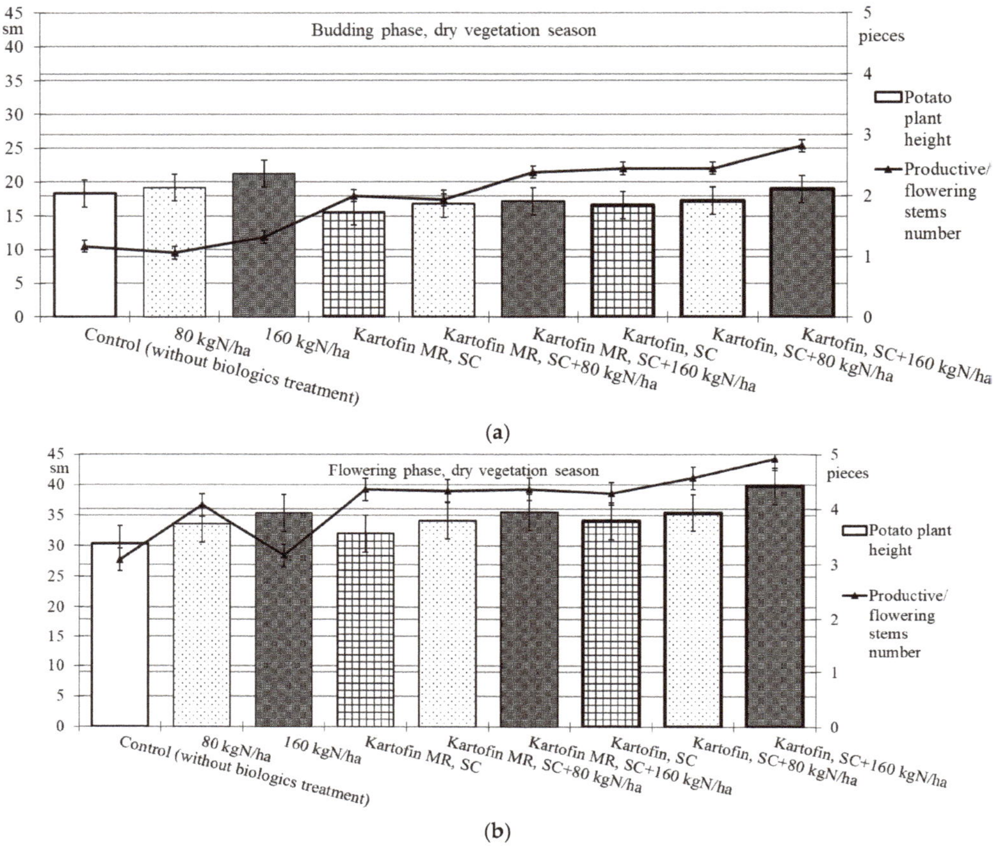

Figure 3. Effect of application polyfunctional biologicals (Kartofin, SC; Kartofin MR, SC) and orga fertilizer BIAGUM in doses 80; 160 kg N ha^{-1} on biometric indicators (plant height—sm; product stems number—pieces) of the potato plants' development in organic growing during dry vegetat season: (**a**)—in budding phase; (**b**)—in flowering phase.

Under the influence of the studied doses of biologicals Kartofin, SC and Karto MR, SC and also against the background of the organic fertilizer BIAGUM, a significa ($p \leq 0.01$) increase by 1.6–2.5 times in the number of productive stems by the buddi phase was observed, regardless of the growing season conditions (Figures 2 and 3). Th under optimal hydrothermal conditions, by the budding phase, the number of producti stems significantly increased ($p \leq 0.01$) when using Kartofin, SC together with BIAGUM doses of 80 and 160 kg N ha^{-1}. In the flowering phase, the maximum number of stems w observed in the trial variants applying Kartofin, SC together with BIAGUM in a dose 160 kg N ha^{-1}. Significant ($p \leq 0.05$) differences in the number of productive stems we noted in the test options applying Kartofin, SC and BIAGUM in a dose of 160 kg N ha and in Kartofin, SC together with BIAGUM in a dose of 80 kg N ha^{-1} relative to the cont The productive stems number increased 1.2–2 times (Figure 2).

In a dry vegetation season, an increase in the potato plants' assimilating surface in test options was achieved by increasing the number of productive stems (peduncles) p bunch. In all trial runs, the number of stems in the flowering culture phase was 1.3–2.3 tim higher than the control (Figure 3). Under arid development conditions (June to July 202 the number of productive stems in the options using biologicals was 1.5–1.8 times high which may indicate a significant increase in potato plant stress resistance under their action. In the budding phase, the maximum stimulating effect on this indicator was observed when using Kartofin, SC together with BIAGUM in a dose of 160 kg N ha^{-1}. Significant ($p \leq 0.05$) differences compared to the control were noted in the test options applying Kartofin, SC and Kartofin, SC together with BIAGUM in a dose of 80 kg N ha^{-1} and Kartofin MR, SC together with BIAGUM in a dose of 160 kg N ha^{-1} (Figure 3). In the flowering phase, the maximum efficiency was noted in the options applying Kartofin, SC together with BIAGUM in doses of 80 and 160 kg N ha^{-1}. Significant ($p \leq 0.05$) differences in comparison with the number of stems in the control were also recorded in the following test options: BIAGUM in a dose of 80 kg N ha^{-1} and Kartofin MR, SC and Kartofin MR, SC together with BIAGUM in doses of 80 and 160 kg N ha^{-1}. The productive stems number increased 1.5 to 2 times (Figure 3). There were no significant differences in the phytoregulatory activity between biologicals Kartofin, SC and Kartofin MR, SC (Figures 2 and 3).

The application of biologicals Kartofin, SC and Kartofin MR, SC as well as organic fertilizer BIAGUM in doses of 80 and 160 kg N ha^{-1} in the organic cultivation of the Udacha potato cultivar led to a significant ($p \leq 0.01$) increase in the biological yield of tubers by 1.6–1.8 times compared with the control, regardless of the growing season conditions (Figure 4). Organic compost BIAGUM in both doses significantly increased the potato crop compared to the control in 2018 to 2020 under optimal hydrothermal conditions. The highest potato productivity under these conditions was observed in the test options applying Kartofin, SC together with BIAGUM in a dose of 80 kg N ha^{-1} and especially in a dose of 160 kg N ha^{-1}. On the contrary, in 2021 dry growing season, the potato yields significantly increased only in options applying organic compost BIAGUM in a dose of 160 kg N ha^{-1} and biologicals together with compost at the same application rate of latter. Thus, under extreme growing conditions, the potato productivity increase was affected only by the application of organic fertilizer BIAGUM (Figure 4).

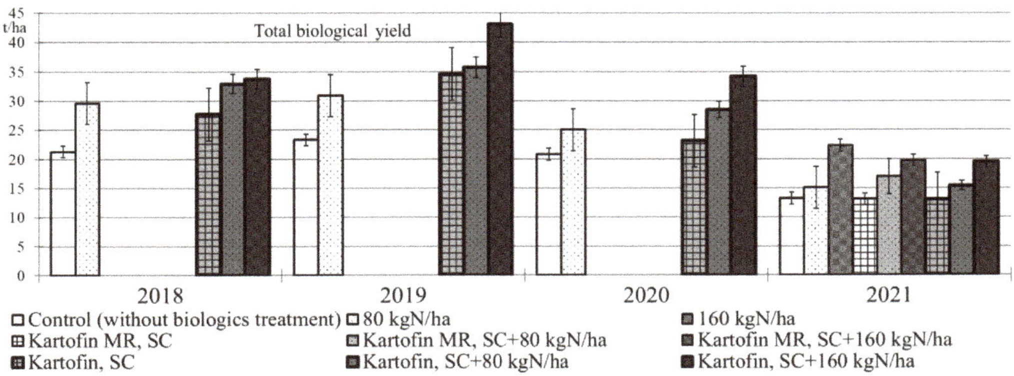

Figure 4. Effect of application polyfunctional biologicals (Kartofin, SC; Kartofin MR, C) and organic fertilizer BIAGUM in doses of 80 and 160 kg N ha^{-1} on potato yield during its organic cultivating in 2018–2021 growing seasons.

The harmfulness of the dominant fungal disease combinations on the organic Udacha potato variety during the research period is shown the Table 2. Disease incidence and disease development as harmfulness indicators under new biorationals influence are provided for the optimal and dry vegetation seasons (2020 and 2021, respectively).

Table 2. The disease incidence and disease development of the dominant fungal diseases com plex on organic Udacha potato variety under new biorationals' influence during 2020–202 vegetation seasons.

Treatment	Disease Incidence/Disease Development, %			
	Optimal Vegetation Season 2020		Dry Vegetation Season 2021	
	Flowering Phase	Before Harvesting	Flowering Phase	Before Harvesting
Control (without treatment)	29.4/14.2	48.1/33.1	36.8/16.2	97.8/54.1
BIAGUM 80 kg N ha^{-1}	16.5/7.3	36.2/24.6	23.6/7.1	68.6/39.2
BIAGUM 160 kg N ha^{-1}	-	-	28.5/10.2	89.7/57.2
Kartofin, SC	3.9/1.6	20.6/11.2	19.5/5.9	41.5/16.2
Kartofin, SC + 80 kg N ha^{-1}	3.1/1.1	22.6/10.8	17.3/6.3	65.8/28.1
Kartofin, SC+ 160 kg N ha^{-1}	-	-	19.9/5.8	43.8/23.4
Kartofin MR, SC	-	-	23.4/7.7	51.1/22.4
Kartofin MR, SC+ 80 kg N ha^{-1}	-	-	16.1/5.6	47.3/23.8
Kartofin MR, SC+ 160 kg N ha^{-1}	-	-	13.2/4.3	33.8/15.1

Polyfunctional biologicals have shown high biological efficacy in reducing the inc dence of dominant fungal diseases (Table 2, Figures 5 and 6).

Under optimal conditions during the flowering phase, the biological efficacy in dom nant fungal diseases incidence reduction was 90 % in all test options, decreasing up to 50 by the end of the growing season (Table 2, Figure 5). Drought at the start of the growin season had a negative effect on this indicator, and the biological efficacy in reducing diseas incidence for the test options did not exceeded 38 to 65%. Nevertheless, by the end of th growing season in pre-harvest, the dominant fungal diseases incidence in all test option was 74–85% lower than in the control (Table 2, Figure 6).

The maximum efficiency for reducing the incidence of dominant fungal disease during the flowering phase was noted in the options using Kartofin MR, SC together wit BIAGUM in doses of 80 and 160 kg N ha^{-1} and at the end of the growing season in th option applying Kartofin, SC together with BIAGUM in a dose of 160 kg N ha^{-1} (Table Figures 5 and 6).

The new biologicals have also shown high biological efficacy in suppressing the deve opment of dominant fungal diseases (Table 2, Figures 5 and 6). Under optimal condition during the flowering phase, the biological efficiency in reducing disease development wa 90% in all test options, decreasing up to 60% by the end of the growing season (Figure The maximum efficacy in reducing disease development at the beginning of the growin season was noted in the treatments by Kartofin, SC and Kartofin, SC together with B AGUM in doses of 80 and 160 kg N ha^{-1} and at the end of the growing season in the optio applying Kartofin, SC together with BIAGUM in the dose of 160 kg N ha^{-1} (Figure Drought at the beginning of the growing season somewhat reduced the effectiveness o biologicals in disease development decreased up to 60 to 75%. Nevertheless, by the end o the growing season, the biological efficiency in reducing the development of potato diseas complexes was as high as in optimal growing conditions, amounting up to 50–70% in th pre-harvest phase. The maximum efficiency in reducing disease development at the start o the growing season was noted in the option using Kartofin, SC together with BIAGUM in dose of 160 kg N ha^{-1} and at the end of the season in the options applying Kartofin, S and Kartofin, SC together with BIAGUM in a dose of 160 kg N ha^{-1} (Figure 6).

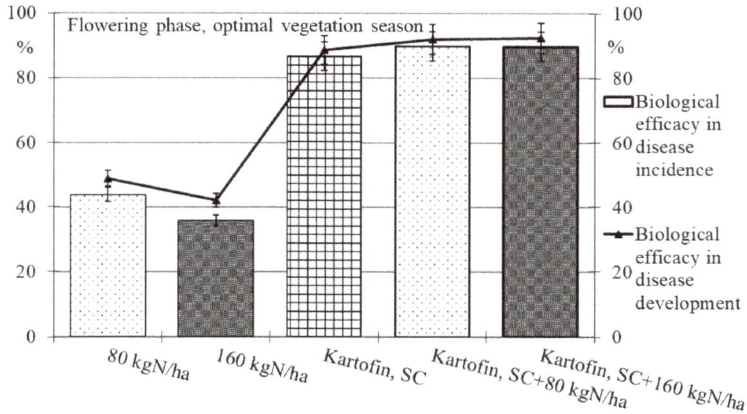

(a)

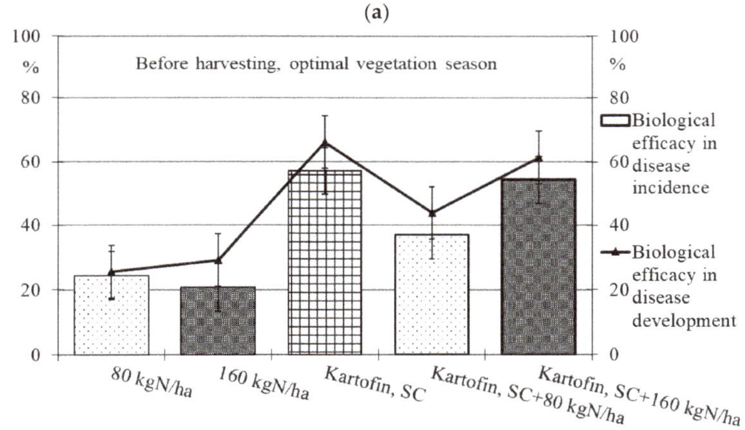

(b)

Figure 5. Biological efficacy of application polyfunctional biologicals (Kartofin, SC; Kartofin MR, SC) and organic fertilizer BIAGUM in doses of 80 and 160 kg N ha^{-1} in the protection of the organic potatoes from dominant fungal diseases complex during optimal vegetative season: (**a**)—in flowering phase; (**b**)—before harvesting.

Organic compost BIAGUM at the studied doses to some extent reduced the incidence and development of fungal potato diseases. Under optimal hydrothermal conditions of the growing season, the biological efficiency of the compost reduced the incidence and development of diseases during the flowering phase by 42% and 45%, respectively, and decreased up to 20–25% and 25–30%, respectively, in pre-harvest (Figure 4). In arid conditions, the efficiency with respect to the incidence and development of disease was 22–30% and 35–55%, respectively, in the flowering phase and 10–30% and 5–30%, respectively, according to test options in pre-harvest (Figure 6). The data obtained indicate the positive impact of organic fertilizers (special organic compost) in studied doses to increase the potato plants' disease resistance.

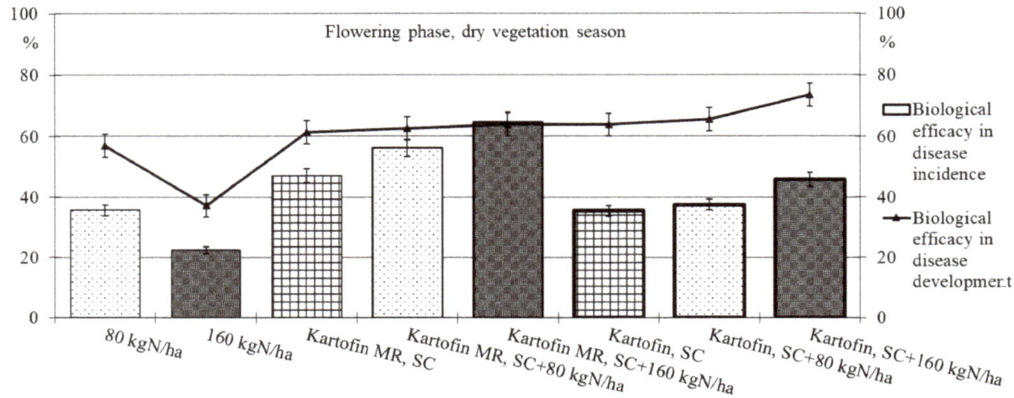

(a)

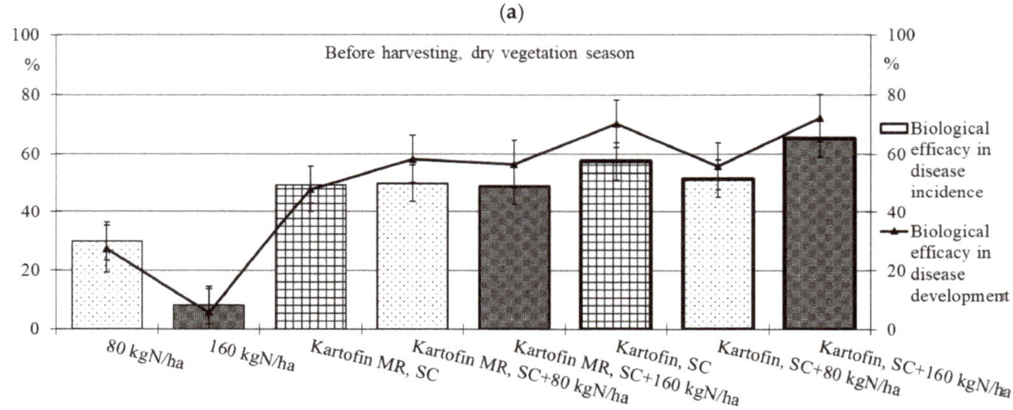

(b)

Figure 6. Biological efficacy of application polyfunctional biologicals (Kartofin, SC; Kartofin MR, and organic fertilizer BIAGUM in doses of 80 and 160 kg N ha^{-1} in the protection of the orga potatoes from dominant fungal diseases complex during dry vegetative season: (**a**)—in floweri phase; (**b**)—before harvesting.

4. Discussion

Useful microorganisms from the rhizo- and phyllospheres are capable of synthesizi complex active BAS, including antibiotics of various chemical classes, enzymes and meta lites with signaling and hormonal functions [81–84]. Auxins, gibberellins, cytokin-i abscisic (ABA), salicylic and jasmonic acids synthesized by microbial strains are fou to be natural growth regulators [85,86]. Phytohormones have significant effects on p tosynthesis, growth and plant productivity. It has been shown that gibberellin enhanc photosynthetic phosphorylation processes, while the intensity of chlorophyll unit utili tion and assimilation number increases. Many bacterial strains from the genera *Bacill* *Azospirillium*, *Pseudomonas*, etc., were found capable of synthesizing auxins, which stimul the root system's development for a more active uptake of water and nutrients by plar These processes in combination increase the plants resistance to diseases and allow th to pass the development stages when they are most susceptible to pathogens [87]. C tokinins can be produced by *Bacillus*, *Rhizobium*, *Arthtrobacter*, *Azotobacter*, *Azospirillium* a *Pseudomonas*. For example, the content of chlorophyll and cytokinins increased in pla when they were inoculated with cytokinin-producing *B. subtilis* strains, which subsequently caused an increase in the biomass of the root system and vegetative part. The *Bacillus*, *Brevibacterium*, *Azospirillum*, *Pseudomonas* and *Lysinibacillus* strains were found capable of ABA synthesis (especially under stress, salinity, drought, etc.), which optimized their endogenous hormonal balance [87,88]. The new polyfunctional biologicals based on microbial antagonists to pathogens possess protective properties which are due to a combination of microbes' antagonistic activity with the ability to increase the disease resistance in plants.

The compound compost formatting is due to the mineral and organic colloids complexing, new biogenic cycles organizing, increasing the enzymatic activity of organic matter and the respiration of living organisms. Such compound compost improves the water regime and nutrition conditions to each organism living in its structure [89,90]. The introduction of compound compost into the soil expands the ecological niches possibilities of cultivated plants in the soil cover system. Multicomponent compost is a good environment for the development of a significant number of species and populations of living organisms producing enzymes, vitamins and other active substances. In terms of chemical and physical properties, compound composts are heterogeneous and multi-dispersed temporary systems, and in terms of living organisms' gene pools, they represent a rich, very complex substrate. Incorporating compound compost into the topsoil affects the populations of virtually all major microbial groupings, with oligotrophs accounting for up to 50% of the total microbial community. The dominant position is occupied by the prokaryotic complex, which includes a large number of species that ensure soil suppressiveness (in particular, genera *Bacillus* and *Pseudomonas*). The increased activity of nitrogen-fixing and especially cellulose-destroying groups is noteworthy. The increase in actinomycetes number especially genus *Streptomyces* representatives was also observed during the compost application. *Streptomyces* species turned out to be the most numerous in the studied composts; they are members of the microbial complex which decomposes complex organic substances and is characterized by a wide enough range of species. Analysis of the data obtained showed that actinomycetes belonging to the *Cinereus* section were the most common, while the *Chromogenes*, *Violaceus* and *Aureus* series were less common. Increasing the biodiversity and density of microbial populations provided soil suppressiveness and caused a reduction in phytopathogenic species' incidence and development. As an example, the preplanting of sawdust and manure compost (10 t ha^{-1}) in a 1:1 ratio enhanced the antagonistic soil activity, reduced rhizoctoniosis development by 40–44%, increased the tuber yield by 1.5–2.1 t ha^{-1}, or by 6–10%, and increased the yield of healthy tubers by 89 to 137%. [91]. The above studies show that applying compost not only provides plants with available nutrition sources but also effectively protects against a variety of diseases.

Thus, our research results have shown that various action mechanisms of biologicals' Kartofin producer strain—*B. subtilis* I-5-12/23—in combination with the available sources of organic nutrition provided by BIAGUM compost can optimize the physiological state of plants by stimulating their growth and development and can significantly increase stress and disease resistance and potato yields.

5. Conclusions

1. The joint use of special compost BIAGUM and new polyfunctional biologicals Kartofin, SC and Kartofin MR, SC provided reliable 1.2-fold increases in the tubers' total yield by 4.3 t/ha under optimal conditions of the growing season and by 2.2 t/ha under dry conditions compared to the control without the application of biologicals. The improvement in yield quality was revealed in increasing total nitrogen, dry matter and starch content 1.2-fold compared with the control regardless of the growing season conditions.
2. The increase in yield under optimal hydrothermal conditions of potato growing was due to the combined use of the polyfunctional biologicals Kartofin, SC; and Kartofin MR, SC with special compost BIAGUM. Under extreme growing conditions,

the potato's productivity was provided by organic fertilizer BIAGUM use in th dose–effect format.
3. New polyfunctional biologicals Kartofin, SC and Kartofin MR, SC and special compo BIAGUM in the studied doses had a stimulating effect on the growth and developme of the Udacha variety potato plants until the flowering phase compared with th control, regardless of the growing season conditions. The potato plants' growth ra and foliage increase 1.2- and1.3-fold, respectively, and the increase in the quantity productive stems (pedicels) increased 1.6-fold regardless of which vegetation seaso conditions were observed.
4. The degree of the stimulating effect on potato plants' growth and development wa caused by the joint application of biorationals (Kartofin, SC and Kartofin MR, SC an special compost BIAGUM) and was proportional to an increase in compost dose.
5. Polyfunctional biologicals Kartofin, SC and Kartofin MR, SC have shown high biolog cal efficacy in relation to the incidence rate and development of potato fungal disease complex at its organic cultivation: 90% under optimal hydrothermal conditions; 65 and 75%, respectively, under drought growing conditions.
6. The biological effectiveness of organic compost BIAGUM in the studied doses i reducing the incidence and development of fungal diseases in Udacha variety potatoe in organic cultivation reached 45% and 50%, respectively, during the tuber yiel formation, regardless growing season conditions.

Author Contributions: Conceptualization, I.N., V.M. and J.T.; methodology I.N., V.M., J.T. and A.Z validation, I.N. and A.Z.; investigation, I.N., V.M., J.T., A.Z., I.B., I.K. and E.M.; data curation, I.N V.M., J.T., I.K., I.B. and E.M.; writing—original draft preparation, J.T., I.K. and E.M.; writing—revie and editing, I.N., V.M., J.T. and I.K.; supervision, I.N. and A.Z.; funding acquisition, A.Z. All autho have read and agreed to the published version of the manuscript.

Funding: The research was undertaken within the framework of Russian Finnish project KS 17 "Environmentally Friendly Smart Organic Agriculture" South-East Finland-Russia CBC Programm 2014–2020.

Institutional Review Board Statement: Not applicable.

Informed Consent Statement: Not applicable.

Data Availability Statement: The data is contained within this article.

Conflicts of Interest: The authors declare no conflict of interest.

References

1. FAOSTAT. Food Balance Sheet. Available online: http://www.fao.org/faostat/en/#data/FBS (accessed on 19 February 2022)
2. *Potato Industry: Handbook*, 2nd ed.; Starovoitova, V.I. (Ed.) All-Russian Research Institute of Potato Farming, Rosselkhozakademiy Moscow, Russia, 2013; p. 272.
3. Loginov, G.A.; Fomin, I.N.; Klein, V.F.; Varlamov, A.G. *Optimization of Technical and Technological Solutions in Potato Growin* SZNIIMESH: SPb-Pavlovsk, Russia, 2009; p. 192.
4. Devaux, A.; Goffart, J.-P.; Kromann, P.; Andrade-Piedra, J.; Polar, V.; Hareau, G. The Potato of the Future: Opportunities an Challenges in Sustainable Agri-food Systems. *Potato Res.* **2021**, *64*, 681–720. [CrossRef] [PubMed]
5. Komarova, O.V.; Beresneva, R.I. The development of the organic agriculture market in the Russian Federation. *Sci. Quest. Inno Eng. Technol.* **2019**, *1*, 89–94. (In Russian)
6. Nesterenko, N.Y.; Pakhomova, N.V.; Richter, K.K. Sustainable development of organic agriculture: Strategies of Russia an its regions in context of the application of digital economy technologies. *St Petersburg Univ. J. Econ. Stud.* **2020**, *36*, 217–24 [CrossRef]
7. Shchukin, S.V.; Trufanov, A.M.; Latsko-Bartoshova, M.; Dorokhova, V.I. *The State, Prospects and Problems of the Development Organic Agriculture in Slovakia and Russia*; Innovation Management, Entrepreneurship and Sustainability (IMES): Prague, Czec Republic, 2020. [CrossRef]
8. Abbott, L.K.; Manning, D.A.C. Soil health and related ecosystem services in organic agriculture. *Sustain. Agric. Res.* **2015**, *4*, 11 [CrossRef]

Korshunov, S.A.; Lyubovedskaya, A.A. Biological Plant Protection in Compliance with the Standards of Organic Agriculture—The Necessary Directions of Research Activities. In *Biological Plant Protection Is the Basis for the Stabilization of Agroecosystems, Proceedings of the 10th International Scientific and Practical Conference, Krasnodar, Russia, 11–14 September 2018*; IOBC-EPRS: Moscow, Russia, 2018; pp. 527–532. (In Russian)
Korshunov, S.A.; Asaturova, A.M.; Khomyak, A.I.; Volkova, G.V. Formation and prospects of organic farming in Russia (review). *Potatoes Veg.* **2018**, *11*, 2–7. (In Russian) [CrossRef]
Shabanov, A.A.; Romanovsky, C.A.; Autko, A.A.; Zen, A.V.; Taranda, N.I. The Use of Biological Products in Organic Crop Production. In *Agriculture—Problems and Prospects*; Pestis, V.K., Ed.; Vikas Pub: Grodno, Belarus, 2018; pp. 140–146. (In Russian)
Syed, A.B.; Rahman, S.F.; Singh, E.; Pieterse, C.M.J.; Schenk, P.M. Emerging microbial biocontrol strategies for plant pathogens. *Plant. Sci.* **2018**, *267*, 102–111. [CrossRef]
Novikova, I.I. Polyfunctional biological products for phytosanitary optimization of agroecosystems in biological agriculture. *Tekhnologii Tekhnicheskie Sredstva Mekhanizirovannogo Proizv. Prod. Rastenievod. Zhivotnovodstva* **2019**, *2*, 183–194. (In Russian) [CrossRef]
Titova, J.A.; Krasnobaeva, I.L. Multirecycled biologics for plant protection and the possibility of their use in organic farming. *Tekhnologii Tekhnicheskie Sredstva Mekhanizirovannogo Proizv. Prod. Rastenievod. Zhivotnovodstva* **2019**, *2*, 164–183. (In Russian) [CrossRef]
Finckh, M.; Schulte-Geldermann, E.; Bruns, C. Challenges to organic potato farming: Disease and nutrient management. *Potato Res.* **2006**, *49*, 27–42. [CrossRef]
Adesemoye, A.O.; Kloepper, J.W. Plant-microbes interactions in enhanced fertilizer use efficiency. *Appl. Microbiol. Biotechnol.* **2009**, *85*, 1–12. [CrossRef]
Calvo, P.; Nelson, L.; Kloepper, J.W. Agricultural uses of plant biostimulants. *Plant Soil* **2014**, *383*, 3–41. [CrossRef]
Luna-García, H.A.; Martínez-Hernández, J.L.; Ilyina, A.; Segura-Ceniceros, E.P.; Aguilar, C.N.; Ventura-Sobrevilla, J.M.; Flores-Gallegos, A.C.; Chávez-González, M.L. Production of Unicellular Biomass as a Food Ingredient from Agro-Industrial Waste. In *Valorisation of Agro-Industrial Residues—Volume I: Biological Approaches. Applied Environmental Science and Engineering for Sustainable Future*; Zakaria, Z.A., Boopathy, R., Dib, J.R., Eds.; Springer: Berlin/Heidelberg, Germany, 2020; Volume 1, pp. 219–238. [CrossRef]
Adesemoye, T.O. *Introduction to Biological Products for Crop Production and Protection*; University of Nebraska: Lincoln, NE, USA, 2017; p. EC3019.
Treuer, T.; Choi, J.; Janzen, D.; Hallwachs, W.; Peréz-Aviles, D.; Dobson, A.P.; Powers, J.S.; Shanks, L.C.; Werden, L.K.; Wilcove, D.S. Low-cost agricultural waste accelerates tropical forest regeneration. *Restor. Ecol.* **2017**, *26*, 275–283. [CrossRef]
Pergola, M.; Persiani, A.; Palese, A.M.; Di Meo, V.; Pastore, V.; D'Adamo, C.; Celano, G. Composting: The way for sustainable agriculture. *Appl. Soil Ecol.* **2018**, *123*, 744–750. [CrossRef]
Othman, N.Z.; Sarjuni, M.N.H.; Rosli, M.A.; Nadri, M.H.; Yeng, L.H.; Ying, O.P.; Sarmidi, M.R. Spent Mushroom Substrate as Biofertilizer for Agriculture Application. In *Valorisation of Agro-Industrial Residues—Volume I: Biological Approaches. Applied Environmental Science and Engineering for Sustainable Future*; Zakaria, Z.A., Boopathy, R., Dib, J.R., Eds.; Springer: Berlin/Heidelberg, Germany, 2020; Volume 1, pp. 37–58. [CrossRef]
Siddiqui, Z.A. *PGPR: Biocontrol and Biofertilization*; Siddiqui, Z.A., Ed.; Springer: Dordrecht, The Netherlands, 2006; p. 318, ISBN 1402040024. [CrossRef]
Parnell, J.; Berka, R.; Young, H.; Sturino, J.; Kang, Y.; Barnhart, D.; Dileo, M. From the lab to the farm: An industrial perspective of plant beneficial microorganisms. *Front. Plant Sci.* **2016**, *7*, 1110. [CrossRef]
De los Santos-Villalobos, S.; Díaz-Rodríguez, A.M.; Ávila-Mascareño, M.F.; Martínez-Vidales, A.D.; Parra-Cota, F.I. COLMENA: A culture collection of native microorganisms for harnessing the agro-biotechnological potential in soils and contributing to food security. *Diversity* **2021**, *13*, 337. [CrossRef]
Rana, K.L.; Kour, D.; Nath, A.; Yadav, A.N.; Yadav, N.; Saxena, A.K. Agriculturally Important Microbial Biofilms: Biodiversity, Ecological Significances, and Biotechnological Applications. In *New and Future Developments in Microbial Biotechnology and Bioengineering. Microbial Biofilms: Current Research and Future Trends*; Yadav, M.K., Singh, B.P., Eds.; Elsevier Inc.: Amsterdam, The Netherlands, 2020; pp. 221–265. [CrossRef]
De los Santos-Villalobos, S.; Parra-Cota, F.I. Current trends in plant growth-promoting microorganisms research for sustainable food security. *Curr. Res. Microb. Sci.* **2020**, *2*, 100016. [CrossRef]
Mohanram, S.; Kumar, P. Rhizosphere microbiome: Revisiting the synergy of plant-microbe interactions. *Ann. Microbiol.* **2019**, *69*, 307–320. [CrossRef]
Gupta, T.; Chakraborty, D.; Sarkar, A. Structural and Functional Rhizospheric Microbial Diversity Analysis by Cutting-Edge Biotechnological Tools. In *Omics Science for Rhizosphere Biology*; Pudake, R.N., Sahu, B.B., Kumari, M., Sharma, A.K., Eds.; Springer Nature: Singapore, 2021. [CrossRef]
Salwan, R.; Sharma, V.; Saini, R.; Pandey, M. Identification of plant beneficial *Bacillus* spp. for Resilient agricultural ecosystem. *Curr. Res. Microb. Sci.* **2021**, *2*, 100046. [CrossRef]
He, D.-C.; He, M.-H.; Amalin, D.M.; Liu, W.; Alvindia, D.G.; Zhan, J. Biological control of plant diseases: An evolutionary and eco-economic consideration. *Pathogens* **2021**, *10*, 1311. [CrossRef]
Wasai, S.; Minamisawa, K. Plant-associated microbes: From rhizobia to plant microbiomes. *Microbes Environ.* **2018**, *33*, 1–3. [CrossRef] [PubMed]

33. Khan, N.; Bano, A.; Ali, S.; Babar, M.A. Crosstalk amongst phytohormones from planta and PGPR under biotic and abic stresses. *Plant Growth Regul.* **2020**, *90*, 189–203. [CrossRef]
34. Glick, B.R.; Gamalaro, E. Recent developments in the study of plant microbiomes. *Microorganisms* **2021**, *9*, 1533. [CrossR [PubMed]
35. Haroon, U.; Khizar, M.; Liaquat, F.; Ali, M.; Akbar, M.; Tahir, K.; Batool, S.S.; Kamal, A.; Chaudhary, H.J.; Munis, M.F Halotolerant plant growth-promoting rhizobacteria induce salinity tolerance in wheat by enhancing the expression of SOS ge *J. Plant Growth Regul.* **2021**, *27*, 1–4. [CrossRef]
36. Nivetha, N.; Lavanya, A.K.; Vikram, K.V.; Asha, A.D.; Sruthi, K.S.; Bandeppa, S.; Annapurna, K.; Paul, S. PGPR-Media Regulation of Antioxidants: Prospects for Abiotic Stress Management in Plants. In *Antioxidants in Plant-Microbe Interaction*; Sin H.B., Vaishnav, A., Sayyed, R., Eds.; Springer Nature: Singapore, 2021. [CrossRef]
37. Lengai, G.; Muthomi, J. Biopesticides and their role in sustainable agricultural production. *J. Biosci. Med.* **2018**, *6*, 7–41. [CrossR
38. Zhang, J.; Cook, J.; Nearing, J.T.; Zhang, J.; Raudonis, R.; Glick, B.R.; Langille, M.G.I.; Cheng, Z. Harnessing the plant microbio to promote the growth of agricultural crops. *Microbiol. Res.* **2021**, *245*, 126690. [CrossRef]
39. Pavlyusin, V.A.; Novikova, I.I.; Boikova, I.V. Microbiological control in phytosanitary optimization technologies for agroecos tems: Research and practice (review). *Agric. Biol.* **2020**, *55*, 421–438. [CrossRef]
40. Köhl, J.; Kolnaar, R.; Ravensberg, W.J. Mode of action of microbial biological control agents against plant diseases: Relevar beyond efficacy. *Front. Plant Sci.* **2019**, *10*, 845. [CrossRef]
41. Sasirekha, B.; Srividya, S. Siderophore production by Pseudomonas aeruginosa FP6, a biocontrol strain for Rhizoctonia solani a Colletotrichum gloeosporioides causing diseases in chilli. *Agric. Nat. Resour.* **2016**, *50*, 250–256. [CrossRef]
42. Ghazy, N.; El-Nahrawy, S. Siderophore production by Bacillus subtilis MF497446 and Pseudomonas koreensis MG209738 a their efficacy in controlling Cephalosporium maydis in maize plant. *Arch. Microbiol.* **2021**, *203*, 1195–1209. [CrossRef]
43. Santoyo, G.; Orozco-Mosqueda, M.C.; Govindappa, M. Mechanisms of biocontrol and plant growth-promoting activity in s bacterial species of *Bacillus* and *Pseudomonas*: A review. *Biocontr. Sci. Technol.* **2012**, *22*, 855–872. [CrossRef]
44. Yu, C.; Fan, L.; Gao, J.; Wang, M.; Wu, Q.; Tang, J.; Li, Y.; Chan, J. The platelet activating factor acetyl hydrolase gene derived fr *Trichoderma harzianum* induces maize resistance to *Curvularia lunata* through the jasmonic acid signaling pathway. *J. Environ. Health* **2015**, *50*, 708–717. [CrossRef]
45. Singh, A.; Chauhan, P.S. Ecological Significance of Soil-Associated Plant Growth-Promoting Biofilm-Forming Microbes for Str Management. In *Biofilms in Plant and Soil Health*; Ahmad, I., Husain, F.M., Eds.; John Wiley & Sons: Hoboken, NJ, USA, 2C Volume 14, p. 291. [CrossRef]
46. Sehrawat, A.; Sindhu, S.S. Potential of biocontrol agents in plants disease control for improving food safety. *Def. Life Sci. J.* **2019** 220–225. [CrossRef]
47. Jiao, X.; Takishita, Y.; Zhou, G.; Smith, D.L. Plant associated rhizobacteria for biocontrol and plant growth enhancement. *Fr Plant Sci.* **2021**, *12*, 1–8. [CrossRef]
48. Zehra, A.; Raytekar, N.A.; Meena, M.; Swapnil, P. Efficiency of microbial bio-agents as elicitors in plant defense mechanism un biotic stress: A review. *Curr. Res. Microb. Sci.* **2021**, *2*, 1–14. [CrossRef]
49. Wang, Y.; Liu, H.; Shen, Z.; Miao, Y.; Wang, J.; Jiang, X.; Shen, Q.; Li, R. Richness and antagonistic effects co-affect plant grov promotion by synthetic microbial consortia. *Appl. Soil Ecol.* **2022**, *170*, 104300. [CrossRef]
50. Sendi, Y.; Pfeiffer, T.; Koch, E.; Mhadhbi, H.; Mrabet, M. Potential of common bean (*Phaseolus vulgaris* L.) root microbiome in biocontrol of root rot disease and traits of performance. *J. Plant Dis. Prot.* **2020**, *127*, 453–462. [CrossRef]
51. Mohanty, P.; Singh, P.K.; Chakraborty, D.; Mishra, S.; Pattnaik, R. Insight into the role of PGPR in sustainable agriculture a environment. *Front. Sustain. Food Syst.* **2021**, *5*, 667150. [CrossRef]
52. Pathak, E.; Sanjyal, A.; Regmi, C.R.; Paudel, S.; Shrestha, A. Screening of potential plant growth promoting properties of Bacil species isolated from different regions of Nepal. *Nepal J. Biotechnol.* **2021**, *9*, 79–84. [CrossRef]
53. Sehrawat, A.; Sindhu, S.S.; Glick, B.R. Hydrogen cyanide production by soil bacteria: Biological control of pests and promotior plant growth in sustainable agriculture. *Pedosphere* **2022**, *32*, 15–38. [CrossRef]
54. Nugmanova, T.A. Biopreparations in vegetable and potato growing. *Potatoes Veg.* **2017**, *6*, 2–6.
55. Chakraborty, T.; Akhtar, N. Biofertilizers: Prospects and Challenges for Future. In *Biofertilizers*; Wiley: Hoboken, NJ, USA, 2C pp. 575–590. [CrossRef]
56. Plekhanova, L.P.; Buldakov, S.A. Effectiveness of action of biological preparation and fungicidal agents against plant disea potatoes tuber and their influence on productivity of land. *Int. Res. J.* **2019**, *87*, 28–33. [CrossRef]
57. Gaynatulina, V.V.; Khasbiullina, O.I. Efficiency of application of biopreparations and fungicides in potato rhizoctoniosis cont *Vestn. Far East. Branch Russ. Acad. Sci.* **2020**, *4*, 93–99. [CrossRef]
58. Anand, A.; Chinchilla, D.; Tan, C.; Mene-Sarane, L.; L'Haridon, F.; Weisskopf, L. Contribution of hydrogen cyanide to antagonistic activity of *Pseudomonas* strains against *Phytophthora Infestans*. *Microorganisms* **2020**, *8*, 1144. [CrossRef]
59. Ali, A.A.; Awad, M.Y.M.; Hegab, S.A.; Abd El Gawad, A.M.; Eissa, M.A. Effect of potassium solubilizing bacteria (*Bacillus cer* on growth and yield of potato. *J. Plant. Nutr.* **2021**, *44*, 411–420. [CrossRef]
60. Fasusi, O.A.; Cruz, C.; Babalola, O.O. Agricultural sustainability: Microbial biofertilizers in rhizosphere management. *Agricult* **2021**, *11*, 163. [CrossRef]

Kumar, S.; Diksha; Sindhu, S.S.; Kumar, R. Biofertilizers: An ecofriendly technology for nutrient recycling and environmental sustainability. *Curr. Res. Microb. Sci.* **2022**, *3*, 1–26. [CrossRef]

Valenzuela-Aragón, B.; Parra-Cota, F.I.; Santoyo, G.; Arellano-Wattenbarger, G.L.; de los Santos-Villalobos, S. Plant-assisted selection: A promising alternative for in vivo identification of wheat (*Triticum turgidum* L. subsp. *durum*) growth promoting bacteria. *Plant Soil* **2019**, *435*, 367–384. [CrossRef]

Basu, A.; Prasad, P.; Das, S.N.; Kalam, S.; Sayyed, R.Z.; Reddy, M.S.; El Enshasy, H. Plant growth promoting rhizobacteria (PGPR) as green bioinoculants: Recent developments, constraints, and prospects. *Sustainability* **2021**, *13*, 1140. [CrossRef]

Etesami, H.; Jeong, B.R.; Glick, B.R. Contribution of arbuscular mycorrhizal fungi, phosphate–solubilizing bacteria and silicon to P uptake by plant. *Front. Plant Sci.* **2021**, *12*, 699618. [CrossRef]

Wang, Y.; Peng, S.; Hua, Q.; Qiu, C.; Wu, P.; Liu, X.; Lin, X. The long-term effects of using phosphate-solubilizing bacteria and photosynthetic bacteria as biofertilizers on peanut yield and soil bacteria community. *Front. Microbiol.* **2021**, *12*, 693535. [CrossRef] [PubMed]

Santoyo, G.; Guzmán-Guzmán, P.; Parra-Cota, F.I.; de los Santos-Villalobos, S.; Orozco-Mosqueda, M.D.C.; Glick, B.R. Plant Growth Stimulation by Microbial Consortia. *Agronomy* **2021**, *11*, 219. [CrossRef]

Bryukhanov, A.Y.; Vasilev, E.V.; Shalavina, E.V.; Uvarov, R.A. Methods for solving environmental problems in animal and poultry farming. *Sel'skokhozyaystvennye Mashiny Tekhnologii* **2019**, *13*, 32–37. (In Russian) [CrossRef]

Qu, G.; Cai, Y.; Lv, P.; Ma, X.; Xie, R.; Xu, Y.; Ning, P. Effect of EM microbial agent on aerobic composting for dairy cattle manure. *Int. J. Environ. Sci. Technol.* **2019**, *16*, 6945–6958. [CrossRef]

Shan, G.C.; Li, W.G.; Gao, Y.J.; Tan, W.B.; Xi, B.D. Additives for reducing nitrogen loss during composting: A review. *J. Clean. Prod.* **2021**, *307*, 127308. [CrossRef]

Bryukhanov, A.Y.; Subbotin, I.A.; Uvarov, R.A.; Vasilev, E.V. Method of designing of manure utilization technology. *Agron. Res.* **2017**, *15*, 658–663.

Uvarov, R.A.; Shalavina, E.V.; Vasilev, E.V. Manure utilisation technologies in the Baltic Sea Region: Analysis and emerging trends. *AgroEcoEngineeriya* **2021**, *3*, 117–128. (In Russian)

Titova, J.A. Spent Mushroom Substrates Valorization via Brand New Multirecycled Polyfunctional Biologics Producing on Them. *Waste Biomass Valor.* **2021**, *13*, 1089–1100. [CrossRef]

Experimental Mycology Methods: A Handbook; Bilai, V.N. (Ed.) Nauk. Dumka: Kiev, Ukraine, 1982. (In Russian)

Shabanov, A.E.; Zhevora, S.V.; Anisimov, B.V.; Kiselev, A.I.; Dolgova, T.I.; Malyuta, O.V.; Zebrin, S.N. *Parameters of Potential Yield of Potato Varieties of the Selection Center VNIIKH, (Reference)*; FSBIU VNIIKH: Moscow, Russia, 2016; p. 13. (In Russian)

Methodological Guidelines for Registration Testing of Fungicides in Agriculture; FSBSI VIZR: Saint-Petersburg, Russia, 2009. (In Russian)

Guidelines for Registration Trials of Plant Growth Regulators, Defoliant Agents and Desiccants in Agriculture: Production and Practical Publication; FSBSI Rosinformagrotech: Moscow, Russia, 2016. (In Russian)

Methodological Guidelines for Registration Tests of Pesticides in Terms of Biological Efficacy; FSBSI Rosinformagrotech: Moscow, Russia, 2019. (In Russian)

Minin, V.B.; Popov, V.D.; Maksimov, D.A.; Ustroev, A.A.; Papushin, E.; Melnikov, S.P. Developing of modern cultivation technology of organic potatoes. *Agron. Res.* **2020**, *18*, 1359–1367. [CrossRef]

Peacock, J.L.; Peacock, P.J. *Oxford Handbook of Medical Statistics*; Oxford University Press: Oxford, UK, 2011. [CrossRef]

Petrie, A.; Sabin, C. *Medical Statistics at a Glance*; Wiley Blackwell: Hoboken, NJ, USA, 2019; p. 139.

Tojo, S.; Tanaka, Y.; Ochi, K. Activation of Antibiotic Production in *Bacillus spp.* by Cumulative Drug Resistance Mutations. *Antimicrob. Agents Chemother.* **2015**, *59*, 7799–7804. [CrossRef]

Hamdache, A.; Lamarti, A.; Aleu, J.; Collado, I.G. Non-peptide metabolites from the genus *Bacillus*. *J. Nat. Prod.* **2011**, *74*, 893–899. [CrossRef]

Quardros, C.P.; Duarte, M.C.T.; Pastore, G.M. Biological activities of a mixture of biosurfactant from *Bacillus subtilis* and alkaline lipase from *Fusarium oxysporum*. *Braz. J. Microbiol.* **2011**, *42*, 354–361. [CrossRef]

Dodd, I.C.; Zinovkina, N.Y.; Safronova, V.I.; Belimov, A.A. Rhizobacterial mediation of plant hormone status. *Ann. Appl. Biol.* **2010**, *157*, 361–379. [CrossRef]

Kudoyarova, G.R.; Melentiev, A.; Martynenko, E.; Timergalina, L.; Arkhipova, T.; Shendel, G.; Kuz'mina, L.; Dodd, I.; Veselov, S.Y. Cytokinin producing bacteria stimulate amino acid deposition by wheat roots. *Plant Physiol. Biochem.* **2014**, *83*, 285–291. [CrossRef]

Kilian, M.; Steiner, U.; Krebs, B.; Junge, H.; Schmeiedeknech, T.G.; Hain, R. FZ B 24® *Bacillus subtilis*—Mode of action of microbial agent enhancing plant vitality. *Pflanzenschutz-Nachr. Bayer.* **2000**, *1*, 72–93.

Porcel, R.; Zamarreño, A.M.; Garna-Mina, J.M.; Aroca, R. Involvement of plant endogenous ABA in *Bacillus megaterium* PGPR activity in tomato plants. *BMC Plant Biol.* **2014**, *14*, 36. [CrossRef]

Belimov, A.A.; Dodd, I.C.; Safronova, V.I.; Dumova, V.A.; Shaposhnikov, A.I.; Ladatko, A.G.; Davies, W.J. Abscisic acid metabolizing rhizobacteria decrease ABA concentrations in planta and alter plant growth. *Plant Physiol. Biochem.* **2014**, *74*, 84–91. [CrossRef]

Belyuchenko, I.S. Complex compost as an important source of enrichment the soil cover with nutrients. *Sci. J. KubGAU* **2014**, *97*, 1–21. (In Russian)

90. Belyuchenko, I.S. The role of living organisms in the development of complex compost. *Sci. J. KubGAU* **2014**, *96*, 232–24 (In Russian)
91. Petukhov, A.V. Potato Diseases and Measures to Control Them in the North Amur Region and Buryatia Conditions. Ph.D. Thes Novosibirsk State University, Novosibirsk, Russia, 2002. (In Russian).

sustainability

MDPI

·icle

omparison of Vegetables of Ecological and Commercial roduction: Physicochemical and Antioxidant Properties

cnicté Olguín-Hernández [1], Quinatzin Yadira Zafra-Rojas [1,*], Nelly del Socorro Cruz-Cansino [1], ·e Alberto Ariza-Ortega [1], Javier Añorve-Morga [2], Deyanira Ojeda-Ramírez [3], Reyna Nallely Falfan-Cortes [2], ·e Arias-Rico [4] and Esther Ramírez-Moreno [1,*]

[1] Área Académica de Nutrición, Centro de Investigación Interdisciplinario, Instituto de Ciencias de la Salud, Circuito Actopan-Tilcuautla s/n. Ex-hacienda La Concepción, San Agustín Tlaxiaca 42160, Hidalgo, Mexico; olguinh@uaeh.edu.mx (Z.O.-H.)

[2] Área Académica de Química, Instituto de Ciencias Básicas e Ingeniería, Ciudad del Conocimiento, Carretera Pachuca-Tulancingo Km. 4.5 Col. Carboneras, Mineral de la Reforma 42184, Hidalgo, Mexico

[3] Área Académica de Medicina Veterinaria y Zootecnia, Instituto de Ciencias Agropecuarias, Universidad Autónoma del Estado de Hidalgo, Av. Universidad Km 1, Ex-Hda. de Aquetzalpa, Tulancingo 43600, Hidalgo, Mexico

[4] Área Académica de Enfermería, Centro de Investigación Interdisciplinario, Instituto de Ciencias de la Salud, Circuito Actopan-Tilcuautla s/n. Ex-hacienda La Concepción, San Agustín Tlaxiaca 42160, Hidalgo, Mexico

* Correspondence: quinatzin_zafra@uaeh.edu.mx (Q.Y.Z.-R.); esther_ramirez@uaeh.edu.mx (E.R.-M.)

tion: Olguín-Hernández, Z.; ·a-Rojas, Q.Y.; Cruz-Cansino, S.; Ariza-Ortega, J.A.; ·rve-Morga, J.; Ojeda-Ramírez, D.; ·an-Cortes, R.N.; Arias-Rico, J.; ·írez-Moreno, E. Comparison of ·tables of Ecological and ·mercial Production: ·sicochemical and Antioxidant ·erties. *Sustainability* **2023**, *15*, ·. https://doi.org/10.3390/ ·065117

·demic Editors: Miklas Scholz, ·o A. Pagnotta, Panayiotis ·itrakopoulos and ·iya Noorani

·ived: 7 February 2023
·ised: 7 March 2023
·pted: 9 March 2023
·ished: 14 March 2023

Abstract: This research aimed to compare some physicochemical and antioxidant properties in vegetables (chard, beet, coriander, spinach, lettuce, radish, carrot, and tomato) of ecological and commercial production. The ecological products were cultivated and obtained from three harvests in an ecology garden with standardized methodologies for implementation while the commercial samples were obtained from a local supplier. On the same purchase or harvest day, the color, texture, moisture, and ashes parameters were measured in the fresh produce without unpeeling. In the lyophilized samples, bioactive compounds (total phenolic compounds, ascorbic acid, chlorophyll *a* and *b*, β-carotenes, anthocyanins, betalains, and lycopene) were determined, and antioxidant activity was found using the 2,2-azino-bis-3-ethylbenzothiazoline-6-sulfonic acid ($ABTS^{\bullet+}$), 2,2′-diphenyl-1-picrylhydrazyl (DPPH) and ferric-reducing antioxidant power (FRAP) assays, and chelating activity. The ecological vegetables presented better color (high luminosity and intensity) than commercial samples, and, according to the value of ΔE, this is a difference that can be perceived by the human eye. In the same way, the ecological vegetables were more turgid than the commercial samples ($p < 0.05$). The content of bioactive compounds was found in higher concentrations in ecologically produced vegetables and this was correlated positively with antioxidant capacity. It is important to carry out more studies to determine the effect on health of these vegetables when they are integrated into the diet and thus to be able to recommend their inclusion in the diet as a sustainability strategy in the production of vegetables for self-consumption.

Keywords: ecological vegetables; phenolic; bioactive compounds; antioxidant activity

1. Introduction

A food security system is based on economic, social, and environmental factors to provide safe food for the entire population. In addition, it is completely profitable, with benefits for society, and has a positive or neutral impact on the environment (economic, social, and environmental sustainability) [1]. Therefore, a food security system requires carefully diversified food sources in terms of land (diversification of crops, organic fertilizers) and environmental sustainability (production of food in a way that does not affect the environment) to satisfy the nutritional requirements of the people [2]. The consumption of vegetable food can play an important role in improving the nutritional status of the population and preventing non-communicable and deficiency diseases, among oth due to its well-recognized nutritional and medicinal value [3,4]. Starting in the year 20 certification processes for food production were established in the United States w methodologies that respect ecosystems and that do not use fertilizers or pesticides t compromise the environment; this process is called "organic production" which imp administrative procedures and an economical cost that the producer absorbs [5]. Co quently, the concept of organic farming was implemented in agricultural systems cente on the need to develop technologies and practices that do not have adverse effects the environment (goods and services), but that are accessible and effective for farm which leads to improvements in food productivity. In this process, it may be underst that everyone follows the standards of an organic system but with the term of "ecologi production" [6]. This ecological process does not have a certification; it is only validated the ethical practices of the responsible producer [7].

A sustainability strategy is the production of food for self-consumption, using verti orchards, green roofs, and ecological school gardens when feasible; vegetables for frequ consumption can be produced in them, allowing access to these vegetable products the time [8,9]. These strategies take care of the environment, avoid food waste thro compost production, contribute to the economy of families, and provide better food qua (higher bioactive compounds, dietary fiber, minerals, and others) [10].

This food production system turns out to be more attractive to consumers [11]. can also contribute to the diminished food supply chain and storage time, and, theref preserve the hygienic and nutritional qualities of the vegetables, including those that perceived by the senses (taste, smell, color, texture, shape, and appearance) related freshness and better flavor [12,13].

The physical characteristics of vegetables are the result of the chemical properties the plant and are associated with the stability of their cell membranes, the distributi of moisture in the tissues, and the synthesis and integrity of bioactive compounds su as phenolic compounds, vitamins, minerals, and pigments [14]. These have been ass ated with biological functions in the human body that favor health effects, especially antioxidants [15].

Consequently, vegetables that are produced using organic methods may have hig concentrations of these compounds [16]. The synthesis of these compounds occurs w the plants are exposed to controlled stress conditions (as occurs in organic products wh mineralization or inorganic fertilization process is not carried out) [17,18]. So, the veg bles that are produced in this way may have a greater presence of these compounds regional production. As a consequence, these protection mechanisms promote an incre in bioactive compounds during plant growth in ecological production systems. If the h dling and storage conditions (time, humidity, temperature, and pH) are controlled duri postharvest, the tissues of the plants are affected, which will modify the concentration the bioactive compounds due to losses or modification of their natural form. For this reas variations in production systems and storage conditions could explain the differences pigment concentration and humidity between ecological and commercial vegetables [1

Considering how production systems influence the physical and chemical qua of vegetables, and the little information there is on the quality of vegetables that produced in an ecological system, this study aims to compare some physical propert (color and texture) and antioxidant properties in vegetables in an ecological garden aga commercial products.

2. Materials and Methods

2.1. Obtaining Vegetable Material

The commercial samples were obtained from a local supplier of Pachuca Hidal México, choosing the color and texture by considering commercial maturity without erations or external lesions, as would be selected by a local consumer. The ecologi production was obtained in the same city, with a dry temperate climate and an ann average of 15 °C with winds blowing from the northeast throughout the year [20]. It was carried out as follows: Homemade compost was generated, which was left to mature; the sowing substrate prepared with compost and plant soil (50:50) was placed in wooden containers, and 6 plants per container were sown; irrigation was performed as needed. Pest management was done culturally (garlic, chili and pepper as natural insecticides for aphids and whiteflies). The harvest was carried out according to the maturity of the vegetable. For the analysis, greens from 3 harvests in different ecological gardens were used. The analyzed species were chard (*Beta vulgaris var. cicla*), beetroot (*Beta vulgaris*), coriander (*Coriandrum sativum*), spinach (*Spinacia oleracea*), lettuce (*Lactuca sativa*), radish (*Raphanus sativus*), carrot (*Daucus carota sativus*), and tomato (*Solanum*). These vegetables were analyzed between September 2021 and February 2022 based on their natural maturity, and all the plant materials were obtained according to their similar sensory characteristics (appearance, color, and texture), including the commercial samples. All the samples were carried to the laboratory in a thermal container. The color, texture, and moisture parameters were measured in the fresh produce without unpeeling the same purchase or harvest day. In another part of the vegetables, the edible parts of the vegetables (20–50 g) were cut into small pieces (<0.5 cm) and lyophilized (LABCONO WWR26671-581, Kansas City, MO, USA) at −52 °C for 48 to 72 h. Lyophilisates samples were ground (analytical mill A11 2,900,001, EE. UU.), homogenized, and sieved into particles less than 500 μm. The powder was stored in a resealable plastic bag until analysis.

2.2. Physical Properties

2.2.1. Color Measurement

The color was recorded with a Chroma meter CR-300 series (CE Minolta, Osaka, Japan). In the leaves, color was measured at three points and on the back. In the vegetables of beet, tomato, radish, and carrot, two external points and one internal were taken. The CIE-Lab parameters were determined by luminosity (L^*), a^*, and b^* coordinates.

The data of L^*, a^*, and b^* were converted to color difference (ΔE) using the following equation [21]:

$$\Delta E = \sqrt{(\Delta L^*)^2 + (\Delta a^*)^2 + (\Delta b^*)^2} \quad (1)$$

ΔE: Color difference;
$\Delta L^{*2} = (L^*_1 - L^*_2)^2$ where L^*_1 is commercial vegetable and L^*_2 is ecological vegetable
$\Delta a^{*2} = (a^*_1 - a^*_2)^2$ where a^*_1 is commercial vegetable and a^*_2 is ecological vegetable
$\Delta b^{*2} = (b^*_1 - b^*_2)^2$ where b^*_1 is commercial vegetable and b^*_2 ecological vegetable

2.2.2. Texture, Moisture, and Ashes

The parameters of texture to assess turgor in vegetables were determined using a texture analyzer (Texture Analyzer, TA Plus, Stable Microsystems Co., Surrey, UK). A Kramer shear cell with a five-blade probe was used to analyze leafy vegetables, performed in 10 replicates per sample (placing 20 g of chopped vegetables in a load cell for a Kramer probe with 5 cutting blades). For the carrot, radish, beet, and tomato, a puncture probe P/2N was used to assess the skin breaking strength and the pulp firmness. The test was performed at a 10 mm distance using a test speed of 2 mm/s of force. The sample was analyzed and reported as maximum force (N/g) [22].

The moisture was determined according to AOAC procedure 945.15 by desiccation at 105 °C until constant weight and the ashes were obtained according to AOAC procedure 962.09.

2.3. Bioactive Compounds Analysis

2.3.1. Extraction of Antioxidant Compounds

To carry out the extraction process, 250 mg of lyophilized samples was weighed and extracted by shaking at room temperature with 50 mL of a mixture of methanol–water (50:50 *v*/*v*, 60 min, room temperature; constant shaking) and 50 mL of mixture of acetone–

water (70:30 v/v, 60 min, room temperature; constant shaking) [23]. After centrifugatio (15 min, 25 °C, 3000 rpm), the supernatants were brought to a total volume of to 25 mL, wi a mixture (1:1) of the solution of solvents and used to determine total phenolic compound ascorbic acid, anthocyanin, betalains, and antioxidant activity ($ABTS^{\bullet+}$, FRAP, DPPH, ar chelating activity of ferrous ions).

2.3.2. Total Phenolic Compounds

The total phenolic content was estimated based on the Folin–Ciocalteau procedure [2 An aliquot of the extract solution (100 µL) was mixed with 0.5 mL of Folin–Ciocalteu reage (2.5 mL, previously diluted with water 1:10 v/v) and sodium carbonate (75 g/L). Aft adding 400 µL of sodium carbonate (7.5%), samples were allowed to stand for 30 mi at room temperature, and the absorbance was read at 765 nm using a microplate read (Power Wave XS UV-Biotek, software KC Junior, Winooski, VT, USA). The gallic acid wa used as a reference standard, and the results were expressed as milligrams of gallic aci equivalents per 100 g of dry weight (mg GAE/100 g dw).

2.3.3. Ascorbic Acid

The ascorbic acid content in the vegetable extract was determined according to Dürü et al. [25]. The sample was diluted 1:10 in a solution of 0.4% oxalic acid. Briefly, 100 µL of th extract was mixed with 100 µL of acetate buffer and 800 µL of 2,6-dichlorophenolindophen (DCPI). The absorbance of the mixture was measured at 520 nm in the microplate read (Power Wave XS UV-Biotek, software KC Junior, USA), and ascorbic acid was used as reference standard, and the results were expressed as mg ascorbic acid per 100 g of dr weight (mg AA/100 g dw).

2.3.4. Total Anthocyanins Content (TAC)

The spectrophotometric pH differential method mentioned in the literature [26] wa employed to measure the total anthocyanin content (TAC) in the extracts. Two dilutions of the same sample were made in 0.025 M potassium chloride pH 1.0 and 0.4 M sodium aceta pH 4.5. The pH was adjusted by adding concentrated HCl respectively. The absorbance of each dilution was measured a 520 and 700 nm using a Power Wave XS UV-Biotek, softwa KC Junior, USA. The total anthocyanin content (TAC) was determined as follows:

$$Abs = (Abs_{510} - Abs_{700})pH_{1.0} - (Abs_{510} - Abs_{700})pH_{4.5}$$

$$TAC\left(\frac{\text{mg}}{\text{L}}\right) = \frac{Abs * Mw * DF * 1000}{\varepsilon * 0.52} \quad (\text{:}$$

where:

Abs: Absorbance

Mw: Molecular weight of cyanidin-3-O-glucoside (449.2 g/mol).

DF: Dilution factor of the sample.

ε: Molar extinction coefficient of cyanidin-3-O-glucoside 26,900 L mol^{-1} cm^{-1}.

0.52 cm: Path length

2.3.5. Total Betalain Content (TBC)

The total betalain content (TBC) of the betacyanins and betaxanthins content wa determined according to the method described by Fernández-López et al. [27] with som adaptations. An aliquot of 100 µL of the extract was determined by spectrophotomet UV/VIS (Power Wave XS UV-Biotek, software KC Junior, USA) at 480 nm for betacyani and 535 nm for betaxanthin [28]. The betalain content was determined as follows and th results were expressed in mg.

$$TBC\left(\frac{\text{mg}}{\text{g}}\right) = \frac{A * FD * MM * 1000}{E * 0.29} \quad (\text{:}$$

A: Absorbance
FD: Dilution factor
0.29 mL: Tucket length
E: Molar extinction coefficient
The following values were used:
MM = 550 g/mol, ε = 60.0 L/mol cm in H_2O for betacyanin determination
MM = 308 g/mol, ε = 48.0 g/mol cm of H_2O for betaxanthin determination

2.3.6. β-Carotene, Chlorophylls, and Lycopene
Extraction Method

To the determination of pigments, 500 mg of the lyophilized sample was homogenized with 10 mL of an acetone–hexane (2:3) mixture for 2 min. Immediately, the samples were sonicated (sonicator Bandelin HD3100 Sonopuls, Hamburg, Germany) for 3 min (5 cycles: Puls 30 s, pause 10 s). Homogenates were centrifuged (Eppendorf Hamburg, Germany) at 5000 rpm for 10 min at 20 °C. The sample was frozen until used and it was maintained in an ice-water bath to prevent overheating [29].

The pigments were determined according to the method of Nagata and Yamashita [29]. The absorbance spectrum of each supernatant was measured and the absorption maxima were read at 453, 505, 645, and 663 nm (Power Wave XS UV-Biotek, software KC Junior, EE. UU.). The β-carotene and chlorophyll (Chl *a*, Chl *b*) content were calculated (Microsoft Office Excel 97-2003) from the following equations:

$$\beta - carotene\left(\frac{\text{mg}}{100\ \text{mL}}\right) = (0.216 * A663) - (1.22 * A645) - (0.304 * A505) + (0.452 * A453) \quad (4)$$

$$Chlorophyll\ a\left(\frac{\text{mg}}{100\ \text{mL}}\right) = (0.999 * A663) - (0.0989 * A645) \quad (5)$$

$$Chlorophyll\ b\left(\frac{\text{mg}}{100\ \text{mL}}\right) = (0.328 * A663) + (1.77 * A645) \quad (6)$$

The lycopene was determined according to the method of Nagata and Yamashita [30]. The determination of lycopene was performed only on tomato samples. The absorbance spectrum of the extract of each supernatant was measured, and the absorption maxima was read at 453, 505, 645, and 663 nm (Power Wave XS UV-Biotek, software KC Junior, EE. UU.). The lycopene content was calculated from the following equation:

$$Lycopene\left(\frac{\text{mg}}{100\ \text{mL}}\right) = (0.0485 * A663) - (0.204 * A645) + (0.372 * A505) - (0.0806 * A453) \quad (7)$$

2.4. *Antioxidant Activity*

2.4.1. 2,2-Azino-bis-3-ethylbenzothiazoline-6-sulfonic Acid ($ABTS^{\bullet+}$)

The antiradical capacity was measured according to Kuskoski et al. [31]. Briefly, the radical cation ($ABTS^{\bullet+}$) was produced by reacting 7 mmol/L $ABTS^{\bullet+}$ stock solution with 2.45 mmol/L potassium persulfate under dark conditions and room temperature for 16 h before use. The $ABTS^{\bullet+}$ solution was diluted with deionized water to an absorbance of 0.70 ± 0.10 at 754 nm. Then, 20 μL of the extract of the sample was added to 980 μL of diluted $ABTS^{\bullet+}$ solution; absorbance readings (754 nm) were taken in a microplate reader (Power Wave XS UV-Biotek, software KC Junior, EE. UU.) after incubation for 7 min at room temperature. The antioxidant capacity was expressed as micromole of Trolox equivalents per 100 g of dry weight (μmol TE/100 g dw).

2.4.2. 2,2′-Diphenyl-1-picrylhydrazyl (DPPH)

The antiradical activity was measured with 1,1-diphenyl-2-picrylhydrazylradical (DPPH) [30]. An ethanolic solution (7.4 mg/100 mL) of the stable DPPH radical was prepared. Then, the extracts (100 μL) were taken into vials and 500 μL of DPPH solution were added, and the mixture was for to stand 1 h at room temperature. Finally, absorbance was measured at 520 nm using a microplate reader (Power Wave XS UV-Biotek, software

KC Junior, Winooski, VT, USA). The standard curve was concentrations of 0, 50, 100, 2 and 300 µmol of Trolox. Free radical scavenging activity was expressed as micromole Trolox equivalents per 100 g on a dry weight (µmol TE/100 g dw).

2.4.3. Ferric-Reducing Antioxidant Power (FRAP)

The FRAP method was assayed according to Benzie and Strain [31]. FRAP soluti with 2,4,6-tripyridyl-s-triazine (TPTZ) 10 mM solution in 40 mM HCl, $FeCl_3$ 20 mM, a acetate buffer 300 mM (sodium acetate anhydrous and glacial acetic acid, pH 3.6). aliquot of 30 µL of the sample was mixed with 90 µL of distilled water and 900 µL of FR solution in the dark room. Absorbance was measured at 593 nm using a microplate rea (Power Wave XS UV-Biotek, software KC Junior, EE. UU). The results were express as micromoles of Fe (II) per 100 g of dry weight (µmol Fe (II)/100 g dw) of each one the vegetables.

2.4.4. The Chelating Activity of Ferrous Ions

The chelating activity (CA) was determined as described by Gulcin et al. [32]. Brie 100 µL of the undiluted sample was placed in vials and 50 µL of ferric (II) chloride soluti (2 mM) and 450 µL of methanol were added. The solution was vortexed and left for 5 mir room temperature before adding 400 µL of ferrozine (5 mM). The mixture was vortexed a then allowed to settle for 10 min at room temperature. The absorbance was read at 562 in a microplate reader (Power Wave XS UV-Biotek, software KC Junior, EE. UU.) usi EDTA (0.1 M) [33], and deionized water as control. The chelating activity was calcula using the following equation:

$$\%\, CA \frac{A0 - A1}{A0} * 100$$

$A0$: Absorbance of the control;
$A1$: Absorbance of the sample.

2.5. Statistical Analysis

All data were reported as mean ± standard deviation for at least three repetitions each treatment. To determine differences between samples of ecological and commerc vegetables to levels of statistical significance a Student *t*-test was conducted. In bc statistical tests, the significance level was set at $p = 0.05$. The analysis of the correlati between the results of antioxidant activity and bioactive compounds, also between color results and concentration of pigments, was carried out using the Spearman test. D were analyzed with program SPSS (System for WINTM version 25, Institute Inc., Ca NC, USA).

3. Result and Discussion

3.1. Physical Properties

3.1.1. Color

The color of foods is an important attribute and it is associated with consumer ceptability and preference [34]. Color measurements on the front and back of the le vegetables (chard, coriander, spinach, and lettuce) were performed, as well as in the p carp and pulp of whole-piece vegetables (beet, tomato, radish, and carrot) (Tables 1 anc respectively). In the L^* parameter, the ecologically produced vegetables were significan ($p < 0.05$) brighter than the commercial vegetables. This behavior occurred both in le vegetables and in those evaluated in the whole piece, except for the tomato sample in external part (Table 1). Concerning the a^* and b^* coordinates, all green vegetables were the green–yellow quadrant, while red vegetables were in the red–green quadrant, ap from the radish in the inside part (Table 2).

Table 1. Color of the adverse/external part of vegetables from the ecological garden and from commercial production.

Vegetable	Type	L*	a*	b*	ΔE
Chard	Com	36.32 ± 3.64	−6.32 ± 0.81	46.32 ± 3.48	23.49 ± 2.20
	Eco	54.88 ± 3.62 *	−9.52 ± 0.81 *	53.16 ± 4.32 *	
Coriander	Com	34.60 ± 1.64	−7.75 ± 0.88	48.82 ± 4.27	13.08 ± 0.92
	Eco	46.69 ± 3.52 *	−10.65 ± 0.75 *	56.55 ± 3.03 *	
Spinach	Com	30.25 ± 2.91	−6.72 ± 1.25	42.45 ± 4.47	11.81 ± 0.79
	Eco	37.08 ± 2.58 *	−10.25 ± 0.42 *	50.19 ± 3.18 *	
Lettuce	Com	53.76 ± 3.70	−5.46 ± 1.05	22.89 ± 2.21 *	9.57 ± 0.32
	Eco	59.00 ± 5.32 *	−5.88 ± 0.81	21.37 ± 2.17	
Beet	Com	28.71 ± 2.87	5.58 ± 0.87	23.77 ± 2.24	12.91 ± 1.10
	Eco	34.73 ± 2.65 *	16.77 ± 4.86 *	28.15 ± 5.85	
Carrot	Com	48.64 ± 2.38	33.33 ± 1.36	59.99 ± 4.31	10.49 ± 1.09
	Eco	52.92 ± 2.48 *	27.35 ± 3.75 *	62.50 ± 3.77	
Tomato	Com	36.27 ± 1.54	23.72 ± 2.30	45.61 ± 1.21	2.79 ± 0.40
	Eco	34.48 ± 2.09 *	26.97 ± 3.62	45.42 ± 5.32	
Radish	Com	29.07 ± 2.09	39.19 ± 3.61	24.98 ± 3.21	4.67 ± 0.03
	Eco	34.47 ± 1.25 *	51.81 ± 2.74 *	47.16 ± 1.98 *	

COM: Commercial and ECO: Ecological. * significantly different at $p < 0.05$ in the two-sided test of equality for column means. Cells without a subscript are not included in the test. The tests assume equal variances.

Table 2. Color of the back/inside part of vegetables from the ecological garden and from commercial production.

Vegetable	Type	*L**	*a**	*b**	ΔE
Chard	Com	36.32 ± 2.46	−6.32 ± 0.92	39.32 ± 2.24	5.7 ± 0.62
	Eco	44.50 ± 2.87 *	−9.91 ± 0.63 *	48.09 ± 2.67 *	
Coriander	Com	33.24 ± 3.20	−8.60 ± 0.54	40.50 ± 4.03	8.64 ± 0.19
	Eco	47.94 ± 3.85 *	−10.14 ± 0.26 *	45.75 ± 4.52 *	
Spinach	Com	37.72 ± 2.21	−7.15 ± 0.96	47.78 ± 3.78	12.70 ± 0.27
	Eco	44.97 ± 3.08 *	−8.39 ± 5.71	50.34 ± 4.82	
Lettuce	Com	61.16 ± 4.76	−7.00 ± 1.15	33.28 ± 3.40 *	8.76 ± 0.69
	Eco	65.33 ± 3.47 *	−7.00 ± 0.66	27.19 ± 2.63	
Beet	Com	11.25 ± 0.75	36.58 ± 2.59	22.53 ± 3.88	4.63 ± 0.40
	Eco	15.09 ± 1.44 *	38.92 ± 1.69	23.14 ± 2.38	
Carrot	Com	53.80 ± 3.80	34.00 ± 5.67	45.91 ± 2.96	12.7 ± 1.06
	Eco	59.06 ± 2.24 *	35.85 ± 2.32	58.95 ± 6.24 *	
Tomato	Com	50.15 ± 6.50	15.77 ± 2.79	23.27 ± 2.60	8.56 ± 0.67
	Eco	55.11 ± 9.90	17.68 ± 1.20 *	23.19 ± 2.64	
Radish	Com	74.36 ± 2.36	0.59 ± 0.21	6.66 ± 0.97	2.8 ± 0.49
	Eco	75.77 ± 1.57	−0.44 ± 0.14 *	7.92 ± 0.67 *	

COM: Commercial and ECO: Ecological. * significantly different at $p < 0.05$ in the two-sided test of equality for column means. Cells without a subscript are not included in the test. The tests assume equal variances.

All green vegetables in ecological production were significantly different in the parameter *a** (green–red) compared to commercial samples, showing a greener color trend, excluding the lettuce (ecological and commercial samples), which had similar values i both parts of the studied vegetables, while the spinach was similar between the vegetabl samples only at the rear. Regarding red vegetables, the ecological samples showed a te dency to be red, placing the values in the red–yellow quadrant; these samples did nc present significant differences, and only the commercial carrot in the external value of *a* was significantly higher with a tendency to be orange in color.

About the parameter *b** (yellow–blue), the ecological vegetables such as chard, coria der, spinach, and radish had a higher yellow color than commercial products ($p < 0.05$ omitting lettuce in the commercial vegetables (Table 2).

The ΔE value indicates the perception of color difference by the human eye an according to Simunovic [35], values greater than 3 indicate that the color differences ar perceptible to the naked eye, and the consumer associates this difference in color wit greater freshness [36]. The ΔE values at all points of the evaluated vegetables in this stud were higher than 3, ranging from 4.67 to 23.49, (Tables 1 and 2), without counting the oute part of the tomato and the internal part of the radish, which have values slightly less than (2.79 and 2.8 respectively).

3.1.2. Texture, Moisture, and Ashes

The results of texture, moisture, and ashes are shown in Table 3. It can be observed tha the texture reported less shear force in the vegetables from the ecological garden, which i associated with better turgidity. This is an indicator of freshness and quality in vegetable because it is maintained when water is found in plant tissues, in the correct compartmen such as vacuoles, cytosol, and intermembrane space [32,37].

Table 3. Texture, moisture, and ashes of vegetables from the ecological garden and from commerci production.

Vegetable	Type	Texture (N/100 g o Piece)	Moisture (%)	Ashes (%)
Chard	Com	3684 ± 312	88 ± 0.04	0.11 ± 0.001
	Eco	3146 ± 289 *	91 ± 0.03	0.12 ± 0.001
Coriander	Com	4825 ± 228	87 ± 0.01	0.10 ± 0.01
	Eco	3537 ± 228 *	87 ± 0.01	0.22 ± 0.01 *
Spinach	Com	4884 ± 343	89 ± 0.01 *	0.018 ± 0.03
	Eco	2878 ± 171 *	84 ± 0.02	0.10 ± 0.01
Lettuce	Com	2510 ± 241	93 ± 0.01	0.03 ± 0.00
	Eco	2210 ± 213 *	91 ± 0.01	0.08 ± 0.02 *
Beet	Com	15.85 ± 0.84	86 ± 0.02	0.08 ± 0.002
	Eco	15.08 ± 0.66 *	85 ± 0.04	0.13 ± 0.023
Carrot	Com	18.99 ± 1.72	97 ± 0.45	0.09 ± 0.019
	Eco	16.51 ± 0.73 *	89 ± 0.03	0.97 ± 0.017
Tomato	Com	1.66 ± 0.08	93 ± 0.0	0.02 ± 0.00
	Eco	0.60 ± 0.09 *	94 ± 0.0 *	0.06 ± 0.02 *
Radish	Com	6.85 ± 0.70	94 ± 0.01 *	0.03 ± 0.00
	Eco	3.69 ± 0.47 *	91 ± 0.06	0.02 ± 0.01

COM: Commercial and ECO: Ecological. * significantly different at $p < 0.05$ in the two-sided test of equality f column means. Cells without a subscript are not included in the test. The tests assume equal variances.

Most of the samples did not show significant differences in moisture, except fc commercial spinach and radish which had higher moisture, and the ecological toma had higher moisture than the commercial ones. This behavior may be due to postharve storage, handling conditions, and the sale period [37]. An usual commercial practice of th merchants is to keep the vegetable products submerged in water or with constant spraying so that the vegetables maintain their fresh appearance [38]. The water in the vegetable is absorbed by the hemicellulose networks of plants; however, this water content is placed on the surface and not in the proper cell compartments [39]. This type of management was not carried out on ecological vegetables since these were harvested and immediately transferred for physical analysis. Because this process generates hydration of the hemicellulose network of the cell wall, increasing its elasticity, it can be verified with the evaluation of texture [40].

The ash values showed significant statistical differences in ecological coriander, tomato, and lettuce in comparison with commercial vegetables. According to Butnariu et al. [41], edible vegetables contain minerals, mainly K, Ca, Mg, P, and Fe, and oligo-element traces (Cr, Cu, I, F, Zn, Mg, Mo, and Se). The mineral and trace element content of plants is known to be affected by the cultivar of the plant, soil, weather conditions, use of fertilizers, and by the state of the plant maturity during the growing season and after the harvest [42].

3.2. *Bioactive Compounds*

Antioxidant compounds have an essential role in human health due to these compounds preventing different types of chemical damage caused by free radicals, avoiding cell oxidation, and therefore reducing the risk of non-communicable diseases [43,44].

In general, all ecologically produced vegetables showed a higher concentration of antioxidant compounds than commercial samples as presented below.

3.2.1. Total Phenolics Compounds (TPC)

The vegetables of ecological production had a concentration of TPC between 553 to 1429 mg GAE/100 g dw, and the samples with a high concentration were coriander (1429.35 mg GAE/100 g dw) and radish (1120.26 mg GAE/100 g dw), while the carrot presented the lowest concentration (102.22 mg GAE/100 g dw). Specifically, the TPC concentration in ecological vegetable products was significantly higher ($p < 0.05$) than in commercial products, except in tomatoes (Table 4). According to Oliveira [45], the synthesis of these compounds occurs when the plants are exposed to controlled stress conditions (as occurs in organic products where mineralization or inorganic fertilization process is not carried out), or during ripening, especially with a major activity (up to 140%) of phenylalanine ammonia-lyase (PAL) enzyme activity caused by physiological alterations [46].

3.2.2. Ascorbic Acid

The studied vegetables had a content of ascorbic acid between 932 to 1310 mg AA/100 g dw; however, the ecological vegetables with a high concentration were chard, coriander, tomato, lettuce, and radish. The lettuce of ecological production was the vegetable with the highest concentration of ascorbic acid (1310 mg AA/100 g dw), followed by chard with 1276.51 mg AA/100 g dw (Table 4). Considering that these vegetables in their fresh form contribute on average of 12.18 mg/100 g fw (fresh weight) of ascorbic acid, this amount could cover more than 100% of the recommended RDI (75 mg/day) for ascorbic acid in adults [47]. The vegetable with the lowest concentration of ascorbic acid was the commercial tomato with 940.16 mg AA/100 g dw. It is possible that the low availability of nitrogen in organic or ecological products decreases the synthesis of biomass in the vegetable, but increases the concentration of ascorbic acid [48], as acclaimed by the behavior observed in vegetables such as spinach, tomato, lettuce, and radish according to other studies [46,49].

Table 4. Total phenolic compounds, ascorbic acid, β-carotene, chlorophyll *a* and *b* in vegetables fro the ecological garden and from commercial production (mg/100 g dw).

Vegetable	Type	TPC (GAE)	Ascobic Acid (AA)	β-Carotene	Chlorophyll *a*	Chlorophyll *b*
Chard	Com	426.76 ± 6.68	1219.37 ± 7.27	50.68 ± 0.27	509.24 ± 8.21	1286.05 ± 8.27
	Eco	485.25 ± 2.38 *	1276.51 ± 7.27 *	50.34 ± 0.13	525.10 ± 3.06 *	1302.07 = 0.81 *
Coriander	Com	1151.11 ± 27.74	1035.24 ± 8.25	50.71 ± 0.08	524.33 ± 2.39	1233.34 ± 1.94
	Eco	1429.35 ± 10.61 *	1070.16 ± 7.21 *	50.61 ± 0.20	529.14 ± 0.83 *	1306.03 = 1.31 *
Spinach	Com	443.73 ± 14.82	1208.41 ± 40.50	49.21 ± 0.29	508.18 ± 3.03	1261.31 ± 3.19
	Eco	592.50 ± 5.46 *	1222.71 ± 90.02	49.47 ± 0.14	508.48 ± 6.92	1271.62 ± 8.08
Lettuce	Com	205.93 ± 3.67	1262.38 ± 17.17	11.61 ± 0.29	79.57 ± 5.53	128.39 ± 4.69
	Eco	754.23 ± 44.41 *	1310.00 ± 12.60 *	47.20 ± 0.28 *	508.64 ± 3.29 *	854.94 ± 12.76 *
Beet	Com	277.99 ± 8.29	1284.44 ± 7.27 *	1.44 ± 0.04	4.47 ± 0.36	11.08 ± 0.86
	Eco	553.15 ± 33.22 *	1043.17 ± 11.68	2.55 ± 0.16 *	44.13 ± 0.27 *	68.44 ± 1.65 *
Carrot	Com	88.18 ± 7.77	1278.25 ± 13.75	52.05 ± 0.09	12.71 ± 2.40	34.7 ± 1.65
	Eco	102.22 ± 3.96 *	1310 ± 26.51	53.23 ± 0.04 *	40.05 ± 5.38 *	140.16 ± 5.11 *
Tomato	Com	873.59 ± 2.56 *	940.16 ± 11.98	18.04 ± 0.04	6.53 ± 0.30	21.49 ± 1.01
	Eco	506.89 ± 1.31	1038.57 ± 20.76 *	19.33 ± 0.16 *	12.61 ± 0.53 *	13.39 ± 2.71 *
Radish	Com	596.64 ± 4.26	932.22 ± 11.00	14.69 ± 0.59	19.99 ± 0.91	46.46 ± 0.45
	Eco	1120.26 ± 7.57 *	1105.24 ± 19.05 *	16.71 ± 0.19 *	21.45 ± 0.16	45.74 ± 0.80

COM: Commercial and ECO: Ecological. * significantly different at $p < 0.05$ in the two-sided test of equality column means. Cells without a subscript are not included in the test. The tests assume equal variances.

3.2.3. β-Carotene and Chlorophyll

The concentration of β-carotene in studied samples was between 47 to 53 mg/100 g c except for beets, radishes, and commercial lettuce which had lower values of this parame The results showed that the ecological production had a higher concentration of β-carote than commercial samples, mainly in carrots, radishes, lettuce, tomatoes, and beets (Table Considering fresh weight in the samples (1.52 to 5.04 mg/100 g fw), this concentration of carotene covers more the 50% of the RDI (2–4 mg/day) in the diet of an adult [50], with t exception of beet which only covers 10% of these recommendations. The lower values we in tomato, radish, and commercial lettuce (between 11.61 to 19.33 mg/100 g dw) and be (1.44 to 2.55 mg/100 g dw). This difference in β-carotene concentration can be explained the production methods (ecological and commercial) and the maturity of the plant at t time of harvest [51], but there is more scientific evidence that handling and postharv storage are determining factors in the concentration of these compounds [52]. After 3 4 days postharvest, the loss of bioactive compounds becomes faster if the humidity a temperature conditions in storage are not adequate [53]. This could be the factor that decisively influencing the results of the analysis carried out in this study.

Chlorophyll is a compound characteristic of green vegetables and it can be found different chemical structures that contribute to several tones of green [54]. The conte of chlorophylls in all the samples of the ecological type were higher than in comm cial vegetables. The highest concentration of chlorophyll *b* (chemical structure leads colors yellow-green) was observed in chard, coriander, spinach, and lettuce (1306.03 1271.62 mg/100 g dw), and therefore, chlorophyll *b* (blue-green) was lower in these produ (508–524 mg/100 g dw) (Table 4).

3.2.4. Betalains, Lycopene, and Anthocyanins

The presence of anthocyanins in radishes, betalains in beets, and lycopene in tomato is characteristic and contributes to the color attributes of each vegetable, and it is associat with their freshness [55,56]. The concentration of anthocyanins in ecological vegetables was 604.39 mg in radish, 5114 mg of betacyanins, and betaxanthins of 5567 mg in beet, and lycopene 2927 mg in tomato considering 100 g of dry mass while the commercial vegetables had of 25 to 48% lower concentration (Figure 1). Statistically, an association was found with color parameters (b^* and a^* coordinate) and concentration of these pigments in ecological vegetables (r^2 = 0.6 to 0.9*). Also, there was a correlation between color and pigments in commercial vegetable samples, but with lower correlation values (r^2 = 0.4 to 0.6*).

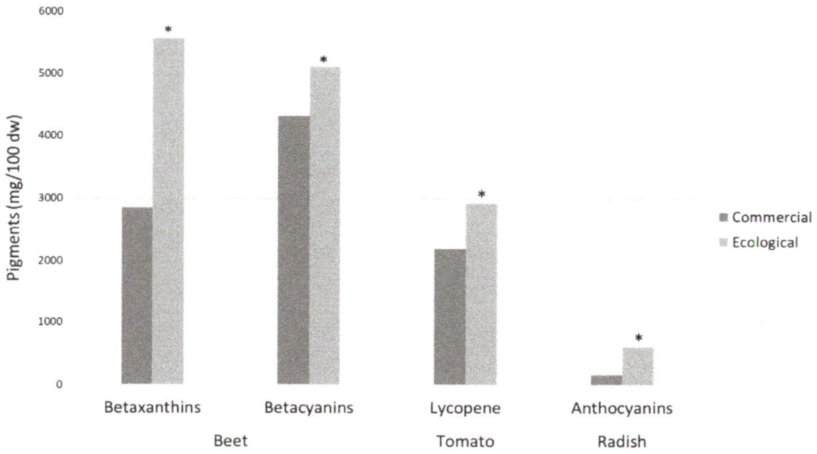

Figure 1. Concentrations of betalains in beet, lycopene in tomato, and anthocyanins in radish. * The asterisk expresses the statistically significant difference at $p < 0.05$ in the two-sided test of equality between the commercial and ecological samples.

In general, commercial products could have more manipulations during processing, for instance practices such as washing and soaking [51,57]. Therefore, due to the high solubility of anthocyanins and betalains in water, the losses of these compounds could significantly increase in radishes and beets during storage and commercial distribution [53,54].

In particular, it has been observed that lycopene is higher in tomatoes that ripen on the plant in open environments, while in tomatoes produced in a greenhouse, lycopene synthesis could be higher during postharvest [46].

3.3. Antioxidant Activity

The presence of these compounds is related to antimicrobial, antiproliferative, anti-inflammatory, and antioxidant bioactivities [58,59]. Many postharvest conditions such as harvesting techniques, storage conditions, temperature, atmosphere, and pH could influence oxidative stress [60]. Some symptoms of oxidative stress in vegetables are inhibition of chloroplast development, core browning, superficial scald, disruption in membrane integrity, inactivation of protein due to the action of proteases, and bleaching of pigmentation [61]. Therefore, the presence of bioactive compounds contributes to the reduction of the oxidative state and the quality of vegetable products.

The Table 5 shows the results of antioxidant activity and chelating activity. The values of antioxidant activity measured as $ABTS^{\bullet+}$ were between 4137.38 to 6565.53 µmol TE/100 g dw. The samples of tomato and carrot had the lowest values (between 23 to 120.12 µmol TE/100 g dw) in ecological and commercial vegetables. In this study, this methodology reported high values of antioxidant activity compared to DPPH, FRAP, and chelating activity in vegetables, and it can be explained by the affinity with hydrophilic (polyphenols and ascorbic acid) and lipophilic (β-carotene, chlorophyll, and lycopene) compounds. A moderate association (r^2 = 0.567*) was found between the TPC concentration and the antioxidant activity measured by ABTS•+ in the vegetables from the ecological garde[n]. These characteristics of the ABTS•+ method makes it possible to record the antioxidan[t] activity of various compounds in the samples [62].

Table 5. Antioxidant activity, ascorbic acid, and iron chelating activity of vegetables from th[e] ecological garden and from commercial production (100 g dw).

Vegetable	Type	ABTS•+ (µmol TE)	DPPH (µmol TE)	FRAP (µmol Fe (II))	Chelating Activity (%)
Chard	Com	3961.95 ± 342.37	55.44 ± 3.11	1056.35 ± 1.24	85.57 ± 1.87
	Eco	3737.05 ± 159.54	108.52 ± 2.00 *	3446.17 ± 207.77 *	90.19 ± 0.03 *
Coriander	Com	3596.63 ± 235.40	27.54 ± 1.07	1063.44 ± 374.00	79.34 ± 0.14
	Eco	3560.29 ± 226.21	73.00 ± 1.28 *	1793.67 ± 15.48 *	89.73 ± 0.78 *
Spinach	Com	2465.32 ± 181.42	0.12 ± 0.00	124.03 ± 3.23	81.50 ± 0.42
	Eco	3395.38 ± 52.81 *	0.14 ± 0.01 *	176.94 ± 3.15 *	92.75 ± 0.13 *
Lettuce	Com	4951 ± 194	6390.2 ± 268.7	263.49 ± 1.22	21:01 ± 0.74
	Eco	6623 ± 639 *	7600.32 ± 97.28 *	865.05 ± 8.77 *	49.56 ± 0.13 *
Beet	Com	4362.63 ± 382.41	30.03 ± 2.78	197.49 ± 3.73	44.45 ± 1.00
	Eco	4231.84 ± 132.64	58.23 ± 1.93 *	259.38 ± 6.12 *	52.12 ± 0.09 *
Carrot	Com	89 ± 6.01	—	62.28 ± 4.61	23.86 ± 0.30
	Eco	120.12 ± 8.88 *	—	94.15 ± 3.49 *	53.01 ± 4.50 *
Tomato	Com	41.28 ± 1.62 *	0.04 ± 0.0	37.99 ± 0.92	83.98 ± 0.15
	Eco	23.29 ± 1.17	0.07 ± 0.0 *	57.27 ± 1.15 *	92.35 ± 0.09 *
Radish	Com	4137.38 ± 302.41	275 ± 14	135.79 ± 5.71	21.27 ± 0.85
	Eco	6565.53 ± 56.29 *	1857 ± 101 *	684.75 ± 13.79 *	32.14 ± 2.56 *

COM: Commercial and ECO: Ecological. * significantly different at $p < 0.05$ in the two-sided test of equality f[or] column means. Cells without a subscript are not included in the test. The tests assume equal variances —: N[ot] performed analysis.

The DPPH radical evaluates the free radical scavenging capacity of antioxidan[ts] according to their hydrogen-donating capacity. And it can evaluate the mechanisms of th[e] transfer of hydrogen atoms and electrons [63]. As a result of this test, it can be observe[d] that the vegetables produced in the ecological garden showed higher antioxidant activi[ty] than commercial samples. The ecological lettuce with 7600.32 µmol TE/100 g dw was th[e] one that best captured free radicals (Table 5). On the other hand, in spinach and tomat[o] the values were lower than 0.07 µmol TE/100 g dw, while in carrots this assay was n[ot] detected. This activity was correlated with ascorbic acid concentrations with statistic[al] significance ($r^2 = 0.714$*).

As in DPPH, in FRAP analysis the antioxidant activity registered by all samples [of] ecological vegetables was significantly ($p < 0.05$) higher than the commercially produce[d] samples. The samples of ecological chard were the ones that reported the best antioxida[nt] activity (3446.17 µmol Fe (II)/100 g dw) and the samples that recorded the lowest antio[x]idant activity were commercial tomatoes (37.99 µmol Fe (II)/100 g dw) (Table 5). In th[e] vegetables from the ecological garden, the correlation was given from the values of th[e] FRAP test with chlorophyll *b* ($r^2 = 0.71$*).

Table 5 also showed the results of the metal chelating assay. All the samples [of] the vegetables produced in the ecological garden have higher chelating activity than th[e] samples of commercial production. The ecological spinach sample reported the highe[st] percentage of chelation (92.35%), while commercial lettuce had the lowest value (21.01%). This result could be explained due to the strong iron-binding properties of polypheno[ls] whether the iron-chelating ability of catechol or polyphenols plays a key role in the antioxidant activity and anti-lipid peroxidation by blocking the Fenton reaction [64,65]. In addition, a moderate correlation was reported between the chelating assay with TP[C] content ($r^2 = 0.657$*).

4. Conclusions

The ecological production vegetables presented better physical characteristics in colo[r] (luminosity and intensity), which were perceptible by the human eye (ΔE). In additio[n] they were more turgid compared to the commercial samples. Furthermore, the ecological products had higher bioactive compounds (TPC, ascorbic acid, β-carotenes, chlorophylls, anthocyanins, betalains, and lycopene) than the commercial products, and the content was positively correlated with antioxidant capacity.

However, it is essential to carry out more studies to clarify whether the differences found in favor of ecological products are reflected in the health of the people who consume them and thus promote the establishment of urban and peri-urban organic gardens as a strategy of sustainability for healthy eating.

Author Contributions: All the authors have contributed equally to this paper. All authors have read and agreed to the published version of the manuscript.

Funding: This research received no external funding.

Data Availability Statement: All data from the analyses performed are available from the corresponding authors of this publication.

Conflicts of Interest: The authors declare no conflict of interest.

ferences

FAO. *Sustainable Food Systems, Concept and Framework*; Food and Agriculture Organization of the United Nations: Rome, Italy, 2015.

De Brito, C.; Gonçalves, Q.; Schlindwein, M.M. Agroforestry Systems: A Systematic Review Focusing on Traditional Indigenous Practices, Food and Nutrition Security, Economic Viability, and the Role of Women. *Sustainability* **2021**, *13*, 11397.

Brandt, K.; Leifert, C.; Sanderson, R.; Seal, C.J. Agroecosystem management and nutritional quality of plant foods: The case of organic fruits and vegetables. *CRC. Crit. Rev. Plant Sci.* **2011**, *30*, 177–197. [CrossRef]

Lairon, D. Nutritional quality and safety of organic food. A review. *Agron. Sustain. Dev.* **2010**, *30*, 33–41. [CrossRef]

Browne, H.R.B.A.W.; Harris, P.J.C.; Cadoret, K. Smallholder Farmers and Organic Certification: Accessing the EU Market from the Developing World. *Biol. Agric. Hortic.* **2001**, *19*, 183–199. [CrossRef]

Olesen, J.E.; Bindi, M. Consequences of climate change for European agricultural productivity, land use and policy. *Eur. J. Agron.* **2002**, *16*, 239–262. [CrossRef]

Willer, H.; Yussefi-Menzler, M.; Sorensen, N. *The World of Organic Agriculture: Statistics and Emerging Trends 2008*; IFOAN: Bonn, Germany, 2008; ISBN 9781849775991.

Williams, D.R.; Dixon, P.S. Impact of Garden-Based Learning on Academic Outcomes in Schools: Synthesis of Research between 1990 and 2010. *Rev. Educ. Res.* **2013**, *83*, 211–235. [CrossRef]

Amaya-Castellanos, C.; Shamah-Levy, T.; Escalante-Izeta, E.; del Morales-Ruán, C.M.; Jiménez-Aguilar, A.; Salazar-Coronel, A.; Uribe-Carvajal, R.; Amaya-Castellanos, A. Development of an educational intervention to promote healthy eating and physical activity in Mexican school-age children. *Eval. Program Plann.* **2015**, *52*, 159–168. [CrossRef]

Luciana, N.G.; da Cassiano, S.O.; Lorena, G.S.; Renata, M.T.D.; Cristiane, B.; Erika, M.M.T. Nutritional composition of vegetables grown in organic and conventional cultivation systems in Uberlndia, MG. *Afr. J. Agric. Res.* **2017**, *12*, 1848–1851. [CrossRef]

Schifferstein, H.N.J.; Wehrle, T.; Carbon, C. Consumer expectations for vegetables with typical and atypical colors: The case of carrots. *Food Qual. Prefer.* **2018**, *72*, 98–108. [CrossRef]

Arce-lopera, C.; Masuda, T.; Kimura, A.; Wada, Y.; Okajima, K. Model of vegetable freshness perception using luminance cues. *Food Qual. Prefer.* **2015**, *40*, 279–286. [CrossRef]

Thomsen, M.G.; Riley, H.; Borge, G.I.A.; Lea, P.; Gunnar, B.; Division, A.C. Effects of soil type and fertilization on yield, chemical parameters, sensory quality and consumer preference of swede (*Brassica napus* L. ssp. rapifera). *Eur. J. Hortic. Sci.* **2018**, *82*, 1611–4434. [CrossRef]

Huyskens-Keil, S.; Schreiner, M. Quality Dynamics and Quality Assurance of Fresh Fruits and Vegetables in Pre- and Postharvest. *Prod. Pract. Qual. Assess. Food Crop.* **2006**, *3*, 401–449. [CrossRef]

Tena, N.; Mart, J. State of the Art of Anthocyanins: Antioxidant Activity, Sources, Bioavailability, and Therapeutic E ff ect in Human Health. *Antopxidants* **2020**, *9*, 451. [CrossRef]

Divya, P.; Puthusseri, B.; Neelwarne, B. Carotenoid content, its stability during drying and the antioxidant activity of commercial coriander (*Coriandrum sativum* L.) varieties. *Food Res. Int.* **2012**, *45*, 342–350. [CrossRef]

Oliveira, F.; Dos Santos, R.; De Souza Rosa, L.; Anderson-Junger, T. Organic and conventional vegetables: Comparison of the physical and chemical characteristics and antioxidant activity. *Afr. J. Biotechnol.* **2016**, *15*, 1746–1755. [CrossRef]

Hattab, S.; Bougattass, I.; Hassine, R.; Dridi-Al-Mohandes, B. Metals and micronutrients in some edible crops and their cultivation soils in eastern-central region of Tunisia: A comparison between organic and conventional farming. *Food Chem.* **2019**, *270*, 293–298. [CrossRef]

Sharma, M.; Usmani, Z.; Gupta, V.K.; Bhat, R. Critical Reviews in Biotechnology Valorization of fruits and vegetable wastes and by- products to produce natural pigments. *Crit. Rev. Biotechnol.* **2021**, *41*, 535–563. [CrossRef]

INEGI. *Compendio de Información Geográfica Municipal 2010 Pachuca de Soto Hidalgo*; INEGI: Aguascalientes, Mexico, 2010.

21. Johnson, G.M.; Fairchild, M.D. A Top Down Description of S-CIELAB and CIEDE2000. *Color Res. Appl.* **2003**, *28*, 425–4 [CrossRef]
22. Medina, M.S.; Tudela, J.A.; Marín, A.; Allende, A.; Gil, M.I. Short postharvest storage under low relative humidity improv quality and shelf life of minimally processed baby spinach (*Spinacia oleracea* L.). *Postharvest. Biol. Technol.* **2012**, *67*, 1–9. [CrossR
23. Viñas, P.; Campillo, N. *Gas Chromatography: Mass Spectrometry Analysis of Polyphenols in Foods*; 2019; ISBN 9780128137630.
24. Muñoz-Bernal, Ó.A.; Torres-Aguirre, G.A.; Núñez-Gastélum, J.A.; de la Rosa, L.A.; Rodrigo-García, J.; Ayala-Zavala, J.F.; Álvar Parrilla, E. Nuevo Acercamiento a La Interacción Del Reactivo De Folin-Ciocalteu Con Azúcares Durante La Cuantificación Polifenoles Totales. *Tip* **2017**, *20*, 23–28. [CrossRef]
25. Dürüst, N.; Sümengen, D.; Dürüst, Y. Ascorbic Acid and Element Contents of Foods of Trabzon (Turkey). *J. Agric. Food Ch* **1997**, *45*, 2085–2087. [CrossRef]
26. Giustu, M.; Wrolstad, R.E. Characterization and Measurement of Anthocyanins by UV-Visible Spectroscopy. *Curr. Protoc. F Anal. Chem.* **2001**, *1*, 1–13. [CrossRef]
27. Ferndndez-López, J.A.; Castellar, R.; Obón, J.M.; Alrnela, L. Screening and Mass-Spectral Confirmation of Betalains in Cac Pears. *Chromatographia* **2002**, *55*, 591–595.
28. Righi, H.; Camila, S.; Bolanho, B.C. Ultrasonic-assisted extraction of betalains from red beet (*Beta vulgaris* L.). *J. Food Process E* **2018**, *41*, e12833. [CrossRef]
29. Braniša, J.; Jenisová, Z.; Porubská, M.; Jomová, K.; Valko, M.; Klaudia, R. Spectrophotometric Determination of Chlorophylls a Carotenoids. An Effect of Sonication and Sample Processing. *J. Microbiol. Biotechnol. Food Sci.* **2014**, *3*, 61–64.
30. Nagata, M.; Yamashita, I. Simple method simultaneous determination of Chrorophyll and carotenoids in tomato fruit. *Nipp Shokuhin Kogyo Gakkaishi* **1992**, *39*, 925–928. [CrossRef]
31. Kuskoski, E.M.; Asuero, A.G.; Troncoso, A.M.; Mancini-Filho, J.; Fett, R. Aplicación de diversos métodos químicos para determir actividad antioxidante en pulpa de frutos. *Ciência Tecnol. Aliment.* **2005**, *25*, 726–732. [CrossRef]
32. Benzie, I.F.F.; Strain, J.J. Ferric reducing/antioxidant power assay: Direct measure of total antioxidant activity of biologi fluids and modified version for simultaneous measurement of total antioxidant power and ascorbic acid concentration. *Meth Enzymol.* **1999**, *299*, 15–27. [CrossRef]
33. Gulcin, I.; Sat, I.; Beydemir, S.; Kufrevioglu, O.I. Evaluation of the in vitro antioxidant properties of broccoli extracts (*Brass oleracea* L.). *Ital. J. Food Sci.* **2004**, *16*, 17–30.
34. Kamga, R.T.; Kouamé, C.; Atangana, A.R.; Chagomoka, T.; Ndango, R. Nutritional Evaluation of Five African Indigenc Vegetables. *J. Hortic. Res.* **2013**, *21*, 99–106. [CrossRef]
35. Simunovic, M.P. Acquired color vision deficiency. *Surv. Ophthalmol.* **2016**, *61*, 132–155. [CrossRef]
36. Moreira, M.D.R.; Roura, S.I.; Del Valle, C.E. Quality of Swiss chard produced by conventional and organic methods. *LWT—F Sci. Technol.* **2003**, *36*, 135–141. [CrossRef]
37. Chen, X. Main Factors Affecting Post-Harvest Grain Loss during the Sales Process: A Survey in Nine Provinces of Chi *Sustainability* **2018**, *10*, 661. [CrossRef]
38. Rajapaksha, L.; Gunathilake, D.M.C.C.; Pathirana, S.M.; Fernando, T.N. Reducing post-harvest losses in fruits and vegetables ensuring food security—Case of Sri Lanka. *MOJ Food Process Technols* **2021**, *9*, 7–16. [CrossRef]
39. Jayalath, M.M.; Perera, H.N. Mapping Post-Harvest Waste in Perishable Supply Chains through System Dynamics: A Sri Lank Case Study. *J. Agric. Sci. Sri Lanka* **2021**, *16*, 526–543. [CrossRef]
40. Masarirambi, M.T.; Mavuso, V.; Songwe, V.D.; Nkambule, T.P. Indigenous post-harvest handling and processing of traditio vegetables in Swaziland: A review. *Afr. J. Agric. Res.* **2010**, *5*, 3333–3341. [CrossRef]
41. Butnariu, M.; Butu, A. Chemical Composition of Vegetables and their Products. In *Handbook of Food Chemistry*; Spring Berlin/Heidelberg, Germany, 2014; ISBN 9783642416095.
42. Kapusta-duch, J.; Leszczy, T.; Florkiewicz, A.; Filipiak-florkiewicz, A. Ecology of Food and Nutrition Comparison of Calcium a Magnesium Contents in Cruciferous Vegetables Grown in Areas around Steelworks, on Organic Farms, and Those Available Retail. *Ecol. Food Nutr.* **2011**, *50*, 37–41. [CrossRef]
43. Działo, M.; Mierziak, J.; Korzun, U.; Preisner, M.; Szopa, J.; Kulma, A. The potential of plant phenolics in prevention and thera of skin disorders. *Int. J. Mol. Sci.* **2016**, *17*, 160. [CrossRef]
44. González-Barraza, L.; Díaz-Godínez, R.; Castillo-Guevara, C.; Nieto-Camacho, A.; Méndez-Iturbide, D. Phenolic compoun Presence, identification and antioxidant activity in plants and fruits Compuestos fenólicos: Presencia, identificación y propieda antioxidantes en plantas y frutos. *Mex. J. Biotechnol.* **2017**, *2017*, 46–64. [CrossRef]
45. Oliveira, A.B.; Moura, C.F.H.; Gomes-Filho, E.; Marco, C.A.; Urban, L.; Miranda, M.R.A. The Impact of Organic Farming Quality of Tomatoes Is Associated to Increased Oxidative Stress during Fruit Development. *PLoS ONE* **2013**, *8*, e56354. [CrossR
46. Pinheiro, J.; Alegria, C.; Abreu, M.; Gonçalves, E.M.; Silva, C.L.M. Kinetics of changes in the physical quality parameters of fr tomato fruits (Solanum lycopersicum, cv. 'Zinac') during storage. *J. Food Eng.* **2013**, *114*, 338–345. [CrossRef]
47. Granger, M.; Eck, P. *Dietary Vitamin C in Human Health*, 1st ed.; Elsevier Inc.: Amsterdam, The Netherlands, 2018; Volume 83.
48. Koh, E.; Charoenprasert, S.; Mitchell, A.E. Effect of organic and conventional cropping systems on ascorbic acid, vitamin flavonoids, nitrate, and oxalate in 27 varieties of spinach (*Spinacia oleracea* L.). *J. Agric. Food Chem.* **2012**, *60*, 3144–3150. [CrossR [PubMed]

49. Adalid, A.M.; Roselló, S.; Nuez, F. Evaluation and selection of tomato accessions (Solanum section Lycopersicon) for content of lycopene, β-carotene and ascorbic acid. *J. Food Compos. Anal.* **2010**, *23*, 613–618. [CrossRef]
50. 50. Mangione, C.M.; Barry, M.J.; Nicholson, W.K.; Cabana, M.; Chelmow, D.; Coker, T.R.; Davis, E.M.; Donahue, K.E.; Doubeni, C.A.; Jaén, C.R.; et al. Review, Vitamin, Mineral, and Multivitamin Supplementation to Prevent Cardiovascular Disease and Cancer US Preventive Services Task Force Recommendation Statement. *JAMA* **2023**, *23*, 2326–2333. [CrossRef]
51. Pintea, A.; Andrei, S.; Bunea, C.; Pop, R.; Bele, C. Carotenoid and fatty acid profiles of bilberries and cultivated blueberries from Romania. *Chem.Pap.* **2012**, *66*, 935–939. [CrossRef]
52. Barrett, D.M.; Beaulieu, J.C.; Shewfelt, R. Color, flavor, texture, and nutritional quality of fresh-cut fruits and vegetables: Desirable levels, instrumental and sensory measurement, and the effects of processing. *Crit. Rev. Food Sci. Nutr.* **2010**, *50*, 369–389. [CrossRef]
53. Walsh, R.; Bartlett, H.; Eperjesi, F. Variation in carotenoid content of kale and other vegetables: A review of pre and post-harvest effects. *J. Agric. Food Chem.* **2015**, *63*, 9677–9682. [CrossRef]
54. Delgado-Vargas, F.; Jiménez, A.R.; Paredes-López, O. Natural Pigments: Carotenoids, Anthocyanins, and Betalains—Characteristics, Biosynthesis, Processing, and Stability. *Crit. Rev. Food Sci. Nutr.* **2000**, *40*, 173–289. [CrossRef]
55. Nemzer, B. Betalainic and nutritional profiles of pigment-enriched red beet root (*Beta vulgaris* L.) dried extracts. *Food Chem.* **2011**, *127*, 42–53. [CrossRef]
56. Habibi, F.; García-Pastor, M.E.; Puente-Moreno, J.; Garrido-Auñón, F.; Serrano, M.; Valero, D. Anthocyanin in blood oranges: A review on postharvest approaches for its enhancement and preservation. *Crit. Rev. Food Sci. Nutr.* **2022**, *6*, 1–13. [CrossRef]
57. Benítez-Estrada, A.; Villanueva-Sánchez, J.; González-Rosendo, G.; Alcántar-Rodríguez, V.E.; Puga-Díaz, R.; Quintero-Gutiérrez, A.G. Determinación de la capacidad antioxidante total de alimentos y plasma humano por fotoquimioluminiscencia: Correlación con ensayos fluorométricos (ORAC) y espectrofotométricos (FRAP). *TIP Rev. Espec. Cienc. Químico Biológ.* **2020**, *23*, 1–9. [CrossRef]
58. Gandía-herrero, F.; Escribano, J.; García-carmona, F. Biological Activities of Plant Pigments Betalains. *Crit. Rev. Food Sci. Nutr.* **2014**, *56*, 37–41. [CrossRef]
59. Rodriguez-amaya, D.B. *Natural Food Pigments and Colorants*; Springer: Cham, Switzerland, 2019; ISBN 9783319780306.
60. Li, H.; Deng, Z.; Wu, T.; Liu, R.; Loewen, S.; Tsao, R. Microwave-assisted extraction of phenolics with maximal antioxidant activities in tomatoes. *Food Chem.* **2012**, *130*, 928–936. [CrossRef]
61. Solovchenko, A.E.; Merzlyak, M.N. Screening of Visible and UV Radiation as a Photoprotective Mechanism in Plants. *Russ. J. Plant Physiol.* **2008**, *55*, 719–737. [CrossRef]
62. Miller, N.J.; Rice-Evans, C.; Davies, M.J.; Gopinathan, V.; Milner, A. A novel method for measuring antioxidant capacity and its application to monitoring the antioxidant status in premature neonates. *Clin. Sci.* **1993**, *84*, 407–412. [CrossRef]
63. Moon, J.-K.; Shibamoto, T. Antioxidant assays for plant and food components. *J. Agric. Food Chem.* **2009**, *57*, 1655–1666. [CrossRef]
64. Bouaziz, A.; Abu, M.; Abdalla, S.; Baghiani, A.; Charef, N. Phytochemical analysis, hypotensive effect and antioxidant properties of *Myrtus communis* L. growing in Algeria. *Asian Pac. J. Trop. Biomed.* **2015**, *5*, 19–28. [CrossRef]
65. Djenidi, H.; Khennouf, S.; Bouaziz, A. Antioxidant activity and phenolic content of commonly consumed fruits and vegetables in Algeria. *Prog. Nutr.* **2020**, *22*, 224–235. [CrossRef]

land

MDPI

ticle

Iicrobiological Analysis and Metagenomic Profiling of the acterial Community of an Anthropogenic Soil Modified from ypic Haploxererts

etro Barbaccia [1], Carmelo Dazzi [1], Elena Franciosi [2], Rosalia Di Gerlando [1], Luca Settanni [1,*] d Giuseppe Lo Papa [1]

1 Dipartimento Scienze Agrarie, Alimentari e Forestali, Università degli Studi di Palermo, Viale delle Scienze 4, 90128 Palermo, Italy; pietro.barbaccia@unipa.it (P.B.); carmelo.dazzi@unipa.it (C.D.); rosalia.digerlando@unipa.it (R.D.G.); giuseppe.lopapa@unipa.it (G.L.P.)

2 Research and Innovation Centre, Fondazione Edmund Mach (FEM), 38098 San Michele all'Adige, Italy; elena.franciosi@fmach.it

* Correspondence: luca.settanni@unipa.it

Abstract: This work aimed to characterize the microbial communities of an anthropogenic soil originating from application of pedotechniques to Vertisols in a Mediterranean environment. Bare soil profiles were sampled at three depths (0–10 cm, 10–30 cm, and 30–50 cm) and compared with the original soil not transformed at the same depths. The anthropogenic soils were characterized by a higher $CaCO_3$ concentration (360–640 g/kg) than control soil (190–200 g/kg), while an opposite trend was registered for clay, where control soil showed a higher concentration (465 g/kg on average) than anthropogenic soil (355 g/kg on average). Organic carbon content was much higher in the untransformed soil. All samples were microbiologically investigated using a combined culture-dependent and -independent approach. Each pedon displayed a generally decreasing level with soil depth for the several microbial groups investigated; in particular, filamentous fungi were below the detection limit at 30–50 cm. To isolate bacteria actively involved in soil particle aggregation, colonies with mucoid appearance were differentiated at the strain level and genetically identified: the major groups were represented by *Bacillus* and *Pseudomonas*. MiSeq Illumina analysis identified Actinobacteria and Firmicutes as the main groups. A high microbial variability was found in all the three anthropogenic pedons and the microorganisms constitute a mature community.

Keywords: anthropogenic soil; applied soil ecology; extracellular polymeric substances; MiSeq Illumina; viable bacteria

tion: Barbaccia, P.; Dazzi, C.; nciosi, E.; Di Gerlando, R.; anni, L.; Lo Papa, G. robiological Analysis and agenomic Profiling of the erial Community of an hropogenic Soil Modified from c Haploxererts. *Land* **2022**, *11*, https://doi.org/10.3390/ 11050748

demic Editor: Krish chandran

eived: 15 April 2022
epted: 16 May 2022
lished: 18 May 2022

lisher's Note: MDPI stays neutral regard to jurisdictional claims in lished maps and institutional affil- ns.

1. Introduction

In recent decades, ex novo soil formation by human action has become increasingly important to crop-linked activities [1]. Human activity is recognized as one of the most important factors involved in soil generation in the 21st century [2,3]. Generally, the term "Anthrosol" refers to soil formed or strongly modified by human activity such as deep tillage, intensive fertilization, addition of organic waste, irrigation with water rich in sediments or creation of rice fields [4].

Studies focusing on the genetic peculiarities, characteristics and properties of soils created in urban areas, mines, forested and agricultural areas evidence that the main issue of the pedogenetic process is the loss of the soil structure [2,3,5–7]. In particular, the agricultural areas of southern Italy are characterized by a continuous loss of structure as a result of deep tillage and displacement of material. This phenomenon, combined with semi-arid conditions and poor vegetation cover, determines runoff and wind erosion during seasonal rainfall, and in the long term, reduces soil fertility [8].

Organic and inorganic input, cover crops and deep tillage have a negative influence on the aggregation status of the soil particles, while microorganisms play an important role in soil aggregation [9,10]. Aside from improving the structure of soil, the soil microbior plays a key role in several functional processes such as production of substances for pla growth promotion, degradation of pollutants derived from agrochemical treatment a control of all type of pests [11]. Soil microbial diversity is extremely complex; hundre of thousands of microbial taxa live together in a gram of soil [12], with a total number microorganisms exceeding 10 billion [13].

Prokaryotic organisms, both bacteria and archaea, can live and develop within biofil consisting of extracellular polymeric substances (EPS). These substances are mainly form by exopolysaccharides structural proteins, enzymes, biopolymeric substances such as lipi and other constituents [14,15]. In natural environments, most microorganisms live in gregates such as floccules or biofilms, of which EPSs represent the fundamental structu component [16]. In a biofilm, the microbial cells are embedded within a self-produc EPS matrix. This organization allows microorganisms to adhere to each other and/ to a surface. Specifically, "a biofilm is a fixed system that can be adapted internally environmental conditions by its inhabitants" [17]. This structure represents a real selecti ecological advantage [16]. Thanks to biofilms, several bacteria adhere to and communica with other microorganisms and plants. A given biofilm is characterized by water chann that allow for the passage of nutrients and other agents throughout the bacterial comm nity [18]. The water channels within biofilms are considered as a primitive circulato system which protects bacteria against the accumulation of toxic metabolites by their moval [19]. Biofilms constitute real barriers that protect microorganisms from predati anti-microbial substances and heavy metals, and adverse environmental stresses [15,20,2

The formation of structural aggregates in soil can be greatly improved by the bacter EPS. These substances positively affect the structure, porosity, fertility and productiv of the soil systems [22,23]. EPSs can act as glues due to their viscous texture and io charges that allow for anchoring to soil clays [24], thus improving the aggregation of poo aggregated soil particles [16].

The current study was undertaken (1) to assess the soil microbial community a (2) to investigate specifically EPS-producing populations of an anthropogenic soil pr to its first cultivation cycle. Specifically, the microbial composition of the sampled s was approached by culture-dependent and -independent methods. The soil was a characterized for its physical and chemical properties.

2. Materials and Methods

2.1. Study Area

The study area (Figure 1) is located in the countryside of Palma di Montechiaro— Giordano district in the province of Agrigento (AG) in southern Sicily—and is characteriz by a Mediterranean climate. This hilly agricultural area is also characterized by the prese of several horticultural greenhouse farms, mainly cultivating table grapes, obtained throu anthropogenic interventions on soils. The lithology dates back to the Pliocene and Mioce periods and consists of gypsum, marls, limestones and alluvial deposits. The soils classified as Entisols, Inceptisols, Mollisols and Vertisols.

Soil sampling occurred in March 2019. Climatic data for the reference year w obtained from the Agrigento meteorological station. The average annual temperature w 17.7 °C, with the highest thermometric value (41.5 °C) registered in July and the low (−1.0 °C) in January. The average annual precipitation was 497 mm.

2.2. Applied Pedotechnics

A new soil was generated by pedotechnics in 2012. The entire area is characteriz by Typic Haploxererts. This soil was covered with calcareous marls in October 2012 a levelled using a caterpillar machine during the following summer. In Autumn 2013, s was ploughed to an approx. depth of 90–100 cm using a one mouldboard, single-furr plough. Thereafter, it was left uncultivated.

Figure 1. Location of the study area.

2.3. Sampling

Soil samples for physicochemical and microbiological analyses were collected from the following pedons (Figure 2): Mant 0 (37°10′08.1″ N–13°46′56.8″ E), Mant 1 (37°10′09.3″ N–13°47′01.4″ E), Mant 2 (37°10′09.4″ N–13°47′00.8″ E) and Mant 3 (37°10′09.7″ N–13°47′00.1″ E). Mant 0 refers to the original Vertisol not affected by pedotechnics (control soil), while Mant 1, Mant 2 and Mant 3 concern profound transformations due to the pedotechnics used and represent three technical repeats. Each pedon was sampled in duplicate at 3 different depths: I, 0–10 cm; II, 10–30 cm; and III, 30–50 cm, corresponding to part of Ap and ˆA horizons, in Vertisol and anthropogenic soils, respectively.

Figure 2. Sampling points. Red circles are soil profiles.

2.4. Physicochemical Analyses

Soil samples were air-dried and sieved by means of a 2 mm stainless steel siev (Laboratory Test Sieve, London, UK). Texture was determined with the sedimentatio method [25]. pH was measured on soil–water mixtures 1:2.5 (*w*/*v*) with a pH mete (XS instruments, Carpi, Italy). The electrical conductivity (EC) was determined on eac soil–water mixture 1:5 with a portable conductivity-meter Cond 7 (XS instruments, Carp Italy). The cation exchange capacity (CEC) was determined by barium chloride an triethanolamine method (MIPAF, 2000). Total carbonates ($CaCO_3$ Tot) were determine by gas volumetric method using a Dietrich–Fruehling calcimeter and HCl, while organi carbon (C_{org}) was determined with the wet oxidation method of Walkley and Black [26].

2.5. Culture-Dependent Microbiological Analysis

Soil samples for microbiological analysis were collected in aseptic conditions: a steril area was obtained using a portable Bunsen burner, and each sample (approximately 200 g was picked up with a sterile stainless-steel spatula and transferred into a sterile BagLight 400 MultilayerR bag (Interscience, Saint Nom, France). All samples were transporte under refrigeration in a portable fridge to the Laboratory of Agricultural Microbiology University of Palermo.

All samples, according to Thomson et al., were dried for 24 h under a laminar hoo to avoid environmental contaminations and sieved with the 2 mm sieve after autocla ing (121 °C for 15 min) [27]. All sieved samples were placed into 9 cm diameter Pet dishes, sealed with Parafilm, and kept at room temperature until plate counting and tot DNA extraction.

Plate counts were performed to determine the levels of the main soil microbial pop lations such as total mesophilic aerobic (TMA) microorganisms, total mesophilic anaerobi (TMAn) microorganisms, filamentous fungi (FF), Actinobacteria (AB), N-fixing (NF) ba teria, and the main EPS and glue-producing bacteria, *Sphingomonas* and *Caulobacter*. So samples (20 g) were put into conical flasks and diluted (1:10) with sodium pyrophospha (0.16% *w*/*v*) solution (180 mL) by homogenization with an orbital shaker for 10 min a 150 rpm [28]. Further decimal dilutions of soil samples were performed in test tube containing Ringer's solution (Sigma-Aldrich, Milan, Italy) subjected to homogenizatic by vortexing.

The microbial suspensions were inoculated in different culture media and incubated i the optimal conditions (temperature and time). TMA were inoculated in bacteria mediur incubated at 30 °C for 48 h, while FF were inoculated in fungi medium at 30 °C for 7 d; bo media were prepared as described by Zhang et al. [29]. TMAn were inoculated in soil extra medium (SEM), prepared as described by Deutsche Sammlung von Mikroorganismen un Zellkulturen (https://www.dsmz.de/microorganisms/medium/pdf/DSMZ_Mediur 2.pdf, accessed on 19 March 2019), incubated at 30 °C for 48 h in hermetically sealed ja containing the AnaeroGen AN25 system (Oxoid, Milan, Italy). *Caulobacter* and *Sphingomon* were grown in two selective media, *Caulobacter* medium (CM) [30] and NK medium [3 prepared, respectively, and both incubated at 30 °C for 48 h. NF bacteria were inoculated i Blue Green Medium (BG-11) prepared as described by Rippka et al., and the plates we incubated at 30 °C for 48 h [32]. All microbiological analyses were carried out in duplica and the results expressed as Log colony-forming units (CFU) per g of dry weight (g dw determined after drying at 105 °C until constant weight.

White/yellowish colonies with mucoid appearance, presumptive *Caulobacter* an *Sphingomonas*, were harvested from CM and NK, grown in Nutrient Broth (NB) (Oxoid) an subjected to purification by consecutive streaking on Nutrient Agar (NA) until reachin morphology homogeneity of colonies. The isolates were first subjected to a phenotyp investigation under an optical microscope to analyze cell morphology. Only rod bacter were further processed by being genetically investigated, such as *Caulobacter* and ro shaped *Sphingomonas*.

DNA was extracted after overnight growth of the pure cultures in NB at 30 °C for 48 h. The cells were harvested from 1 mL of broth cultures by centrifugation at 10,000× *g* for 5 min, and the pellets were subjected to repeated washing steps with sterile distilled H_2O and finally re-suspended in 1 mL of the same diluent. Cell lysis was performed by Instagene Matrix (Bio-Rad, Hercules, CA, USA) following the protocol provided by the supplier. Crude cell extract from each culture was used as template DNA for the genotypic characterization.

All isolates were differentiated at the strain level to reduce the number of bacteria to be genetically identified. The typing procedure was carried out by random amplification of polymorphic DNA (RAPD)-PCR as described by Gaglio et al. [33]. RAPD profiles were visualized under UV light after electrophoresis of 2% (*w*/*v*) agarose gels in 1 × TBE buffer (Sigma-Aldrich, Milan, Italy). The isolates that showed diverse RAPD profiles were considered different strains and analyzed by sequencing the ribosomal 16S rRNA gene for species identification, applying the protocol described by Weisburg et al. [34]. PCR products of about 1600 bp were purified using the QIAquick kit (Quiagen S.p.a., Milan, Italy) and sequenced at the center for innovation of quality systems, traceability and certification of agri-food—AGRIVET (University of Palermo), with the same oligonucleotides used for PCR analysis. The identity of each sequence was obtained by comparing the sequences acquired with those available in GenBank/EMBL/DDBJ (http://www.ncbi.nlm.nih.gov, accessed on 19 March 2021) [35] and Ez-Taxon (http://eztaxon-e.ezbiocloud.net/ accessed on 19 March 2021) [36] databases. The latter database compares the sequences obtained with those of the type strains only.

2.6. Culture-Independent Analysis

The amount of soil for total genomic DNA extraction from each sample ranged between 0.4 and 0.6 g. DNA extraction was performed by the Gene MATRIX Soil DNA Purification Kit (EURx, Gdansk, Poland) according to the manufacturer's instructions, and quantified using a Nanodrop 8800 Fluorospectrometer (Thermo Fisher Scientific, Waltham, MA, USA).

2.6.1. Amplicon Library Preparation

The pooled libraries and pair-end sequencing quality and quantification were carried out using the sequencing platform of Edmund Mach Foundation (FEM, San Michele a/Adige, Italy). Soil genomic DNAs were amplified with primers specific to the V3–V4 region [37,38] of the 16S rRNA gene of *Escherichia coli* corresponding to the positions 341 to 805. The PCR reaction volume of 25 µL contained 1 µM of each primer. PCR results were obtained through the GeneAmp PCR System 9700 (Thermo Fisher Scientific). Amplicons were checked on 1.5% agarose gel and then purified through the system Agencourt AMPure XP (Beckman Coulter, Brea, CA, USA). A further PCR was necessary to apply dual indices and Nextera XT Index Primer (Illumina, San Diego, CA, USA), which are Illumina sequencing adapters. The resulting libraries were purified as reported above. The libraries were also checked for quality on the Typestation 2200 platform (Agilent Technologies, Santa Clara, CA, USA). Barcoded libraries were pooled in an equimolar ratio and sequenced on the MiSeq Illumina® (PE300) platform (MiSeq Control Software 2.5.0.5 and Real-Time Analysis software 1.18.54.0).

2.6.2. Illumina Data Analysis and Sequences Identification by QIIME2

FASTQ files containing raw paired-end sequences were demultiplexed by idemp (https://github.com/yhwu/idemp/blob/master/idemp.cpp, accessed on 17 December 2021) and imported into Quantitative Insights Into Microbial Ecology, Qiime2, version 2020.11 [39]. The DADA2 program was used for quality-filtering, trimming, de-noising, and merging of sequences [40]. All chimeric sequences were removed through the consensus method in DADA2. Sequence alignment was performed with MAFFT and phylogenetic reconstruction of the aligned sequences occurred in FastTree using the plugins alignment and phylogeny [41]. Taxonomic and compositional analyses were conducted w the plugins feature classifier (https://github.com/qiime2/q2-feature-classifier, access on 17 December 2021). A pre-trained Naive Bayes classifier based on the Greenger gg_13_5_otus.tar.tgz Operational Taxonomic Units (OTUs) database (http://greengen secondgenome.com/?prefix=downloads/greengenes_database/gg_13_5/, accessed on December 2021/), which had been previously trimmed to the V4 region of 16S rDN bound by the 341F/805R primer pair, was applied to paired-end sequence reads to generate taxonomy tables.

Data generated by Illumina sequencing were uploaded in the NCBI Sequence Re Archive (SRA) and are available under Ac. PRJNA809221.

2.7. Alpha Diversity Analysis

Alpha diversity helps us to understand the bacterial community structure in terms the number of taxonomic groups (richness) and/or the distribution of group abundar (evenness). Richness and evenness of the sites investigated were analyzed through Shann Wiener diversity index (H_{Sh}) and Gini–Simpson diversity index (H_{Si}) [42,43]:

$$H_{Sh} = -\sum_{i=1}^{N_s} p_i \times \ln(p_i)$$

$$H_{Si} = 1 - \sum_{i=1}^{N_S} p_i^{\,2}$$

In these equations, p_i is the ratio between the number of sequences of the OTUs i (as on the total sequences) and the richness N_s (total number of OTUs identified).

The results of the sites with different N_s were compared by equitability (the ev distributed values) of the indices H_{Sh} and H_{Si} determined as follows:

$$E_{Sh} = \frac{H_{Sh}}{\ln N_s}$$

$$E_{Si} = \frac{H_{Si}}{1 - \frac{1}{N_S}}$$

where E_{Sh} and E_{Si} represent the even-distributed values of H_{Sh} and H_{Si}, respectively.

2.8. Statistical Analysis

Microbiological count data were subjected to One-Way Variance Analysis (ANOV using XLStat software version 7.5.2 for Excel (Addinsoft, New York, NY, USA). Tukey's t was applied only for comparison between different pedons at the same depth. Statisti significance was attributed to p values of $p < 0.05$ and are marked with different letters.

Principal component analysis (PCAn) was carried out to evaluate the relationsh between the bacterial phyla and the physicochemical of each sample. Kaiser criterion w used to select the number of principal factors with an eigen value >1.00 [44]. A check of statistical significance within the data set was performed using Barlett's sphericity test [Data were processed using the XLStat software reported above. In addition, agglomerati hierarchical clustering (AHC) (joining, tree clustering) was performed to group data soils based on their mutual dissimilarity, measured by Euclidean distances. A clus aggregation was obtained, applying the single linkage method of Todeschini (1998) [46

In order to better correlate bacterial taxa and soil properties, a correlation analy between bacterial phyla and all physicochemical characteristics of the soil samples w performed by means of Pearson's correlation coefficient calculation within the XLS software reported above. The results were plotted in a heat map.

3. Results

3.1. Physicochemical Characteristics of Soils

The results of physicochemical analysis of the studied soil plot are reported in Table 1. The concentration of $CaCO_3$ in the control soil Mant 0 (190–200 g/kg) was consistently lower than the values registered in the anthropogenic soil samples Mant 1–Mant 3 (360–640 g/kg). In particular, the last samples showed highly variable data, with the highest levels detected for the deepest aliquots of Mant 1 and Mant 3. On the contrary, clay levels of control soils were higher than those characterizing the anthropogenic soil samples. The EC the values of Mant 0 were lower than values registered for the other pedons and an increasing trend was observed with depth. Regarding C_{org}, the values of all anthropogenic pedons were lower than those displayed by the control soil Mant 0. Among anthropogenic sites, the highest value of C_{org} was reached by the sample Mant 2-II (6.97 g/kg), while the highest value in Mant 0 was shown by the most superficial sample Mant 0-I (15.11 g/kg) and decreased with depth until 8.52 g/kg.

Table 1. Main physicochemical features of investigated anthropic soils profiles in Giordano area.

Profile	Depth	$CaCO_3$ Tot (g/kg)	Clay (g/kg)	Silt (g/kg)	Sand (g/kg)	pH (H_2O)	EC (μS/cm)	CEC ($Cmol_{(+)}$/kg)	C_{org} (g/kg)
Mant 0	I	200	446	269	285	7.83	82.0	33.75	15.11
Mant 0	II	190	452	304	244	7.89	93.2	25.00	8.72
Mant 0	II	200	498	259	243	7.76	95.5	31.25	8.52
Mant 1	I	550	371	308	321	7.87	147.6	22.50	3.10
Mant 1	II	640	295	328	377	7.90	145.5	20.00	2.13
Mant 1	III	640	369	342	289	7.89	192.5	18.75	2.90
Mant 2	I	550	363	418	219	7.87	172.5	21.30	5.03
Mant 2	II	360	402	317	281	7.88	150.0	25.00	6.97
Mant 2	III	420	390	307	303	7.90	160.5	21.30	3.87
Mant 3	I	460	363	299	338	7.86	178.9	22.50	5.62
Mant 3	II	640	333	373	294	7.88	154.9	20.00	5.23
Mant 3	III	640	314	313	373	7.88	141.1	21.30	6.00

Abbreviations are as follows: $CaCO_3$ Tot, total content of carbonate; EC, electrical conductivity; CEC, cation exchange capability; C_{org}, content of organic carbon. Soil samples: Mant 0, control soil (original Vertisol not affected by pedotechnics), while Mant 1, Mant 2 and Mant 3 represent the anthropogenic soils of sites 1 to 3. Depths: I, 0–10 cm; II, 10–30 cm; and III, 30–50 cm.

3.2. Analysis of Microbial Communities

The results of the microbiological counts are reported in Table 2. TMA cell densities were in the range 6.06–7.56 CFU per gram of soil. Each pedon displayed a significant decreasing level for TMA with soil depth; TMAn were also at six orders of magnitude per gram of soil and the highest cell densities were registered for the superficial (A) samples. TMAn levels for the samples B and C for all three anthropogenic pedons were consistently lower than the corresponding pedons of the undisturbed soil.

FF and AB groups were characterized by cell densities of 1–3 Log cycles lower than those displayed by TMA. In general, at the highest depth, FF were below the detection limit, with the exception of pedon Mant 2 for which barely 2.0 Log CFU per g dw were found. A higher variability was indicated by AB levels, which were undetectable from B sample for pedon Mant 2, while around 104 CFU per g dw for the other pedons, control soil included.

Bacteria grown on synthetic media CM and NK were isolated to better investigate in future on their role in soil particle aggregation. About 10 colonies sharing the same mucoid appearance and white/yellow color were picked up from agar plates and promptly purified. A total of 834 bacterial cultures presumptively belonging to *Caulobacter* and *Sphingomonas* genera were then microscopically processed, and barely 423 isolates showed a rod-shape.

Table 2. Results of plate count agar of the microbial groups.

Soil	TMA	FF	AB	TMAn	NF	*Sphingomonas*	*Caulobacter*
Control soil							
Mant 0-I	6.78 ± 0.10 b	5.74 ± 0.04 a	4.65 ± 0.05 a	6.61 ± 0.11 a	6.90 ± 0.18 ab	6.42 ± 0.06 c	6.87 ± 0.07 ab
Mant 0-II	6.34 ± 0.08 b	5.30 ± 0.00 a	4.39 ± 0.09 a	6.19 ± 0.05 a	6.30 ± 0.08 a	6.20 ± 0.10 a	6.52 ± 0.20 ab
Mant 0-III	6.14 ± 0.02 c	<2 b	4.00 ± 0.00 b	6.12 ± 0.03 a	5.78 ± 0.17 b	6.00 ± 0.08 b	6.31 ± 0.01 b
Statistical significance	***	***	***	***	***	**	**
Anthropogenic soil							
Mant 1-I	7.00 ± 0.07 b	4.74 ± 0.04 b	4.60 ± 0.20 a	6.45 ± 0.09 a	6.59 ± 0.30 b	6.23 ± 0.03 d	6.64 ± 0.28 b
Mant 1-II	6.22 ± 0.01 bc	4.30 ± 0.30 b	4.30 ± 0.20 a	5.71 ± 0.13 b	6.03 ± 0.30 a	5.97 ± 0.15 b	6.20 ± 0.09 ab
Mant 1-III	6.15 ± 0.05 c	<2 b	4.45 ± 0.15 a	5.75 ± 0.04 b	5.97 ± 0.11 b	5.76 ± 0.14 c	6.29 ± 0.05 b
Statistical significance	***	***		***	*	**	*
Anthropogenic soil							
Mant 2-I	7.09 ± 0.09 ab	4.69 ± 0.09 b	5.00 ± 0.25 a	6.52 ± 0.06 a	7.08 ± 0.04 a	6.63 ± 0.1 b	6.78 ± 0.00 ab
Mant 2-II	6.06 ± 0.01 c	4.24 ± 0.24 b	<2 b	5.43 ± 0.05 c	5.44 ± 0.05 b	5.64 ± 0.00 c	6.5 ± 0.30 b
Mant 2-III	6.37 ± 0.06 b	2.00 ± 0.30 a	<2 c	5.77 ± 0.01 b	6.00 ± 0.02 b	6.10 ± 0.01 b	6.31 ± 0.08 b
Statistical significance	***	***	***	***	***	***	*
Anthropogenic soil							
Mant 3-I	7.56 ± 0.35 a	5.70 ± 0.00 a	4.59 ± 0.11 a	6.66 ± 0.03 a	7.29 ± 0.04 a	6.81 ± 0.07 a	7.06 ± 0.01 a
Mant 3-II	6.77 ± 0.17 a	5.39 ± 0.09 a	4.30 ± 0.25 a	6.06 ± 0.09 a	6.39 ± 0.16 a	6.33 ± 0.03 a	6.59 ± 0.07 a
Mant 3-III	6.74 ± 0.00 a	<2 b	4.00 ± 0.00 b	6.14 ± 0.02 a	6.43 ± 0.17 a	6.52 ± 0.01 a	6.82 ± 0.13 a
Statistical significance	**	***	*	***	***	***	**
Statistical significance I	**	***			**	***	*
Statistical significance II	***	***	***	***	***	***	*
Statistical significance III	***	***	***	***	**	***	***

Abbreviations are as follows: TMA, total mesophilic count in aerobic condition; FF, filamentous fungi; A actinobacteria; TMAn, total mesophilic count in anaerobic condition; NF, nitrogen fixing bacteria. Mant 0. contr soil (original Vertisol not affected by pedotechnics), while Mant 1, Mant 2 and Mant 3 represent the anthropogen soils of sites 1 to 3. Depths: I, 0–10 cm; II, 10–30 cm; and III, 30–50 cm. Results indicate mean values ± S.D. four microbiological counts (carried out in duplicate for two independent sample collections). Unit of measu CFU/100 g of dried soil. Soil samples: Data within a column followed by the same letter are not significant different according to Tukey's test. *. $p \leq 0.05$; **. $p \leq 0.01$; ***. $p \leq 0.001$.

About 40% of the 423 isolates were subjected to RAPD analysis to differentiate th cultures at strain level, and 63 different profiles (Figure 3) indicated a certain biodive sity within putative EPS-producing bacteria. The bacteria showing different RAPD pa terns were considered distinct strains. The dendrogram showed that the majority c strains grouped according to their genus and species, even though the two species c *Stenotrophomonas*, St. *indicatrix* and St. *rhizophila*, clustered quite distantly, the first withi *Variovorax* and the second within *Pseudomonas* and *Lysobacter*. All strains were furth processed by 16S rRNA gene sequencing and a remarkable species diversity emerged fro these results. The isolates were allotted into three phyla (Firmicutes, Actinobacteria an Protobacteria) consisting of 19 genera; the most numerous bacterial groups were repr sented by *Bacillus* and *Pseudomonas*, which were found in all pedons analyzed. Surprising none of the 63 strains were identified as *Sphingomonas* or *Caulobacter*.

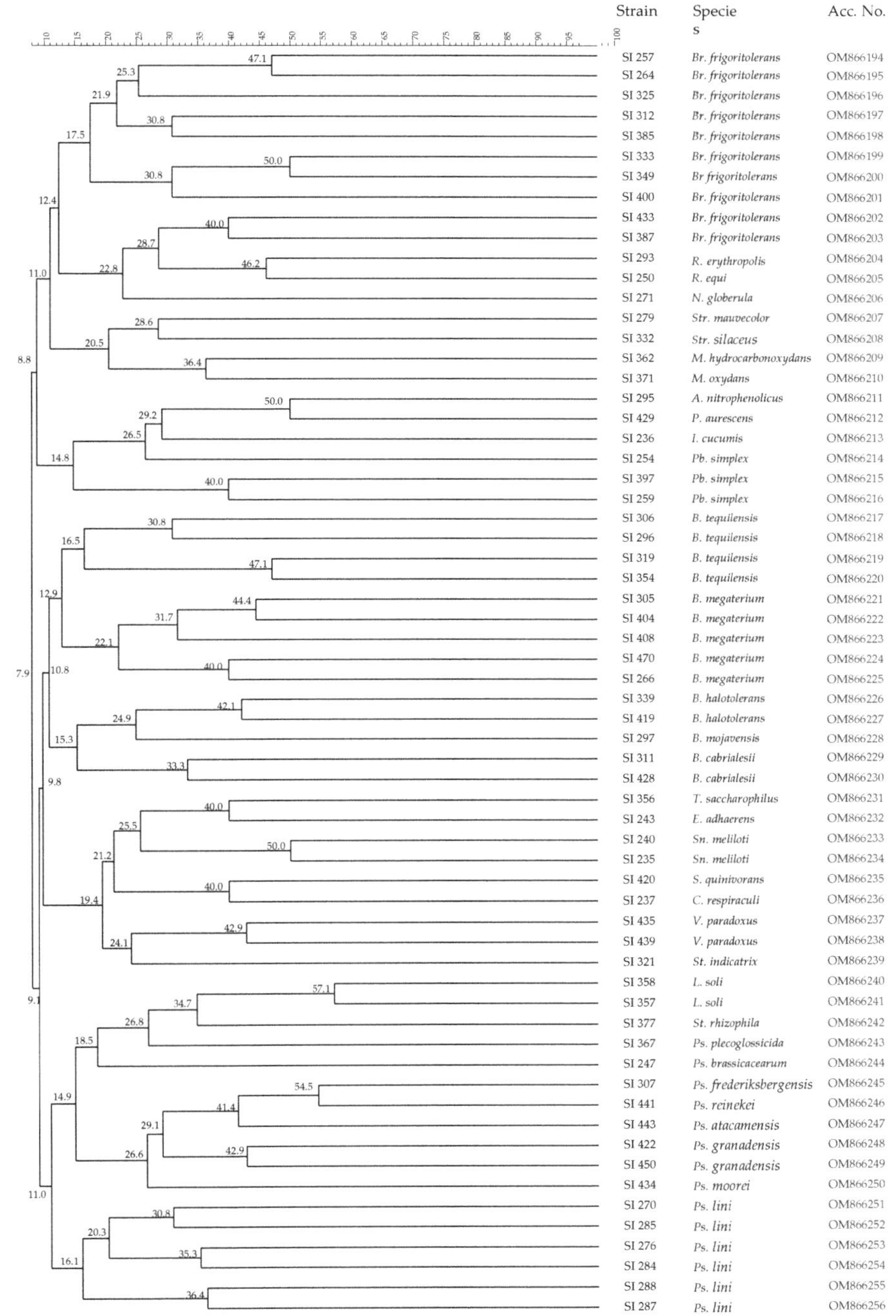

Figure 3. Dendrogram obtained from combined RAPD-PCR patterns generated with three primers (M13, AB106 and AB111) of the dominant bacteria identified from anthropogenic soils. Abbreviations: A., *Arthrobacter*; B., *Bacillus*; Br., *Brevibacterium*; C., *Cupriavidus*; E., *Ensifer*; I., *Isoptericola*; L., *Lysobacter*; M., *Microbacterium*; N., *Nocardia*; P., *Paenarthrobacter*; Pb., *Peribacillus*; Ps., *Pseudomonas*; R., *Rhodococcus*; S., *Serratia*; Sn., *Sinorhizobium*; St., *Stenotrophomonas*; Str., *Streptomyces*; T., *Terribacillus*; V., *Variovorax*.

3.3. Culture-Independent Analysis

In order to deeply analyze the bacterial communities of control and anthropogenic s samples, a next-generation sequencing (NGS) approach was also applied. This approa was necessary because the great majority of microorganisms (>99%) of natural ecosyste cannot be detected by means of the culture-dependent tools alone [47]. In this study, all s samples were investigated using MiSeq Illumina technology, and the results are reported Figure 4 (at phylum level) and Figure S1 (all taxonomic levels mixed). The total bacterial versity of the soil under evaluation was composed of 11 phyla—Actinobacteria, Chlorofle Firmicutes and Proteobacteria were present in all pedons at the three depths investigat

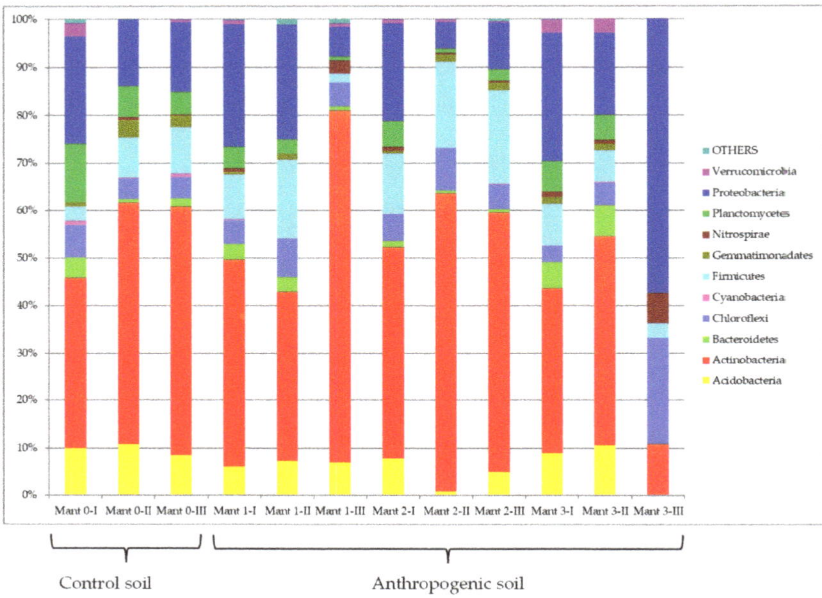

Figure 4. Relative abundances (%) of bacterial phyla identified by MiSeq Illumina in soil samp Abbreviations are as follows: Soil samples: Mant 0, control soil (original Vertisol not affected pedotechnics), while Mant 1, Mant 2 and Mant 3 represent the anthropogenic soils of sites 1 t Depths: I, 0–10 cm; II, 10–30 cm; and III, 30–50 cm.

With the exception of pedon Mant 3-III, Actinobacteria represented the most abunda phylum in almost all samples analyzed, reaching the maximum level (74.04%) in Ma 1-III. Among this phylum, only members of Gaiellaceae family were found in all sampl with percentages ranging from 4.46% in Mant 3-I and 33.06% in Mant 1-III. Within t phylum, radiotolerant and halotolerant bacteria belonging to the *Rubrobacter* genus w quite ubiquitous, even though their presence was particularly variable (0.5–23.5%).

The presence of Firmicutes, especially within the genus *Bacillus*, was detected in alm all pedons of both control and anthropogenic soils; in particular, this genus was detec at very high percentages in all Mant 2 pedons (12.63–19.06%). Additionally, the order Clostridiales was found in almost all pedons, reaching the maximum abundance (9.08%) Mant 1-II.

Six families of Proteobacteria (Bradyrhizobiaceae, Rhodospirillaceae, Syntrophobac aceae, Enterobacteriaceae, Moraxellaceae and Pseudomonadaceae) were also ubiquitous the soil samples, but at percentages lower than those of Actinobacteria. Only in Mant 3- did Proteobacteria account for 57.45% of total OTUs. *Erwinia* and *Acinetobacter* were most consistent genera, with 18.34 and 16.81%, respectively.

Two main classes within Chloroflexi phylum were present in our samples, in particular, Anaerolineae and Ktedonobacteraceae. The first group was detected in all soil samples, with percentages ranging from 0.03% in Mant 3-I to 8.40% in Mant 3-III, while the second was found only in Mant 3-III, but the relative abundance (14.01%) was quite high.

3.4. Analysis of Biodiversity Indexes (Alpha Diversity)

The intra-site diversity and richness were assessed by alpha diversity considering the phyla of the bacterial community. The distribution of phyla among soils samples was studied and the results are reported in Table 3.

Table 3. Quantitative data analysis of bacterial phyla (alpha diversity).

Samples	No. of Phyla	H_{Sh}	E_{Sh}	H_{Si}	E_{Si}
Mant 0-I	11	1.84	0.74	0.79	0.86
Mant 0-II	11	1.58	0.64	0.70	0.76
Mant 0-III	12	1.58	0.64	0.68	0.75
Mant 1-I	12	1.64	0.66	0.73	0.79
Mant 1-II	9	1.73	0.69	0.78	0.85
Mant 1-III	10	1.06	0.42	0.44	0.48
Mant 2-I	12	1.64	0.66	0.73	0.80
Mant 2-II	11	1.20	0.48	0.56	0.61
Mant 2-III	12	1.44	0.58	0.65	0.71
Mant 3-I	11	1.82	0.73	0.78	0.85
Mant 3-II	11	1.77	0.71	0.75	0.82
Mant 3-III	5	1.17	0.47	0.60	0.66
Reference value for a perfect even community		2.48	1	0.92	1

Abbreviations: H_{Sh}, Shannon–Wiener diversity index; E_{Sh}, equitability of H_{Sh}; *his*, Gini–Simpson diversity index; E_{Si}, equitability of H_{Si}.; Soil samples: Mant 0, control soil (original Vertisol not affected by pedotechnics), while Mant 1, Mant 2 and Mant 3 represent the anthropogenic soils of sites 1 to 3. Depths: I, 0–10 cm; II, 10–30 cm; and III, 30–50 cm.

The total number of phyla (n. = 12) was found only in four samples, while only one soil sample (Mant 3-III) was characterized by a very low number of phyla (n. = 5). The samples collected at the highest depths showed the lowest biodiversity since the samples Mant 1-III, Mant 3-III and Mant 2-II were characterized by a H_{Sh} of 1.06, 1.17 and 1.20, respectively. Shannon indexes increased until 1.82 and 1.84 in samples Mant 3-I and Mant 0-I, respectively, indicating that the most superficial soil horizon is characterized by a high bacterial biodiversity. This trend was also confirmed by Simpson's index, since the lowest H_{Si} (0.44) was registered for the sample Mant 1-III and the highest (0.79) for sample Mant 0-I. Regarding this last biodiversity index, except at site Mant 2, the samples collected at the first two depths (0–10 cm and 10–30 cm) from the anthropogenic plot were highly comparable. Both biodiversity indexes and their equitability showed that the bacterial community of control and anthropogenic soils are not particularly different.

3.5. Multivariate Data Analysis

All pedons were subjected to AHC (Figure 5) to obtain their classification in accordance with their mutual dissimilarity and relationship based on physicochemical parameters and relative abundance (%) of microbiota. Anthropogenic soil samples clustered almost randomly considering their depth. In particular, the only two samples that showed a very low dissimilarity were Mant 1-I and Mant 2-I, while Mant 3-I clustered with Mant 3-II, and Mant 2-II with Mant 2-III. Considering a dissimilarity level of 30%, all Mant 0 samples formed a single mega-cluster distant from the other clusters. The results showed that the mega-cluster of Mant 0 samples is significantly different from the other two clusters of anthropogenic soils with values of dissimilarity higher than 60.65% and values of within-class variance lower than 13.20%. This analysis clearly demonstrated high dissimilarit among anthropogenic soil samples.

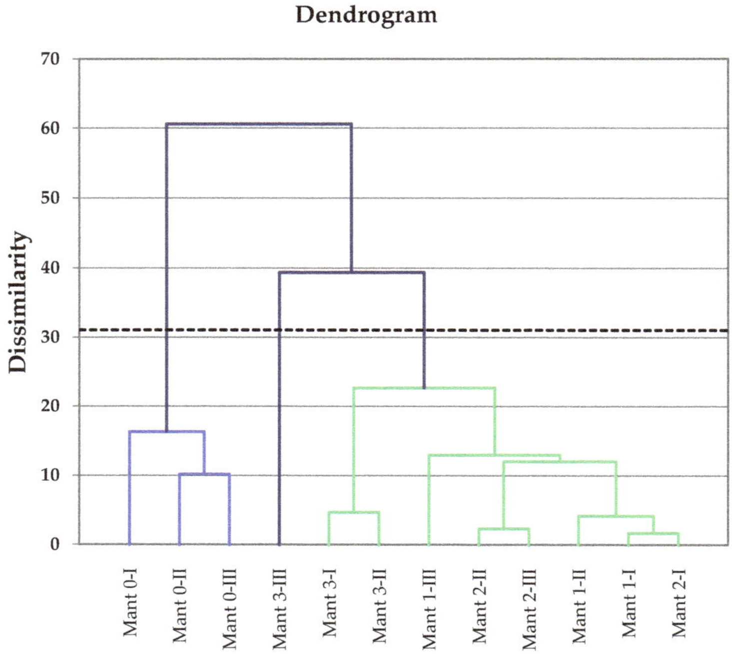

Figure 5. Dendrogram obtained from the hierarchical cluster analysis (AHC). Abbreviations are follows: Soil samples: Mant 0, control soil (original Vertisol not affected by pedotechnics), while Ma 1, Mant 2 and Mant 3 represent the anthropogenic soils of sites 1 to 3. Depths: I, 0–10 cm; II, 10–30 c and III, 30–50 cm.

Furthermore, PCA was also carried out with the aim to discriminate a very high nur ber of physicochemical and microbiological parameters in a few factors (Figure 6). Facto 1 and 2 had an eigen value higher than 3 and explained 58.71% of total variability (38.56 and 20.14%. for F1 and F2., respectively). Biplot representation shows that CEC and C_o strongly influence the F1. Considering microbial groups, Acidobacteria, Planctomycet and Cyanobacteria showed a strong influence on the factor F1, while Verrucomicrobia ar Bacteroidetes only showed a weak influence on F1, and this contributed to grouping contr soil samples into a mega cluster. On the contrary, Proteobacteria and Chloroflexi strong influence the F2 factor.

To examine the reciprocal influence of bacterial community and soil characteristics, correlation analysis was performed between bacterial phyla and physicochemical parar eters of the samples investigated (Figure 7). The most significant positive correlations Planctomyces, Gemmatimonadates, Cyanobacteria and Acidobacteria were found for cl CSC and C_{org}. A highly positive correlation with clay was also showed by Actinobacter with C_{org} by Bacteroidetes, and with CSC and C_{org} by Verrucomicrobia. Chloroflexi, N trospirae, Proteobacteria. Non-identified members of the community showed the highe positive influence on limestone and sand. A positive influence on the last characterist was also registered for Bacteroidetes. Actinobacteria and Firmicutes were mainly impact soil conductivity. Finally, Firmicutes and other members influenced soil pH.

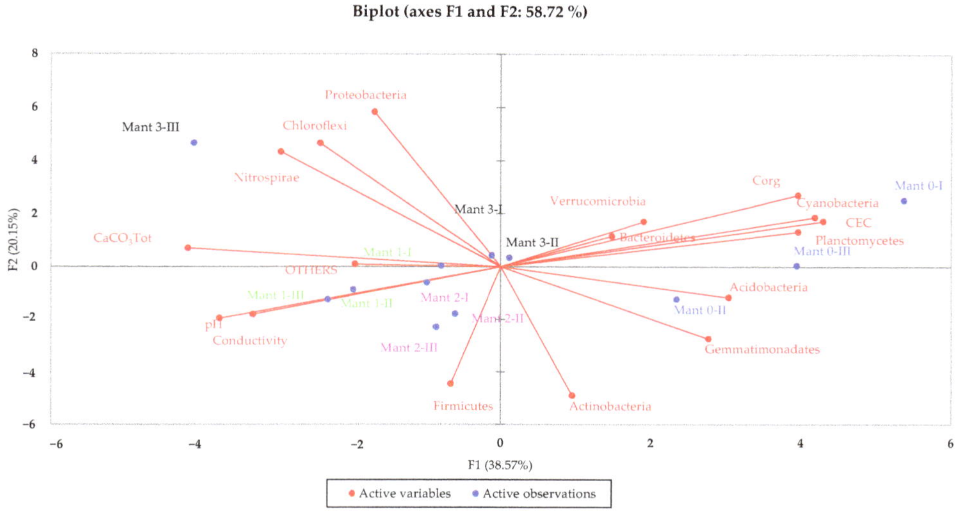

Figure 6. Principal component analysis (PCA) among the microbiological and physicochemical characteristics of soil sample. Abbreviations are as follows: $CaCO_3$ Tot; total content of carbonate; EC, electrical conductivity; CEC, cation exchange capability; C_{org}, content of organic carbon. Soil samples: Mant 0, control soil (original Vertisol not affected by pedotechnics), while Mant 1, Mant 2 and Mant 3 represent the anthropogenic soils of sites 1 to 3. Depths: I, 0–10 cm; II, 10–30 cm; and III, 30–50 cm.

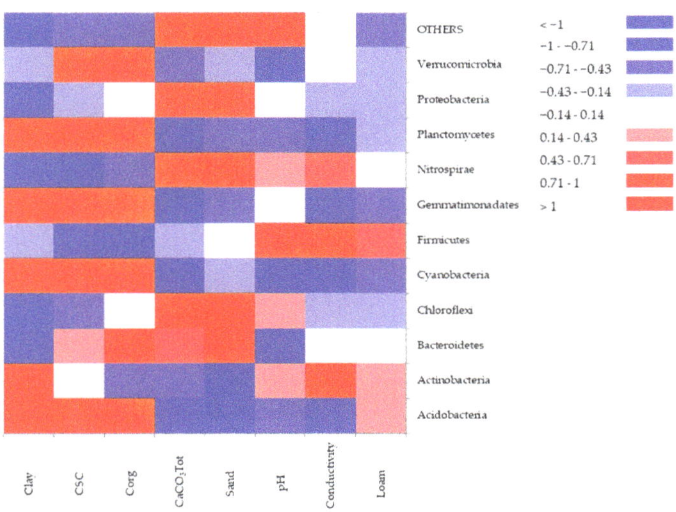

Figure 7. Pearson's correlation between bacterial phyla and physicochemical characteristics of soil samples. Colour intensity indicates the level of association. Abbreviations are as follows: $CaCO_3$ Tot; total content of carbonate; EC, electrical conductivity; CEC, cation exchange capability; C_{org}, content of organic carbon.

4. Discussion

The success of anthropogenic soils is mainly due to the high economic incomes deriving from high yields through high labor inputs [48]. Since the 1970s, the entire agricultural area of Palma di Montechiaro (Sicily) has been subjected to a change in land use for

economic purposes. Olive yards and almond yards were converted into vineyards [
Moreover, the cultivated soils of this area were transformed by native farmers into
thropogenic soils, adding consistent amounts of calcareous marls in order to improve
organoleptic traits of table grape.

The chemical parameters of the anthropogenic soils under investigation were co
pared to those of control soil. The concentration of $CaCO_3$ of the control soil was betwe
190 and 200 g/kg, comparable to that characterizing other Haploxerert soils located in
same area [50]. The concentration of calcium carbonate of the anthropogenic soils w
higher, in the range 420–640 g/kg. These high results have been also registered for sc
from Quaternary limestone [51]. The values of pH of anthropogenic soils were around
an average value displayed by soils with calcareous marls [6,50–52]. These high pH val
are due to the high concentrations of $CaCO_3$ that exerts a buffer capacity [53]. Furtherm
CEC of control soil was different from that of the anthropogenic soils analyzed. Regardi
this parameter, the average value of the anthropogenic soils was lower than that of cont
soil because of the lower clay percentage after intervention [51,54–57]. The concentrati
of C_{org} Tot was highest in control soil; this parameter ranged between 8.52 and 15.11 g/
commonly registered for Vertisols [50]. The level of C_{org} Tot in anthropogenic soils w
consistently low (2.13 to 6.97 g/kg) and this is due to the dilution effect determined by
addition of calcareous-marl material.

To our knowledge, anthropogenic soil microbiology in Sicily is almost unexplor
and the basic relationships among soil modifications and bacterial evolution merit furt
investigation. In general, the anthropic actions applied to soil might cause microb
biodiversity reduction in the short term [58]. Thus, in this work, microbial imaging analy
was performed after 7 years from soil modification, just before the first cycle of pl
cultivation. The microbial communities of the soils located in Palma di Mantechiaro w
first investigated by a polyphasic culture-dependent approach to provide direct evide
on the viable populations. Plate count analysis carried out on soil samples showed t
TMA levels were, on average, in the range 6.7–7.5 Log CFU per g dw. These lev
characterize commonly uncultivated soils [59,60]. In general, TMA decreased with de
and this finding is imputed to nutrient and oxygen limitations [61]. FF levels were 1
2 orders of magnitude lower than those of TMA. These results are due to the fact t
bacteria fill an ecological niche composed of aerobic, facultative anaerobic and N-fi
microorganisms [56], which is much larger than that occupied by fungi. The same lev
of microscopic fungi were reported by numerous authors in the superficial parts of bu
soils [60,61]. Generally, in agricultural soils, FF decrease at undetectable levels already
20–30 cm, but in our anthropogenic soil samples, the absence of fungi was only recorc
at a 30–50 cm depth. This could be explained by a greater penetration of oxygen into
deeper layers of soils enriched with calcareous material. In all cases, the viable levels
the microbial groups investigated decreased with depth. For Actinomycetes, the co
range was close to that of FF, settling on values ranging between 4.0 and 5.0 Log C
per g dw, confirming what has been observed by other authors [62,63]. The same auth
also reported that the highest Actinomycetes densities are mostly found in the superfi
soil horizon.

In order to identify bacteria playing a direct role in soil aggregate formation, the gen
more typically associated with EPS production were investigated after colony isolati
Bacteria grown on media used to count *Sphingomonas* and *Caulobacter* were purified a
cultured in broths in order to perform 16S rRNA gene sequencing. Despite growth on
and NK, none of the 63 presumptive *Sphingomonas* and *Caulobacter* isolates were confirm
to belong to these genera. These findings are imputable to the scarce selectivity of the t
agar media used for these genera. The isolates were confirmed to represent 63 differ
strains by RAPD investigation. The comparison of the 16S rRNA gene sequences w
those available in GenBank/DDBJ/EMBL and EZtaxon databases allotted the 63 stra
into 19 genera of three phyla: Actinobacteria, Firmicutes and Protobacteria. All three ph
generally represent the most abundant soil bacterial groups [64,65]. Bacteria of the ge

Bacillus have been found in all pedons. Five strains were identified as *Bacillus megaterium*, which is known for its ability to produce EPS [66]. Furthermore, eight *Pseudomonas* strains have been also identified. Bacteria of this genus are widely known to produce EPS [16]. They have been used singly or in association with Stenotrophomonas to stabilize aggregates of artificial soils [67]. Among presumptive EPS-producing bacteria, some strains were identified as *Cupriavidus* and *Sinorhizobium*, both N-fixing bacteria. Besides *Bacillus* and *Pseudomonas*, some of the other bacteria identified, such as *Streptomyces* and *Serratia*, are considered plant-growth-promoting bacteria [68].

In order to deepen the information about the bacterial community, MiSeq Illumina technology was applied to total bacterial DNA extracted from all soil samples. The superficial layer of control soil was mainly characterized by the presence of Actinobacteria, Acidobacteria, Protobacteria and Planctomycetes. Apart from Planctomycetes, the other bacterial orders are generally detected in cultivated and uncultivated soils [69,70]. The dominant phyla of bacteria for the anthropogenic soils were almost comparable to those of control soil. However, some exceptions were observed; Mant 3 was characterized by Proteobacteria, Actinobacteria and Chloroflexi as the bacteria found at the highest percentages. These groups are generally detected in soils subjected to conventional and minimum tillage [71]. The high relative abundance of Ktedonobacteraceae in Mant 3-III is unsurprising, since this family uses sugars and peptides as carbon sources [72]. Among these bacteria, Acidobacteria are known as ecological indicators because they are oligotrophic and are present in soils with a scarce presence of nutrients, while, on the contrary, being detected at very low levels in fertilized soils [73]. On the other hand, Proteobacteria are considered copiotrophs [74] and are generally commonly detected in environments with greater nutritional opportunities [75]. In the current study, Proteobacteria was widely represented, for example in soil sample Mant 3-III, where their presence accounted for 57.45% of relative abundance. The dominance of this phylum over the bacterial community of soil has been reported by several authors [76–79]. These bacteria play important roles in soil; for example, Bradyrhizobiaceae and Rhodospirillaceae found almost in all samples are N-fixers [80,81], while Oxalobacteraceae, found in A and B layers of control and anthropogenic soils, are the main degraders of soil organic matter and they are also involved in production of phytohormones, such as auxin, gibberellin and siderophores [82]. These families of Protobacteria are also indicators of soil fatigue [70]. Copiotrophs consistently contribute to carbon mineralization [83].

Actinobacteria and Firmicutes cannot be considered copiotrophic nor oligotrophic [59], but rather as ubiquitous bacteria capable of living in completely different conditions such as forest and grassland soils [78,79,84] and even in polluted soils [77,85,86]. Specifically, *Microbacterium* detected in almost all soils are able to degrade polycyclic aromatic hydrocarbons with a high molecular weight [87]. In particular, the culture-dependent approach confirmed that *Microbacterium oxydans* and *Microbacterium hydrocarbonoxydans* were present in viable form in the superficial horizon.

An abundance of Chloroflexi are identified among soil bacterial communities [88–91]. The primary reason for their ubiquity is due to their metabolic diversity; this phylum includes heterotrophs, lithotrophs and phototrophs adapted to both oxic and anoxic environments [92]. All soil samples analyzed in this study displayed bacteria within this phylum in a range similar to those previously documented by other authors, even though, surprisingly, Mant 3-III showed 22.42% of relative abundance allotted into Chloroflexi. With regard to *Sinorhizobium*, *Pseudomonas*, *Bacillus* and *Lysobacter* genera, the culture-independent approach showed some inconsistencies with the culture-dependent approach, since their sequences were not amplified from DNAs extracted from samples that showed their presence in viable form. This is not surprising, since these genera might have been allotted into unassigned OTUs or their sequences were damaged by nucleases or were below 0.1% abundance and not considered during classification.

Diversity indexes are useful to evaluate differences among environmental samples [93]. In this study, Shannon and Simpson indexes were calculated to compare control and anthropogenic soils at the depths 0–10, 10–30 and 30–50 cm in terms of bacterial taxa diversi at the phylum level. Due to the dominance of given phyla, the bacterial community wa basically uneven in all samples investigated. Alpha diversity analysis clearly showed tha bacterial diversity decreases with depth, and that among anthropogenic and control soi the differences were quite negligible.

Finally, all data were subjected to a multivariate statistical analysis. Both PCA an AHC have been largely used to study the behavior of microorganisms in soil, the influenc of different external variables in the distribution of microorganisms or to study diffe ent anthropogenic soils [94,95]. The main chemical parameters implicated on microbia diversity were pH and C_{org}. These two parameters are generally used as predictors o bacterial diversity in a huge number of sites around the world [96,97]. In those studies, soil pH around 7 correlated with a high biodiversity, while a decrease in pH determine a reduction in bacterial biodiversity. Regarding the main bacterial groups found in ou study, Actinobacteria were not significantly correlated with pH and C_{org}, and this coul be explained by the ubiquity and capacity of these bacteria to live in different terrestria environments, even the most extreme soils characterized by nutrient deficiencies [98–10 Furthermore, Proteobacteria, Chloroflexi and Firmicutes were negatively correlated wi C_{org} value. Regarding the direct influence of bacterial taxa on soil characteristics, th correlation analysis confirmed the trend found for PCA. The spatial position of the thre pedons could not be a factor driving any variance in the microbial profile, since the onl factor dominant and regulating all soil variables is the man action in generating suc soils by transformation of the original Vertisol using the described pedotechnique. Th effects of any natural soil genetic factors were canceled at the moment of the creation of th anthropogenic soil.

5. Conclusions

In conclusion, plate count results showed a significant decrease in almost all microbi groups with increasing sample depth. The genetic analysis of the strains highlighte the presence of 19 different genera, denoting a remarkable biodiversity within the pl analyzed. Only two genera (*Bacillus* and *Pseudomonas*) were found in all pedons. Th results obtained from the isolation and sequencing of putative EPS-producing bacter identified the presence of *Bacillus*, *Pseudomonas* and *Stenotrophomonas*. Further studies wi be performed to test their potential to improve the aggregation of soils in a Mediterranea environment. Next-generation sequencing indicated that the most represented phyla i anthropogenic soils are Actinobacteria and Proteobacteria. Viable dominant populatio indicated that the anthropogenic soil hosted a mature microbial community useful fo land cultivation.

Supplementary Materials: The following supporting information can be downloaded at: http //www.mdpi.com/article/10.3390/land11050748/s1, Figure S1: Relative abundances (%) of bacte identified by MiSeq Illumina in soil samples.

Author Contributions: Investigation, P.B., E.F. and R.D.G.; data curation, P.B. and R.D.G.; visu ization, P.B. and G.L.P.; writing—original draft preparation, P.B.; conceptualization, C.D. and L. resources, C.D.; supervision, C.D.; writing—review and editing, L.S. and G.L.P.; funding acquisitic L.S.; project administration, G.L.P. All authors have read and agreed to the published version the manuscript.

Funding: This research has received co-funding from the European Commission's ERASMU Programme under grant agreement No 2017-1-SE01-KA203-034570.

Institutional Review Board Statement: Not applicable.

Informed Consent Statement: Not applicable.

Conflicts of Interest: The authors declare no conflict of interest.

ferences

Dudal, R.; Nachtergaele, F.O.; Purnell, M.F. Human factor of soil formation. In Proceedings of the 17th World Congress Soil Science, Bangkok, Thailand, 14–21 August 2002.

Bockheim, J.G.; Gennadiyev, A.N. The role of soil-forming processes in the definition of taxa in Soil Taxonomy and the World Soil Reference Base. *Geoderma* **2000**, *95*, 53–72. [CrossRef]

Bryant, R.B.; Galbraith, J.M.; Eswaran, H. *Incorporating Anthropogenic Processes in Soil Classification*; CRC Press: New York, NY, USA, 2003.

Dazzi, C.; Lo Papa, G.; Palermo, V. Proposal for a new diagnostic horizon for WRB Anthrosols. *Geoderma* **2009**, *151*, 16–21. [CrossRef]

Andrés-Abellán, M.; Del Álamo, J.B.; Landete-Castillejos, T.; López-Serrano, F.R.; García-Morote, F.A.; del Cerro-Barja, A. Impacts of visitors on soil and vegetation of the recreational area "Nacimiento del Rio Mundo" (Castilla-La Mancha, Spain). *Environ. Mon. Assess.* **2005**, *101*, 55–67.

Dazzi, C.; Monteleone, S. Anthropogenic processes in the evolution of a soil chronosequence on marly-limestone substrata in an Italian Mediterranean environment. *Geoderma* **2007**, *141*, 201–209. [CrossRef]

Hearing, K.C.; Daniels, W.L.; Galbraith, J.M. Mapping and classification of southwest Virginia mine soils. *Soil Sci. Soc. Am. J.* **2005**, *69*, 463–472. [CrossRef]

Salvati, L.; Zitti, M.; Ceccarelli, T. Integrating economic and environmental indicators in the assessment of desertification risk: A case study. *Appl. Ecol. Environ. Res.* **2008**, *6*, 129–138. [CrossRef]

Morgan, R.P.C. *Soil Erosion and Conservation*; John Wiley & Sons: Hoboken, New Zealand, 2009.

Peña-Angulo, D.; Nadal-Romero, E.; González-Hidalgo, J.C.; Albaladejo, J.; Andreu, V.; Bagarello, V.; Barhi, H.; Batalla, R.J.; Bernal, S.; Bienes, R.; et al. Spatial variability of the relationships of runoff and sediment yield with weather types throughout the Mediterranean basin. *J. Hydrol.* **2019**, *571*, 390–405. [CrossRef]

Pankhurst, C.E.; Ophel-Keller, K.; Doube, B.M.; Gupta, V.V.S.R. Biodiversity of soil microbial communities in agricultural systems. *Biodivers. Conserv.* **1996**, *5*, 197–209. [CrossRef]

Allison, S.D.; Martiny, J.B.H. Resistance, resilience, and redundancy in microbial communities. *Proc. Natl. Acad. Sci. USA* **2008**, *105*, 11512–11519. [CrossRef]

Torsvik, V.; Øvreås, L. Microbial diversity and function in soil: From genes to ecosystems. *Curr. Opin. Microbiol.* **2002**, *5*, 240–245. [CrossRef]

Flemming, H.C.; Wingender, J. Relevance of microbial extracellular polymeric substances (EPSs)-Part I: Structural and ecological aspects. *Water Sci. Technol.* **2001**, *43*, 1–8. [CrossRef] [PubMed]

Wingender, J.; Jaeger, K.E.; Flemming, H.C. Interaction between extracellular polysaccharides and enzymes. In *Microbial Extracellular Polymeric Substances*; Wingender, J., Neu, T.R., Flemming, H.C., Eds.; Springer: Berlin, Germany, 1999; pp. 231–251.

Costa, O.Y.; Raaijmakers, J.M.; Kuramae, E.E. Microbial extracellular polymeric substances: Ecological function and impact on soil aggregation. *Front. Microbiol.* **2018**, *9*, 1636. [CrossRef] [PubMed]

Vert, M.; Doi, Y.; Hellwich, K.H.; Hess, M.; Hodge, P.; Kubisa, P.; Rinaudo, M.; Schué, F. Terminology for biorelated polymers and applications (IUPAC Recommendations 2012). *Pure Appl. Chem.* **2012**, *84*, 377–410. [CrossRef]

Saini, R.; Saini, S.; Sharma, S. Biofilm: A dental microbial infection. *J. Nat. Sci. Biol. Med.* **2011**, *2*, 71. [CrossRef]

Vlastarakos, P.V.; Nikolopoulos, T.P.; Maragoudakis, P.; Tzagaroulakis, A.; Ferekidis, E. Biofilms in ear, nose, and throat infections: How important are they? *Laryngoscope* **2007**, *117*, 668–673. [CrossRef]

Vardharajula, S.; Ali, S.Z. The production of exopolysaccharide by *Pseudomonas putida* GAP-P45 under various abiotic stress conditions and its role in soil aggregation. *Microbiology* **2015**, *84*, 512–519.

Wang, X.; Sharp, C.E.; Jones, G.M.; Grasby, S.E.; Brady, A.L.; Dunfield, P.F. Stable-isotope probing identifies uncultured Planctomycetes as primary degraders of a complex heteropolysaccharide in soil. *Appl. Environ. Microbiol.* **2015**, *81*, 4607–4615. [CrossRef]

Bronick, C.J.; Lal, R. Soil structure and management: A review. *Geoderma* **2005**, *124*, 3–22. [CrossRef]

Dinel, H.; Levesque, P.E.M.; Jambu, P.; Righi, D. Microbial activity and long-chain aliphatics in the formation of stable soil aggregates. *Soil. Sci. Soc. Am. J.* **1992**, *56*, 1455–1463. [CrossRef]

Chenu, C. Extracellular polysaccharides: An interface between microorganisms and soil constituents. In *Environmental Impact of Soil Component Interactions*, 1st ed.; Huang, P.M., Berthelin, J., Bollag, J.M., McGill, W.B., Eds.; CRC Press: New York, NY, USA, 1995; pp. 217–233.

Taubner, H.; Roth, B.; Tippkötter, R. Determination of soil texture: Comparison of the sedimentation method and the laser-diffraction analysis. *J. Plant Nutr. Soil Sci.* **2009**, *172*, 161–171. [CrossRef]

Walkley, A.; Black, I.A. An examination of the Degtjareff method for determining soil organic matter, and a proposed modification of the chromic acid titration method. *Soil Sci.* **1934**, *37*, 29–38. [CrossRef]

Thomson, B.C.; Ostle, N.J.; McNamara, N.P.; Whiteley, A.S.; Griffiths, R.I. Effects of sieving, drying and rewetting upon soil bacterial community structure and respiration rates. *J. Microbiol Meth.* **2010**, *83*, 69–73. [CrossRef] [PubMed]

Settanni, L.; Miceli, A.; Francesca, N.; Moschetti, G. Investigation of the hygienic safety of aromatic plants cultivated in soil contaminated with *Listeria monocytogenes*. *Food Control.* **2012**, *26*, 213–219. [CrossRef]

29. Zhang, X.; Ma, L.; Gilliam, F.S.; Wang, Q.; Li, C. Effects of raised-bed planting for enhanced summer maize yield on rhizosph soil microbial functional groups and enzyme activity in Henan Province, China. *Field Crops Res.* **2012**, *130*, 28–37. [CrossRef]
30. Sun, L.N.; Yang, E.D.; Wei, J.C.; Tang, X.Y.; Cao, Y.Y.; Han, G.M. *Caulobacter flavus* sp. nov., a stalked bacterium isolated fr rhizosphere soil. *Int. J. Syst. Evol. Microbiol.* **2015**, *65*, 4374–4380. [CrossRef]
31. Huang, H.D.; Wang, W.; Ma, T.; Li, G.Q.; Liang, F.L.; Liu, R.L. *Sphingomonas sanxanigenens* sp. Nov., isolated from soil. *Int. J. S Evol. Microbiol.* **2009**, *59*, 719–723. [CrossRef]
32. Rippka, R.; Waterbury, J.B.; Stanier, R.Y. Isolation and purification of cyanobacteria: Some general principles. In *The Prokaryo* Starr, M.P., Stolp, H., Trüper, H.G., Balows, A., Schlegel, H.G., Eds.; Springer: Berlin, Germany, 1981; pp. 212–220.
33. Gaglio, R.; Francesca, N.; Di Gerlando, R.; Mahony, J.; De Martino, S.; Stucchi, C.; Moschetti, G.; Settanni, L. Enteric bacteria food ice and their survival in alcoholic beverages and soft drinks. *Food Microbiol.* **2017**, *67*, 17–22. [CrossRef]
34. Weisburg, W.; Barns, S.M.; Pelletier, D.A.; Lane, D.J. 16S ribosomal DNA amplification for phylogenetic study. *J. Bact.* **1991**, 697–703. [CrossRef]
35. Altschul, S.F.; Madden, T.L.; Schäffer, A.A.; Zhang, J.; Zhang, Z.; Miller, W.; Lipman, D.J. Gapped BLAST and PSI-BLAST: A n generation of protein database search programs. *Nucleic Acids Res.* **1997**, *25*, 3389–3402. [CrossRef]
36. Chun, J.; Lee, J.H.; Jung, Y.; Kim, M.; Kim, S.; Kim, B.K.; Lim, Y.W. EzTaxon: A web-based tool for the identification of prokaryo based on 16S ribosomal RNA gene sequences. *Int. J. Syst. Evol. Microbiol.* **2007**, *57*, 2259–2261. [CrossRef]
37. Baker, G.C.; Smith, J.J.; Cowan, D.A. Review and re-analysis of domain-specific 16S primers. *J. Microbiol. Methods* **2003**, 541–555. [CrossRef] [PubMed]
38. Claesson, M.J.; Wang, Q.; O'sullivan, O.; Greene-Diniz, R.; Cole, J.R.; Ross, R.P.; O'Toole, P.W. Comparison of two next-generat sequencing technologies for resolving highly complex microbiota composition using tandem variable 16S rRNA gene regio *Nucleic Acids Res.* **2010**, *38*, e200. [CrossRef] [PubMed]
39. Bolyen, E.; Rideout, J.R.; Dillon, M.R.; Bokulich, N.A.; Abnet, C.C.; Al-Ghalith, G.A.; Alexander, H.; Alm, E.J.; Arumugam, Asnicar, F.; et al. Reproducible, interactive, scalable and extensible microbiome data science using QIIME 2. *Nat. Biotechnol.* **2** *37*, 852–857. [CrossRef] [PubMed]
40. Callahan, B.J.; McMurdie, P.J.; Rosen, M.J.; Han, A.W.; Johnson, A.J.A.; Holmes, S.P. DADA2: High-resolution sample infere from Illumina amplicon data. *Nat. Methods* **2016**, *13*, 581–583. [CrossRef]
41. Katoh, K.; Standley, D.M. MAFFT multiple sequence alignment software version 7: Improvements in performance and usabi *Mol. Biol. Evol.* **2013**, *30*, 772–780. [CrossRef]
42. Spellerberg, I.F.; Fedor, P.J. A tribute to Claude Shannon (1916–2001) and a plea for more rigorous use of species richness, spe diversity and the 'Shannon–Wiener'Index. *Global Ecol. Biogeog.* **2003**, *12*, 177–179. [CrossRef]
43. Chao, A.; Jost, L.; Chiang, S.C.; Jiang, Y.H.; Chazdon, R.L. A two-stage probabilistic approach to multiple-community simila indices. *Biometrics* **2008**, *64*, 1178–1186. [CrossRef]
44. Jolliffe, I.T. *Principal Component Analysis for Special Types of Data*; Springer: New York, NY, USA, 2002.
45. Mazzei, P.; Francesca, N.; Moschetti, G.; Piccolo, A. NMR spectroscopy evaluation of direct relationship between soils a molecular composition of red wines from Aglianico grapes. *Anal. Chim. Acta* **2010**, *673*, 167–172. [CrossRef]
46. Todeschini, R. *Introduzione Alla Chemiometria: Strategie, Metodi e Algoritmi Per L'analisi e il Modellamento di Dati Chimici, Farmacolo e Ambientali*; Edises: Naples, Italy, 1998.
47. Wen, C.; Wu, L.; Qin, Y.; Van Nostrand, J.D.; Ning, D.; Sun, B.; Xue, K.; Liu, F.; Deng, Y.; Liang, Y.; et al. Evaluation of reproducibility of amplicon sequencing with Illumina MiSeq platform. *PLoS ONE* **2017**, *12*, e0176716. [CrossRef]
48. Moran, E.F. Socio-economic aspects of acid soil management. In *Plant-Soil Interactions at Low pH: Principles and Management*; D R.A., Grundon, N.J., Rayment, G.E., Probert, M.E., Eds.; Springer: Berlin, Germany, 1995; pp. 663–669.
49. Lo Verde, F.M. *Agricoltura e Mutamento Sociale*; Éditions L'Harmattan: Paris, France, 1995.
50. Lo Papa, G.; Antisari, L.V.; Vianello, G.; Dazzi, C. Soil interpretation in the context of anthropedogenic transformations a pedotechniques application. *Catena* **2018**, *166*, 240–248. [CrossRef]
51. Asio, V.B.; Cabunos, C.C., Jr.; Chen, Z.S. Morphology, physiochemical characteristics, and fertility of soils from Quatern limestone in Leyte, Philippines. *Soil Sci.* **2006**, *171*, 648–661. [CrossRef]
52. Loeppert, R.H.; Suarez, D.L. Carbonate and gypsum. In *Methods of Soil Analysis Part 3. Chemical Methods*; Sparks, D.L., Page, A Helmke, P.A., Loeppert, R.H., Eds.; John Wiley & Sons: Hoboken, NJ, USA, 1996; pp. 437–474.
53. Ulrich, B. Natural and anthropogenic components of soil acidification. *J. Plant Nutr. Soil Sci.* **1986**, *149*, 702–717. [CrossRef]
54. Ghorbani, H.; Kashi, H.; Hafezi Moghadas, N.; Emamgholizadeh, S. Estimation of soil cation exchange capacity using mult regression, artificial neural networks, and adaptive neuro-fuzzy inference system models in Golestan Province, Iran. *Comm Soil Sci. Plant Anal.* **2015**, *46*, 763–780. [CrossRef]
55. Khaledian, Y.; Brevik, E.C.; Pereira, P.; Cerdà, A.; Fattah, M.A.; Tazikeh, H. Modeling soil cation exchange capacity in multi countries. *Catena* **2017**, *158*, 194–200. [CrossRef]
56. McBratney, A.B.; Minasny, B.; Cattle, S.R.; Vervoort, R.W. From pedotransfer functions to soil inference systems. *Geoderma* **2** *109*, 41–73. [CrossRef]
57. Ulusoy, Y.; Tekin, Y.; Tümsavas, Z.; Mouazen, A.M. Prediction of soil cation exchange capacity using visible and near infra spectroscopy. *Biosyst. Eng.* **2016**, *152*, 79–93. [CrossRef]

8. Dazzi, C.; Galati, A.; Crescimanno, M.; Lo Papa, G. Pedotechnique applications in large-scale farming: Economic value, soil ecosystems services and soil security. *Catena* **2019**, *181*, 104072. [CrossRef]
9. Rech, M.; Pansera, M.R.; Sartori, V.C.; Ribeiro, R.T.D.S. Microbiota do solo em vinhedos agroecológico e convencional e sob vegetação nativa em Caxias do Sul. RS. *Rev. Bras. Agroecol.* **2013**, *8*, 141–151.
0. Silva, J.G.D.; da Luz, J.M.R.; Henrique, J.; de Carvalho, J.J.; da Silva, J.E.C. Domestic wastewater for forage cultivation in Cerrado soil. *J. Agric. Sci.* **2018**, *10*, 248. [CrossRef]
1. Carvalho, J.J.; da Luz, J.M.R.; Henrique, J.; Silva, J.G.D.; Silva, J.E.C.; Santos, E.A. Biofertigation of forage with effluents of green line of a cattle slaughterhouse: Microbial diversity and leaf dry mass productivity. *J. Agric. Sci.* **2018**, *10*, 353. [CrossRef]
2. Neeraj, K.S. Organic amendments to soil inoculated arbuscular mycorrhizal fungi and *Pseudomonas fluorescens* treatments reduce the development of root-rot disease and enhance the yield of *Phaseolus vulgaris* L. *Eur. J. Soil Biol.* **2011**, *47*, 288–295. [CrossRef]
3. Smith, S.E.; Facelli, E.; Pope, S.; Smith, F.A. Plant performance in stressful environments: Interpreting new and established knowledge of the roles of arbuscular mycorrhizas. *Plant Soil.* **2010**, *326*, 3–20. [CrossRef]
4. Fierer, N.; Bradford, M.A.; Jackson, R.B. Toward an ecological classification of soil bacteria. *Ecology* **2007**, *88*, 1354–1364. [CrossRef] [PubMed]
5. Lennon, J.T.; Aanderud, Z.T.; Lehmkuhl, B.K.; Schoolmaster, D.R., Jr. Mapping the niche space of soil microorganisms using taxonomy and traits. *Ecology* **2012**, *93*, 1867–1879. [CrossRef] [PubMed]
6. Chowdhury, S.R.; Manna, S.; Saha, P.; Basak, R.K.; Sen, R.; Roy, D.; Adhikari, B. Composition analysis and material characterization of an emulsifying extracellular polysaccharide (EPS) produced by *Bacillus megaterium* RB-05: A hydrodynamic sediment-attached isolate of freshwater origin. *J. Appl. Microb.* **2011**, *111*, 1381–1393. [CrossRef]
7. Caesar-TonThat, T.C.; Caesar, A.J.; Gaskin, J.F.; Sainju, U.M.; Busscher, W.J. Taxonomic diversity of predominant culturable bacteria associated with microaggregates from two different agroecosystems and their ability to aggregate soil in vitro. *Appl. Soil Ecol.* **2007**, *36*, 10–21. [CrossRef]
8. Fiodor, A.; Singh, S.; Pranaw, K. The contrivance of plant growth promoting microbes to mitigate climate change impact in agriculture. *Microorganisms* **2021**, *9*, 1841. [CrossRef]
9. Wolińska, A.; Kuźniar, A.; Zielenkiewicz, U.; Izak, D.; Szafranek-Nakonieczna, A.; Banach, A.; Błaszczyk, M. Bacteroidetes as a sensitive indicator of agricultural soil usage revealed by culture-independent approach. *Appl. Soil Ecol.* **2017**, *119*, 128–137. [CrossRef]
0. Wolińska, A.; Kuźniar, A.; Zielenkiewicz, U.; Banach, A.; Błaszczyk, M. Indicators of arable soils fatigue–Bacterial families and genera: A metagenomic approach. *Ecol. Ind.* **2018**, *93*, 490–500. [CrossRef]
1. Legrand, F.; Picot, A.; Cobo-Díaz, J.F.; Carof, M.; Chen, W.; Le Floch, G. Effect of tillage and static abiotic soil properties on microbial diversity. *Appl. Soil Ecol.* **2018**, *132*, 135–145. [CrossRef]
2. Chang, Y.J.; Land, M.; Hauser, L.; Chertkov, O.; Glavina Del Rio, T.; Nolan, M.; Copeland, A.; Tice, H.; Cheng, J.-F.; Lucas, S.; et al. Non-contiguous finished genome sequence and contextual data of the filamentous soil bacterium *Ktedonobacter racemifer* type strain (SOSP1-21T). *Stand. Genomic Sci.* **2011**, *5*, 97–111. [CrossRef] [PubMed]
3. Koyama, A.; Wallenstein, M.D.; Simpson, R.T.; Moore, J.C. Soil bacterial community composition altered by increased nutrient availability in Arctic tundra soils. *Front. Microbiol.* **2014**, *5*, 1–16. [CrossRef] [PubMed]
4. Aislabie, J.; Deslippe, J.R. Soil microbes and their contribution in soil services. In *Ecosystem Services in New Zealand-Conditions and Trends*; Dyamond, J.R., Ed.; Manaaki Whenua Press: Lincoln, New Zeland, 2013; pp. 143–161.
5. Koch, A.L. Oligotrophs versus copiotrophs. *Bioessays* **2001**, *23*, 657–661. [CrossRef] [PubMed]
6. dos Santos, H.F.; Cury, J.C.; do Carmo, F.L.; dos Santos, A.L.; Tiedje, J.; van Elsas, J.D.; Rosado, A.S.; Peixoto, R.S. Mangrove bacterial diversity and the impact of oil contamination revealed by pyrosequencing: Bacterial proxies for oil pollution. *PLoS ONE* **2011**, *6*, e16943. [CrossRef] [PubMed]
7. Militon, C.; Boucher, D.; Vachelard, C.; Perchet, G.; Barra, V.; Troquet, J.; Peyretaillade, E.; Peyret, P. Bacterial community changes during bioremediation of aliphatic hydrocarbon-contaminated soil. *FEMS Microbiol. Ecol.* **2010**, *74*, 669–681. [CrossRef]
8. Nacke, H.; Thürmer, A.; Wollherr, A.; Will, C.; Hodac, L.; Herold, N.; Schöning, I.; Schrumpf, M.; Daniel, R. Pyrosequencing-based assessment of bacterial community structure along different management types in German forest and grassland soils. *PLoS ONE* **2011**, *6*, e17000. [CrossRef]
9. Roesch, L.F.; Fulthorpe, R.R.; Riva, A.; Casella, G.; Hadwin, A.K.M.; Kent, A.D.; Daroub, S.H.; Camargo, F.A.O.; Farmerie, W.G.; Triplett, E.W. Pyrosequencing enumerates and contrasts soil microbial diversity. *ISME J.* **2007**, *1*, 283–290. [CrossRef]
0. Anderson, C.R.; Condron, L.M.; Clough, T.J.; Fiers, M.; Stewart, A.; Hill, R.A.; Shrelock, R.R. Biochar induced soil microbial community change: Implications for biogeochemical cycling of carbon, nitrogen and phosphorus. *Pedobiologia* **2011**, *54*, 309–320. [CrossRef]
1. Reis, V.M.; Teixeira, K.R. Nitrogen fixing bacteria in the family Acetobacteraceae and their role in agriculture. *J. Basic Microbiol.* **2015**, *55*, 931–949. [CrossRef]
2. Ofek, M.; Hadar, Y.; Minz, D. Ecology of root colonizing *Massilia* (Oxalobacteraceae). *PLoS ONE* **2012**, *7*, e40117. [CrossRef]
3. Hu, S.J.; Van Bruggen, A.H.C.; Grünwald, N.J. Dynamics of bacterial populations in relation to carbon availability in a residue-amended soil. *Appl. Soil Ecol.* **1999**, *13*, 21–30. [CrossRef]
4. Jones, R.T.; Robeson, M.S.; Lauber, C.L.; Hamady, M.; Knight, R.; Fierer, N. A comprehensive survey of soil acidobacterial diversity using pyrosequencing and clone library analyses. *ISME J.* **2009**, *3*, 442–453. [CrossRef] [PubMed]

85. Allen, J.P.; Atekwana, E.A.; Duris, J.W.; Werkema, D.D.; Rossbach, S. The microbial community structure in petroleum contaminated sediments corresponds to geophysical signatures. *Appl. Environ. Microbiol.* **2007**, *73*, 2860–2870. [CrossRef]
86. Saul, D.J.; Aislabie, J.M.; Brown, C.E.; Harris, L.; Foght, J.M. Hydrocarbon contamination changes the bacterial diversity of so from around Scott Base, Antarctica. *FEMS Microbiol. Ecol.* **2005**, *53*, 141–155. [CrossRef] [PubMed]
87. Gauthier, E.; Déziel, E.; Villemur, R.; Juteau, P.; Lépine, F.; Beaudet, R. Initial characterization of new bacteria degrading hig molecular weight polycyclic aromatic hydrocarbons isolated from a 2-year enrichment in a two-liquid-phase culture system. *Appl. Microbiol.* **2003**, *94*, 301–311. [CrossRef] [PubMed]
88. Delgado-Baquerizo, M.; Oliverio, A.M.; Brewer, T.E.; Benavent-González, A.; Eldridge, D.J.; Bardgett, R.D.; Maestre, F.T Singh, B.K.; Fierer, N. A global atlas of the dominant bacteria found in soil. *Science* **2018**, *359*, 320–325. [CrossRef]
89. Mehrshad, M.; Salcher, M.M.; Okazaki, Y.; Nakano, S.I.; Šimek, K.; Andrei, A.S.; Ghai, R. Hidden in plain sight—highly abundan and diverse planktonic freshwater Chloroflexi. *Microbiome* **2018**, *6*, 176. [CrossRef]
90. Sunagawa, S.; Coelho, L.P.; Chaffron, S.; Kultima, J.R.; Labadie, K.; Salazar, G.; Djahanschiri, B.; Zeller, G.; Mende, D.R Alberti, A.; et al. Structure and function of the global ocean microbiome. *Science* **2015**, *348*, 1261359. [CrossRef]
91. Thompson, L.R.; Sanders, J.G.; McDonald, D.; Amir, A.; Ladau, J.; Locey, K.J.; Prill, R.J.; Tripathi, A.; Gibbons, S.M.; Ackerman G.; et al. A communal catalogue reveals Earth's multiscale microbial diversity. *Nature* **2017**, *551*, 457–463. [CrossRef]
92. Whitman, W.B. *Bergey's Manual of Systematics of Archaea and Bacteria*; Wiley Online Library: New York, NY, USA, 2015.
93. Guida, G.; Palmeri, V.; Settanni, L.; Gaglio, R.; Tolone, M.; Ferro, V. Ability of soil bacterial composition as an indicator of levels soil erosion in a badland. *Int. J. Sediment Res.* **2022**, *37*, 493–504. [CrossRef]
94. Golobočanin, D.D.; Škrbić, B.D.; Miljević, N.R. Principal component analysis for soil contamination with PAHs. *Chemometr. Inte Lab.* **2004**, *72*, 219–223. [CrossRef]
95. Wu, T.; Chellemi, D.O.; Graham, J.H.; Martin, K.J.; Rosskopf, E.N. Comparison of soil bacterial communities under divers agricultural land management and crop production practices. *Microb. Ecol.* **2008**, *55*, 293–310. [CrossRef] [PubMed]
96. Lauber, C.L.; Hamady, M.; Knight, R.; Fierer, N. Pyrosequencing-based assessment of soil pH as a predictor of soil bacteria community structure at the continental scale. *Appl. Environ. Microbiol.* **2009**, *75*, 5111–5120. [CrossRef]
97. Tripathi, B.; Kim, M.; Singh, D.; Lee-Cruz, L.; Lai-Hoe, A.; Ainuddin, A.N.; Go, R.; Abdul Rahim, R.; Husni, M.H.A.; Chun, J et al. Tropical soil bacterial communities in Malaysia: pH dominates in the equatorial tropics too. *Microb. Ecol.* **2012**, *64*, 474–48 [CrossRef]
98. Dhakal, D.; Pokhrel, A.R.; Shrestha, B.; Sohng, J.K. Marine rare actinobacteria: Isolation, characterization, and strategies fo harnessing bioactive compounds. *Front. Microbiol.* **2017**, *8*, 1106. [CrossRef]
99. Goodfellow, M.; Nouioui, I.; Sanderson, R.; Xie, F.; Bull, A.T. Rare taxa and dark microbial matter: Novel bioactive actinobacteri abound in Atacama Desert soils. *Antonie Van Leeuw.* **2018**, *111*, 1315–1332. [CrossRef]
100. Qin, S.; Xing, K.; Jiang, J.H.; Xu, L.H.; Li, W.J. Biodiversity, bioactive natural products and biotechnological potential o plant-associated endophytic actinobacteria. *Appl. Microbiol. Biotechnol.* **2011**, *89*, 457–473. [CrossRef]

sustainability MDPI

:icle

arriers and Levers for the Implantation of Sustainable ˙ature-Based Solutions in Cities: Insights from France

loé Duffaut [1,†], Nathalie Frascaria-Lacoste [2] and Pierre-Antoine Versini [1,*]

1 Hydrology Meteorology and Complexity Laboratory, École des Ponts ParisTech, 77455 Champs-sur-Marne, France
2 Université Paris-Saclay, CNRS, AgroParisTech, Ecologie Systématique Evolution, 91190 Gif-sur-Yvette, France
* Correspondence: pierre-antoine.versini@enpc.fr
† Chloé Duffaut Left the Hydrology Meteorology and Complexity Laboratory, École des Ponts ParisTech, on 1 September 2021.

Abstract: The challenges of the 21st century, namely, climate change and loss of biodiversity, especially present in heavily populated areas, should be addressed. Nature-based Solutions (NBS) seem to offer a suitable answer to these challenges. However, this new concept is not always easy to implement in a sustainable manner. In an effort to identify the barriers and levers for the implementation in cities of these sustainable NBS, several professionals working on them in France were interviewed. The first analysis with the most quoted words shows that the constraints would be mainly related to technique, management, biodiversity and people. The levers would be related to projects, roofs, people, legislation and services. Further analysis shows that most of the interviewees feel that the main barriers are the lack of knowledge, political will, financial resources and regulations. There are also cultural limitations, such as the use of exotic horticultural plants rather than wild local species. According to them, the technical problems should be easy to solve. To address these issues, the interviewees propose the development of transdisciplinary research disciplines, as well as on-field collaboration between all NBS actors in cities. They also recommend specific funds for NBS and their implication in related regulations. Demonstrative examples of urban NBS highlighting their multiple benefits are also needed to encourage their replication or upscaling. Education and communication are essential to shift the traditional points of view on nature in the city.

Keywords: nature-based solution; city; barriers; levers; interview

tion: Duffaut, C.; caria-Lacoste, N.; Versini, P.-A. iers and Levers for the lantation of Sustainable ıre-Based Solutions in Cities: ghts from France. *Sustainability* , *14*, 9975. https://doi.org/ 390/su14169975

lemic Editors: Miklas Scholz, ıyiotis Dimitrakopoulos, Mario agnotta and Arshiya Noorani

ived: 9 May 2022
pted: 21 July 2022
ished: 12 August 2022

lisher's Note: MDPI stays neutral regard to jurisdictional claims in ished maps and institutional affil- ns.

1. Introduction

Today, urban environments are facing many challenges, including global warming, loss of biodiversity and all their consequences. Within this context, Nature-Based Solutions (NBS) appear to be the ideal tool. Indeed, they can act on both fronts. The International Union for Conservation of Nature (IUCN) defines NBS as follows: "actions to protect, sustainably manage and restore natural or modified ecosystems that address societal challenges effectively and adaptively, simultaneously providing human well-being and biodiversity benefits" [1]. The European Commission defines NBS as the following: "Solutions that are inspired and supported by nature, which are cost-effective, simultaneously provide environmental, social and economic benefits and help build resilience. Such solutions bring more, and more diverse, nature and natural features and processes into cities, landscapes and seascapes, through locally adapted, resource-efficient and systemic interventions." [2]. These two definitions are often used as references, but others also exist. Indeed, there are many definitions of NBS, but most agree that these solutions are beneficial to the environment and to humans rather than just focusing on the restoration of nature and its conservation [3]. The following examples can be listed: restoration of mountainous slope forests, which help avoid erosion and landslides; preservation of mangroves to limit the risk of submersion; revegetation of buildings to improve their thermal properties.

The NBS concept is mainly European and is often associated with climate chan urbanization, water management, urban heat islands, air pollution, well-being, hum health and sustainability [4]. The concept is recent and is still emerging [5,6], with ma related scientific publications. In a search performed by the authors on 10 June 202_, Scop found 975 documents with titles, abstracts and keywords containing the term "natu based solutions". Under the topic of the urban environment, the usual cited NBS are follows: urban forests [7,8], green roofs and walls [9,10], ecological corridors [11], or gre swales [9,12]. They are usually studied to manage stormwater issues or to mitigate urb heat islands.

It is worth noting that some scientists, especially in the field of nature conservati treat NBS as "yet another buzzword" [13]. Even if the concept is currently a hot topic, i challenged by its physical implementation, which is often complicated by many factors [These difficulties are particularly noticeable in the urban environment due to its typi dense nature. Indeed, the implementation of NBS in urban settings raises the followi number of operational questions: How to adequately integrate them into the concern context? How to evaluate their costs and benefits? How to design solutions to meet t different challenges? etc. [15]. In addition to these issues, there are many other barri facing the implementation of sustainable NBS in cities. With the help of a dozen peo interviewed working on NBS in France, this study aims to identify these barriers a highlight the levers that can be used to overcome them. By responding to this proble answers for subsequently upscaling NBS in cities can be obtained.

2. Materials and Methods

For this study, several actors in France working on NBS were interviewed in a se structured way. The interviewees were selected because the authors knew that they their organizations/companies were working on urban NBS in France for many ye Hence, they were able to share relevant and eclectic discourses, experiences and analy The authors are aware that this selection may have led to a bias, but as this work rather qualitative, this bias may be negligible. This study allows for the identification barriers and levers by "experts" in the field and is not a quantitative study on the le of knowledge of the general public. In order to cover the whole spectrum of specialt interviewees from academic, institutional and operational circles were chosen, ensuri the same number of people in each category. The profiles of these interviewees are lis herewith the following: Ecologist and project leader at the Regional Biodiversity Ager researchers in Hydrology or Ecology at the Centre for Studies and Expertise on Risks, Environment, Mobility and Urban Planning (CEREMA), the French Research Instit for Development (IRD) and the Museum of Natural History, general manager and he of the planning and natural environment department of the Intercommunal Union Hydraulic Development of the Croult and Petit Rosne Valleys, project manager at IUCN, independent agro-economist and project manager at the Gally design office, pro manager in a design study at Topager, technical manager at SOPRANATURE (gre roofs and facades of SOPREMA, building, waterproofing and insulation company), an manager of the EcoQuartier mission (EcoQuartier is an approach supported by the Fre Ministry of Ecological Transition, which promote new ways of designing, building a managing the city sustainably. It includes 20 commitments, a label, a club to meet and to to train). More details are provided in Table 1.

Table 1. List of interviewees with their reference number, function, organization, French acronym and website of their organization, sector of activity and the date they were interviewed. In the rest of the article, the letters in brackets correspond to the reference letters in this table to indicate, which interviewee is cited.

Letter	Function	Organization	Acronym	Website	Sector	Date
A	Head of studies	Regional Biodiversity Agency	ARB	https://www.arb-idf.fr/ (accessed on 20 July 2022)	Institutional	17 November 2020
B	Regional animator	Regional Biodiversity Agency	ARB	https://www.arb-idf.fr/ (accessed on 20 July 2022)	Institutional	27 November 2020
C	Researcher in hydrology	Center for studies and expertise on risks, environment, mobility and development	CEREMA	https://www.cerema.fr/fr (accessed on 20 July 2022)	Academic	4 December 2020
D	Head of the urban planning and natural environment department	Mixed syndicate for the hydraulic development of valleys	SIAH	https://www.siah-croult.org/ (accessed on 20 July 2022)	Operational	8 December 2020
E	NBS project manager	International Union for Conservation of Nature	UICN	https://uicn.fr/ (accessed on 20 July 2022)	Institutional	14 December 2020
F	General manager	Mixed syndicate for the hydraulic development of valleys	SIAH	https://www.siah-croult.org/ (accessed on 20 July 2022)	Operational	16 December 2020
G	Research director in ecology	Institute of Research for Development	IRD	https://www.ird.fr/node/8 (accessed on 20 July 2022)	Academic	7 January 2021
H	Project manager	Gally's design office (design, plants, urban biodiversity and agriculture)	GALLY	https://www.lesjardinsdegally.com/agence/le-bureau-detudes-de-gally (accessed on 20 July 2022)	Operational	20 January 2021
I	Project manager/Head of mission	Topager (Edible and wild urban landscape)/Museum of Natural History	TOPAGER/MNHN	http://topager.com/ (accessed on 20 July 2022) https://www.mnhn.fr/fr (accessed on 20 July 2022)	Operational/Academic	3 February 2021
J	Researcher in ecology	Center for studies and expertise on risks, environment, mobility and development	CEREMA	https://www.cerema.fr/fr (accessed on 20 July 2022)	Academic	4 February 2021
K	Technical manager	SOPRANATURE (Vegetation system)	SOPREMA	https://www.soprema.fr/fr/nos-produits/vegetalisation/sopranature (accessed on 20 July 2022)	Operational	19 February 2021
L	Responsible of EcoQuartier	Ministry of Ecological Transition	MTE	http://www.ecoquartiers.logement.gouv.fr/ (accessed on 20 July 2022)	Institutional	4 March 2021

A questionnaire was especially prepared for this purpose. Most of the questions wer developed by the authors; however, the first Author adapted some of the questions to f the profile of each interviewee. The questionnaire was divided into the following six part (i) context of the interview, (ii) personal information, (iii) NBS in the urban environmen (iv) biodiversity and the ecosystem functions of these solutions, (v) constraints, barriers an levers for the implementation and the sustainability of these solutions, and (vi) perspective The main questions about barriers and levers are the following: "In your opinion, wha could be the obstacles to the implementation and sustainability of NBS in cities?" an "What are the levers that could be used to promote the establishment and sustainabili of NBS in cities?". For more details on the questions, see the "Questionnaire used durin interviews" in the Supplementary Material.

The semi-structured interviews were conducted in French by the first author via videoco ference techniques using the Meet Jitsi website (https://meet.jit.si/; accessed on 20 July 202 at the following address: https://meet.jit.si/EntretiensSolutionsFondeesNature (accessed o 20 July 2022). Each interview lasted between 45 and 90 min. They were carried out betwee 17 November 2020 and 4 March 2021. The first hour of the interviews was recorded (imag and sound) directly via the Meet Jitsi site and saved on the Dropbox cloud. The comple interviews were recorded (sound only) via a voice recorder on a digital tablet. In addition handwritten notes taken during the interviews, full transcripts were made from the video an audio recordings.

First, word clouds were created with the answers translated into English to the que tions on constraints and barriers on the one hand and on levers on the other hand. Th enabled us to highlight the words most used by the interviewees in their answers (the mo often a word is quoted the bigger it appears in the cloud). The word clouds were made wit R and Iramuteq software using the active forms of the words (verbs, nouns, adjectives) cite at least 5 times for barriers and 3 times for levers. Words that did not provide informatic were removed (put, lot, thing, good, case, addition, false, feel, today, small, real, relate, bac subject, start, true, set, show, fact, begin, end, sense, remain, bite, area, bring, percentag type, part, talk, necessarily). Words of the same family were merged.

Then, a deep analysis of the results of the interviews was conducted to explain in deta the main identified constraints and levers. In order to illustrate the different argumen put forward by the interviewees, verbatim their answers are inserted in the presentatio of the results, and the related discussion. Illustrating a personal opinion, these verbatim shed light on a specific topic and provide relevant information that can be confronted wi bibliographic references. The results of the interviews are divided into two parts. Th first part lists the constraints and difficulties encountered by the actors in implementin Nature-Based Solutions. The second part gives the levers for effective implementatio under a framework of sustainability.

3. Results and Discussion about Barriers to the Implementation of Sustainable NBS

According to the word cloud (Figure 1a), the constraints to urban NBS are various. Th most highlighted words can be related to categories discussed in the following Sub-Sectior technique (Section 3.1.2), management (Section 3.1.3), biodiversity (Section 3.3.2), peop (Section 3.3.3), etc.

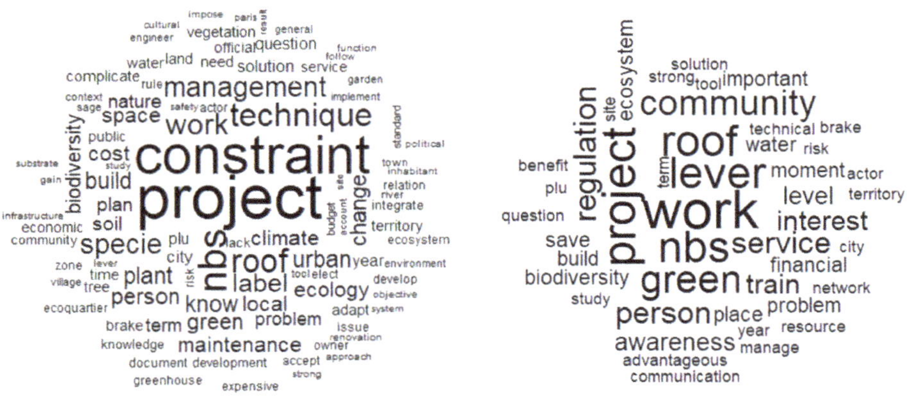

(a) (b)

Figure 1. (**a**) Word clouds with answers to the questions on constraints and barriers to the implementation and sustainability of nature-based solutions (NBS) in cities; (**b**) word clouds with answers to the questions on levers to the implementation and sustainability of NBS in cities. The larger a word appears, the more it has been quoted.

3.1. Knowledge and Technical Barriers

3.1.1. Confusion Due of Multiplicity of Terms and Lack of Knowledge

The interviewed ecologist points out a problem, which is the *"multitude of concepts that designate more or less the same thing, that come together to designate nature-related elements: nature-based solutions, green and blue solutions, green and blue webs, nature in the city, ecological engineering, urban restoration, green infrastructure, etc."* (A). The same person adds the following: *"I think what you call NBS, is not called so by the people who implement them, especially in urban areas. NBS is a new concept, not very common in communities, they'll talk more about a landscape project, a wetland or a retention pond, a park or a garden, etc."* (A). For him, this term is mainly used by scientists. The problem of NBS-connected terms, even synonymous ones, has already been raised in the scientific literature [16]. The multiplicity of terms can be counterproductive, as it makes research more difficult to conduct. Accordingly, there is a risk of misunderstanding due to contested nomenclature that might lead to a replication of efforts. Moreover, the IUCN project manager said that the term NBS *"[is] not always well-used (one of my roles is to re-define it to ensure that it is used well)"* (E). It is therefore necessary to recall the adopted definition to homogenize the discourse among different stakeholders.

Several interviewees pointed out that there is still a lack of data and knowledge on NBS in urban settings. Even if studies on the subject have multiplied in recent years, there are still grey zones that need to be filled [17,18]. Indeed, ecosystems and the urban environment are complex, and their interrelationship is difficult to analyze. Thus, soils are often underrepresented and tools enhancing soil quality are needed, according to one interviewee. An interviewed researcher in ecology said that it was necessary to *"increase scientific knowledge on the components of biodiversity that allow to maximize ecosystem functions and services"* (J). According to the literature, there is little evidence of the multiple benefits of NBS [3,19].

A project leader on NBS said the following: *"There is a lack of indicators that reveal the efficiency of NBS. [...] It is difficult to evaluate the gains resulting from NBS, especially when taking into account all the costs and benefits, and not just the targeted problem. It is accordingly hard to mobilize stakeholders, due to the lack of concrete proof on returns."* (B). According to Kabisch et al. [18], there is a need to develop indicators that incorporate environmental performance, health and well-being, citizen engagement, and maintenance/transferability. People who implement NBS need to be reassured that these solutions will deliver the expected benefits; otherwise, investing in them might be useless. It is also important mention that long-term studies (i.e., over large time spans) on NBS in cities are still too f especially those dealing with the longevity of the advantages brought by these solutio

3.1.2. Technical Problems That Can Often Be Overcome

Despite a general lack of knowledge, many interviewees reported that technical issu were not the most difficult elements to solve. In fact, most of the time, technicians kn how to eliminate these types of constraints thanks to their own knowledge and to their fi expertise, as illustrated by an ecologist as follows: *"Yes, there are technical constraints, they can be easily overcome, if you do things in the right order with specialists it can be done."* (There is even a technical handbook specific to NBS that can help people work on them [However, not all technical constraints are easily overcome. While this might be the c for technical constraints tied up to buildings, the technical constraints linked to ecolo are harder to solve. Indeed, as observed, there is a definite lack of knowledge and skills this particular field. As a guideline, a hydrologist told the authors that simplicity shou be sought as follows: *"Simpler solutions should be developed. Currently, NBS comprise of a of things: some things are very technical and others are more elementary. In my views, the m elementary solutions are closer to NBS and are better than the very technical aspects."* (C).

3.1.3. Lack of Maintenance and Durability

The ecologist of the Regional Biodiversity Agency said, *"There is usually a lack monitoring over time, that is to say a before/after assessment. Also, there is a lack of understand on whether a gain for biodiversity has been attain and whether certain ecological functions u affected, [. . .] to make an annual follow-up for instance is difficult, [. . .] Similarly, the fear f part of communities exists based on not knowing what works over a long term."* (A). The hydrau union manager for the natural environment (D) gave the example of a watercourse t was reopened in 2014, where only a single fauna and flora inventory had been carried According to the manager, it would have been better to carry out a survey every two years least. She also explained that long-term monitoring is important to determine, for instar *"the behavior of the stream after dry weather, or rainy weather, and see how the vegetation mana to recover."* (D). She added that it was tricky to find *"a good boundary between the desire f natural stream and the reality of urban area."* (D). NBSs are partly made up of living beings a therefore are in constant evolution, thus requiring continuous maintenance. However, t maintenance is not always taken into consideration in NBS projects [18]. The problem lack of monitoring and/or maintenance may be related to the fact that administrations h a short-term view when a long-term one should be favored for NBS [21,22]. The problem maintenance raises the following number of questions: *"Who will do the maintenance? H do we make sure we have sustainable funding for this maintenance?"* (E).

Although maintenance can be seen as a barrier to NBS, it can also represent an portunity, as one person from the water development union (D) explained. Indeed, projects on which they work must manage rainwater at the plot parcel level. Currently the case of underground water basins, maintenance is performed only in the first few ye Then, because the basins are not visible, maintenance tends to drop. So, the union fav the management of open-air rainwater, for which it is easier to implement a perenn maintenance program and is prone to the creation of NBS such as wetlands.

3.2. Contextual Barriers

3.2.1. The Challenge of Adapting NBS to Local Climate and Climate Change

In the questionnaire, the climatic factors are considered constraints because th require a significant adaptation of NBS (e.g., choice of plant species). Such a task is always easy and requires considerable knowledge. Ideally, it is better to use native pla adapted to the local climate and to adapt the plant range on a case-by-case basis. Howe for practical reasons, this is not always possible.

In addition, climate change and its consequences must be taken into account, and species that can cope with them must be found in the near future [16,23]. However, this is not an obvious task, as the project manager of the Gally design office said the following: *"If I plant trees, they must still be there in 20 or 30 years. Hence, how do I select them?"* (H). A researcher asked the following similar question: *"In anticipation of a change in climate, shouldn't we start considering species that are not necessarily local, but more adapted to periods of drought (e.g., Mediterranean species)?"* (J). Even if certain people try to find solutions to answer this problem, it is not yet the norm and many more efforts are still needed.

To limit greenhouse gases and therefore mitigate climate change, it might be interesting to consider short-term actions, for example, trying to use local materials for the substrate of green roofs. In this regard, a seller of green roofs explained that this is something that can be very complicated to do in practice. Indeed, it is necessary to *"[extract] the soil at the right time during the construction phase, [characterize it] in the laboratory, [use it] partly on a roof [after having it mounted]."* (K). Mounting the soil is just as complicated, as follows: *"You have to put it in big bags, which are very expensive. This means that you have to set up a bagging workshop on site or you have to move the local material onto a platform with semi-trailer trucks to use this material, bag it as is or slightly rework it with lightening or structural elements, to bring it back to the site for use."* (K). This leads to having to rethink the implantation/construction methods.

3.2.2. Too Little Space in the City and Land Prices Are Often High

Something that has not been mentioned much, but which is a characteristic reality of the urban environment, is the lack of space and the price of land which can be very costly in some cities such as Paris. Indeed, the average built density ((footprint × building height)/study area) in Paris is equal to 2; in comparison, the built density of an individual housing operation is about 0.3 [24]. In April 2021, the average price of a square meter in Paris was EUR 10,780 [25]. This problem does not concern green roofs as they are not in direct competition with other infrastructures because they can be inserted on the top of buildings. However, the land characteristic is a major problem for ground NBS. The city is a built environment that can be very dense and NBS can require large spaces that are difficult to find in such an environment [26]. This lack of space can even lead to the redesign of an urban project (D). On this subject, an interviewee points out that NBS can take up a little more space than traditional solutions. He added the following: *"Today, we are already struggling to get people to accept more open space in development projects, more green spaces or land. Projects are often far too dense. So, there's an issue of creating space for these NBS."* (A). The price of land can also be a disadvantage for the implementation of NBS and sometimes conventional infrastructures can be more profitable, especially in the short term. This competition for land use has already been discussed in the literature [18].

3.3. Cultural Barriers

3.3.1. Greenwashing: NBS Are Not as Eco-Friendly as Portrayed and Failed Cases of NBS Give the Impression That NBS Are Not Efficient Solutions

One of the risks of environmental projects is greenwashing. Indeed, sometimes the ecological aspect of certain projects can be oversold and overused by the NBS concept term [27]. The project leader at the regional biodiversity agency thus refers to *"NBS can be used indiscriminately, for example referring to false ecological engineering"* (B). One researcher explains it the following way: *"The risk with green roofs is that they oversell ecological interests [...] services and benefits that might never be provided at the end."* (G). In this context, green roofs can be considered mono-specific lawns that represent little interest in biodiversity. Thus, many private NBS commitments are presented as offsets, which often imply greenwashing [28].

Failed examples are also a major issue. The general manager of the water management union said that there were bad examples in which NBS *"brought disadvantages and increased risks"* (F). He made particular reference to bad practices in a reopened river zone, such as *"uncontrolled picnics spots, quads, and motocross activities"* (F) even the accumulation of waste, and the excessive consumption of alcohol. When NBS planners observe such effects, the might reconsider the implementation of their projects.

3.3.2. Biodiversity Is Not in the Foreground of Our Societies

Generally, biodiversity and its erosion are not in the foreground of our societies. Fc example, in the United States, most people prioritize other issues such as terrorism, healt or the economy [29]. Presumably, these concerns are the same in Western societies suc as France. The recent health crisis due to COVID-19 has undoubtedly increased healt concerns and consequently pushed the biodiversity crisis into the background [30]. Th economy is also often at the center, leaving little space for environmental issues [31]. Th person responsible for urban planning and the natural environment of the hydraulic unio confirms that in urban development, biodiversity is not the priority, as follows: *"Designing housing project with a focus on biodiversity as a priority can be complicated, since it comes second* (D). Even within environmental issues, the project manager of the Regional Biodiversit Agency told the authors that protecting biodiversity is not a priority as follows: *"[th Ministry is really focused] on adapting to climate change, climate change in general and renewab energies."* (B). Biodiversity often takes a back seat behind aesthetics (or agronomy), a one ecologist noted the following regarding NBS substrates: *"Today, substrates are mac by landscapers who want to give an agronomic or ornamental aspect to their project. In any cas it does not consider the type of plants* i.e., *wild plants, or local plants,* etc. *[. . .] And that's a issue in NBS: to move towards more local plants."* (A). Another ecologist noted, *"The fact th there are neither landscapers, nor ecologists on the project, can raise the question of how someor (i.e., the architect) who does not necessarily have this eco-friendly culture is going to put in plac everything."* (I). Indeed, the implementation of NBS in cities is often performed through a opportunistic approach (A).

According to someone from the hydraulic union, despite what was presented, biod versity seems to be increasingly taken into consideration, particularly in the legislations c urban planning. However, as she explained, this approach is often more oriented towarc humans as follows: *"Concerns today about the impact of projects, the long-term vision of a si the integration of nature, having green corridors, having less concrete, and more vegetation remai within the framework of a vision of an environment adapted to humans: "You are going to be in neighborhood with lots of trees"."* (D).

3.3.3. Social and Cultural Barriers Are Often Predominant

An ecologist declared the following: *"I would say the first constraint to implementir NBS is cultural. It's the fear of using nature versus grey infrastructure. [. . .] We're afraid of the solutions because we don't find them reliable."* (A). The hydrologist (C) also mentioned the fe that NBS may not work, which has been documented in literature such as the fear of low performance of green infrastructure versus grey infrastructure [32]. This cultural barri seems to be present at all levels, whether from the general public, communities, urba planners, etc. Thus, there is a lack of citizen awareness, support and interest in NBS [22,3

Later, the ecologist added the following: *"People are afraid of nature. [. . .] People st have a rather negative relationship with nature, even if it is changing. We see it in the case of t wetland in Vignois, there are complaints from inhabitants about mosquitoes. It's a daily job to t and get people to accept it."* (A). This fear of nature may relate to the fear of the unknow discussed by Kabisch et al. [18] in the face of uncertainties and risks of implementing NE in cities, as well as the changes these may induce in urban planning. This fear of natu can also relate to real problems called ecosystem disservices, such as the mosquito bit mentioned hereabove. Indeed, ecosystem disservices are inconveniences caused by natu and they can be diverse in cities [34].

The manager of the urban planning and natural environment in the hydraulic unic mentions many human-related barriers as follows: *"Often statements such as "Biodiversity very good, but not in my place!" are common. From the moment when vegetation is allowed to gro having a height of 50 cm of vegetation [. . .] in an urban environment is not something acceptab because it does not look clean. [...] We try to make something beautiful so that it becomes better accepted. [...] We can also have complaints because of pollen and its associated allergies"* (D).

Another aspect of social barriers is the cost of organizing services or changing roles within communities (C).

3.4. Institutional Barriers

3.4.1. Too Little Funding for NBS

There are also economic constraints, including the uncertainty about the cost of NBS. For example, one hydrologist (C) said that NBSs are perceived as more costly than conventional solutions. Such observations are also found in the literature on green infrastructures [32]. The technical manager of SOPRANATURE told us, *"The economic constraints of the project mean that green spaces are the fifth wheel. When there are savings to be made, they are found in green spaces. The same applies for building, when savings are to be made, it is often vegetation roofing that will suffer."* (K). The ecologist from the Biodiversity Agency told us the following on this topic: *"We are afraid of having to manage ecosystems for too long. Whereas with good grey infrastructure, we know the cost, at least for the short term."* (A). His colleague added the following: *"Big problem to obtain financing complicates the mobilization of the public on NBS, in particular because NBS are very broad, [...]. Many terms already used before. So, funders are wondering what NBS have to add."* (B). She also says it is hard to obtain a quick return on investment. According to the European Commission's 2015 report [35], the cities' budgets for green spaces are very small. Specifically, the lack of dedicated funding for NBS implementation in cities has already been highlighted [36], and financial incentives to use NBS are also missing [37]. According to Toxopeus and Polzin [38], the main financial barriers to urban NBSs are lack of coordination between public and private funding and a lack of integration of NBS benefits into valuation and accounting methods.

3.4.2. Too Little Space for NBS in the Regulations

Regulations can be a constraint for the development of NBS in the city because, as one interviewee points out, *"planning documents today do not have a space for these NBS."* (A). He cites the PLU (i.e., the Local Urbanism Plan) and the SCoT (i.e., the Territorial Coherence Scheme). An NBS officer (B) gave an example that an NBS was to be implemented in a zone labeled as "to be urbanized" in the PLU, and that the project in which she was involved could fund NBS only in zones classified in the document as natural. This example shows how regulations can become obstacles to the implementation of NBS in cities. For now, French regulations do not sufficiently encourage these practices. Kabisch et al. noted back in 2016 [18] that urban administrations may lack information on legal instruments and requirements for implementing NBS.

3.4.3. Lack of Political Will

Another important obstacle facing NBS in cities is the lack of political will. Indeed, the development of such solutions requires political initiatives from elected officials, but that is not always the case. This lack of political will as a major constraint to the implementation of NBS in cities was also identified by Sarabi et al. [22]. Moreover, the representative of the Biodiversity Agency (B) told that even if the political will was to be there, municipalities and elected officials can change quickly and be replaced by some people less concerned by NBS. This can compromise projects undertaken during previous mandates. In France, municipal councils and mayors have a six-year mandate [39]. In comparison, the implementation of an urban project takes at least ten years, often much longer than a political mandate [40]. The problem of changing administration was also mentioned by Kabisch et al. [18]. Davies et al. [35] also discussed the long-term vision for green spaces that must be modified due to policy changes. *"It would be necessary for municipalities or agglomerations to have a real desire, to be the driving force, in the PLU to reinforce the presence of biodiversity component on their territories"* (D), said a person from the hydraulic union.

4. Results and Discussion about Levers to the Implementation of Sustainable NBS

According to the word cloud (Figure 1b), the levers for urban NBS are numerous. T most highlighted words can be related directly to the categories discussed in the followi next sub-sections: projects and roofs (Section 4.1.3), communities and people (Section 4.2 regulations (Sections 4.3.3 and 4.3.4), services (Section 4.2.1), etc.

4.1. Raising Awareness through Knowledge

4.1.1. To Address the Lack of Knowledge, Research and Diagnostic Efforts

To respond to the lack of knowledge of NBS, more research is required. The risks m be diagnosed in detail in order to identify the sectors that are in most need of NBS. It co also be interesting to develop *"territorial diagnostic tools, for example, to know where the risk flooding or heat waves lay, and/or to target the different risk levels."* (A). An ecologist cites ot avenues of research as follows: *"defining the NBS scale, or what is grouped behind it, and position of biodiversity within it."* (A). To maximize the contribution of biodiversity, m knowledge of species ecology is required in order to create for them favorable conditic It would also be interesting to evaluate the real monetary costs and benefits of an N implementation in the city. According to Kabisch et al. [18], the areas of knowledge be developed are the effectiveness of NBS, the relationship between NBS and society, NBS design, and its implementation. There are already several research projects on urb NBS, such as the European Horizon 2020 "REGREEN" project. The main research axes this project are the following: improving knowledge of NBS, the development of mappi and modeling tools and the study of the links between well-being, health and nature cities [41]. There is also the GROOVES study (Green ROOfs Verified Ecosystem Servic carried out by the Regional Biodiversity Agency, which applies to green roofs in the Pa region. In this study, inventories of flora and fauna were carried out and ecosystem servi such as water retention were studied. In general, more weight should be given to N studies [42]. To address the lack of follow-up indicators on the NBS, IUCN has develope Global Standard with 3–5 indicators per criterion in the form of a traffic light (E).

Beyond research, the operational people who set up NBS in cities should mak diagnostic effort, taking into account the context as a whole (local climate, context a coherence). For example, the flora in the vicinity where NBS is to be set up should studied, and possibly plant the same into the NBS. One of the ecologists (A) associates t with ecological engineering, which should be a reflex when setting up NBS.

4.1.2. Formation and Education on Nature and NBS at All Levels

To go against the cultural barriers, the director of the hydraulic union explains follows: *"We are in a logic of training and not of communication; this is not enough to fi prejudices, caused by for instance excessive services of green spaces (over shaving of banks un the responsibility of the union). Afterwards, we must not be surprised to witness erosion on banks"* (F). One of the ecologists (A) interviewed insisted that elected officials, techni services, schools, communities and private sector individuals must be trained on N issues and ecology, in general, to become well informed on these topics and therefore implement successfully NBS. In this dynamic, the director of the hydraulic union talk about the following program, initiated by them: *"an educational program that we introdu this year into riparian schools"* (F). The ecologist (A) also believes that engineering fir need more technical training. Indeed, he noted that in most cases ecological compensat measures were badly implemented by engineering companies as they lacked the necess skills. According to this ecologist, landscapers and architects must be trained in NBS t so that they can also implement them correctly. Indeed, it would be good if profession working on urban infrastructure were trained in NBS and not just in grey infrastructure [On this subject, a researcher in ecology adds the following: *"[We must] rethink the train of public works which are still under the influence of large groups historical lobbying for such the "all pipe" for stormwater management lobby, "all to the sewer" lobby or any classic sche that manage the networks."* (J). None of the interviewees specified whether these educat programs should be developed at the pre-service or in-service level, but presumably, they should exist at both levels. Education programs on NBS and related topics at many levels have already been mentioned as a lever for their implementation [3,43]. Thus, through education, people can be motivated to protect the environment [44]. This education can take many forms; for example, in Poland (in Katowice), a local community that trains youth in sustainability was the main actor in organizing and networking citizens and the city for a food festival where the term "nature-based" solutions was explained [10].

4.1.3. Demonstrator Examples That Can Be Replicated

It is important to have good examples to replicate. Indeed, the following is one of the seven lessons about urban NBS that Frantzeskaki [10] drew from her study: the need to learn about NBS and replicate them over the long term. An ecologist stated the following: *"I think we need large demonstrators today, such as large wetlands, or experiments like in Lyon, on rue Garibaldi, on the cooling effect of trees, which show us concretely what benefits these NBS bring, and to be able to quantify them."* (A). According to another interviewee, it is necessary *"to bring the elected representatives on the site"* or *"project owners should come and see what is being done"* (F). One NBS project manager summarized it the following way: *"You need to have examples, [which is] needed by communities. Knowing that another community had the same problem, seeing what has been done on their site and the positive effects of NBS application and seeing that it works, makes you want to do the same. [We need to] get the momentum going."* (B) A hydrologist said the following: *"[We must] manage to show that NBS work, that they are good at absorbing rain, that in places with nature, people are happier and to show that NBS are sustainable"* (C). People likely to uptake NBS need to be reassured. If they see NBS functioning properly and providing ecosystem services, they may wish to replicate them. For the Life ARTISAN project (Increasing the Resilience of Territories to Climate Change by Encouraging Nature-Based Adaptation Solutions, https://www.ofb.gouv.fr/le-projet-life-integre-artisan; accessed on 20 July 2022, there are 10 demonstration sites, including two in the Ile-de-France region (Les Mureaux and AQUI'Brie), which have been selected on the basis of the NBS project, to demonstrate the implementation of nature-based adaptation solutions. Demonstration sites provide an opportunity to evaluate NBS in practice and adapt their management approach [18]. The European Union has invested heavily in such NBS demonstration projects, notably with the Horizon 2020 research and innovation program [45]. Beyond demonstration sites, there are initiatives such as EKLIPSE that aim to evaluate the performance and benefits of NBS [46].

4.2. Multiplicity of Services and Supports

4.2.1. A Major Advantage of NBS Compared to Traditional Solutions Is Their Multifunctionality

As mentioned by the definitions of IUCN [1] and the European Commission [2], NBS are multifunctional since they must simultaneously have benefits for humans and biodiversity. One interviewee puts it the following way: *"Often, the objective of NBS is to be multi-functional, and therefore should not meet only one environmental challenge. Otherwise, it is not called an NBS. At the very least, it must meet the objectives of adapting to climate change and bringing benefits to biodiversity. If, in addition, it can cool a place or store water, store carbon or be of recreational value to people, it is all for the better."* (A). This aspect is very important as the multifunctional aspect of NBS in cities can bring benefits to many fields such as (micro-)climate, ecology, hydrology, socioeconomics, land use planning, architecture, etc. In consequence, NBS can mitigate many risks such as flooding, heat waves or coastal erosion, etc. [47].

The hydrologist thinks that it is necessary to put forward the multifunctionality of the NBS for promoting them as follows: *"[there is not just one] service but several [...]. For example, a pipe handles water better than a green roof, but the roof has more benefits."* (C). He cited the following many functions delivered by NBS: *"Cooling, fighting against climate change, water management, human well-being, biodiversity conservation, city renaturation, and reconnecting with nature. For water management, an example of advantages is treatment at the source reduction of reject volumes in the networks, maximizing infiltration, minimizing runoff*

in urban surfaces to avoid pollution transfer, limiting impacts on the natural environment." (C
Other functions provided by NBS in the city, such as the case of wetlands, were mentione
as follows: *"living environment, depollution, shelter for aquatic fauna, maintenance of banks."* (D
The multifunctional aspect of NBS is one of its strongest points. For example, in a series o
interviews conducted in Australia, 18 of 27 interviewees mentioned this aspect, and it wa
often compared to the single-benefit nature of grey infrastructures [27].

4.2.2. NBS Arch across Many Fields, therefore Transdisciplinarity and the Establishment o Network of Actors Must Be Encouraged

As suggested by their multifunctional aspect, the work on NBS in the city calls fo many disciplinary fields. It is therefore essential to promote transdisciplinarity and t create networks between the different stakeholders. *"The importance of transversal governanc is that it is very broad and that it associates the locals with the actors of the concerned territory i order to have a common co-constructed project"* (E). This collaborative governance is also on of the seven lessons put forward by Frantzeskaki [10] to improve the implementation o NBS in the city.

"It is also necessary to make more co-constructions between the actors of the city and actors o biodiversity and of water. Because in fact, we often have projects done separately. We have our visio of things while the people working on the city have their own. Maybe creating a multidisciplinar working group on NBS would be interesting." (A). The fact that professionals from differen fields do not have the same ways of thinking refers to "silo thinking" and has already bee identified as a barrier to green and blue infrastructure development [48].

It is important that actors working on urban NBS collaborate [19] and, in particula make the link between operational needs and applied research, or the public and priva sectors [22]. Public-private partnerships can even facilitate investments in NBS [36]. Th need for transdisciplinary work, as well as the need to co-design NBS, have alread been identified as a lever for NBS development [16,49]. Sarabi et al. [3] even identifie stakeholder partnerships as an enabling factor for NBS in 27 papers. There are alread collaborations between research and the operational field, such as one between École de Ponts ParisTech and the company SOPREMA, which works on insulation, waterproofin and roofs and offers a wide range of green roofs and facades.

To facilitate transdisciplinarity, it is important that professionals from various field connect with each other [22]. On the brand "Végétal local" [50] (i.e., local vegetation), whic sells wild local plants in France, an NBS representative stressed that it would be necessar *"to link the different initiatives like this one and to do everything together, while consulting eac other and maintaining a good understanding between each other"* (B). The representative of th Regional Agency for Biodiversity explained her role as follows: *"To act as a link between a the actors, to help local authorities interested in carrying out NBS: they contact me and I gui them in finding technical support or consultancy firms that can carry out their field studies; I ma sure that there is a follow-up of the projects, encourage them to find funding, and provide the mea to carry out these projects. Interface role, lobbying for NBS."* (B) Such approaches should b developed in the future to better promote the NBS in the city. This interviewee also referre to the ARTISAN project already mentioned above and in which she is involved. Unde ARTISAN and at the national scale, there is a *"resource network consisting of working grou of different themes for the production of resources and tools to support actors on NBS, develop a w interface, provide support for actors, support the mobilization of funding, initiate training progran and develop studies for building indicators needed for the implementation of NBS, etc."* (B) At th regional level, the actors *"decide together to build a roadmap over 5 years period, and design strategy for the development of NBS."* (B).

4.3. Institutional and Financial Support

4.3.1. What If NBS Could Save Money?

The ecologist from the Regional Biodiversity Agency had the opportunity to start study to compare the costs of green infrastructures to those of grey infrastructures. Th study could not be completed due to a lack of data, but the preliminary results from a small sample showed that *"green infrastructures, in terms of investment and management, are less expensive than grey infrastructures. For example, underground concrete tanks for rainwater management are much more expensive than systems of swales, ponds or gardens."* (A). A more in-depth study is needed to confirm these results, but NBS are potential sources of financial savings. For example, a green roof can last twice as long as a traditional roof [51], thus saving money on roof renewal [52]. According to Fan et al. [53], the presence of NBS can even attract investments, while improving the image of the city that supports them. Green spaces that are part of NBS could also help attract knowledge professionals and participate in the city's economic development [54].

4.3.2. Have Special Funding for NBS

To compensate for the lack of financial means necessary for the development of NBS, one interviewee mentioned the following several ideas: *"Economic levers to facilitate investments in these NBS, at the community level are important. Perhaps, mechanisms like payments for ecosystem services, or favorable taxation on NBS are a plausible idea. Advantageous loans for communities willing to implement NBS are also a good idea."* (A). He also referred to the possibility of having funds from the European Commission for the development of NBS or at least for the restoration or the creation of ecosystems. In any case, it is very important that public and private funds become available for NBS in cities [55]. In the literature, there are different financial instruments that can encourage the use of NBS, such as the modification of user fees for ecosystem services, limiting impacts on natural areas and setting fiscal measures [36]. There may also be grants such as the one given by the European Social Fund to an NGO in Szeged, Hungary, for community gardens [56]. There are already initiatives to finance NBS, such as "Nature 2050" [57] (a program designed by CDC Biodiversity), which allows companies to fund NBS (some of which are in cities). The IUCN, in its 2018 French NBS committee brochure, indicates other ways to fund NBS, at least in France, such as funding for "climate" projects, "natural risk prevention" projects, calls for projects from water agencies, etc. [58].

4.3.3. Labels and Certifications That Endorse and Promote NBS

Labels and certifications can promote NBS in cities by enhancing their value. In this regard, an ecologist expressed his point of view about labels that, in his opinion, are closest to NBS, as follows: *"the BiodiverCity label on buildings. There are Ecojardin, EVE (Espace Végétal Écologique) labels for green spaces [. . .]. Labels are an opportunity for the development of NBS."* (A). The BiodiverCity label, supported by cibi (International Council on Biodiversity and Real Estate), concerns all urbanization projects in urban, peri-urban or natural sites, and displays the performance of real estate projects that take biodiversity into account [59]. Two of the goals referred to by NBS and set by this label are to "Maximize useful biotopes and ecological functionality" and to "Provide nature services for building users". One interviewee commented the following on this label: *"BiodiverCity provides a number of criteria to be fulfilled, among which the indigeneity of the plant palette, which demands to have a certain percentage of indigenous plants"* (I). A person working on the EcoQuartier label indicated that this label is obtained at the urban or rural project level. It includes the following four dimensions: approach and process, living environment and uses, territorial development, environment and climate. These dimensions contain 20 commitments. The following two of these commitments are linked to NBS: "Proposing urban planning to anticipate and adapt to climate change and risks" and "Preserve, restore and enhance biodiversity, soils and natural environments". These two commitments can therefore be used to implement NBS in the city. The presence of a green roof on a site can allow obtaining labels such as BiodiverCity, Effinature ("certification devoted entirely to the consideration of biodiversity in construction, renovation and development projects" [60]), or HQE (high environmental quality). However, it can also work the other way, as follows: Wanting one of these labels on a project can push them to implement a green roof. Currently, in French urban projects, the label BiodiverCity seems to be the one that supports the most biodiversity and NBS cities. However, according to several interviewees, this label does not demand enough fr an ecological point of view. Experts [61] agree that other certifications also (BREEAM a LEED) incorporate NBS but not in sufficient detail, particularly with respect to vegetati Therefore, it would be necessary to develop the existing labels and create new ones. specific label for NBS could be beneficial.

4.3.4. Include NBS in the Regulations to Make Their Use Mandatory and Sustainable

Regulations are very important to promoting NBS [22]. According to the head of town planning and natural environment department of the union for hydraulic devel ment, it is even the best lever for setting up NBS in cities. She mentioned the Local Urb Plan (PLU; see Table 2) *"to have more nature in the city. This document allows for clearly defi orientations"* (D). This document has the power to impose NBS in cities. This union hydraulic development also participates in the PLU by working with the municipalities free up as much space as possible. An ecologist specified on this subject: *"with an urb planning document, one can make a zoning plan, let's say, for a wetland; one can impose the green of roofs in the articles of the regulation; one can suggest the management of rainwater on the p in green spaces; one can also suggest vegetated swales in front of buildings. In fact, you can a lot of things with PLU."* (A). His colleague from the Regional Agency for Biodivers explained as follows: *"We are trying to introduce often as we can in the PLU, PCAET (Territo Climate-Air-Energy Plan) and in all other planning aspects the words NBS and measures NBS."* (B). Water management at the parcel level can also be written into the PLU a hence promote NBS. Regarding NBS affecting water management, there are other Fren documents. One is called SDAGE (Water Development and Management Master Pla Another one is named SAGE, which is a variation of SDAGE. This SAGE focuses mo at the local scale, on catchment areas and their watercourses. Since 2020, the hydrau development union has been in charge of integrating the regulations of the SAGE with *"implementation of protection measures, promoting non-imperviousness around the streets, and infiltration of the first 8 mm of rain."* (D). Such regulations may encourage the use of NBS urban water management. The registration of NBS in documents, such as the PLU ensu the durability of these solutions, beyond political changes (B). A hydrologist also cited regulatory constraints by the communities that are in charge of sanitation such as the c of Paris (the departments, the water management union or the water agency), which can *"either discharge limitation, or an abatement of so many mm of rainfall per event, rather than at scale of a development or combination of several techniques, several NBS"* (C). Such docume can also promote the use of NBS in cities. In any case, it would be preferable for regulati to take into account the multi-benefit aspect of NBS [62].

Table 2. Names of French regulatory documents cited by interviewees (with their acronyms, names in French and English, and definitions).

Acronym	Full Name in French	Full Name in English	Definitions
PLU(i)	Plan Local d'Urbanisme (intercommunal)	Local (intermunicipal) Urban Plan	Main urban planning document at municipal or inter-municipal level.
PCAET	Plan Climat-Air-Energie Territorial	Territorial Climate-Air-Energy Plan	A planning tool that aims to mitigate climate chang develop renewable energy and control energy consumption.
SDAGE	Schéma Directeur d'Aménagement et de Gestion des Eaux	Water Development and Management Master Plan	Main tool for the implementation of the Community's water policy and set the guidelines f 6 years to achieve the objectives of "good water status".
SAGE	Schéma d'Aménagement de Gestion des Eaux	Water Management Plan	A more local version of the SDAGE, to reconcile th satisfaction and development of uses and the protection of aquatic environments.

4.4. Appropriation by the General Public

4.4.1. Climatic Hazards Can Encourage the Use of NBS

Although climate can be considered a constraint for NBS, several interviewees said that they saw it more as an opportunity. Indeed, NBSs are often intended to manage the effects of climate change [23,63]. An ecologist declared the following: *"[the communities] will try to appropriate the subject. [. . .] if they see that there is flooding, they will want to deal with it", "People are realizing that we should put back trees and vegetation, and that we should uncover rivers which have been buried in the recent past and so that they can play the role of sponges."* (A). This sponge notion refers to the "sponge city" concept that was brought forward by the Chinese government in 2013. It "describes an urban environment that is devoted to finding ecologically suitable alternatives to transform urban infrastructures into green infrastructures so these could capture, control and reuse precipitation in a useful, ecologically sound way." [64]. There are climate-related initiatives, such as the Life ARTISAN project already mentioned above, which deal only with the climate change adaptation facet of NBS. The fight against climate change may also represent an opportunity to include NBS in the regulations, as developed in the previous paragraph.

4.4.2. Raising Awareness and Communication to Re-Educate on Nature and Mainstream NBS

According to a water union official, *"It takes a lot of communication, for re-educating people about nature."* (D). *"There is also a need in governance, to have people in communities who are responsible for these NBS issues. That this concept must also be carried out at national level."* (A). There is a necessity to raise awareness in order to move away from the classical landscape approach that has subsisted for years and from the bad habit of using exotic plants (B). The representative of the Agency of Biodiversity proposed *"to go looking for the communities, to inform them, and to mainstream the term NBS"* (B). To mainstream the concept of NBS, media such as the internet, television, radio, and newspapers can be helpful [3]. It is also important to raise awareness about the effectiveness of NBS [45]. To have political and public support, awareness about the links between climate, health and the benefits of NBS is required [65]. The person responsible for the natural environment of the water union gave another piece of advice to educate the general public toward increased integration of nature into the city as follows: *"Create stages that would allow the eye to get used to the changes. For example, if you demolish a building, it attracts the eye, but if you remove the floors one by one, then you don't necessarily see that the building is being destroyed. With nature it's kind of the same, by going step by step, it can allow the uninitiated eye to better accept situations gradually rather than shocking it by a single event."* (D).

5. Conclusions

NBSs are considered an efficient tool to develop sustainable cities. For this reason, they are promoted by the UICN and the EU. By conducting several interviews with different professionals working on NBS (of academic, institutional and operational backgrounds), this study analyzed the barriers and levers related to their implementation in France. Despite belonging to different professional categories, the respondents seem to agree on common barriers affecting NBS implementation. The lack of scientific and cultural knowledge and the absence of financial, political, and institutional support appear to be the main reasons for the current low use of these solutions, whereas the technical problems raised by NBS seem to be more manageable.

To address these issues while putting more emphasis on the conservation of biodiversity, several avenues have been proposed by the interviewees. First, in order to improve the current knowledge of the functioning of NBS, some significant research efforts should be undertaken (essentially promoted by the academic and operational categories). They have to be carried out within a multidisciplinary framework. This seems necessary to better understand and assess the different ecosystem services (regulation, supply, cultural) provided by NBS, as they refer to the following several different disciplines: ecology, hydrology, mechanics, social sciences, urban planning, microclimate . . . It is also important to have concrete demonstrative examples of NBS in cities that work well in order t highlight this multifunctional aspect (promoted by all categories). The follow-up and mon toring of these pilot sites will also contribute to producing quantitative data that will fee research activities.

The second track, which is also deeply related to the development of research, i education (which is promoted by all categories). The concept of NBS must be better an more widely introduced in training programs. The enrichment of knowledge will allow the consolidation of higher education courses and the promotion of their abilities to solv operational problems. The higher education framework will also facilitate the adoption o the multidisciplinary approach mentioned above and will promote the networking of th different actors involved in urban planning. The link between academic knowledge an its operational applications can also be performed by certifications and labeling that ar commonly adopted by stakeholders. Obtaining a label or certification will eliminate an suspicion of "greenwashing" and also justify the implementation of an NBS rather than traditional solution (essentially promoted by operational category). Note that this wor has begun with the definition of the IUCN standard. These quantification tools could valu NBS performances and also benefit from monitored pilot sites.

NBSs are also currently considered as climate change adaptation tools. In this applic tion, they could be renamed NBAS (for Nature-Based Adaptation Solutions). In such a evolving climatic context, the sustainability of their ecosystem services and their associate performances over time is poorly known. This raises many questions about the choic of species to implement, their evolution in an urban environment strongly impacted b climate change, and their possible need for maintenance ... Here again, experimentatio and the development of knowledge will provide answers that must be taken into account we do not want to create new barriers to the implementation of NBS.

Finally, in addition to these scientific requirements, more social-societal initiatives ca be proposed. Indeed, most of the seven lessons for planning NBS in cities proposed b Frantzeskaki [10] refer to consultation and co-construction approaches. The definition new common goods, the involvement of the different local authorities, the appropriatic of urban space by citizens, and the construction of a relationship of trust between thes different entities, are all avenues to be favored to facilitate the implementation of NBS. Th societal reorganization of urban planning, coupled with the democratization of knowledg surely represents the best levers for overcoming the various obstacles listed in this articl

Supplementary Materials: The following supporting information can be downloaded at: https:/ www.mdpi.com/article/10.3390/su14169975/s1, Questionnaire used during the interviews.

Author Contributions: Conceptualization, C.D., N.F.-L. and P.-A.V.; methodology, C.D., N.F.-L. ar P.-A.V.; formal analysis, C.D.; investigation, C.D.; writing—original draft preparation, C.D.; writing review and editing, N.F.-L. and P.-A.V.; supervision, N.F.-L. and P.-A.V.; project administratio P.-A.V.; funding acquisition, P.-A.V. All authors have read and agreed to the published version the manuscript.

Funding: This research was funded by Agence Nationale de la Recherche, grant number AN 17-CE22-0002-01 EVNATURB project dealing with the evaluation of ecosystem performances f re-naturing urban environment. The APC was also funded by Agence Nationale de la Recherch grant number ANR-17-CE22-0002-01 EVNATURB.

Institutional Review Board Statement: Ethical review and approval were waived for this study du to the interviews require only the information and consent of the interviewees. Informed conse was obtained verbally before participation.

Informed Consent Statement: Informed consent was obtained from all subjects involved in the stu

Data Availability Statement: Data sharing not applicable.

Acknowledgments: The authors would like to thank the different interviewees that accept to answ the questionnaire for their kind and constructive collaboration. We also want to thank Mario Sayah and Ghislaine Mangeney for correcting the English.

Conflicts of Interest: The authors declare no conflict of interest. The funders had no role in the design of the study; in the collection, analyses, or interpretation of data; in the writing of the manuscript, or in the decision to publish the results.

ferences

IUCN. *Guidance for Using the IUCN Global Standard for Nature-Based Solutions. User-Friendly Framework for the Verification, Design and Scaling up of Nature-Based Solutions*, 1st ed.; IUCN: Gland, Switzerland, 2020; p. 1.

European Commission, Research and Innovation Branch. *Biodiversity and Nature-Based Solutions: Analysis of EU-Funded Projects*; Publications Office: Luxembourg, 2020; p. 3.

Sarabi, S.E.; Han, Q.; Romme, A.G.L.; de Vries, B.; Wendling, L. Key enablers of and barriers to the uptake and implementation of nature-based solutions in urban settings: A review. *Resources* **2019**, *8*, 121–140. [CrossRef]

Templier, F.; (École des Ponts ParisTech, Paris, France). Les solutions fondées sur la nature: De quelle nature parle-t-on lorsqu'elles sont développées en milieu urbain? Personal Communication (Internship Report), 2020.

Pauleit, S.; Zölch, T.; Hansen, R.; Randrup, T.B.; van den Bosch, C.K. Nature-based solutions and climate change–four shades of green. In *Nature-Based Solutions to Climate Change Adaptation in Urban Areas*; Kabisch, N., Korn, H., Stadler, J., Bonn, A., Eds.; Springer: Cham, Switzerland, 2017; pp. 29–49.

Dorst, H.; van der Jagt, A.; Raven, R.; Runhaar, H. Urban greening through nature-based solutions—Key characteristics of an emerging concept. *Sustain. Cities Soc.* **2019**, *49*, 101620. [CrossRef]

Tomao, A.; Quatrini, V.; Corona, P.; Ferrara, A.; Lafortezza, R.; Salvati, L. Resilient landscapes in Mediterranean urban areas: Understanding factors influencing forest trends. *Environ. Res.* **2017**, *156*, 1–9. [CrossRef]

Yao, X.; Zhao, M.; Escobedo, F.J. What Causal Drivers Influence Carbon Storage in Shanghai, China's Urban and Peri-Urban Forests? *Sustainability* **2017**, *9*, 577. [CrossRef]

Xing, Y.; Jones, P.; Donnison, I. Characterisation of nature-based solutions for the built environment. *Sustainability* **2017**, *9*, 149. [CrossRef]

Frantzeskaki, N. Seven lessons for planning nature-based solutions in cities. *Environ. Sci. Policy* **2019**, *93*, 101–111. [CrossRef]

Giannakis, E.; Bruggeman, A.; Poulou, D.; Zoumides, C.; Eliades, M. Linear parks along urban rivers: Perceptions of thermal comfort and climate change adaptation in Cyprus. *Sustainability* **2016**, *8*, 1023. [CrossRef]

Scott, M.; Lennon, M.; Haase, D.; Kazmierczak, A.; Clabby, G.; Beatley, T. Nature-based solutions for the contemporary city/Re-naturing the city/Reflections on urban landscapes, ecosystems services and nature-based solutions in cities/Multifunctional green infrastructure and climate change adaptation: Brownfield greening as an adaptation strategy for vulnerable communities?/Delivering green infrastructure through planning: Insights from practice in Fingal, Ireland/Planning for biophilic cities: From theory to practice. *Plan. Theory Pract.* **2016**, *17*, 267–300.

Eggermont, H.; Bailan, E.; Azevedo, J.M.N.; Beumer, V.; Brodin, T.; Claudet, J.; Fady, B.; Grube, M.; Keune, H.; Lamarque, P.; et al. Nature-based solutions: New influence for environmental management and research in Europe. *Ecol. Perspect. Sci. Soc.* **2015**, *24*, 243–248. [CrossRef]

Kumar, P.; Debele, S.E.; Sahani, J.; Aragão, L.; Barisani, F.; Basu, B.; Bucchignani, E.; Charizopoulos, N.; Di Sabatino, S.; Domeneghetti, A.; et al. Towards an operationalisation of nature-based solutions for natural hazards. *Sci. Total Environ.* **2020**, *731*, 138855. [CrossRef] [PubMed]

Raymond, C.M.; Frantzeskaki, N.; Kabisch, N.; Berry, P.; Breil, M.; Nita, M.R.; Geneletti, D.; Calfapietra, C. A framework for assessing and implementing the co-benefits of nature-based solutions in urban areas. *Environ. Sci. Policy* **2017**, *77*, 15–24. [CrossRef]

Nesshöver, C.; Assmuth, T.; Irvine, K.N.; Rusch, G.M.; Waylen, K.A.; Delbaere, B.; Haase, D.; Jones-Walters, L.; Keune, H.; Kovacs, E.; et al. The science, policy and practice of nature-based solutions: An interdisciplinary perspective. *Sci. Total Environ.* **2017**, *579*, 1215–1227. [CrossRef]

Fernandes, J.P.; Guiomar, N. Nature-based solutions: The need to increase the knowledge on their potentialities and limits. *Land Degrad. Dev.* **2018**, *29*, 1925–1939. [CrossRef]

Kabisch, N.; Frantzeskaki, N.; Pauleit, S.; Naumann, S.; Davis, M.; Artmann, M.; Haase, D.; Knapp, S.; Korn, H.; Stadler, J.; et al. Nature-based solutions to climate change mitigation and adaptation in urban areas: Perspectives on indicators, knowledge gaps, barriers, and opportunities for action. *Ecol. Soc.* **2016**, *21*, art39. [CrossRef]

Short, C.; Clarke, L.; Carnelli, F.; Uttley, C.; Smith, B. Capturing the multiple benefits associated with nature-based solutions: Lessons from a natural flood management project in the Cotswolds, UK. *Land Degrad. Dev.* **2019**, *30*, 241–252. [CrossRef]

Eisenberg, B.; (Technische Universität München, Munich, Germany). Nature Based Solutions—Technical Handbook. Personal Communication, 2019.

Burch, S. Transforming barriers into enablers of action on climate change: Insights from three municipal case studies in British Columbia, Canada. *Glob. Environ. Chang.* **2010**, *20*, 287–297. [CrossRef]

Sarabi, S.; Han, Q.; Romme, A.G.L.; de Vries, B.; Valkenburg, R.; den Ouden, E. Uptake and implementation of Nature-Based Solutions: An analysis of barriers using Interpretive Structural Modeling. *J. Environ. Manag.* **2020**, *270*, 110749. [CrossRef] [PubMed]

23. Hobbie, S.E.; Grimm, N.B. Nature-based approaches to managing climate change impacts in cities. *Philos. Trans. R. Soc.* **2020**, *3* 20190124. [CrossRef] [PubMed]
24. Intermezzo. Available online: http://www.intermezzo-coop.eu/fr/actualites/densite-batie-a-paris-opendata#:~:text=La% densit%C3%A9%20b%C3%A2tie%20est%20un,approximative%20de%20sa%20densit%C3%A9%20b%C3%A2tie (accessed 12 July 2021).
25. EffiCity. Available online: https://www.efficity.com/prix-immobilier-m2/v_paris_75/ (accessed on 12 April 2021).
26. O'Donnell, E.C.; Lamond, J.E.; Thorne, C.R. Recognising barriers to implementation of Blue-Green Infrastructure: A Newcas case study. *Urban Water J.* **2017**, *14*, 964–971. [CrossRef]
27. Moosavi, S.; Browne, G.R.; Bush, J. Perceptions of nature-based solutions for Urban Water challenges: Insights from Austral researchers and practitioners. *Urban For. Urban Green.* **2021**, *57*, 126937. [CrossRef]
28. Seddon, N.; Smith, A.; Smith, P.; Key, I.; Chausson, A.; Girardin, C.; House, J.; Srivastava, S.; Turner, B. Getting the message ri on nature-based solutions to climate change. *Glob. Chang. Biol.* **2021**, *27*, 1518–1546. [CrossRef]
29. Novacek, M.J. Engaging the public in biodiversity issues. *Proc. Natl. Acad. Sci. USA* **2008**, *105*, 11571–11578. [CrossRef] [PubM
30. Goymann, W. What the Corona (SARS-CoV 2) pandemic, climate change, and the biodiversity crisis teach us about human nat *Ethology* **2020**, *126*, 593–594. [CrossRef] [PubMed]
31. Giddings, B.; Hopwood, B.; O'Brien, G. Environment, economy and society: Fitting them together into sustainable developm *Sustain. Dev.* **2002**, *10*, 187–196. [CrossRef]
32. Dhakal, K.P.; Chevalier, L.R. Managing urban stormwater for urban sustainability: Barriers and policy solutions for gr infrastructure application. *J. Environ. Manag.* **2017**, *203*, 171–181. [CrossRef]
33. Wamsler, C.; Wickenberg, B.; Hanson, H.; Alkan Olsson, J.; Stålhammar, S.; Björn, H.; Falck, H.; Gerell, D.; Oskarsson, Simonsson, E.; et al. Environmental and climate policy integration: Targeted strategies for overcoming barriers to nature-ba solutions and climate change adaptation. *J. Clean. Prod.* **2020**, *247*, 119154. [CrossRef]
34. Lyytimäki, J.; Petersen, L.K.; Normander, B.; Bezák, P. Nature as a nuisance? *Ecosystem services and disservices to urban lifest* *Environ. Sci.* **2008**, *5*, 161–172. [CrossRef]
35. Davies, C.; Hansen, R.; Rall, E.; Pauleit, S.; Lafortezza, R.; De Bellis, Y.; Santos, A.; Tosics, I. Green infrastructure planning a implementation. *The status of European green space planning and implementation based on an analysis of selected European cityregi* *Green Surge Proj. Deliv.* **2015**, *5*, 1–134. [CrossRef]
36. Dorste, N.; Schröter-Schlaack, C.; Hansjürgens, B.; Zimmermann, H. Implementing nature-based solutions in urban ar Financing and governance aspects. In *Nature-Based Solutions to Climate Change Adaptation in Urban Areas*; Kabisch, N., Korn, Stadler, J., Bonn, A., Eds.; Springer: Cham, Switzerland, 2017; pp. 307–321.
37. Li, C.; Peng, C.; Chiang, P.C.; Cai, Y.; Wang, X.; Yang, Z. Mechanisms and applications of green infrastructure practices stormwater control: A review. *J. Hydrol.* **2019**, *568*, 626–637. [CrossRef]
38. Toxopeus, H.; Polzin, F. Reviewing financing barriers and strategies for urban nature-based solutions. *J. Environ. Manag.* **2** *289*, 112371. [CrossRef]
39. Demarches.interieur.gouv.fr. Available online: https://www.demarches.interieur.gouv.fr/particuliers/elections-municipal ~:text=28%20juin%202020.-,Les%20conseillers%20municipaux%20sont%20%C3%A9lus%20 (accessed on 12 July 2021).
40. Le Moniteur. Available online: https://www.lemoniteur.fr/article/prendre-en-compte-le-temps-long-du-projet-urbain.286 (accessed on 12 July 2021).
41. ARB-IDF. Available online: https://www.arb-idf.fr/article/projet-europeen-regreen/ (accessed on 13 January 2022).
42. Davies, C.; Lafortezza, R. Transitional path to the adoption of nature-based solutions. *Land Use Policy* **2019**, *80*, 406–409. [CrossR
43. Frantzeskaki, N.; Borgström, S.; Gorissen, L.; Egermann, M.; Ehnert, F. Nature-Based Solutions Accelerating Urban Sustainabi Transitions in Cities. In *Nature-Based Solutions to Climate Change Adaptation in Urban Areas*; Kabisch, N., Korn, H., Stadle Bonn, A., Eds.; Springer: Cham, Switzerland, 2017; pp. 65–88.
44. Lysack, M. Environmental decline, loss, and biophilia: Fostering commitment in environmental citizenship. *Crit. Soc. Work* **2** *11*, 48–66.
45. Faivre, N.; Fritz, M.; Freitas, T.; de Boissezon, B.; Vandewoestijne, S. Nature-Based Solutions in the EU: Innovating with natur address social, economic and environmental challenges. *Environ. Res.* **2017**, *159*, 509–518. [CrossRef] [PubMed]
46. Raymond, C.M.; Breil, M.; Nita, M.R.; Kabisch, N.; de Bel, M.; Enzi, V.; Frantzeskaki, N.; Geneletti, G.; Lovinger, Cardinaletti, M.; et al. An impact evaluation framework to support planning and evaluation of nature-based solutions proj Report prepared by the EKLIPSE Expert Working Group on Nature-Based Solutions to Promote Climate Resilience in Ur Areas. *Cent. Ecol. Hydrol.* **2017**, 1–72. [CrossRef]
47. Andersson, E.; Borgström, S.; McPhearson, T. Double insurance in dealing with extremes: Ecological and social factors for mak nature-based solutions last. In *Nature-Based Solutions to Climate Change Adaptation in Urban Areas*; Kabisch, N., Korn, H., Stadle Bonn, A., Eds.; Springer: Cham, Switzerland, 2017; pp. 51–64.
48. O'Donnell, E.C.; Lamond, J.E.; Thorne, C.R. Learning and Action Alliance framework to facilitate stakeholder collaboration social learning in urban flood risk management. *Environ. Sci. Policy* **2018**, *80*, 1–8. [CrossRef]
49. Albert, C.; Schröter, B.; Haase, D.; Brillinger, M.; Henze, J.; Herrmann, S.; Gottwald, S.; Guerrero, P.; Nicolas, C.; Matzdor Addressing societal challenges through nature-based solutions: How can landscape planning and governance research contrib *Landsc. Urban Plan.* **2019**, *182*, 12–21. [CrossRef]

0. Végétal Local. Available online: https://www.vegetal-local.fr/la-marque (accessed on 12 July 2021).
1. Mann, G. Ansätze zu objektbezogenen Kosten-Nutzen-Analysen. *10Oktober* **2005**, *2002*, 54. Available online: https://www.gruendaecher.de (accessed on 12 July 2021).
2. Claus, K.; Rousseau, S. Public versus private incentives to invest in green roofs: A cost benefit analysis for Flanders. *Urban For. Urban Green.* **2012**, *11*, 417–425. [CrossRef]
3. Fan, P.; Ouyang, Z.; Basnou, C.; Pino, J.; Park, H.; Chen, J. Nature-based solutions for urban landscapes under post-industrialization and globalization: Barcelona versus Shanghai. *Environ. Res.* **2017**, *156*, 272–283. [CrossRef]
4. Florida, R. *Cities and the Creative Class*, 1st ed.; Routledge: New York, NY, USA, 2004. [CrossRef]
5. European Commission. Towards an EU research and innovation policy agenda for nature-based solutions and re-naturing cities. In *Final Report of the Horizon 2020 expert group on "Nature-Based Solutions and Re-Naturing Cities"*; European Commission: Brussels, Belgium, 2015; Volume 74.
6. Van der Jagt, A.P.; Szaraz, L.R.; Delshammar, T.; Cvejic, R.; Santos, A.; Goodness, J.; Buijs, A. Cultivating nature-based solutions: The governance of communal urban gardens in the European Union. *Environ. Res.* **2017**, *159*, 264–275. [CrossRef]
7. Nature 2050. Available online: https://www.nature2050.com/le-programme-nature2050/presentation-du-programme-nature2050/ (accessed on 12 July 2021).
8. UICN. *Les Solutions fondées sur la Nature Pour Lutter Contre les Changements Climatiques et Réduire les Risques Naturels en France*; IUCN: Paris, France, 2018.
9. Cibi. BiodiverCity. Available online: http://cibi-biodivercity.com/biodivercity/ (accessed on 12 July 2021).
0. IRICE. Available online: https://irice.fr/effinature (accessed on 12 July 2021).
1. Enzi, V.; Cameron, B.; Dezsényi, P.; Gedge, D.; Mann, G.; Pitha, U. Nature-Based Solutions and Buildings–The Power of Surfaces to Help Cities Adapt to Climate Change and to Deliver Biodiversity. In *Nature-Based Solutions to Climate Change Adaptation in Urban Areas*; Kabisch, N., Korn, H., Stadler, J., Bonn, A., Eds.; Springer: Cham, Switzerland, 2017; pp. 159–183.
2. Zuniga-Teran, A.A.; Staddon, C.; de Vito, L.; Gerlak, A.K.; Ward, S.; Schoeman, Y.; Hart, A.; Booth, G. Challenges of mainstreaming green infrastructure in built environment professions. *J. Environ. Plan. Manag.* **2020**, *63*, 710–732. [CrossRef]
3. Seddon, N.; Chausson, A.; Berry, P.; Girardin, C.A.J.; Smith, A.; Turner, B. Understanding the value and limits of nature-based solutions to climate change and other global challenges. *Philos. Trans. R. Soc.* **2020**, *375*, 20190120. [CrossRef] [PubMed]
4. Liu, H.; Jia, Y.; Niu, C. "Sponge city" concept helps solve China's urban water problems. *Environ. Earth Sci.* **2017**, *76*, 473. [CrossRef]
5. Santiago Fink, H. Human-nature for climate action: Nature-based solutions for urban sustainability. *Sustainability* **2016**, *8*, 254. [CrossRef]

land

MDPI

icle

lainstreaming Nature-Based Solutions in City Planning: xamining Scale, Focus, and Visibility as Drivers of ıtervention Success in Liverpool, UK

Mell [1,*], Sarah Clement [2] and Fearghus O'Sullivan [1]

1 Department of Planning & Environmental Management, University of Manchester, Manchester M13 9PL, UK
2 College of Science, Australian National University, Canberra, ACT 2601, Australia
* Correspondence: ian.mell@manchester.ac.uk

Abstract: Nature-based solutions (NBS) have been central to the European Union's drive to address climate change, ecological degradation, and promote urban prosperity. Via an examination of the Horizon 2020-funded URBAN GreenUP project in Liverpool, this paper explores mainstreaming NBS in city planning. It uses evidence from pre- and post-intervention surveys with Liverpool residents and interviews with local business, environmental, government, and community sector experts to illustrate how a complex interplay of scale, location, focus, and visibility of NBS influences perceptions of the added value of NBS. This paper highlights the requirement that NBS interventions be bespoke and responsive to the overarching needs of residents and other stakeholders. Moreover, we underscore the importance of expert input into the design, location, and maintenance of NBS and call for these key drivers of successful delivery to be better integrated into work programs. This paper also notes that the type and size of NBS interventions impact perceptions of their value, with smaller projects being viewed as less socially and ecologically valuable compared to larger investments. We conclude that while small-scale NBS can support climatic, health, or ecological improvements in specific instances, strategic, larger-scale, and more visible investments are required to accrue substantive benefits and gain acceptance of NBS as a legitimate and effective planning tool.

Keywords: Nature-Based Solutions; urban planning; community perceptions; urban nature; biodiversity; multi-functionality; climate change; green infrastructure

tion: Mell, I.; Clement, S.; llivan, F. Mainstreaming ıre-Based Solutions in City ning: Examining Scale, Focus, Visibility as Drivers of vention Success in Liverpool, Land 2023, 12, 1371. https:// rg/10.3390/land12071371

demic Editor: Panayiotis trakopoulos

ived: 15 May 2023
sed: 5 July 2023
pted: 6 July 2023
shed: 8 July 2023

1. Introduction

The rise of Nature-based solutions (hereafter NBSs) has been catalyzed by significant investment from the European Union (EU) through its Research and Innovation (R&I) portfolio. EU funding aims to build the evidence base for how nature-focused interventions can address climate change, improve public health and well-being, support economic growth, and promote urban renewal [1]. The Horizon 2020 schemes focusing on NBSs fund research and demonstration projects to showcase the breadth of opportunities available to planners, politicians, the environment, and the public sectors to solve public problems and integrate more ecologically sustainable development into urban development [2]. Through a broad program of micro/singular, street, and area/neighborhood-based interventions, NBSs have been implemented in European cities to test the positive impact that nature-focused interventions can have at multiple scales (micro, e.g., a lamp post; site, e.g., a park or building); street; neighborhood; and across different urban contexts [3]. The following uses "scale" to define the size of a NBS intervention. In addition, the focus of NBS is reflective of their multiple socio-economic and ecological functions and how they aid the delivery of climate change adaptation/mitigation, health and well-being, economic prosperity, and improved quality of life. At the same time, the visibility of NBS relates to the ease with which NBS interventions are seen and interacted with in an urban context. The goal is ultimately transformative, i.e., to provide ecological and socio-economic evidence for NBS, enhance their visibility to potential users, and mainstream them within urb planning and regeneration design and delivery.

To examine the added value of NBS in delivering ecological and socio-econon benefits in cities, we leveraged insights from a six-year project to design, implement, a test the contribution of nature-centric projects to solving urban challenges [4]. The pa uses the Horizon 2020-funded URBAN GreenUP project as a case study, and specifica the interventions delivered in Liverpool (UK), to illustrate the complexity of translating theoretical promise of NBS into the practice of greening highly urbanized environme The paper sets out to answer the following:

1. Which NBS are considered most appropriate by residents, businesses, and ot communities of interest to address a range of sociocultural, economic, and ecologi challenges in the Liverpool case study area?
2. What barriers can hinder the delivery of NBS within a high-density urban area?
3. What are the most appropriate NBS options in terms of scale (micro, site, street, nei borhood), location (urban, urban-fringe, rural), NBS ecological and socio-econor function, visibility, and interactivity that can be used to address the widest range issues impacting high-density urban areas?

The paper presents two sets of interlinked evidence to answer these questio (1) a survey of residents in Liverpool examining the perception of existing GI and URB GreenUP-funded NBS; and (2) insights from development, third, and environmental or nizations. This directly responds to the enthusiastic advocacy for urban NBS in policy a practice, providing evidence from a live demonstration project that tests the promise of N in practice. This evidence bridges professionals, e.g., local government and representati of environmental organizations, and local perspectives on the perceived added ecologi and/or socio-economic value provided by investment in NBS. This allows for a detail commentary on local and strategic considerations for implementing NBS in Liverpool. T is of global interest, considering it was a centerpiece of EU investment and part of a wi effort to leverage the power of NBS in cities as other cities will face similar challen Within this context, successful delivery is framed as an investment that local participa perceive as positively enhancing the quality of quality of life socially, economically, or e logically. Success is not presented as calculable (or quantitatively evidenced) improveme in urban ecosystem functionality, as this is beyond the scope of this paper. Therefore, analysis presented relates to the perceptions of business, environmental, and residen stakeholders of the additional benefits that NBS provides in Liverpool. The paper sh cases where links between climate adaptation, improved access to nature, enhanced quality, and improved health and well-being can be enhanced through NBS investm However, our research also underscores that communication of the benefits of investm in NBS, co-design with residents, the third sector, and environmental organizations, a delivery that meets identified local needs are core factors that must be foregrounded in cities to guide investment.

Overall, the paper argues for a locally grounded appreciation of which NBS may appropriate in urban areas and consideration of what benefits and functions are need to maximize the value of such interventions for resolving complex challenges. Wh NBS has been automatically accepted as a public good, we call for consideration in proposal and design stages of what socio-economic and/or ecological "additionality" a NBS intervention will provide and to whom [5]. This supports the evolving discussi of NBS interventions developed by Kabisch, Frantzeskaki, and Hansen [6], who, as p of the wider discourse supporting NBS, suggest that effective NBS investment sho be structured against five core principles: (a) systematic understanding; (b) benefits people and biodiversity; (c) inclusive solutions that are long-term; (d) context considerat and (e) communication and learning. While the debates presented in this paper ackno edge the value of such framings, the discussion presented in the case of Liverpool pla an increased emphasis on considerations of NBS in terms of (i) elements, (ii) functio

(iii) benefits, and (iv) beneficiaries to ensure that locally appropriate investments are delivered.

2. Framing NBS in Urban Planning

NBS emerged internationally as an approach to resolving linked challenges relating to climate change, biodiversity loss, and community livelihoods [7]. Although the concept originally focused on the conservation, restoration, and enhancement of natural ecosystems and broader landscapes, the focus of these debates in Europe shifted to greening cities, where the majority of people live and where environmental challenges are most acute [8]. Approximately 70% of the EU's population lives in urban areas, driving significant changes in the functionality of linked socio-economic and ecological systems. This includes impacts on water quality and quantity, biodiversity loss, and degraded air quality, with consequences for ecological function, human health, and the economic viability of cities. The Horizon 2020 program aims to add to the evidence base to understand how investment in "nature" in its myriad forms (e.g., street trees, green facades, parks, or sustainable drainage) can act as a viable solution to the problems associated with growth and unsustainable development patterns in cities.

By building on conversations about the benefits of nature, e.g., ecosystem services [9] and the connective [10], accessible, and multi-functional principles of green infrastructure (GI) [11,12], NBSs are being promoted as an innovative way to enhance these benefits by integrating ecological thinking into engineered, i.e., built environment, systems [13]. NBS as a term is new, but conceptually, it is built on decades of research discussing the value of an investment in urban nature, for example, in the urban ecology literature [14,15]. However, the promotion of "nature" as the central principle of investment does differ from previous forms of green space and landscape development [16,17]. In such a scenario, GI, or greenspace, could be positioned as an overarching concept, putting a conceptual and thematic structure in place that contextualizes investment in nature as essential infrastructure [18]. Practice-based delivery can subsequently promote using NBS as the action component of a wider, environment-centric framework to deliver ecologically focused development. By working with nature as a means to deliver on the objectives of core goals rather than as an afterthought, investments in NBS can offer cost-effective and responsive forms of urban management that support greener and more sustainable growth in cities [19,20].

The capacity of government decision-makers and the environmental and business sectors to implement NBS varies substantially geographically and across governance levels (e.g., local, regional, and national) [21]. Consequently, although advocates in environmental organizations and academia have argued convincingly for investment in NBS, there remain diverse views of the added value that NBS can provide [22]. Current debates on NBS are starting to address this issue to ensure that the technical, legal, and political challenges faced by practitioners, scientists, and decision-makers working in cognate sustainability disciplines are more effectively integrated into urban development practice. Consequently, despite the relative infancy of NBS as an academic subject, it has emerged as core terminology within urban nature debates in Europe. This shift is visible in the catalogs of NBS typologies and investment options being proposed in the literature. These include the use of urban forestry, sustainable drainage, increased biodiverse planting, green wall/roof/facade technology, parklets and parks, urban agriculture, and roadside verge pollinator enhancements [1,18,23]. More technological solutions are also being debated, utilizing sensors to examine heat, pollution, water, and biodiversity change linked to variations in the form and function of each of the types of NBS noted above. The breadth of interventions available highlights the need for an understanding of NBS to be informed by examples of their value in practice.

The benefits of ecologically centered investment include the delivery of comparable functionality to engineered solutions but at lower costs and more responsiveness to changes in the fabric of an urban environment [24]; thus, they offer a 'dynamic mutability' to the pressures placed on urban landscapes [19]. Therefore, taking a wide perspective on what can be considered NBS is a useful approach to addressing such variability. Furthermor the research literature suggests that NBS optimize the benefits of ecological systems withi the built environment, promoting a more nuanced appreciation of "nature" within praxis t ensure different stakeholders can more effectively engage with NBS compared to tradition engineered solutions. When aligned with a consideration of NBS elements, function benefits, and beneficiaries, it is possible to apply a more holistic framing to NBS, examinin what is delivered, what change they are promoting, and how local people can engage wit and benefit from them [18].

3. The URBAN GreenUP Project and Greenspace Planning in Liverpool

These assertions can be tested by examining the design, delivery, and monitorin of NBS interventions associated with the URBAN GreenUP project. URBAN GreenU (https://www.urbangreenup.eu/, accessed on 7 July 2023) is a 5-year EU-funded R&I proje testing the value of integrating innovative NBS in urban areas. Due to delays in the delive of NBS caused by COVID-19, URBAN GreenUP was extended by 12 months to become 6-year program. The project comprised a consortium of 27 universities and research ins tutions, Small–Medium Enterprises (SMEs), local governments, and environmental orgar zations located in eight countries (China, Columbia, Germany, Italy, Spain, Turkey, the U and Vietnam).

Liverpool (UK) is one of three front-runner cities, along with Izmir (Turkey) an Valladolid (Spain), that led the delivery of NBS interventions and the development (a transferable methodology for planning and implementing NBS. The project also i volved five follower cities (Chengdu in China, Ludwigsburg in Germany, Mantova in Ita Medellin in Columbia, and Quy Nhon in Vietnam), which tested the project's innovatio and methods. Each partner city brings a wealth of local government, environmental, an technological expertise to the project that has been integrated to aid in the identificatic of solutions to a range of sociocultural, economic, and ecological problems. The thre front-runner cities developed portfolios of NBS interventions, which have subsequent been expanded into a series of renaturing strategies for cities. The project also facilitate a more holistic understanding of the complexity associated with NBS delivery, as it h worked across varied geographic, climatic, political, and governance systems.

The URBAN GreenUP project does not sit in isolation but is one of three EU Horizon 202 funded NBS R&I projects. Its sister projects, Connecting Nature (https://connectingnatu eu/, accessed on 7 July 2023) and Grow Green (https://growgreenproject.eu/, accessed c 7 July 2023), are comparable to URBAN GreenUP in terms of their consortium compositic Variations are visible in each project's strategic objectives. Grow Green in Manchester focus on a single large neighborhood-scale project, West Gorton Sponge Park, and the Glasgo (UK)-based components of Connecting Nature, including a range of small- and mediur sized interventions. The breadth of delivery has provided important insights for the E enabling them to examine what works, how it works, and where barriers remain to successf NBS interventions.

While NBS should be delivered at a landscape scale to meet global standards [2 retrofitting NBS within existing 'hard' urban forms makes this challenging. This challen is evident in the portfolio of NBS interventions delivered by URBAN GreenUP, whic were at a site and street scale. Figure 1a–d illustrates examples of the interventions, whic include green walls, pollinator lamp posts, biodiverse pollinator street planting, and i vestment in street trees to add shade, promote pollination, and intercept pollution ar rainfall. In addition, Table 1 provides a profile of each research/investment area ar the NBS interventions located in each. Moreover, ecological floating islands, deculve ing/bioretention and sustainable drainage works, and urban garden bio-filters were a delivered to examine the potential impact that NBS could have on people, the environme and the local economy. A full portfolio of investments can be accessed via the proje webpages: https://www.urbangreenup.eu/solutions/, accessed on 7 July 2023.

(a) (b) (c) (d)

Figure 1. (**a**) NBS interventions in Liverpool—St John's Centre Green Wall, Liverpool City Centre. (**b**) NBS interventions in Liverpool—Pollinator lamp post installation, Baltic Triangle, Liverpool. (**c**) NBS interventions in Liverpool—Biodiverse pollinator planting, Park Lane, Liverpool (out of season). (**d**) NBS interventions in Liverpool—Street tree interventions to address traffic pollution and surface water flooding, The Strand, Liverpool.

Both policy and research on NBS underscore the need for participatory planning and co-design [26,27]. Over its lifespan, URBAN GreenUP worked extensively with local government, the environment and business sectors, technology start-ups/companies, engineers, landscape architects, and, to a lesser extent, local communities. The aim was to ensure that (a) the right NBS is located in the most appropriate place, (b) local stakeholders are aware of the NBS interventions and the associated socio-economic and ecological benefits, and (c) the range of benefits associated with each NBS intervention is grounded in robust evidence. This approach has been crucial in generating a detailed appreciation of what NBS interventions are needed, how they address local needs, and how they support local government policy mandates to address climate change, health and well-being, and economic development issues. The following reflects on the perceptions of local respondents to the changes afforded to the physical environment, their interaction with and valuing of urban nature, and the socio-economic and ecological benefits that investment in NBS can deliver. As a front-runner city within the URBAN GreenUP project, the city of Liverpool is being used as a testing ground for novel approaches to investment in NBS, and the acceptance and critiques of this program of work offer useful insights for cities adopting similar interventions. This study provides an analysis of public acceptance of NBS in terms of size, location, and type that can be used to shape future design and investment. It also provides evidence of the role played by communication, engagement, and co-design investment plans for cities, as well as a more nuanced appreciation of how stakehold make links between alternative environmental and socio-economic factors. Evidence this nature is valuable to cities in different locations and aids in the transferability of b practices (or the identification of poor practices) from which other locations can learn.

Table 1. URBAN GreenUP survey areas and associated NBS interventions.

Survey Location	Description	URBAN GreenUP NBS Interventions
Business Improvement District	City-center business area is characterized by a mixture of commercial, office, and retail space. Limited residential. Serviced via main roads.	St Johns/Parr Street/Chavasse Park Green walls pollinator verges/green spaces, The Strand stree tree investments, Wapping Dock Floating Ecological Island, mobile gardens/mobile forest bathing pods
Baltic Triangle	Area of mixed residential (mostly apartments but some Victorian terrace and social housing), light industrial, creative, and commercial spaces. Located next to the main road (Wapping, Chaloner Street, and Parliament Street) and River Mersey. Limited NBS on-site.	Street trees, pollinator lamp posts, pollinator verges/green space, forest school and church activities, sustainable drainage systems.
Sefton Park	Residential area of south Liverpool is characterized by a mix of apartments, semi-detached houses and converted Victorian townhouses. Some commercial/retail use. Sefton Park is the largest NBS in the area, a Green Flag awarded park, and one of Liverpool's most frequently patronized sites.	Floating Ecological Island, Street/shade trees, pollinator green spaces.
Otterspool	Residential area of south Liverpool is characterized by a mixture of 20th-century terraces and semi-detached houses. Otterspool Promenade and the greenspace system are the largest NBS in the area. Some light industrial use includes a neighborhood recycling center. Located proximate to A561 Aigburth Road and River Mersey.	Pollinator green spaces/verges, sustainable drainage systems, deculverting/bioretention floc work, wood/tree planting.

The delivery of NBS will always be informed by historical debates and local polit and this was certainly the case in Urban GreenUP. In Liverpool, green space is a marke unequal investment and geographic inequalities. The wealthier southern parts of the c have more and higher quality GI and increased engagement with environmental iss while communities in the north face multiple sources of social and economic deprivat and have less (and lower quality) GI [28]. Issues of spatial parity, environmental qual access, and variability of amenities have been extensively debated and documented in Liverpool Green Infrastructure Strategy [29] and the subsequent Liverpool Green and Op Space Review (LG&OSR). These indicated that NBSs in Liverpool are considered valua and that greater investment in environmental management is needed to address clim change, biodiversity loss, a lack of sustainable transport options, and health and well-be inequalities [30]. One direct consequence of the LG&OSR has been the foregrounding nature in the subsequent approach taken by Liverpool City Council to address envir mental issues. It also facilitated Liverpool City Council's engagement with the EU's call partners to join the Horizon 2020 NBS R&I projects and the city's inclusion in the URB GreenUP project.

A political willingness to engage with urban greening has thus been developed o several years. The relationships that Liverpool City Council has with regionally innovat environmental partners working on ecological issues are key to this. These instituti aided Liverpool City Council in engaging expertise to shape their environmental think by facilitating technical, academic, and knowledge exchange. However, despite the visi relationships between the city, environmental organizations, developers, and the public, there remain critiques of Liverpool's environmental policies. These focus on the equitable provision of green space and the limited emphasis placed on tackling environmental quality issues compared to achieving economic development objectives. There is also a perceived lack of accountability associated with the City Council by some communities, which view all development as negatively impacting the city's natural environment [31]. The URBAN GreenUP project attempted to redress these issues by implementing a program of NBS interventions.

4. Materials and Methods

To evaluate the impact of NBS interventions, each frontrunner city developed indicators and a monitoring framework. In Liverpool, social indicators focused on understanding the knowledge, perceptions, and engagement with NBS among businesses, SMEs, environmental organizations and charities, residents, local communities of interest, i.e., church groups or friends of groups, and elected officials. Data were collected over an extended period (2019–2022) to ensure that a range of stakeholders who stand to benefit from the interventions were included.

The following sections analyze the evidence generated from two primary forms of data: interviews with communities of interest and the results of a postal survey (undertaken ex-ante and ex-post of the program of NBS interventions). Both approaches to data collection focused on communities of interest located proximate to URBAN GreenUP NBS interventions, as these respondents were considered to have a more detailed understanding of the local environmental context prior to and post-intervention (see Tables 1 and 2). However, the paper acknowledges that the sample size of the interviews and ex-ante/ex-post surveys is not statistically representative of the population of Liverpool or communities proximate to each intervention (see Table 3 for demographic profiles of proximate wards). The ex-ante survey aimed to establish a baseline position on perceptions of NBS in Liverpool, while the ex-post survey asked respondents to consider the URBAN GreenUP interventions and how these informed their perceptions of urban nature and its benefits. Data collection was influenced by COVID-19, which limited opportunities to engage with respondents face-to-face or on-site (see Section 4.3 for further details).

4.1. Interviews with Communities of Interest

A total of 22 semi-structured interviews were conducted with businesses, SMEs, social enterprises, non-governmental organizations' workers, members of 'Friends of' groups, and elected officials/councilors in Liverpool. Participants were selected due to their prior engagement with urban development and/or environmental issues prior to the commencement of URBAN GreenUP. Interviewees were located proximate to the intervention areas of Sefton Park, Otterspool, the Baltic Triangle, and the Central Business Improvement District (BID) or held a role of responsibility for their development, i.e., elected officials.

Practically, interviewees were provided information about the project, the nature of their engagement, and information regarding consent and anonymization prior to agreeing to engage. Interviews lasted between 30–90 min and were recorded and transcribed. Consent was obtained from all interviewees, enabling the project team to use their commentary in the public domain in an anonymized form. Each interview focused on the following:

(1) Perceptions of the present provision of NBS/green space in Liverpool,
(2) Perceived impacts URBAN GreenUP NBS interventions on the city's natural environment,
(3) URBAN GreenUP governance structure and approach to urban greening, and
(4) The perceived legacy of URBAN GreenUP.

The transcripts were analyzed thematically to illustrate where links were made between policy and governance, co-design, and engagement of local communities, the added socio-economic and/or ecological value of investing in NBS, and the potential for longer-term benefits to Liverpool via the URBAN GreenUP interventions. Issues of design, e gagement, focus of project interventions, long-term maintenance, and the delivery benefits are key themes noted in the research literature focusing on NBS [8,32–34]. Dire commentary has been used from the fifteen participants (along with their organizatior and areas of work—although twenty-two interviews were undertaken in total) shown i Table 2. Commentaries from the remaining seven interviewees corroborated the inform tion presented but did not provide additional examples to extend the debates presented the following sections. It is also acknowledged that the sample size of interviewees do not represent the wider body of professional organizations in Liverpool. However, th proximity of each organization and their knowledge provided interviewees with a great understanding of the local urban and ecological context and the potential added value URBAN GreenUP-funded NBS interventions. The project attempted to engage a larg number of organizations in the interview process. However, due to unforeseen logistic issues, organizations were not able to be involved in the data collection process.

Table 2. Interview profile and stakeholder type.

Interviewee Profile	Stakeholder Type
Business-owner—Planning Consultancy	Business
Manager—Retail Organisation	Business
Owner—Environmental Consultancy Business	Business
Owner—Environmental Consultancy Business	Business
Business-owner—Hospitality	Business
Chair—Residents Organization	SME/Social Enterprise
Managing Director—CIC	SME/Social Enterprise
Manager—Natural Heritage NGO	NGO/NFP
Religious Leader	NGO/NFP
Head of Economic Non-Profit	NGO/NFP
CEO—Civil Society Organisation	NGO/NFP
Head—Parks Organisation	'Friends of' group
Head—Parks Organisation	'Friends of' group
Labour Councillor	Councillor
Green Party Councillor	Councillor

4.2. *Ex-Ante and Ex-Post Postal Survey*

To ensure that a cross-section of local responses was generated, a resident's surv and expert/professional interviews were developed to assess local knowledge and u of NBS proximate to URBAN GreenUP interventions. The survey focused on responde perceptions and relationships with nature in urban environments, their use of these spac and positive and negative assessments of NBS. It also reflected on how NBS could p vide benefits to climate change mitigation, pubic/personal health and well-being, soci inclusion, community engagement, the livability of the area, property values, crime, ar local business opportunities. The questionnaire survey was constructed to provide r spondents with opportunities to respond quantitatively via Likert/preference scales 5-point scale was used—strongly agree, mostly agree, neutral/neither agree nor disagre mostly disagree, strongly disagree) and activity/issue lists, i.e., activities/uses of NBS, ar qualitatively through open-ended questions. Both qualitative and quantitative questio were used to provide respondents with opportunities to detail their understanding of loc NBS (Tables 4 and 5 provide indicative results of these types of questions).

Table 3. Ward Profiles of areas proximate to URBAN GreenUP NBS interventions.

	Population	Male/Female	IMD (1 = Most Deprived to 30 = Least Deprived)	Unemployment Rate	Housing (Most Significant Tenure)
			Wards closet to Sefton Park and Otterspool NBS Investment Areas		
Church	13,722	48.02% (M)/51.98% (F)	30	3.7%	79.92% Owner Occupier
Cressington	15,182	49.64% (M)/50.36% (F)	24	5.2%	69.2% Owner Occupier
Greenbank	15,731	47.73% (M)/52.27% (F)	22	6.1%	48.73% Private Rented
Mossley Hill	13,463	49.91% (M)/50.09% (F)	29	3.5%	73.9% Owner Occupied
Princes Park	20,529	47.67% (M)/52.33% (F)	8	12.8%	49.31% Registered Social Housing
St. Michaels	12,724	47.33% (M)/52.67% (F)	20	7.0	47.46% Owner Occupier
			Wards closest to BID and Baltic Triangle NBS investment area		
Central	33,468	46.92% (M)/53.08% (F)	26	2.3%	78.67% Private Rented
Princes Park	20,529	47.67% (M)/52.33% (F)	8	12.8%	49.31% Registered Social Housing
Riverside	23,498	47.77% (M)/52.23% (F)	15	7.3%	54.89% Private Rented

Table 4. Survey perceptions of NBS quality, quantity, and accessibility located proximate to URBAN GreenUP interventions (ex-ante and ex-post results). Boxes in red denote an overall negative response and those in green an overall positive response.

	Sefton Park (Ex-Anti Intervention)	Sefton Park (Ex-Post Intervention)	Change in Response (Positive/Negative)	Otterspool (Ex-Anti Intervention)	Otterspool (Ex-Post Intervention)	Change in Response (Positive/Negative)
ow would you rate your eighbourhood BS in terms of quantity	93.7% Good/Very Good	92.1% Good/Very Good	Negative	100% Good/Very Good	97.6% good/very good	Negative
ow would you rate your eighbourhood BS in terms of quality	84.7% Good/Very Good	82.5% Good/Very Good	Negative	90.5% Good/Very Good	89.3% Good/Very Good	Negative
In your eighbourhood, ow would you ate the NBS in terms of accessibility.	90.7% Good/Very Good	89.6% Good/Very Good	Negative	85.7% Very Good/Good	92.8% Good/Very Good	Positive
hinking about the city of iverpool as a whole, how ould you rate NBS in terms of quantity.	69.9% Good/Very Good	64.7% Good/Very Good	Negative	76.2% Good Very/Good	65.9% Very Good/Good	Negative
hinking about the city of iverpool as a whole, how ould you rate NBS in terms of quality.	65.6% Good/Very Good	65.6% Good/Very Good	Negative	54.7% Very Good/Good	75.3% Very Good/Good	Positive
hinking about the city of iverpool as a whole, how ould you rate NBS in terms f accessibility.	61.3% Good/Very Good	59.2% Good/Very Good	Negative	65.9% Very Good/Good	70% Very Good/Good	Positive

Table 5. Survey respondent knowledge of URBAN GreenUP NBS interventions (drawn from post survey only). Boxes in red denote an overall negative response and those in green an ove positive response.

	Sefton Park (Post)	Reaction (Positive/Negative/Neutral)	Otterspool (Post)	Reaction (Positive/Negative/Neutra
Have you seen the green wall at St. John's Shopping Centre?	65.1% No	Negative	82.4% No	Negative
Were you aware that it was an URBAN GreenUP intervention?	84.9% No	Negative	96.3% No	Negative
Have you seen the green wall on Parr's Street?	81% No	Negative	90.4% No	Negative
Were you aware that it was an URBAN GreenUP intervention?	91.8% No	Negative	97.6% No	Negative
Have you seen the floating island in Wapping Dock?	79.1% No	Negative	91.8% No	Negative
Were you aware that it was an URBAN GreenUP intervention?	81.9% No	Negative	97.6% No	Negative
Have you seen the floating island in Sefton Park?	69.8% Yes	Positive	56.5% Yes	Positive
Were you aware that it was an URBAN GreenUP intervention?	74.4% No	Negative	81.5% No	Negative
Have you seen the bio-retention pond in Otterspool?	53.5% yes	Positive	71.8% yes	Positive
Were you aware that it was an URBAN GreenUP intervention?	90.7% No	Negative	89.3% No	Negative

Surveys were distributed to homes within a 300 m radius of the two research si Sefton Park and Otterspool, located in south Liverpool and hosting URBAN GreenUP N Interventions. Two surveys were conducted—the first in 2019 prior to NBS interventions establish a baseline of local perceptions of NBS (hereafter the ex-ante survey) and a secc in 2021 following the completion of NBS interventions (hereafter the ex-post survey). T postal survey was delivered by hand and collected by the research team because (a) pos surveys without interaction with the research team generate lower returns and (b) th provided the research team with an additional level of certainty regarding how ma people/homes had been engaged, how many had responded, and whether additional vi to collect/remind residents to complete the survey were needed. Respondents were a provided with the opportunity to return completed surveys via postal mail in a pre-p envelope. The use of a postal survey was considered to add greater validity to the d collection process, as the research team could guarantee (within some tolerances) t the survey would be completed by people living proximate to existing and new N interventions. Respondents were able to complete the survey by hand for collection by project team and return it by pre-paid envelope. They were also able to complete it online via a designated survey weblink. Details of the online survey (including a weblink) were included in the information provided in the postal survey.

A total of 75 survey responses were received from residents in the ex-ante survey (N: 35 Sefton Park and N:42 Otterspool), and 173 responses (N:86 Sefton Park and N:87 Otterspool) were returned following the ex-post survey. The combination of data collected for the survey provided scope to analyze whether links between preferences, uses, appreciation and/or understandings of NBS in Sefton Park and Otterspool could be made.

4.3. COVID-19

URBAN GreenUP ran from 2017–2023, and the main period of data collection occurred during the 2019–2021 COVID-19 pandemic. Consequently, the approach taken to data collection was modified to adhere to institutional and UK government stay-at-home and social distancing regulations. The outcome was that face-to-face surveying work was deemed inappropriate, and data collection was predominately conducted online, i.e., interviews on Zoom or via postal survey, where interaction with the research team was minimal. The authors acknowledge the implications of such a pivot from face-to-face evidence collection in terms of potentially limiting the number and range of potential participants engaged. However, the alternative methodological framework developed was considered robust, providing an effective approach through which residents and communities of interest could be engaged.

5. Results

The outcome of interviews with local communities of interest and the postal survey with residents illustrate a range of considerations regarding the value of both NBS generally in Liverpool and URBAN GreenUP NBS interventions specifically. Results have been categorized and presented in three distinct areas: Elemental and functional influences, Project and locational influences, and Thematic influences of NBS success to highlight the variation in participant responses related to the breadth of views that need to be made when implementing NBS.

The presentation of the results is not split between interviewee and respondent responses but is presented thematically across the three areas noted above. This is a deliberate choice, as it is considered to show greater complementarity between the different voices engaged with NBS. Sections 5.1–5.3 combine evidence from the interviews and survey results, as both address the thematic framings of each sub-section. This provides the discussion with a more nuanced approach to comparing issues deemed significant to professionals and residents. Direct quotes or specific points are attributed to specific interviewees or noted as being drawn from the resident survey.

The data generated from the interviews are also presented as a set of critical commentary reflecting participant considerations of the additional value that NBS and the URBAN GreenUP interventions can provide in Liverpool. The survey responses were analyzed with descriptive statistics to explore trends in the data. In Tables 4 and 5, for example, the data provided is drawn from Likert Scale analysis and depicts the most frequently used responses to a series of questions focused on perceptions of NBS quality, quantity, distribution, visibility, accessibility, and use. It is also acknowledged that there is a cross-over between both the results derived from the interviewee and survey responses, as investment in NBS does not fall solely into the purview of professionals or residents and, as such, should not be compartmentalized when issues of politics, sociocultural, economic, and environmental issues intersect.

5.1. Elemental and Functional Influences: Location, Scale, Functionality, Interactivity

Significant emphasis was placed on the location, scale, and quality of NBS in Liverpool by both respondents to the postal survey and interviewees. From an interviewee's perspective, the city has a host of high-quality NBS in the form of its Victorian parks and waterfront area; however, it was noted that the distribution in terms of north-south, areas low income/affluence, and proportion of these spaces was inequitable. This was reporte by an elected official, who stated that:

> *"I think we have some wonderful parks in the city, particularly in the south. They are not proportionate. I think we need more green corridors, whether they are pedestrian or bike lanes, that people can use."*

The view that NBSs are not distributed equally was also commented on by a busine respondent when considering access to nature in urban areas: "*I have absolutely no idea wh the people who live in the city are doing for their walks and to get out of the house because they ha not got anything that is remotely connected at the moment.*" However, an alternative and mo positive elected official commented that URBAN GreenUP was facilitating a conversatio about the breadth of options available to Liverpool City Council and partners to invest NBS. They stated that:

> *"Scientifically, I do not know how much of an impact [URBAN GreenUP NBS interventions are] going to have because, obviously, they are fairly small [but] something like that [floating island] I think is useful as a talking point. Then we can start the conversation about biodiversity. My attitude is that you should try and work out what the problem is and what you want to do about it, then how you do it, and then find the money. This [URBAN GreenUP] really started with the money, which is not a very good way to start. It is very difficult for a politician or a senior officer to resist a project that comes with money attached."*

Therefore, although issues of location and scale were reported, URBAN GreenU could act as a catalyst for the city to rethink how it invests in urban nature by reflecting what type of intervention would work in specific locations.

Considering NBS from such a perspective would enable the City Council (and partne to address the pervasive view that there is insufficient NBS across Liverpool. Table highlights this issue: Postal survey respondents had largely positive views of NBS (85 positive responses for local NBS), but views of the city's environmental resources were le positive. Responses to the postal survey vary in their assessment of the quality, quanti and accessibility of NBS locally and at a city scale. Local/neighborhood NBS are perceiv more positively than Liverpool's. This aligns with expert commentary in the city a academic literature and highlights an ongoing issue of environmental governance with Liverpool related to parity of resource allocation.

By contrast, the views expressed in the ex-post survey are also conditioned by sponses to COVID-19 and city-wide lockdown protocols. As such, perceptions of acc sibility, quality, and quantity may have altered from 2020 onward if respondents h (a) limited access or (b) spent more time in green space and thus were more critically aw of quality/quantity issues. Moreover, it could be argued that respondents in Sefton Park more critical of the totality of Liverpool's NBS resource base due to their proximity to o of the city's two Green Flag accredited parks (Sefton Park and Stanley Park). Residents Otterspool may not have placed such an emphasis on local sites in their commentary. The fore, they may have been more accepting of the variation in location, size, and functi due to the greater variation in NBS quality compared to Sefton Park residents. Otterspo residents report improvements in accessibility locally (an increase from 85.7% to 92.8 and across the city (an increase from 54.7% to 75.3%) in the ex-post survey, supporti this view.

The variation in commentary from local expert interviews and residents suppo existing discussions concerning the distribution of NBS in Liverpool. This highlights need for continued engagement with issues of how projects identify sites for developm (as well as financing for maintenance) across the city. It also suggests that although URB GreenUP has value for enhancing ecological resources, the interventions may be inadequ especially for addressing the existing disparity caused by long-term development objecti across the city.

5.2. Project and locational influences: Communication, Visibility

> *"Firstly, I think individuals, businesses, and developers have not gotten a clue. They need something like this [URBAN GreenUP] to give them ideas ... they know about buildings, and they know about commerciality, but they have no clue about how they can do it in a very considered way. There is a lack of creativity in terms of what we have here, which is a bunch of facilities that could have green all over the place in terms of their roofing. There is a real lack of imagination in terms of how green space can actually be created. We are not being individually or collectively clever enough."*

The lack of visibility of the URBAN GreenUP interventions was perceived as undermining their success by interviewees in business, the environment, the SME sector, and local government. The quote above from a local NGO highlights two significant issues with NBS interventions: (a) local awareness of project work and (b) the type of projects delivered. The NGO officer critiqued the lack of innovation by local businesses, community, and environmental organizations in terms of experimenting with NBS beyond planting street trees, street greening, or creating pock parks. This raises questions regarding whose responsibility it is to facilitate innovation and/or investment in NBS and who has the authority to deliver change in urban areas where multiple landowners are present. Even where substantial interventions were delivered, i.e., the 65 m long "living green wall" comprising over 14,000 evergreen plants on the St. Johns Shopping Centre, there was a view from a local Small–Medium Sized Enterprise (SME) that *"[local people] will not see the fringe [the green wall] around St. John's Shopping Centre in any way making up for some of the stuff that is going on."* Although a further SME noted that they considered URBAN GreenUP to have support within the city, this was skewed towards the environment sector rather than from a broad cross-section of stakeholders:

> *"I think URBAN GreenUP as a concept has support, but only among a small group of people in the know, predominantly eco-warriors and university types, and people with responsibility within the council for eco-friendly decisions."*

One issue repeatedly noted by interviewees extending this view is the city's role as the facilitator of URBAN GreenUP and the language used to raise awareness of the project. The term "NBS" is considered overtly technical and thus lacks resonance with local people. As noted by a representative of a 'Friends of' Group, such language makes NBS less tactile, as it is not as common compared to parks, trees, or nature (although each of these terms is also complex). Moreover, within the NGO and environmental sectors, two respondents commented:

> *"I would not naturally have put that phrase [NBS] with URBAN GreenUP unless you are defining the problem as there is not enough green space and we would like it to be more ... I would have gone more for something like nature-based enhancements. If it is more about how we actually want to enhance green corridors and make it a nicer place to live, I would not necessarily think 'solution' was the phraseology for that."* (NGO)

> *"I have a personal reaction to the word 'solution', and it is not a good one because a solution emanates from a problem, and I just do not like problem–solving processes; they are too simplistic. You take a problem and you solve it; you [then] usually create ten more problems, often they are somebody else's, and it flows on from there, and sometimes they come back to bite you. It is an endless task. It is also symptomatic of simplifying quite complex systems."* (Environmental consultant)

Both highlight a potential limitation of NBS terminology if it does not readily translate to all communities. Moreover, the comments illustrate how, even within the environmental sector, interpretations vary regarding the meaning of words. In the case of NBS, the use of "solutions" is viewed as problematic if or when interventions are not linked directly to context-specific issues. In Liverpool, the city's Green Infrastructure Strategy [26] and the URBAN GreenUP project proposals identified a range of core climatic, socio-economic, and health issues that NBS could address. However, although this information was integrated into the framing and project documentation, development, and delivery URBAN GreenUP, it may not have been effectively communicated to all communities in city—especially residents (see Table 5, which presents resident awareness of URB GreenUP interventions).

Consequently, there were concerns among a significant number of participants reg ing the communication of the added value of NBS interventions associated with URB GreenUP. This is also visible in the responses from the ex-ante and ex-post postal surv where awareness of both the project and its specific interventions was limited. Tab outlines the awareness of Sefton Park and Otterspool respondents, noting that the float ecological island in Sefton Park and the bioretention work in Otterspool were the o known interventions. This suggests that the interventions delivered were (a) physic too small to be visible to residents, (b) not well-publicized, or (c) in places that peo do not use. Point (c) is countered by a significant number of respondents in the ex-p postal survey who reported using Sefton Park and Otterspool frequently as their lo NBS/greenspace compared to other city center or waterfront locations and said that URBAN GreenUP interventions in these locations may have been increasingly visi compared to those in the Baltic Triangle or BID area. This was noted by a Friends of Gr located close to Sefton Park, who stated that:

> *"The impact was immediate in terms of literally every person passing us . . . stopping to ask what it was about. We did not receive one comment about what a waste of money 'in these times' [they were], which I was quite surprised at. I did assume you would be receiving, 'Oh, well, why are you spending money on this when, you know, we have poverty and COVID and everything else?' [There has been] lots and lots of genuine interest, which was surprising in terms of impact."*

It could be argued that the communication of each intervention could have b improved to facilitate knowledge of both the URBAN GreenUP project and the NBS in vations within it. The commentary from local environmental organization interview supports this, suggesting that consideration of terminology and engagement with com nities of interest are key to raising awareness. Therefore, projects need to be visible to u to facilitate engagement. Where this was possible (e.g., the water management wor Otterspool), users saw work on-site and a direct impact (less flooding) of the interven Thus, the experiential nature of specific projects in highly visible locations was deer necessary in assessing the value of NBS interventions as "solutions".

5.3. Thematic Influences of NBS Success: Ecological and Socio-Ecological Factors

The postal survey results highlight a positive response to assessments of the envi mental quality associated with the URBAN GreenUP interventions. Although knowle of each intervention was less well defined in responses (see Table 5) when residents v asked to discuss the links between NBS and climate change, biodiversity enhancem and quality of life, respondents in both the Sefton Park and Otterspool surveys provi a positive analysis of the resource base in Liverpool. NBS interventions were conside to have a less direct impact on addressing urban heat island effects and improving quality. Although respondents understood they could make a difference in these probl NBS interventions were perceived as more effective for other environmental challer NBSs were perceived as having the least impact on economic factors, with resident su respondents skeptical of linking nature and local business revenue. The lack of posi responses for economic activity was in line with the general commentary from resid in the postal surveys, as they focused more directly on issues of quality of life and p than economic returns. A reading of Table 6 suggests that NBS are considered to positi contribute to the quality and functionality of ecological resources in Liverpool—note strongly agreeing or somewhat agreeing in responses to postal survey questions. H ever, there was a minor proportion of responses considered neutral or negative ove i.e., instances of respondents strongly disagreeing with links between NBS and impro environmental quality. Drawing on the ex-post postal survey, Table 6 notes the promin of improvements in quality of life and biodiversity (and, to a slightly lesser extent, climate change) as positively associated with NBS interventions.

Table 6. Postal survey responses to the impact of URBAN GreenUP NBS interventions. Green boxes denote positive overall associations between NBS interventions and specific benefits, whilst yellow boxes denote a partial relationship between identified in respondent commentary.

	Sefton Park						Otterspool					
	Impact of URBAN GreenUP Intervention (Respondent Commentary—Green = Positive/ Yellow = Neutral/Red = Negative)						Impact of URBAN GreenUP Intervention (Respondent Commentary—Green = Positive/ Yellow = Neutral/Red = Negative)					
	Climate Change	Air Pollution	Urban Heat Island	Local Business Revenue	Quality of Life	Biodiversity	Climate Change	Air Pollution	Urban Heat Island	Local Business Revenue	Quality of Life	Biodiversity
St Johns Shopping entre green wall												
Parr Street green wall												
Floating Island Wapping Dock												
Floating Island efton Park												
oretention Pond Otterspool												

Absent from Table 6 is a commentary on the economic opportunities associated with NBS. These data were generated from interview respondents in the BID and Baltic Triangle rather than the postal survey. Interviewee respondents reported that businesses responded positively to NBS as a facilitator of economic development opportunities. The commentary noted that greener, interactive, and ecologically diverse environments attracted both businesses and supported additional footfall and returns on investment, justifying relocation costs to "greener areas." Respondents also noted links between higher rental and sales values in locations with more NBS. They supported the view that employee productivity would increase if and where neighborhoods had a higher proportion of NBS. All of these compare favorably to the research literature examining the links between NBS/GI and economic value [32,33,35]. To support these statements, an SME and a local business representative noted:

> *"A city that is greener, with a lot more nature in it ... is a much more attractive place to work and to attract businesses. It is a healthier place to work. Therefore, during breaks, you can get out, and very quickly, you are under a tree. You are looking at flowers; you are sitting on a bench where there is some grass around; there is a water feature, or whatever it is. That, of course, impacts the people who come into the city to work, the visitors to the city, and the people like us who live here; it impacts significantly on our well-being and our health."* (SME)

> *"Inward investors are more likely to choose greener cities for their businesses and employees."* (Local business)

However, the need for caution was also noted in terms of over-extrapolating the impact of urban greening/NBS on city-wide economic development opportunities. A local business reported that:

> *" ... businesses use many criteria when deciding where to locate, and many different factors play a part, with green spaces not necessarily high on the list. Transport, parking, and cost would be far more important."*

Interviews with business respondents also provided a more detailed analysis of the perceived links between NBS and the economic benefits of investment. This included a more nuanced appreciation of NBS, as it was more difficult for respondents to substan- ate claims of direct economic benefits associated with NBS interventions. Alternative business respondents outlined how they considered NBS to aid this process, arguing th for example, 80% of respondents in the BID and 50% of respondents in the Baltic Triang reported that investment in NBS would lead to an uplift in property prices. Increase opportunities for employment (based on NBS providing a more attractive work and i vestment environment) were also noted: 50% agreed/strongly agreed in the BID and 75 agreed/strongly agreed in the Baltic Triangle. There was also consensus that locations wi a higher proportion of high-quality, diverse, and interactive environments, including NI had the following effects: (a) they are attractive to businesses (83% agree/strongly agree the BID and 63% agree/strongly agree in the Baltic Triangle), (b) they promoted investme and relocation these areas (66% agree/strongly agree in the BID), (c) they support increase footfall and time spent leading to potential increases in revenue (83% agree/strongly agr in the BID and 50% agree/strongly agree in the Baltic Triangle), and (d) they are green and more interactive places can facilitate increased productivity (83% agree/strongly agr in the BID and 63% agree/strongly agree in the Baltic Triangle) and employee well-bein However, while the majority of respondents identified positives associated with NBS, the remained concerns that parking, public transport, and the cost of rent were more significa influences on economic development opportunities than investment in NBS. Additional businesses in the Baltic Triangle stated that NBS was not a core factor promoting the relocation to the area (63% disagree/strongly disagree).

Consequently, no singular view supports investment in NBS as a facilitator of soc economic or ecological improvements. What is visible is the complexity of the benef and/or functions respondents (residents and interviewees) associate with NBS, what th highlight as positive influences, and where they view change as having a less significa impact. Therefore, defining success in delivery is difficult, as no singular function intervention was deemed to deliver improvements in socio-economic or ecological benef in all cases. This suggests that when planning NBS, local context is critical to the choic being made, which requires awareness of the perceptions of businesses, the environme community-oriented organizations, and residents to ensure investment responds to lo circumstances. Each of these groups defines success differently, and as such, the city nee to be aware of how scale, focus/function, and visibility influence perceptions of success interventions. Although there was a broad consensus among the respondents engag with the project that NBS can enhance the quality, quantity, and functionality of a lo area, the use of NBS should not be seen as a panacea for all urban problems—especia economic issues.

6. Discussion

Investment in NBS has been framed as a "go-to" approach to addressing cities' co plex socio-economic and ecological problems via co-produced nature-centric plans [NBSs offer a breadth of investment options that can be adopted to address issues includi ecological decline, climate change, unsustainable urban forms, and socio-economic depri tion [21,37]. Investment in NBS by the EU, funded via the Horizon 2020 NBS R&I progra accelerated implementation and facilitated experimentation with NBS, which may not ha occurred or been delivered over an extended timeframe via ES or GI planning. This h been particularly visible in the support for NBS from local government, environmen organizations, and SMEs, who have been at the forefront of these delivery programs. C sequently, we can identify a groundswell of engagement with NBS as potential "solutio to urban problems that are flexible enough to adapt to various contexts.

A significant proportion of the literature supporting this position frames NBS an evolution of other nature/green space planning terms, but one that explicitly lo to deliver change. This is becoming clearer in the literature debating scoping, desi implementation, and monitoring of NBS interventions in various European, Chin and African cities. A further commonality across these debates is the framing of N as a facilitator of more effective human/environmental discussions—one where nature offers direct solutions to human (and ecological) problems [19,34,38]. Within the URBAN GreenUP project, links between the ecological benefits and the associated socio-economic opportunities afforded by NBS interventions have been central, e.g., enhancing recreation, access to nature, and economic development [39]. Against this backdrop, each of the five-year Horizon 2020 R&I programs tested a range of delivery options to examine how NBS can create more sustainable and functional urban environments. This framing can be mapped to the approach taken in Liverpool.

Discussions of the added benefit of NBS require an understanding of (a) NBS form and function, (b) the benefits delivered to people and biodiversity, (c) the role of NBS as inclusive long-term solutions, and (d) the contexts in which they are implemented, and such benefits need to be communicated effectively to all communities of interest [6]. In practice, this requires dynamic, iterative planning processes to ensure effective delivery in the nuanced circumstances of different locations. Liverpool, for example, focused extensively on climate change adaptation, access to and improved quality of landscape functionality, and economic development opportunities—A, B, and C above. However, the approach to communication and context-driven delivery was queried by the respondents to the survey work undertaken for URBAN GreenUP.

However, analysis of NBS investments in Liverpool enables advocates to identify whether these findings can be mapped onto the growing series of frameworks developed for NBS to provide signposts for delivery. The research of Frantzeskaki [40], for example, reported a further seven areas that should be considered when developing NBS: NBS should be attractive, help create new green commons, promote trust between the public and city officials, support collaborative governance, facilitate inclusive and holistic policy formation, and be scalable and transferable between locations. All of these are common to each Horizon 2020-funded NBS project and are identified as key factors influencing the framing of NBS in the literature. The evidence discussed above and the wider reporting on the Horizon 2020 NBS projects provide opportunities to reflect on what best practice for NBS looks like, even if the NBS interventions delivered in Liverpool do not fully align with these goals [19,39–43].

6.1. Locating NBS in Local Development Structures

In the context of Liverpool and the delivery of NBS via the URBAN GreenUP project, a continued reflection on issues of appropriateness in terms of the location of project interventions, collaboration and trust between stakeholders, and the creation of additional ecological resources in urban areas were all considered critical. Critiques can be made of the URBAN GreenUP portfolio of interventions, including whether they were the most effective way of delivering change. While the analysis shown above suggests not, the commentary of professional and residential stakeholders supports the links between environmental and socio-economic improvements associated with the project. This was significant when issues of access and use of NBS and perceptions of quality of life were discussed by respondents. However, NBS advocates must remain cognizant of the local context to ensure effective delivery. Otherwise, concerns about legitimacy could be raised, e.g.,

> *"I get the sense that it is more greenwashing than actually addressing urban design problems, living problems, and urban-ecological problems of the human species."* (Environmental Consultant)

> *"While I am generally supportive and positive about the URBAN GreenUP project, my critique would be that what something like URBAN GreenUP represents is this idea that you have to raise lots of money, spend loads of time consulting and planning, and then spend loads of money implementing a complicated big project to encourage nature. What you need to do is just stop wasting your time fighting nature the whole time and allow nature to flourish. Nature can do a pretty good job on its own."* (Elected official)

URBAN GreenUP may have succeeded in extending the foundations of the ci Green Infrastructure Strategy and the LG&OSR [25–27] by building additional moment for interventions in nature-led planning that can subsequently be used to shape policy such a scenario, NBS can be used to explore new delivery pathways in terms of innova design, the choice of projects and the location they are placed in, and the role of ex imentation in terms of the benefits associated with each NBS, which provides scop rethink environmental planning across the city [44–46]. If this can be effectively achie Liverpool City Council could inspire further interaction with nature while promoting efficient use of existing space within the city for development [47]. The evidence genera by URBAN GreenUP could be beneficial if leveraged to support an increased allocatio local government capacity and financing to facilitate longer-term changes in the governa of the city's landscape.

6.2. Elements, Functions, Benefits, and Beneficiaries

To ensure that NBS interventions deliver their intended objectives, there is a nee consider what elements are designed, what functions, i.e., specific ecosystem services, t deliver, what socio-economic and/or ecological benefits they provide, and who bene from investment in nature. Raising awareness of URBAN GreenUP interventions other greening projects is important as a starting point. Despite the visibility of sc NBS interventions, i.e., the Sefton Park floating island, there remains a lack of connec between policy/local government campaigning for NBS and public acknowledgme the added benefit of the interventions. Project partners in Liverpool and elsewhere co benefit from more attention being placed on the needs and aspirations of the div communities that stand to benefit from NBS interventions. The promise that NBS deliver more democratic and sustainable outcomes is central to arguments for their use The "NBS with and for people" approach can help deliver on such promises, enhan equitable distribution of benefits, minimizing disbenefits, and underpinning succes delivery [48]. It could be argued that URBAN GreenUP did not fully align the need people, place, and nature, leading to critiques of the approach.

Working with communities to respond to locally contextual issues also enables supporters to think more creatively about a given intervention's type, size, function, benefits. Moreover, it provides scope to consider how variations in aesthetics and logical diversity can be more effectively aligned with issues of accessibility to pro use [40,49]. URBAN GreenUP highlights that this is not a simple process and requ ongoing collaboration between multiple sectors. If such relationships can be curated parties can better engage in knowledge exchange activities and arrive at more approp nature-focused solutions [26].

6.3. Aligning Political and Local Needs/Priorities

Evidence from Liverpool suggests that in addition to considerations of location function, advocates of NBS also need to align interventions with local political and plan objectives. The diversity of commentary from local elected officials noted above keeping with historical debates about the environment in the city [25,28]. This highlight critical role of consensus (or lack thereof) in shaping investment plans and political prior Where the breadth of benefits associated with NBS, e.g., enhancing ecological function alongside economic development, can be integrated into policy mandates, the options to the City Council and its partners expand. Achieving this requires the city to examin true cost of 'business as usual' development and reflect on the benefits of shifting p attention from predominately built infrastructure to a more NBS-centric approach. T discussed extensively in the literature and demonstrates a need to debate the variability complexity of alternative governance models to ensure that expertise, innovation, and knowledge are integrated into decision-making [2,10,11,16,19,20,34,50]. Where each ca aligned effectively, we can identify investment options for NBS that are locally approp as well as new funding pathways and alternative management practices that make the use of local government, private and environmental institutions and organizations, and community advocates to deliver investment in NBS [33,34,39,40].

Moreover, by examining NBS interventions as a continuum of options linking local, city, and regional perspectives, cities can better identify the most appropriate approaches for investment [51]. Although the URBAN GreenUP project was focused predominately on the local scale, there are options to explore the delivery of comparable innovations at a larger scale due to the scalability of investment in street trees, sustainable drainage, and pollinator networks. This can even be extended to a national scale if the priorities of such policies, for example, the National Planning Policy Framework in England [52] or EU directives on NBS [50], can be aligned with local delivery. This provides greater scope to promote synergies between policy mandates, identify problems associated with resource change and/or technological redundancy, develop adaptive plans to address maintenance and redevelopment issues and assess the immediate and long-term benefits associated with urban change [53].

6.4. Communication and Visibility

Analysis of the NBS interventions in Liverpool also illustrates the critical role played by communication and visibility in developing successful outcomes for investment. Across all commentary, the lack of communication of NBS projects delivered as part of URBAN GreenUP was noted as problematic for communities of interest. Therefore, effective communication of the proposed NBS elements, their functions and benefits, and the beneficiaries associated with interventions must be clearly communicated to all [6]. This is a critical point, as although NBS continues to deliver ecological services (e.g., flooding mitigation, providing habitat, intercepting pollution) even when people are unaware of these functions, the lack of awareness of the benefits can undermine public acceptance of investment in urban nature. Successful communication can support a greater understanding of the functions associated with NBS where they are not highly visible, which in turn can lead to public support for future interventions. It is critical to develop a successful communication strategy highlighting the location, function, and NBS to increase their visibility.

6.5. The Future Role of Stakeholders as Advocates for NBS Investment in Liverpool

The limited role afforded to environmental experts, local practitioners, businesses, and communities during the delivery of the URBAN GreenUP was reported as limiting the perceived value of the project's investment program. Future work should thus consider engaging with local expertise from experts and residents to help shape considerations of what type and size of NBS and in which locations investments should be delivered. Specifically, environmental stakeholders have expertise in landscape design, funding, and maintenance, which could be integrated into project work. Moreover, business leaders are acutely aware of the links between urban/landscape quality and economic prosperity and could be useful commentators on the appropriateness of future NBS interventions. Therefore, stakeholders engaged with urban development and management may be critical allies for the city council in Liverpool if or when they design, plan, and implement additional NBS. In addition, consultation with residents will be a key activity needed to ensure that the type of NBS, scale of intervention, and location of investment are considered appropriate to the local context. The city of Liverpool is aware of the need to engage more effectively with its stakeholders, as reported in the Liverpool Green and Open Space Review [30]. Still, it has not yet established a clear structure to mee this challenge. It may therefore be appropriate for local stakeholders to be afforded a more prominent role in the design, implementation, and management of NBS across the city of Liverpool. However, to date, the city council does not have a governance framework in place to ensure that such a multi-partner approach to investment is implemented. A key outcome of URBAN GreenUP is the need to rethink the role of a diverse set of stakeholders in the city's environmental planning and management.

6.6. Reflection on Limitations

The public and stakeholders perceived NBS as valuable and useful. Still, the metho were not designed to access a deep understanding of public knowledge, values, ar perceptions of the benefits and drawbacks of NBS. If monitoring had been better resource surveys would have been combined with more deliberative methods, as surveys do n provide scope for respondents to elaborate. If the project were completed again, it wou also aim to generate a more representative sample of the local population constructed by a housing type, employment, and income. The survey would also add additional questio to generate baseline knowledge of participant awareness of benefits and disservices compare responses to the URBAN GreenUP NBS. Due to the limitations placed upon t project by time, staffing, and COVID-19, this was not feasible.

The results need to be contextualized in light of the impact of COVID-19, but t direction of influence is potentially positive and negative. The pandemic saw an crease in the use, considerations of accessibility, and perceptions of value placed on urb nature [54,55]. This could have been reflected in our results, as the majority of responde in the postal surveys and interviews noted a greater appreciation of the sociocultu ecological, and economic value of NBS post-lockdown. Although this may not transla directly into new or revised policy supporting investment in NBS/greenspace, it provid valuable evidence of the relationships between people and their environments. Such outcome compares favorably with the results of analysis by Public Health England [and Natural England [37,38] in their analysis of greenspace use, suggesting that furth funding for nature is key to healthy and successful places.

7. Conclusions

The use of NBS as both a term and as an approach to addressing climate, biodivers and health issues within urban planning is growing. Evidence from the Horizon 2020 N program highlights the added value of delivering interventions focusing on nature a making direct links between people, location/place, and ecological functions. Howev care is needed to ensure that the focus, scale, and location of any individual or progra of NBS investment address identified needs and fulfill the climatic, sociocultural, a economic problems associated with a specific location. Moreover, establishing an effect governance regime to co-design, implement, and manage NBS interventions is key their long-term success. These overarching principles have been partially met from t discussion of the URBAN GreenUP project in Liverpool.

Consequently, the visibility of NBS in Liverpool—and therefore their perceived va to the city—remains variable. Smaller-scale NBS were identified as holding a more limi value due to their size, location, and perceived lack of function, for example, pollina lamppost investments. Visibly larger interventions and those located within existing N i.e., street trees and water-based NBS, were reported as holding an increased value residents and professionals. The size and location of NBS within Liverpool were hi lighted as critical factors promoting the visibility of each intervention and the subsequ recognition (or lack thereof) of its benefits. Therefore, future investment in NBS wo benefit from a more nuanced approach to the choice of interventions with residential a commercial/business communities to ensure greater alignment between strategic and lo needs. This was only partially achieved in Liverpool, with significant variations in reported awareness of NBS. Thus, the lessons to be learned from Liverpool include focus NBS delivery on a more integrated approach to siting, designing, communicating, a evaluating projects at the local scale. These are valuable lessons applicable to many ci and can act as a valuable learning experiences to aid the effective mainstreaming of NBS a best practice in city planning.

Author Contributions: Conceptualization, I.M., S.C. and F.O.; methodology, I.M., S.C. and F.O.; validation, F.O., I.M. and S.C.; formal analysis, I.M. and S.C.; data curation, F.O.; writing—original draft preparation, I.M. and S.C.; writing—review and editing, I.M. and S.C.; project administration, S.C.; funding acquisition, I.M. and S.C. All authors have read and agreed to the published version of the manuscript.

Funding: This project has received funding from the EU's Horizon 2020 Research and Innovation Program under grant agreement number 730426. Support was also received from the University of Manchester through their open access funding agreements.

Data Availability Statement: The data used to populate this paper will be made available following the conclusion and validation of the URBAN GreenUP project via EU repositories. Currently, the location of these data has not been made public.

Acknowledgments: The authors would like to thank all participants for their considerable thoughts on the URBAN GreenUP project.

Conflicts of Interest: The authors declare no conflict of interest.

erences

European Commission. Towards an EU Research and Innovation policy agenda for Nature-Based Solutions & Re-Naturing Cities. In *Final Report of the Horizon 2020 Expert Group on "Nature-Based Solutions and Re-Naturing Cities"*; Directorate-General for Research and Innovation Climate Action Environment Resource Efficiency and Raw Materials; European Commission: Brussels, Belgium, 2015.

O'Sullivan, F.; Mell, I.; Clement, S. Novel Solutions or Rebranded Approaches: Evaluating the Use of Nature-Based Solutions (NBS) in Europe. *Front. Sustain. Cities* **2020**, *2*, 572527. [CrossRef]

Cortinovis, C.; Olsson, P.; Boke-Olén, N.; Hedlund, K. Scaling up nature-based solutions for climate-change adaptation: Potential and benefits in three European cities. *Urban For. Urban Green.* **2022**, *67*, 127450. [CrossRef]

Carvalho, P.N.; Finger, D.C.; Masi, F.; Cipolletta, G.; Oral, H.V.; Tóth, A.; Regelsberger, M.; Exposito, A. Nature-based solutions addressing the water-energy-food nexus: Review of theoretical concepts and urban case studies. *J. Clean. Prod.* **2022**, *338*, 130652. [CrossRef]

Melanidis, M.S.; Hagerman, S. Competing narratives of nature-based solutions: Leveraging the power of nature or dangerous distraction? *Environ. Sci. Policy* **2022**, *132*, 273–281. [CrossRef]

Kabisch, N.; Frantzeskaki, N.; Hansen, R. Principles for urban nature-based solutions. *Ambio* **2022**, *51*, 1388–1401. [CrossRef] [PubMed]

Beumer, C.; Martens, P. IUCN and perspectives on biodiversity conservation in a changing world. *Biodivers. Conserv.* **2013**, *22*, 3105–3120. [CrossRef]

Lafortezza, R.; Sanesi, G. Nature-based solutions: Settling the issue of sustainable urbanization. *Environ. Res.* **2019**, *172*, 394–398. [CrossRef] [PubMed]

Hansen, R.; Pauleit, S. From multifunctionality to multiple ecosystem services? A conceptual framework for multifunctionality in green infrastructure planning for urban areas. *Ambio* **2014**, *43*, 516–529. [CrossRef]

Austin, G. *Green Infrastructure for Landscape Planning: Integrating Human and Natural Systems*; Routledge: New York, NY, USA, 2014.

Mell, I.C. *Global Green Infrastructure: Lessons for Successful Policy-Making, Investment and Management*; Routledge: Abingdon, VA, USA, 2016.

Benedict, M.A.; McMahon, E.T. *Green Infrastructure: Linking Landscapes and Communities*; Conservation Fund: Arlington, VA, USA; Island Press: Washington, DC, USA, 2006; Volume June, ISBN 1-55963-558-4.

Liquete, C.; Udias, A.; Conte, G.; Grizzettia, B.; Masi, F. Integrated valuation of a nature-based solution for water pollution control. *Highlighting Hidden Benefits. Ecosyst. Serv.* **2016**, *22*, 392–401.

Wu, J. Urban ecology and sustainability: The state-of-the-science and future directions. *Landsc. Urban Plan.* **2014**, *125*, 209–221. [CrossRef]

Niemelä, J. Is there a need for a theory of urban ecology? *Urban Ecosyst.* **1999**, *3*, 57–65. [CrossRef]

Nesshöver, C.; Assmuth, T.; Irvine, K.N.; Rusch, G.M.; Waylen, K.A.; Delbaere, B.; Haase, D.; Jones-Walters, L.; Keune, H.; Kovacs, E.; et al. The science, policy and practice of nature-based solutions: An interdisciplinary perspective. *Sci. Total Environ.* **2016**, *579*, 1215–1227. [CrossRef] [PubMed]

Oh, Y. All London Green Grid as Nature-Based Solutions for Urban Resilience. In *The Palgrave Handbook of Climate Resilient Societies*; Brears, R.C., Ed.; Springer International Publishing: Cham, Switzerland, 2021; pp. 989–1011. ISBN 978-3-030-42462-6.

Mell, I.; Scott, A. Definitions and context of blue-green infrastructure. In *ICE Manual of Blue-Green Infrastructure*; Washbourne, C.-L., Wansbury, C., Eds.; ICE Manuals; ICE Publishing: London, UK, 2023; pp. 3–22. ISBN 978-0-7277-6542-0.

19. Eggermont, H.; Balian, E.; Azevedo, J.M.N.; Beumer, V.; Brodin, T.; Claudet, J.; Fady, B.; Grube, M.; Keune, H.; Lamarque, P.; e Nature-based Solutions: New Influence for Environmental Management and Research in Europe. *GAIA—Ecol. Perspect. Sci.* **2015**, *24*, 243–248. [CrossRef]
20. Fan, P.; Ouyang, Z.; Basnou, C.; Pino, J.; Park, H.; Chen, J. Nature-based solutions for urban landscapes under p industrialization and globalization: Barcelona versus Shanghai. *Environ. Res.* **2017**, *156*, 272–283. [CrossRef]
21. Kabisch, N.; Frantzeskaki, N.; Pauleit, S.; Naumann, S.; Davis, M.; Artmann, M.; Haase, D.; Knapp, S.; Korn, H.; Stadler, J.; e Nature-based solutions to climate change mitigation and adaptation in urban areas: Perspectives on indicators, knowledge g barriers, and opportunities for action. *Ecol. Soc.* **2016**, *21*, 39. [CrossRef]
22. Dorst, H.; van der Jagt, A.; Toxopeus, H.; Tozer, L.; Raven, R.; Runhaar, H. What's behind the barriers? Uncovering struct conditions working against urban nature-based solutions. *Landsc. Urban Plan.* **2022**, *220*, 104335. [CrossRef]
23. Al Sayah, M.J.; Versini, P.-A.; Schertzer, D. H2020 projects and EU research needs for nature-based adaptation solutions. *Ur Clim.* **2022**, *44*, 101229. [CrossRef]
24. Van Wesenbeeck, B.; Mulder, J.; Marchand, M.; Reed, D.; de Vries, M.; de Vriend, H.; Herman, P. Damming deltas: A practic the past? Towards nature-based flood defenses. *Estuar. Coast. Shelf Sci.* **2014**, *140*, 1–6. [CrossRef]
25. Clement, S. *Governing the Anthropocene: Novel Ecosystems, Transformation and Environmental Policy*; Palgrave Macmillan: Basingst UK, 2021.
26. Clement, S.; Mell, I.C. Nature, Democracy, and Sustainable Urban Transformations. In *Sustainability Transformations, S Transitions and Environmental Accountabilities*; Edmondson, B., Ed.; Palgrave MacMillan: London, UK, 2023.
27. Raymond, C.M.; Frantzeskaki, N.; Kabisch, N.; Berry, P.; Breil, M.; Nita, M.R.; Geneletti, D.; Calfapietra, C. A framew for assessing and implementing the co-benefits of nature-based solutions in urban areas. *Environ. Sci. Policy* **2017**, *77*, 15 [CrossRef]
28. Mell, I. The impact of austerity on funding green infrastructure: A DPSIR evaluation of the Liverpool Green & Open S Review (LG&OSR), UK. *Land Use Policy* **2020**, *91*, 104284. [CrossRef]
29. Mersey Forest. *Liverpool Green Infrastructure Strategy*; The Mersey Forest: Risley Moss, UK, 2010.
30. Liverpool City Council. *Strategic Green and Open Spaces Review Board: Final Report*; Liverpool City Council: Liverpool, UK, 2
31. Nurse, A.; North, P. A place for climate in a time of capitalist crisis? A case study of low-carbon urban policy making in Liver England. *Town Plan. Rev.* **2020**, *91*, 155–179. [CrossRef]
32. Mell, I.C.; Henneberry, J.; Hehl-Lange, S.; Keskin, B. To green or not to green: Establishing the economic value of g infrastructure investments in The Wicker, Sheffield. *Urban For. Urban Green.* **2016**, *18*, 257–267. [CrossRef]
33. Vandermeulen, V.; Verspecht, A.; Vermeire, B.; Van Huylenbroeck, G.; Gellynck, X. The use of economic valuation to create pu support for green infrastructure investments in urban areas. *Landsc. Urban Plan.* **2011**, *103*, 198–206. [CrossRef]
34. Cohen-Shacham, E.; Walters, G.; Janzen, C.; Maginnis, S. *Nature-Based Solutions to Address Global Societal Challenges*; IUCN: Gl Switzerland, 2016.
35. South Yorkshire Forest Partnership & Sheffield City Council. *The VALUE Project: The Final Report*; South Yorkshire Fo Partnership & Sheffield City Council: Sheffield, UK, 2012.
36. Malekpour, S.; Tawfik, S.; Chesterfield, C. Designing collaborative governance for nature-based solutions. *Urban For. Urban G* **2021**, *62*, 127177. [CrossRef]
37. Dumitru, A.; Frantzeskaki, N.; Collier, M. Identifying principles for the design of robust impact evaluation framework nature-based solutions in cities. *Environ. Sci. Policy* **2020**, *112*, 107–116. [CrossRef]
38. Bush, J.; Doyon, A. Building urban resilience with nature-based solutions: How can urban planning contribute? *Cities* *95*, 102483. [CrossRef]
39. Mell, I.; Clement, S.; O'Sullivan, F. Engineering nature-based solutions: Examining the barriers to effective interventio Proceedings of the Institution of Civil Engineers-Engineering Sustainability; Thomas Telford Ltd.: London, UK, 2022; pp. [CrossRef]
40. Frantzeskaki, N. Seven lessons for planning nature-based solutions in cities. *Environ. Sci. Policy* **2019**, *93*, 101–111. [CrossRef]
41. Croeser, T.; Garrard, G.E.; Thomas, F.M.; Tran, T.D.; Mell, I.; Clement, S.; Sánchez, R.; Bekessy, S. Diagnosing delivery capabi on a large international nature-based solutions project. *NPJ Urban Sustain.* **2021**, *1*, 32. [CrossRef]
42. Frantzeskaki, N.; Vandergert, P.; Connop, S.; Schipper, K.; Zwierzchowska, I.; Collier, M.; Lodder, M. Examining the policy n for implementing nature-based solutions in cities: Findings from city-wide transdisciplinary experiences in Glasgow (UK), ((Belgium) and Poznań (Poland). *Land Use Policy* **2020**, *96*, 104688. [CrossRef]
43. Mayor, B.; Toxopeus, H.; McQuaid, S.; Croci, E.; Lucchitta, B.; Reddy, S.E.; Egusquiza, A.; Altamirano, M.A.; Trumbic, T.; T A.; et al. State of the Art and Latest Advances in Exploring Business Models for Nature-Based Solutions. *Sustainability* *13*, 7413. [CrossRef]
44. Kooijman, E.D.; McQuaid, S.; Rhodes, M.L.; Collier, M.J.; Pilla, F. Innovating with nature: From nature-based solutio nature-based enterprises. *Sustainability* **2021**, *13*, 1263. [CrossRef]
45. McPhearson, T.; Anderson, E.; Elmqvist, T.; Frantzeskaki, N. Resilience of and through urban ecosystem services. *Ecosyst.* **2015**, *12*, 152–156. [CrossRef]
46. Tozer, L.; Bulkeley, H.; van der Jagt, A.; Toxopeus, H.; Xie, L.; Runhaar, H. Catalyzing sustainability pathways: Navigating u nature based solutions in Europe. *Glob. Environ. Chang.* **2022**, *74*, 102521. [CrossRef]

. Sowińska-Świerkosz, B.; García, J. What are Nature-based solutions (NBS)? Setting core ideas for concept clarification. *Nat.-Based Solut.* **2022**, *2*, 100009. [CrossRef]

Seddon, N.; Smith, A.; Smith, P.; Key, I.; Chausson, A.; Girardin, C.; House, J.; Srivastava, S.; Turner, B. Getting the message right on nature-based solutions to climate change. *Glob. Chang. Biol.* **2021**, *27*, 1518–1546. [CrossRef] [PubMed]

Hoyle, H.; Jorgensen, A.; Hitchmough, J. What determines how we see nature? Perceptions of naturalness in designed urban green spaces. *People Nat.* **2019**, *1*, 167–180. [CrossRef]

European Commission Directorate-General for Research and Innovation. *Evaluating the Impact of Nature-Based Solutions: A Summary for Policy Makers*; Office of the European Union: Luxembourg, 2021.

Cohen-Shacham, E.; Andrade, A.; Dalton, J.; Dudley, N.; Jones, M.; Kumar, C.; Maginnis, S.; Maynard, S.; Nelson, C.R.; Renaud, F.G.; et al. Core principles for successfully implementing and upscaling Nature-based Solutions. *Environ. Sci. Policy* **2019**, *98*, 20–29. [CrossRef]

Ministry of Housing Communities and Local Government. *National Planning Policy Framework*; MHCLG: London, UK, 2021.

Markolf, S.A.; Chester, M.V.; Eisenberg, D.A.; Iwaniec, D.M.; Davidson, C.I.; Zimmerman, R.; Miller, T.R.; Ruddell, B.L.; Chang, H. Interdependent Infrastructure as Linked Social, Ecological, and Technological Systems (SETSs) to Address Lock-in and Enhance Resilience. *Earth's Future* **2018**, *6*, 1638–1659. [CrossRef]

Natural England. The People and Nature Survey for England: Adult Data Y1Q1 (April–June 2020) (Experimental Statistics). Available online: https://www.gov.uk/government/publications/the-people-and-nature-survey-for-england-adult-data-y1q1-april-june-2020-experimental-statistics/the-people-and-nature-survey-for-england-adult-data-y1q1-april-june-2020-experimental-statistics (accessed on 3 November 2020).

Natural England & Office of National Statistics. Monitor of Engagement with the Natural Environment—The national survey on people and the natural environment. In *Headline Report 2019*; Natural England: Peterborough, UK, 2019.

Public Health England. *Improving Access to Greenspace: A New Review for 2020*; Public Health England: London, UK, 2020; Available online: https://beyondgreenspace.net/2020/07/29/improving-access-to-greenspace-a-new-review-for-2020/ (accessed on 3 December 2020).

land

MDPI

icle

nplementing Nature-Based Solutions in Urban Spaces in the ontext of the Sense of Danger That Citizens May Feel

bara Vojvodíková [1,*], Iva Tichá [2] and Anna Starzewska-Sikorska [3]

1 IURS—Institute for Sustainable Development of Settlements, 70800 Ostrava, Czech Republic
2 Faculty of Social Studies, University of Ostrava, 70200 Ostrava, Czech Republic
3 Institute for Ecology of Industrial Areas, 40-844 Katowice, Poland
* Correspondence: barbara.vojvodikova@email.cz; Tel.: +420-725-117-244

Abstract: Cities are facing the challenges of climate change. The application of nature-based solutions (NBS) to the urban structure is often mentioned in climate change adaptation strategies. In an effort to ensure the greatest possible well-being of citizens in the form of environmentally positive elements, the opinions of citizens are forgotten. This paper presents the results of research focusing on the feelings of unsafety associated with the application of NBS elements directly into the urban structure. In two pilot areas (Ostrava (CZ) and the part of Upper Silesian agglomeration (PL)) the feelings of the inhabitants and the possible feeling of fear or danger in the application of NBS were investigated. In Ostrava, a questionnaire survey was conducted in relation to specific elements of the NBS without discussion of specific locations. In the Upper Silesian agglomeration, residents' feelings about specific NBS were surveyed at specific locations using guided interviews. Both approaches resulted in the identification of elements of concern. Respondents who discussed a specific location had a better understanding of the urban context and worried less. The two approaches demonstrated the need to communicate with residents before finalizing the design of a particular public space and the desirability of discussing site-specific issues with citizens.

Keywords: nature-based solution; worries; sustainable cities; brownfields

on: Vojvodíková, B.; Tichá, I.; wska-Sikorska, A. menting Nature-Based ons in Urban Spaces in the xt of the Sense of Danger That ns May Feel. *Land* **2022**, *11*, https://doi.org/10.3390/land 712

mic Editors: Panayiotis akopoulos, Mario A. Pagnotta, Scholz and Arshiya Noorani

ed: 29 August 2022
ted: 27 September 2022
hed: 2 October 2022

1. Introduction

Since their creation, cities have undergone a series of changes resulting from the changing needs of their inhabitants (fortification and defortification, industrialization and deindustrialization, etc.) [1]. These changes have been manifested in recent decades through densification of urban cores, consolidation of areas and expansion of cities into suburbs [2–4]. Cities are home to more than half of the world's population and are responsible for three-quarters of global energy consumption and greenhouse gas emissions [5]. Just as cities themselves are changing, [6] the perceptions and needs of city residents are also changing, which may be defined in different ways e.g., [7], but the hierarchy of residents' needs [8] and the importance of feeling safe is still relevant. In [9], it is stated that human wants contribute to making cities "richer, smarter, greener, healthier and happier".

Cities face many challenges in their pursuit of sustainable development. A much-cited challenge today is resilience to challenges associated with climate change [10–12]. The fundamental challenge is to develop cities as holistic systems that cater to human needs in the context of environmental and social justice [13]. When applying measures such as NBS, it is essential to always reduce hazards and promote well-being [14]. Climate change is leading to a gradual increase in urban temperatures, resulting in the creation of more intense heat islands [15–17]. Climate change is also increasing flood risk [18], which is mainly associated with pluvial flooding [19], causing localised damage often unrelated to the overtopping of a permanent water body. The application of green elements has become a key strategy for municipalities in adapting to the climate effect of the urban heat island, stormwater runoff and water pollution, as well as an appropriate solution to improve visual image of the city [20].

Cities also have land in their structure that is contaminated mainly by previous dustrial activities. Soil contamination has a strong environmental impact. For exam phytoremediation, which uses the absorption, accumulation and decomposition of c taminants, can be applied to reduce or limit the mobility of contaminants [21]. Wh considering the application of a green feature in areas where contamination is confirmed is advisable to select suitable plants in terms of their phytoremediation capacity.

Proper management of urban green spaces and their well-planned structure has a itive impact on both reducing heat islands and the consequences of pluvial flooding [22, In urban areas, nature-based solutions play an important role [24]. Many cities have pa and green spaces larger than 2 ha in their urban fabric [25,26]. However, these green spa are usually quite far apart, and their influence does not cover the entire area of the city In a dense urban structure, it is rarely possible to introduce new green spaces of large si such as parks, and other solutions must be sought [28,29]. One such solution may be introduction of small green spaces, for which the name "urban environmental acupunct has recently been adopted and which is closely related to nature-based solutions (NBS

Several requirements for the integration of green spaces into the urban fabric be found in adaptation strategies [30–35]. The negotiation of individual implementat measures of adaptation strategies responding to climate change faces a number of obsta at the municipal level [36]. In addition to environmental aspects, an economic effe also associated with the application of NBS. Especially in the case of established pa and urban forests, the influence of the increase in real estate prices [14], the occupanc housing and the increase in the well-being of the territory [21] are mentioned.

One aspect that influences the adoption of an NBS element is the feeling of sa or the feeling of danger and threat. The relationship between greenness in the city fear of danger was already known in the 19th century [37]. According to [8], safet considered the second most basic human need. Feeling safe refers to a certain leve security perceived by an individual [38].

The level of safety is closely related to the perception of safety in the place where inhabitants are. Since public green space has been mentioned in relation to pedest movement, it is essential to create a safe space for pedestrians [39,40]. Projects inclu safe streets for improving pedestrian include not only traffic infrastructure but also renovation of smaller green or open spaces [41].

Places that are usually associated with negative emotions —most commonly fear [42,43]—as a result of crime or criminal activity [44–47] are those that provide pote hiding places for perpetrators, limit visibility or the ability to escape [48].

Fear can also be associated with various types of phobias. For example, general fe animals, fear of insects and fear of birds. In general, some people may suffer from zooph (zoophobia) fear of animals [49]. Entomophobia (the fear of insects) can be also associ with allergies or traumatic experiences with insects [50]). The flowering elements of may be surrounded by pollinating insects. Should residents suffer from ornithophobia of birds), they might feel threatened if a bird species nests in an element [51].

The aim of the research was to find out which elements of NBS are perceive dangerous (pilot area in Ostrava) and whether personal experience and active consulta can influence these perceptions (part of Upper Silesian agglomeration in Poland) so their application by city representatives is best received by residents. The research based on the assumption that individual NBS elements would be placed on individual in accordance with the definition prepared by the Salute4CE project where a potent suitable place for the application of a nature-based solution is:

- a place that is not maintained, is neglected, or no longer fulfils its function (brownfi
- a smaller site—ideally up to 0.2 ha but not more than 0.6 ha to allow feasible im mentation; or

- a place that spoils the image of its surroundings or even reduces property prices in its vicinity.

Brownfields present an opportunity for new development projects. They are also sites for public green space. In special cases, linear brownfields (railway brownfields) can be used as bio-corridors [52]. Brownfields appear in some research to be very suitable, for example, for the location of urban forests, where they bring cooling effects and increase biodiversity [53]. By the loss of their original (e.g., industrial) function, they are suitable sites for change and the location of some elements of the NBS. These elements can also serve to enhance the process of phytoremediation [21]. However, the research presented in this article was focused on brownfields up to 0.6 ha. This practically excluded former light or heavy industry enterprises which are characterized by a high level of contamination.

Two areas in the Upper Silesian Coal Basin were selected for research: Ostrava (CZ) and part of Upper Silesian agglomeration (PL).

2. Pilot Areas in Upper Silesian Coal Basin

The territory is spread over the area of Poland and the Czech Republic, which are connected by common history of coal mining and heavy industry. For the Czech part, Ostrava as a metropolitan city was chosen and for the Polish part, was selected three cities Chorzów, Ruda Śląska and Świętochłowice (part of Upper Silesian agglomeration).

2.1. Pilot Area—Ostrava (CZ)

The city of Ostrava is part of the Moravian-Silesian Region in the Czech Republic. It has 279,791 (as of 1 January 2022) [54] inhabitants. The city is rich in industrial history—coal mining began in 1787 and ended in 1993 [55], and in 1828, the ironworks in Vítkovice (now one of Ostrava's districts) were founded (part of it still operates today). Ostrava is a typical shrinking city, with an ageing population [56]. In terms of the environment, Ostrava was one of the worst polluted areas in the Czech Republic. Thanks to the decline of coal mining, industrial restructuring and investments aimed at improving Ostrava's environment, the environment and air quality have gradually improved. Nevertheless, in recent years, according to the measurements of the Czech Hydrometeorological Institute, it has become one of the most polluted cities in the Czech Republic in terms of pollution by the carcinogen benzopyrene. The concentrations of dust particles in Ostrava's air have also scored the highest rank in the country [57].

The city tries to maintain larger areas of green space, which is achieved, among other things, thanks to its loose urban structure [58]. The City of Ostrava manages the green spaces located on city-owned land in 23 urban districts. At the end of 2019, the total area of green space on city-owned land (excluding forest areas) was approximately 1867 ha. In the urban district of Moravská Ostrava and Přívoz (city centre), this amounts to 225 ha (most of which is located in Moravská Ostrava). In the administrative district of the municipality of Ostrava (the city itself and several surrounding municipalities), there are currently 5399.9 ha of forests, mostly deciduous and mixed [59].

The city of Ostrava is trying to behave responsibly when dealing with the challenges of the standard of living, air pollution, and climate change. It has therefore developed the Strategic Development Plan of the City of Ostrava 2017–2023 and the Adaptation Strategy—Healthy Ostrava [60].

The Adaptation Strategy 2017–2023 notes that a gradual increase in average annual temperature by approximately 2.5–3 °C (compared to the average for the period 1961–2009), more frequent occurrence of heat waves and an increase in their duration can be expected by 2100. By 2100, can be expected a significant increase in summer temperatures by 3–4 °C and therefore a significant increase in the number of tropical days and a decrease in the number of arctic days. The most affected localities mentioned in the text are the urban districts of Vítkovice, Moravská Ostrava and Přívoz, and Hrabůvka. For the preparation of the Adaptation Strategy, the City of Ostrava organised a mapping survey where residents marked their perceived thermal discomfort on a map. (In the text, the term thermal discomfort is used if the authors are referring to sensible heat. Thermal stress is used in th context of heat and overheating problems that have been measured or modelled).

The Ostrava City Development Strategy [61] also addresses feelings of safety as part the standard of living. Therefore, the strategy includes an emotional map of places whe residents do not feel safe.

As described above, NBS applications should be dominated by sites that are inappr priately used, abandoned, or forgotten. No such inventory of sites exists for the city Ostrava. Therefore, the authors of the article used the database of brownfield sites with th knowledge that regeneration of these areas will not be primarily directed towards the u of NBS; however, with the current setup of the whole society, it is possible to assume th application of some NBS as an integral part of the regeneration process.

For the territory of the city of Ostrava, the brownfields database is managed by MS (a company owned by the region).

Brownfields are defined for the purposes of the database as sites or buildings that ha been affected by previous use. Now, they have lost their function and are only partial used or even completely abandoned. They need some necessary support for further different use [62].

There are 122 brownfields registered in Ostrava. Of these brownfields, 83 sites a smaller or equal to 0.6 ha. Of these 83 sites, 62 are smaller than 0.2 ha. All these sites a listed as sites with no proven contamination. In terms of their past use, these are most areas previously used for housing, office space or some small transport structures such bus stops or garages. The database also includes an assessment of their potential for futu use. The authors are aware that this information is not directly related to the application the NBS on these sites, but it illustrates the overall situation.

To visualize the whole situation, a map of brownfields was linked to maps of perceiv temperature discomfort (Figure 1) and perceived danger (Figure 2).

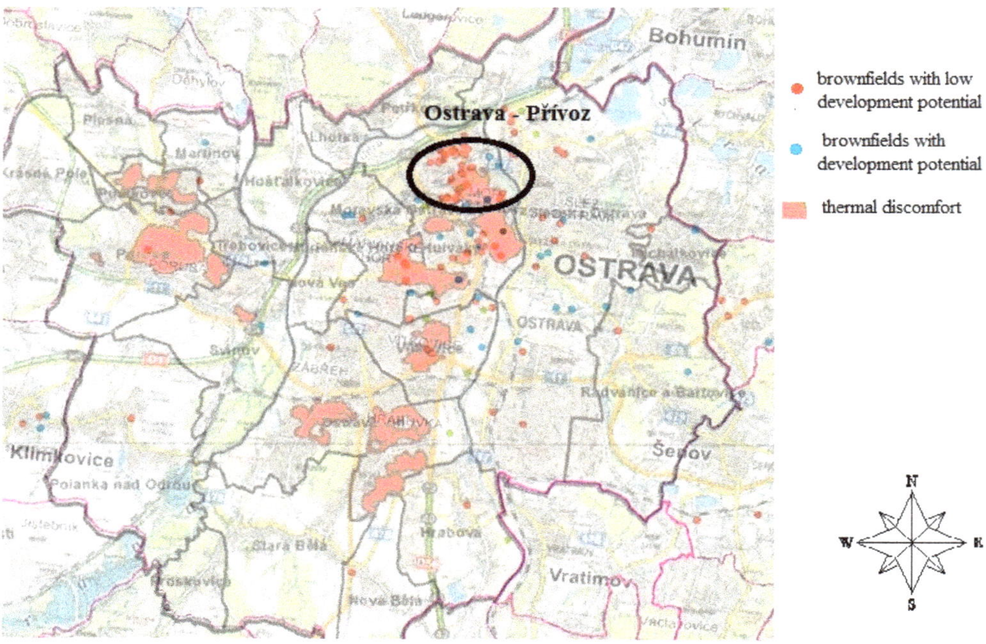

Figure 1. Connecting brownfield sites and places where the residents of Ostrava feel therm discomfort. (source: based on information from [59,60,62] created by authors).

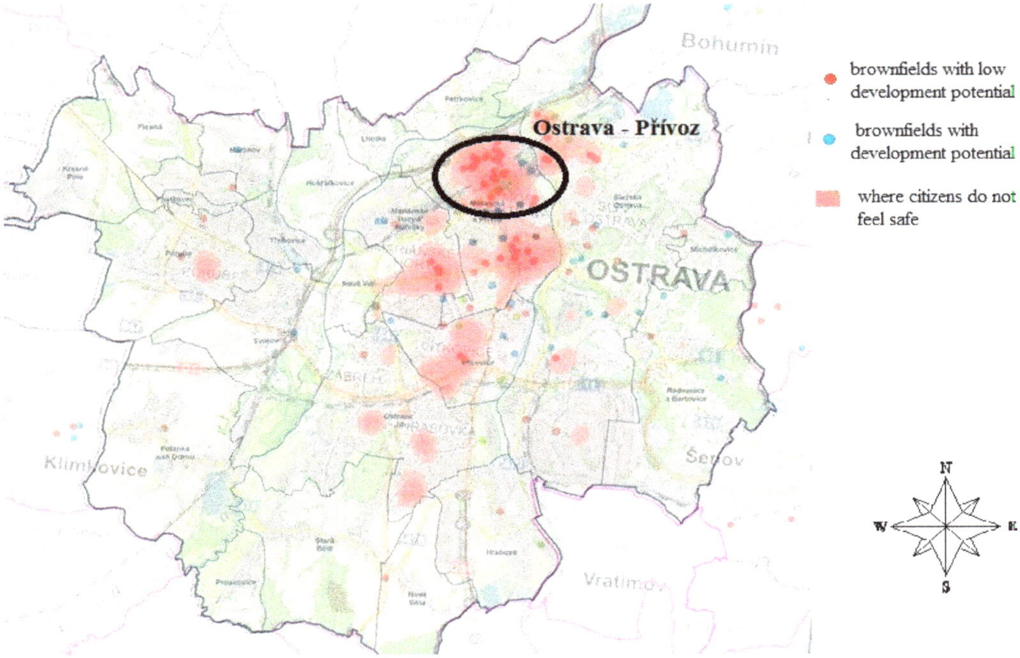

Figure 2. Connecting brownfield sites and places where the residents of Ostrava feel danger. (source: based on information from [59,60,62] created by authors).

The authors are aware that the perception of thermal discomfort is influenced by a number of personal and/or physical characteristics [63,64] and differs from measurement-based temperature maps. For elderly residents, heat can be a health risk or even fatal. [65,66] Similarly, hazard perception and crime maps differ.

Given the targeting of feelings and perceptions of safety towards elements of the NBS, the authors consider the feeling maps to be more appropriate and illustrative.

When linking the feeling of thermal discomfort maps to the brownfields map (Figures 1 and 2), it is clear that there is a high concentration of brownfields (marked by dots) and a feeling of thermal discomfort, especially in the Přívoz district. The same problem site can be observed in the connection between the brownfields map and the feeling of danger (not feeling safe).

2.2. Pilot Area—The Part of Upper Silesian Agglomeration(PL)

Functional Urban Area (FUA) represents three cities: Chorzów, Ruda Śląska and Świętochłowice located within the Metropolitan Association of Upper Silesia and Dąbrowa Basin, usually referred to in Poland as the Silesian Metropolis or Metropolis GZM. The Silesian Metropolis consists of 41 municipalities in the Silesian Voivodeship in Poland. The FUA covers an area of 124.19 km^2 with over 250,000 inhabitants and is located in the centre of Metropolis GZM, which is very important for the development of the area. The peculiar structure of land use is mainly due to the many years of industrial operation within its borders. The industrial landscape was characterised by industrial and post-industrial buildings and areas located in the immediate neighbourhood of city centres as well as by spoil heaps and dumping sites. The proportion of anthropogenic areas reaches over 55% of the whole FUA surface, implying a high level of its transformation.

In the area, two main challenges have been indicated as related to climate char urban resilience: heat island and soil sealing (The core of this article focuses on the probl of thermal stress).

The analysis of land surface temperature (LST) showed the LST distribution Chorzów, Świętochłowice, and Ruda Śląska for the warm seasons period 2016–2020 [After collecting and compiling the statistical data, it can be concluded that the high maximum temperatures (above 39 °C) were recorded in Chorzów City, which is cha terised by a dense urban fabric. The city of Świętochłowice is characterised by the high average temperature (above 27 °C), which is caused by the smallest area of the city a the smallest surface of green areas (including forests). Due to the lack of green spa and forests, Świętochłowice has the highest minimum surface temperature of above 22 among the analysed cities. The city of Ruda Śląska, characterised by separate districts a polycentric structure, recorded the lowest minimum temperature of 21.11 °C, and lowest mean temperature of 26.07 °C. In the entire analysed area, over 60% of the surfac characterised by a temperature of 25–30 °C. 22% of the area of Świętochłowice and 26% the area of Chorzów have an average temperature below 25 °C, 10% of the area of Chorz and Świętochłowice has a temperature of 30–35 °C. The highest percentage of the city a with a temperature above 35 °C was observed in Chorzów (1%), and approximately 0 of the area of Świętochłowice and Ruda Śląska [67].

3. Methods

In Ostrava, the research focused on feelings of danger of implementing the N without targeting the application to a specific location. Respondents, as described bel expressed themselves based on their feelings without having to look for solution their neighbourhood.

In the Silesian agglomeration, respondents were addressing site-specific NBS pro als. As part of proposing specific NBS solutions, structured interviews were conducte determine if they felt there was any danger in applying a particular element

3.1. Methods Applied in Pilote Area Ostrava

3.1.1. Selection of Respondents and Communication Techniques

Three groups of respondents were selected for the questionnaire survey in Ostra

- Group A—respondents living in housing estates in Ostrava. The group consi mostly of residents of retirement age. In the Czech Republic, almost 20% [68], of population is aged 65+. (In this group, there were 3 respondents aged 65–70 5 respondents aged 80–85. The other respondents were in the 70–75 and 75–80 groups). This is a significantly large group of people who tend to be more concer about crime (analysis).
- Group B—respondents were residents of Ostrava, regardless of the urban district type of housing, but their age was up to 26 years. (All respondents in this gr were aged 20–26). This group of respondents was composed mainly of students. age group is the least worried about crime [69], but at the same time, they feel strongly about the problems associated with climate change.
- Group C—respondents were residents of urban districts with a higher crime including residents of the Přívoz district. Respondent age ranged from 26 to 60 y (The age structure of respondents in this group was: 2 respondents 26–30 yea respondents 35–40 years, 3 respondents 40–45 years, 5 respondents 45–50 yea respondent 55–60 years and 1 respondent 60–65 years.) The reason for the selec was that the area consists of a large number of brownfield sites, with a high perce crime rate, and is also a place of perceived thermal discomfort (see Figures 1 and

The selection of groups was not random. From the beginning of the research, there interest in three groups of respondents. The aim was to focus on respondents affected location that is problematic (Ostrava-Přívoz). Respondents have experience with movir public spaces that they do not consider safe. Another significant group was considere be the elderly population who use public spaces throughout the day and at the same time have more concerns. The younger generation was considered to be more environmentally focused. The students were deliberately chosen as they are a homogeneous group in terms of age and education. During the implementation, the presentation and questioning format had to be modified due to the COVID-19 pandemic.

Group A—respondents were asked to complete the questionnaire in person at several meetings. These meetings were organised as part of cultural and social activities for the elderly and their companions. Prior to completing the questionnaire, an introduction to the different types of NBS was given, focusing mainly on a technical description and the NBS impacts on climate change issues. The presenters did not mention anything related to safety or feeling unsafe. The questionnaire was printed in colour. The font size was adapted to this age group. Not all the participants of these presentations were willing to complete the questionnaire.

Group B—this group was approached through their lecturers at the university. These respondents were provided with an online lecture describing each element of the NBS before completing a questionnaire. The presentation included a technical description. Respondents completed the online questionnaire using Google forms.

Group C—reaching this group was the most difficult. The outreach was done through personal contacts and with the help of NGOs operating in the locality. In addition, owners of some brownfields in Ostrava-Přívoz were also asked to cooperate through MSID. Respondents received questionnaires by e-mail or printed out. They had a spoken presentation on google drive where they could learn about the different elements of the NBS.

3.1.2. Questionnaire Structure and Method of Processing

A total of 28 types of suitable NBS were selected for the survey. These NBS were selected based on a search of existing solutions [70–81]. The list was also based on the NBS identified in the SALUTE4CE project. Their list is presented in Figure 3. For better clarity, the NBS were categorized by position (horizontal, vertical, indifferent) and shape (point, line, area).

At the beginning of the questionnaire, the purpose of the survey was briefly explained. Two personal questions were inserted for further processing (gender (female/male/other), age category). The key questions in the questionnaire focused on respondents' feelings about the potential threat of a particular NBS (feeling of danger/neutral feeling/feeling of safety/no comments). The order of individual NBS in the questionnaire was set so that it started with simple horizontal elements such as a lawn. In the middle part of the questionnaire, the NBS was mentioned in relation to rainwater, for example, a rain garden. At the end of the questionnaire, there were the NBS requiring construction solutions placed, for example, on a green wall. A photograph was included with each NBS for illustration. For an example of a question, see Figure 4. Respondents were always given an opportunity to comment with their opinion.

For the self-assessment, respondents' answers were converted into points. In the case of feeling of danger, NBS was rated 3, in the case of neutral feeling, it was rated 2, and if respondents felt safe about it, it scored 1. If respondents had no opinion, NBS was rated 0. The scores from each respondent for each NBS were summed and divided by the number of non-zero responses.

3.2. Methods Applied in Pilot Area—The Part of Upper Silesian Agglomeration

The assessment of feelings of danger in relation to individual NBS for the cities of Chorzów, Ruda Śląska and Świętochłowice (Katowice agglomeration) was linked to public involvement in the search for concrete solutions. The direct involvement of the public was a necessary condition and at the same time enabled the application of structured interviews as one of the research methods [82].

<table>
<tr><th colspan="2"></th><th colspan="3">position</th></tr>
<tr><th colspan="2"></th><th>horizontal</th><th>indifferent</th><th>vertical</th></tr>
<tr><td rowspan="15">largenerss/shape</td><td rowspan="7">area</td><td>Urban meadows</td><td rowspan="7">Urban wilderness / succession area</td><td>Park trees</td></tr>
<tr><td>Ground cover plants</td><td>Green facades with climbing plants</td></tr>
<tr><td>Lawn</td><td>Wall-mounted living walls</td></tr>
<tr><td>Rain gardens (under-drained)</td><td rowspan="2">Hydroponic mobile living walls / vertical gardens</td></tr>
<tr><td>Rockery</td></tr>
<tr><td>Ground crops of vegetables / herbs</td><td rowspan="2">Hanging wall planters (as green street furniture)</td></tr>
<tr><td>Green roof /roof terrace</td></tr>
<tr><td rowspan="4">line</td><td rowspan="2">Green pavements</td><td rowspan="4"></td><td>Street trees</td></tr>
<tr><td>Large shrubs</td></tr>
<tr><td rowspan="2">Road-side swales for retention and infiltration</td><td>Linear wetlands for stormwater filtration</td></tr>
<tr><td>Hedge/hedgerow</td></tr>
<tr><td rowspan="4">point</td><td>Verges / flower beds with native perennials</td><td rowspan="2">VRSS slopes with green fences</td><td>Fruit trees/ shrubs/</td></tr>
<tr><td>Green pergolas/ green arbors</td><td>Herb spiral</td></tr>
<tr><td>Street planters (as green street furniture)</td><td rowspan="2">Rain gardens in planter (=self-contained)</td><td rowspan="2">Vertical vegetable / herb gardens</td></tr>
<tr><td>Green covering shelters</td></tr>
</table>

Figure 3. Selection of nature-based solutions defined by the SALUTE4CE project (source: created by authors).

1. *Lawn – select only ONE option*

1. Trávník – vyberte pouze JEDNU možnost

pocit nebezpečí	O	*feeling of danger*
neutrální pocit	O	*neutral feeling*
pocit bezpečí	O	*feeling of safety*
nemám názor	O	*I have no opinion*

komentář k výběru Vaší odpovědi ....

, Comment on your answer

Figure 4. Questionnaire—a sample question, (English translation inserted for the purposes of the article) (source: create by authors).

Respondents were involved in planning, building, maintaining, and monitoring sites for the NBS application. Respondents were representatives of the local community—

stakeholders. Different types of institutions and organisations were invited to participate in the discussions in the so-called living labs. The living labs were organised at the local level together with citizens and relevant experts. The aim of the living labs was to gain insights from local stakeholders (residents, civic associations, experts), to identify priorities and gain insights on bottom-up needs and opportunities in line with the ideas of public participation, while being in direct contact with respondents [83].

Discussion walks, workshops and a consultation point were organised for respondents.

Respondents—the locals themselves suggested some solutions in concrete localities and commented on feelings of safety.

4. Results

In general, the application of green elements in the city affects the quality of life. A number of authors, e.g., in the book *The green city and social justice* [84], mention a number of examples from the USA or Great Britain, where the expansion, modification and improvement of green areas undoubtedly led to an improvement in the quality of life and even gentrification. However, the following results show that green elements in different forms can be perceived differently.

4.1. Results in Pilot Area Ostrava

A total of 58 respondents provided answers for all 3 groups. The minimum number of included (not 0 or missing) answers about each individual NBS was 50. Most questions were answered by 52–55 respondents.

A total of 23 respondents from Group A were willing to fill in the questionnaire. The group of young respondents (Group B) consisted of a total of 19 respondents. Group C (from Ostrava-Přívoz) consisted of 16 respondents. (Figure 5).

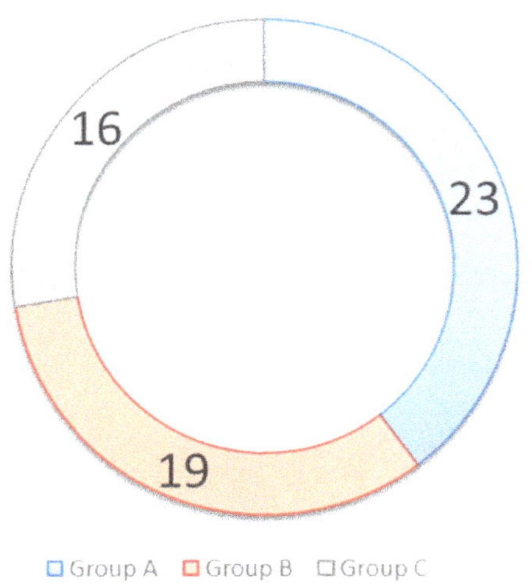

Figure 5. Groups with numbers of respondents (source: created by authors).

For the evaluation, the research was focused mainly on those types of natural-based solutions that have an impact on reducing the heat load. The selection was made on the basis of information [70–81] and the individual NBS were divided into the following groups according to the effect on thermal stress. The breakdown is given in Table 1.

Table 1. The individual NBS were divided into the following groups according to their effect thermal stress.

Dominant Effect	Partial Effect	Additional Effect (Has Another Dominant Function)
Street trees	Urban meadows	Rain gardens (under-drained)
Park trees	Verges/flower beds with native perennials	Road-side swales for retention and infiltration
Fruit trees/shrubs/	Ground cover plants	Linear wetlands for stormwater filtrati
Large shrubs	Lawn	Rockery
Hedge/hedgerow	Green pavements	Herb spiral
Green pergolas/green arbours		Ground crops of vegetables/herbs
Green facades with climbing plants	VRSS slopes with green fences	Rain gardens in planter (=self-containe
Wall-mounted living walls	Green covering shelters	Street planters (as green street furnitu
Hydroponic mobile living walls/vertical gardens		
Vertical vegetable/herb gardens		
Hanging wall planters (as green street furniture)		
Green roof/roof terrace		
Urban wilderness/succession area		

Source: authors.

The first group was a group with a dominant influence. This means that the applica of the nature-based solution from the first group should reduce thermal stress. Given this is primarily a division that will serve as a basis for structured interviews, it has b abstracted from some imperfections. For example, the effect that the tree crown size on the efficiency of reduction of heat stress. Similarly, the height of the green wall and type of plants used were not addressed. In the green roof solution, the distinction betw intensive and extensive was not drawn.

The second group was designated as a partial influence group. These are the N which have other functions such as soil treatment, soil sealing, or infiltration support. Th NBS have a smaller effect on the reduction of thermal stress itself.

The third group were the NBS, which predominantly have a different purpose pecially those that support water infiltration. Even these can partially contribute to reduction of thermal stress, but only marginally (e.g., evaporation from a rain garden

The evaluation showed that opinions on the same type might differ. The results o questionnaire survey in Ostrava are shown in Figure 6.

For example, young people (Group B) felt less fear, especially with NBS associ with the possibility of growing vegetables, whereas Group A and C answered mostly or three about vertical vegetable/herb gardens, indicating that they felt neutral or e unsafe. In the group of young people of the same type, the answers were predomina two and one, indicating that younger people perceive this element as neutral or even

Another NBS type with an interesting result was a pergola. For the pergola, 15 o 16 respondents from Group C gave a rating of three, indicating a feeling of danger, w was again consistent with Group A's answers. In Group B, a response of two (neutral) much more common.

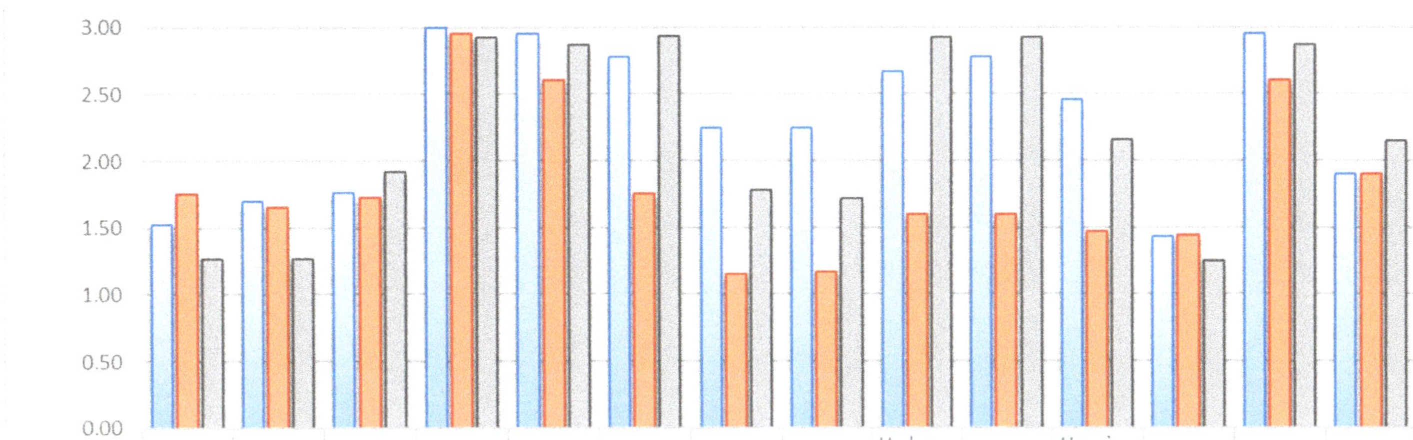

	Street trees	Park trees	Fruit trees/ shrubs/	Large shrubs	Hedge/hedgerow	Green pergolas/ green arbors	Green facades with climbing plants	Wall-mounted living walls	Hydroponic mobile living walls / vertical gardens	Vertical vegetable / herb gardens	Hanging wall planters (as green street furniture)	Green roof /roof terrace	Urban wilderness / succession area	Urban meadows
Group A	1.52	1.70	1.76	3.00	2.95	2.78	2.25	2.25	2.67	2.78	2.45	1.43	2.95	1.90
Group B	1.75	1.65	1.72	2.95	2.60	1.75	1.15	1.17	1.60	1.60	1.47	1.44	2.60	1.90
Group C	1.27	1.27	1.92	2.93	2.87	2.93	1.79	1.71	2.93	2.92	2.15	1.25	2.87	2.14

Figure 6. Evaluation of the feeling of danger associated with the NBS application (points: maximum 3, minimum 1) (source: created by authors).

Overall, there was a consensus on the danger posed by urban wildlife, large shru and hedges, with respondents from all groups expressing feelings of danger associat with these NBS. In all cases, these are visual barriers. Technical features, i.e., green facad and green walls on existing buildings, were perceived negatively for Group C, where green roofs were rated neutrally or positively by all. Urban meadows were considered le safe; respondents showed concerns about ticks or allergies to grass pollen.

The results also showed Group B concerns less (compared with Group A) with tech cal solutions—see Figure 7.

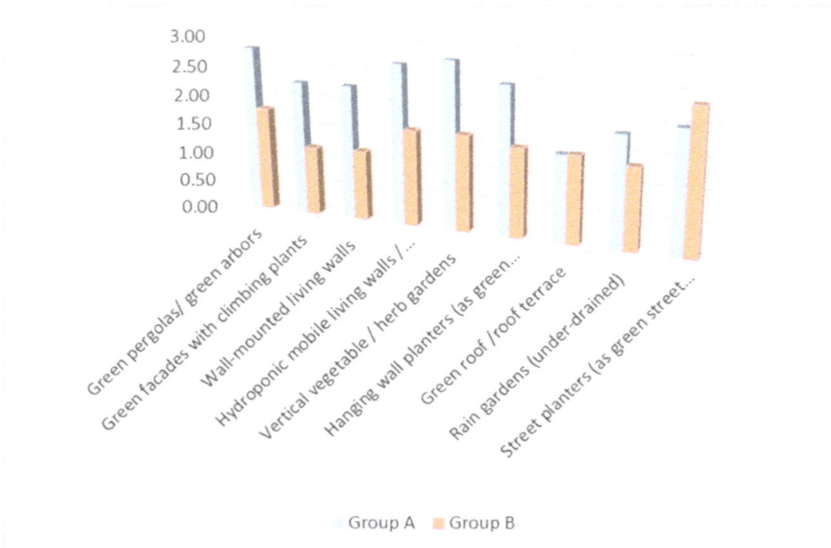

Figure 7. Feelings of respondents (Group A and B) focusing on NBS of technical basis. Source: auth

Respondents were given the opportunity to comment on their answers. Table 2 sho the most recurrent comments on some types of NBS. These comments, together with scores, are the basis for comparison with the situation in Poland.

The authors of the evaluation made minor corrections in the style of the answ leaving the substance unchanged. Comments in the table are only given for NBS wit dominant influence on the temperature of the area.

For some types of NBS, e.g., those designed to hold water, people commented t they were concerned that a child would fall into them and drown. Despite being told w to evaluate, respondents deviated from the essence of the issue, as seen in the example vertical vegetable/herb gardens. Respondents were concerned about damage to the N itself or expressed concern about liability for removing fallen leaves.

The assessment showed that respondents were concerned about those NBS that fo a visual barrier, as there may be hidden threats behind this barrier. Another signific concern was the provision of shelters for drug users. Respondents are concerned t hidden locations may also serve as toilets. Nor do they prefer any features that wo provide a shelter or gathering place for the homeless.

4.2. Results in Pilot Area—The Part of Upper Silesia Agglomeration

As described in Methods (Section 3.2), respondents considered specific locati and defined appropriate NBS solutions for them. Individual proposals were consul with experts. Table 3 shows the respondents' comments, which were noted during consultations (workshops, living labs etc.) (respondents always consulted on the type

NBS in the context of a specific site). Respondents added their feelings about a particular element of the NBS. In total, 48 respondents participated in the events.

Table 2. Respondents' comments regarding different types of NBS.

Types of NBS	Comments
Street trees	Leaves fall from the trees, and it is not clear who will take care of them, but I do not feel any danger.
Park trees	Leaves fall from the trees, and it is not clear who will take care of them. Will junkies gather here?
Fruit trees/shrubs	Children would climb trees and injure themselves.
Large shrubs	In no case, someone can hide behind a shrub, or it will be used as a toilet.
Hedge/hedgerow	There may be a thief or worse hidden behind it.
Green pergolas/green arbours	There will be toilets, and homeless people will gather around.
Green facades with climbing plants	I am worried about the technical solution so that it does not fall on anyone
Wall-mounted living walls	I am worried about the technical solution so that it does not fall on anyone.
Iydroponic mobile living walls/vertical gardens	I am worried about the technical solution so that it does not fall on anyone.
Vertical vegetable/herb gardens	Anyway, someone will destroy it right away and what will grow there will be stolen.
Ianging wall planters (as green street furniture)	It will not last long before someone destroys it.
Green roof/roof terrace	I'I am afraid the roof is not able to carry it.
Urban wilderness/succession area	Junkies would gather there.

Source: authors.

Table 3. The list of results is in the part of Upper Silesian agglomeration.

Type of NBS	Average Value for Ostrava	Respondent Comments of the Part of Upper Silesian Agglomeration
Large shrubs	2.959524	Large bushes are part of, for example, Piastowski Square, forming a noise barrier. Its concept is close to urban wilderness. On this particular site, residents would welcome the retention of shrubs, but landscaped. Some respondents consider the placement of new large shrubs in a vacant gap in the development, for example, to be inappropriate concerns to raise. They are also concerned about large shrubs in a park area and near playgrounds.
Hedge/hedgerow	2.806349	A group of respondents on a particular street (11. November Street) were negative about putting trees on the street because they were concerned about overshadowing but would welcome a green fence which they did not consider risky. A green fence would separate the road from the pedestrian area.
an wilderness/succession area	2.806349	Respondents generally understood the importance of such an element for biodiversity and other positive attributes. The NIMBY syndrome was evident here. For no particular site, this NBS was proposed by residents who were concerned about creating space for troubled citizens.
reen pergolas/green arbours	2.488647	Respondents suggested pergolas for the courtyard area. All yards were private or semi-private spaces. Respondents did not feel apprehensive about pergolas in the courtyards. No pergola was proposed for a purely public space.

Table 3. *Cont.*

Type of NBS	Average Value for Ostrava	Respondent Comments of the Part of Upper Silesian Agglomeration
Vertical vegetable/herb gardens	2.435229	This NBS was proposed by respondents for a backyard area. Al yards were either a private or semi-private space. As none of th sites in the public space discussed were suitable for growing vegetables, this NBS was not discussed by respondents from a safety perspective.
Hydroponic mobile (flexible) living walls/vertical gardens	2.398413	Respondents were positive about mobile living walls. In a few places, they even suggested them as a barrier to traffic. They felt fear in relation to these NBS.
Hanging wall planters (as green street furniture)	2.027359	This NBS was proposed by respondents for a backyard area. Al yards were either a private or semi-private space. Because none the locations in the public space discussed were suitable for growing vegetables or herbs, this NBS was not discussed with respondents from a safety perspective
Urban meadows	1.980952	In several places, respondents suggested this type of NBS. Primarily, these were spaces today consisting of only a lawn. Oft in front of shopping centres or as part of a square or park.
Fruit trees/shrubs	1.800265	Fruit trees were not proposed by respondents for any of the selected sites. Therefore, this NBS was not discussed with the respondents in terms of their safety.
Green facades with climbing plants	1.728571	Respondents in several places suggested a green facade as an appropriate solution. They felt no fear in connection with this element.
Wall-mounted living walls	1.710317	Respondents suggested this NBS for several places as an appropriate solution. They have no worries about this element
Park trees	1.53744	A willow tree was suggested as an appropriate tree by several respondents. Willow is considered by respondents to be a very decorative tree that is also a native species. They have no fear allergens.
Street trees	1.512802	Respondents (as indicated for the large bush section) do not wa to have trees planted in their streets because they do not want reduce the sunshine in their homes. They have no negative feelir about trees in terms of danger.
Green roof/roof terrace	1.376409	According to respondents, green roofs are very costly; therefor they have not been proposed for any site. However, responden reported a positive experience with green roofs in the agglomeration area.

Source: authors.

Table 3 shows only those elements that have been evaluated for Ostrava. The g colour symbolises an agreement in the perception of the inhabitants of both pilot areas. red colour symbolises a significant disagreement between the pilot areas. The yellow co represents a partial disagreement. Grey elements were not proposed by respondents ir Upper Silesian agglomeration; therefore, their perception of danger was also not discu

For the NBS, they were evaluated by the Upper Silesian agglomeration pilot Graphs of response frequencies were prepared for each group of respondents in the area of Ostrava. For the pilot area in Poland, unfortunately, detailed age compositi not available. The authors estimate (based on personal experience from these intervi that this was an age group predominantly in the older working age and younger ser The age structure would be most closely related to Group A and Group C. Accordir Figures 8–10, the difference in the attitude of the young (Group B) is evident. It is also from Figure 9 that Group C is just strongly influenced by the negative experience o area that is problematic. In the pilot area in Poland, none of the sites were left in su significantly negative environment. This is an important experience for further resea

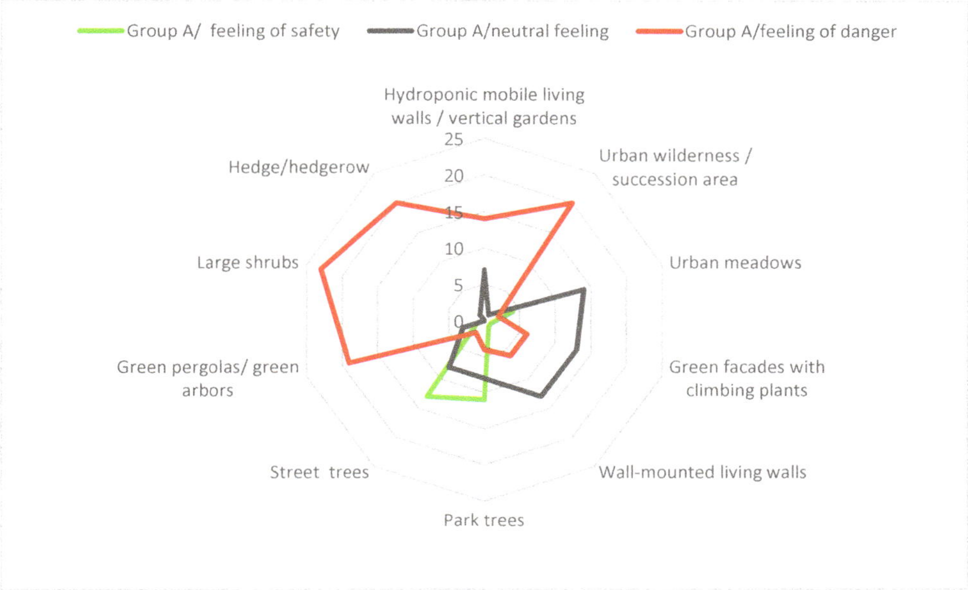

Figure 8. Radar charts with frequency of answers for selected NBS for the 65+ group Source: authors.

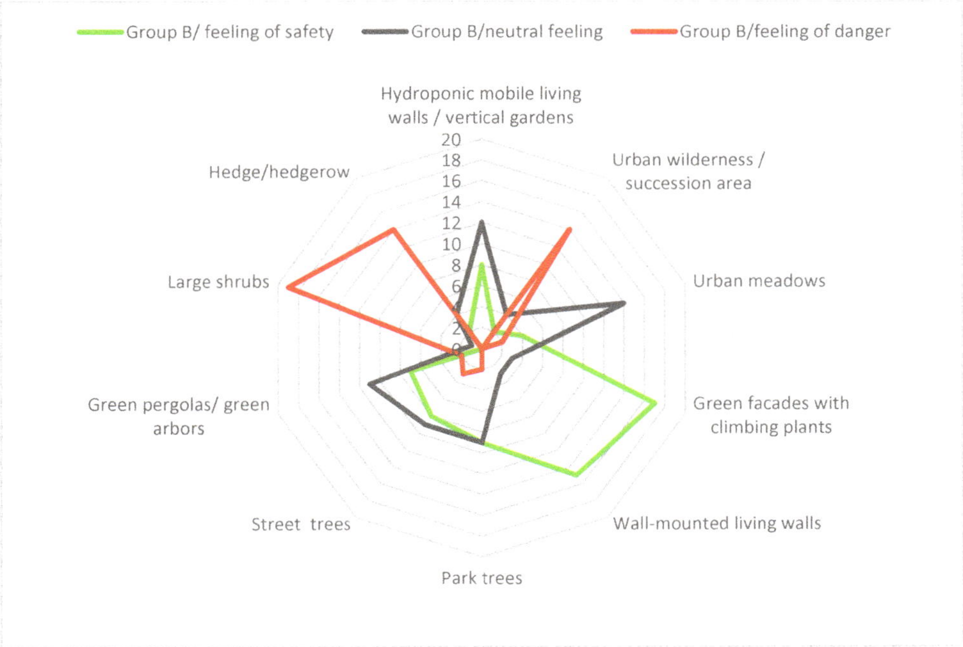

Figure 9. Radar charts with frequency of answers for selected NBS for the student group. Source: authors.

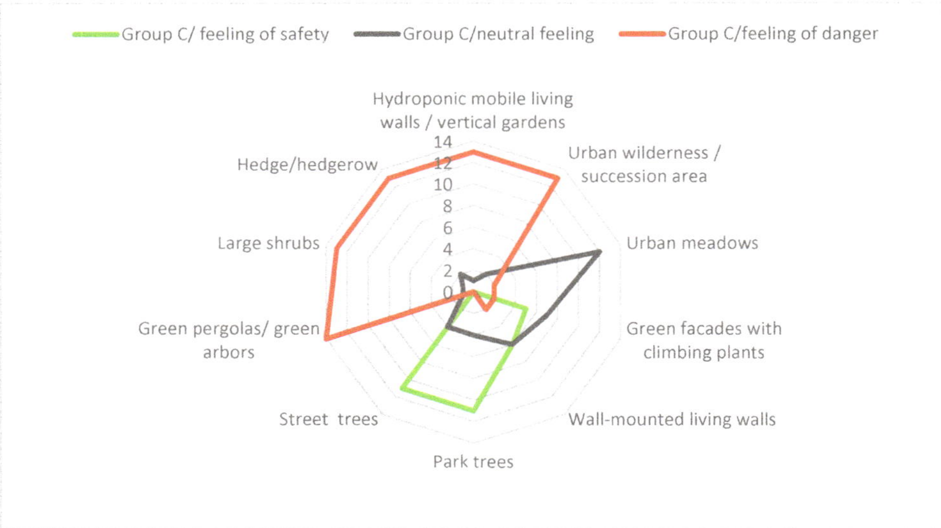

Figure 10. Radar charts with frequency of answers for selected NBS for the group from Přív
Source: authors.

5. Discussion

The relationship between urban greenery and feelings of safety is not a new p nomenon [85]. The research shows that places that provide potential hiding places perpetrators and limited visibility or escape are perceived as dangerous [48]. Large shru and urban wilderness can be considered such features. The result is therefore in accordar with the assumptions. Negative feelings about urban wilderness are also confirmed by [

Discussions can be held about the discrepancy between the research results and already applied implementation of the NBS. An urban forest has been developed brownfield sites in Leipzig with an obvious positive effect on the housing in the a by increasing the occupancy rate [53]. It can be assumed that the main difference is embeddedness in the city structure. In Leipzig, these were former brownfields (size grea than 3 ha), which are now used for recreation but are a separate entity [53]. Residents not forced to enter the forest area in their daily movements (possibly at night). In contr an area of small urban wilderness embedded in a public space that is used for the da movement of residents is a place that residents must encounter in their daily movem This is considered to be the main reason for the difference in residents' feelings.

Hedgerows were perceived differently. Respondents in Ostrava made general cc ments and did not indicate a specific location. Respondents from the Silesian agglomerat indicated a specific site and perceived shrubs more positively. The different approac may be given by the very good experience of the Silesian agglomeration, the different ty of housing development and especially the specific area.

As for the results, urban meadows are interesting, which were evaluated rat negatively compared to the usual assumptions or were not in direct contact with current location. According to [87], urban meadows are perceived positively after th have been established, but [87] also admit that their positive impact on biodiversity ne to be explained to respondents, which confirms the rather positive attitude of responde in the Silesian agglomeration. Biodiversity enhancement in low-mown urban grasslan also addressed by [88], for example

In terms of vertical features, there were differences in results. Green facades w climbing plants and living walls had an average rating of less than two, meaning that m respondents considered them neutral or safe. This result was virtually confirmed in second pilot area. In contrast, elements of hydroponic furniture, living walls/vertical gardens, vertical vegetable/herb gardens and hanging wall planters (as green street furniture) were rated between 2 and 2.4. In the Silesian area, respondents were particularly positive about mobile and flexible walls. It can be assumed that there may have been a lack of perception of the robustness of the design used in the solution in Ostrava, for example, despite the fact that the functionality of such a system in cities has been a topic of an extensive expert discussion for a long time (e.g., [89]). But again, the good experience of the Silesian agglomeration may also have contributed to a different attitude by the respondents. In future research, more attention will need to be paid to vertical elements or should focus on them separately.

A significant limitation of the results for the Ostrava pilot is that there is only a three-level response scale. In further research, it will be necessary to adjust the scale. Respondents should express their feelings on a scale of 1–5 or 1–10. Another limitation that may have affected the result may have been photographs showing a particular element. For further research, it will be necessary to test the neutrality of the feeling evoked by the images in a sample of respondents.

A relatively small number of respondents can be considered a limitation to the research, especially in the Ostrava part. Given the number of respondents, the age composition does not match a demographic curve. Although equal numbers of men and women were approached, efforts to obtain gender-balanced responses were unsuccessful. It is therefore not possible to analyse gender differences in relation to feelings of threat from the data obtained.

A limitation of the research is the actual form of interviewing in the pilot area of Ostrava as mentioned in Chapter 3—Methods. As a result of the COVID-19 pandemic, the method of interviewing had to be modified. The initial assumption of face-to-face meetings in larger groups was not possible. Meeting with the senior generation was practically excluded for almost a year. Another limitation of the results that will need to be taken into account in future research is the likely changes in society as a result of the pandemic, which was a prolonged stay at home and changes in work and social habits.

6. Conclusions

Fear of crime is itself a problem that has costly and long-lasting consequences for a social life of a city, so understanding its causes and why it occurs as a social phenomenon plays a key role in developing the right policies [7].

The city as a living organism is a site of clashes of opinion between different groups of experts, even though they share the common goal of sustainable development. The paths that lead to this goal may be different.

This can be seen in the example of nature-based solutions. A little green acupuncture in the city can have a number of positive effects (reducing not only heat but also noise and dust). However, such a solution cannot be implemented everywhere where it is technically appropriate or advantageous. The acupuncture of an urban environment, especially when associated with brownfield issues, must respect the specificities of the site. Brownfields are often located in areas with higher crime rates. It is therefore necessary to consider and communicate well the development of such NBS that could create a sense of danger to citizens. For example, costly green walls or green roofs are positively received by residents in terms of being unsafe, while having a number of ecosystem services.

If the citizens do not have their own experience with NBS elements or the concept of applying them is not well communicated to them (as was the case in Ostrava—a resistant lecture is not sufficient), fears can arise, which can then lead to non-acceptance of these elements by the city residents.

Another possible explanation is its link to the imagination of the respondents. If they talked about a specific location, they had the opportunity to place a specific element in the context of the environment. However, if they were working with NBS without a predefined location their evaluation was influenced by their level of imagination and the place where they probably subconsciously placed the element. However, this psychological aspect v not the focus of the study but may serve as a stimulus for further research in the fiel urban psychology

This seemingly simple and clear and predictable result is important. In many ca various elements of NBS are being built with the vision of creating sustainable and resili cities. Society is falling into stereotypes that anything green is right and must be enthusia cally embraced by residents. Society is becoming significantly radicalized in environme issues. Which can lead to a deeper dissatisfaction of a part of the population, for wh sustainable and resilient cities become uninhabitable and hostile.

Author Contributions: Conceptualization, B.V., I.T. and A.S.-S.; methodology, B.V., I.T. and A.S validation, B.V., I.T. and A.S.-S.; formal analysis, B.V. and A.S.-S.; investigation, B.V., I.T. and A.S resources, B.V. and I.T., data curation, B.V., I.T. and A.S.-S.; writing—original draft preparation, I.T. and A.S.-S.; writing—review and editing, B.V.; visualization, B.V. and A.S.-S.; supervision, project administration, A.S.-S.; funding acquisition, A.S.-S. All authors have read and agreed to published version of the manuscript.

Funding: This research was funded by SALUTE4CE—Integrated environmental managemer SmALl Green Spots in FUncTional Urban ArEas following the idea of acupuncture, Project in number: CE1472, from Interreg CENTRAL EUROPE Programme.

Institutional Review Board Statement: Ethical review and approval were waived for this study to REASON: The personal data that was provided by the respondents related only to age (age ra and gender. All questionnaires were fully anonymised. The questionnaires were processed onl the authors and the results of the individuals were not published or provided to any third party

Informed Consent Statement: Not applicable.

Data Availability Statement: Not applicable.

Acknowledgments: This research was supported by the programme of the Polish Ministry of ence and Higher Education called: "PMW" in years 2019–2022 (contract no. 5062/INTER CE/19/202/2.

Conflicts of Interest: The authors declare no conflict of interest.

References

1. Anas, A.; Arnott, R.; Small, K. Urban Spatial Structure. *J. Econ. Lit.* **1998**, *36*, 1426–1464. Available online: https://www.jstor. stable/2564805 (accessed on 17 August 2022).
2. Matthews, T. Heat islands: Understanding and mitigating heat in urban areas. *Aust. Plan.* **2012**, *49*, 363–364. [CrossRef]
3. Seto, K.C.; Sánchez-Rodríguez, R.; Fragkias, M. The New Geography of Contemporary Urbanization and the Environment. *A Rev. Environ. Resour.* **2010**, *35*, 167–194. [CrossRef]
4. Xu, R.; Yang, G.; Qu, Z.; Chen, Y.; Liu, J.; Shang, L.; Liu, S.; Ge, Y.; Chang, J. City components—Area relationship and dive pattern: Towards a better understanding of urban structure. *Sustain. Cities Soc.* **2020**, *60*, 102272. [CrossRef]
5. Gouldson, A.; Colenbrander, S.; Papargyropoulou, A.S.E.; Kerr, N.; McAnulla, F.; Hall, S. Cities and climate change mitiga Economic opportunities and governance challenges in Asia. *Cities* **2016**, *54*, 11–19. [CrossRef]
6. Storper, M. Governing the large metropolis. *Territ. Politics Gov.* **2014**, *2*, 115–134. [CrossRef]
7. Doyal, L.; Gough, I. *A Theory of Human Need*; Palgrave Macmillan: Basingstoke, UK, 1991. [CrossRef]
8. Maslow, A.H. A Theory of Human Motivation. *Psychol. Rev.* **1943**, *50*, 370–396. [CrossRef]
9. Glaeser, E.L. *The Triumph of the City: How Our Greatest Invention Makes Us Richer, Smarter, Greener, Healthier, and Happier*; Pen Books: New York, NY, USA, 2011.
10. Carter, J.G.; Cavan, G.; Conelly, A.; Guy, S.; Handley, J.; Kazmierczak, A. Climate change and the city: Building capacity for u adaptation. *Prog. Plan.* **2015**, *95*, 1–66. [CrossRef]
11. Rosezweig, C.; Solecki, W.D.; Hammer, S.A.; Mehrotra, S. *Climate Change and Cities: First Assessment Report of the Urban Cl Change Research Network*, 1st ed.; Cambridge University Press: Cambridge, UK, 2011; p. 312. [CrossRef]
12. Schär, C.; Vidale, P.L.; Lüthi, D.; Frei, C.; Häberli, C.; Liniger, M.A.; Appenzeller, C. The role of increasing temperature varia in European Summer heatwaves. *Nature* **2004**, *427*, 332–336. [CrossRef]
13. Cardoso, R.; Sobhani, A.; Meijers, E. The cities we need: Towards an urbanism guided by human needs satisfaction. *Urban* **2021**, *59*, 2638–2659. [CrossRef]
14. Allam, Z.; Bibri, S.E.; Chabaud, D.; Moreno, C. The '15-Minute City' concept can shape a net-zero urban future. *Humanit. So Commun.* **2022**, *9*, 126. [CrossRef]

Bowler, D.E.; Buyung-Ali, L.; Knight, T.M.; Pullin, A.S. Urban greening to cool towns and cities: A systematic review of the empirical evidence. *Landsc. Urban Plan.* **2010**, *97*, 147–155. [CrossRef]
Vargo, J.; Habeeb, D.; Stone, B., Jr. The importance of land cover change across urban–rural typologies for climate modeling. *J. Environ. Manag.* **2013**, *114*, 243–252. [CrossRef] [PubMed]
Yoon, E.J.; Kim, B.; Lee, D.K. Multi- objective planning model for urban greening based on optimization algorithms. *Urban For. Urban Green.* **2019**, *40*, 183–194. [CrossRef]
Dragoni, W.; Sukhija, B. Climate change and groundwater: A short review. *Geol. Soc. Spec. Publ.* **2008**, *288*, 1–12. [CrossRef]
Houston, D.; Werritty, A.; Bassett, D.; Geddes, A.; Hoolachan, A.; McMillan, M. *Pluvial (Rain-Related) Flooding in Urban Areas: The Invisible Hazard*; Joseph Rowntree Foundation: York, UK, 2011; Available online: https://www.jrf.org.uk/report/pluvial-rain-related-flooding-urban-areas-invisible-hazard (accessed on 20 September 2022).
Shokry, G. Competing risks capes of climate changes, gentrification and adaptation in Philadelphia's hunting park neighborhood. In *Green City and Social Injustice: 21 Tales from North America and Europe*, 1st ed.; Anguelovski, I., Connolly, J.J.T., Eds.; Routledge: London, UK, 2021.
Song, Y.; Kirkwood, N.; Maksimović, Č.; Zheng, X.; O'Connor, D.; Jin, Y.; Hou, D. Nature based solutions for contaminated land remediation and brownfield redevelopment in cities: A review. *Sci. Total Environ.* **2019**, *663*, 568–579. [CrossRef]
Jafari, E.; Soltanifard, H.; Aliabadi, K.; Karachi, H. Assessment of the Effect of Neyshabur Green Spatial Configuration on the Temperature of Land Surface and Heat Islands. *Open J. Ecol.* **2017**, *7*, 554–567. [CrossRef]
Santamouris, M.; Kolokotsa, D. *Urban Climate Mitigation Techniques*, 1st ed.; Routledge: Abingdon, UK, 2016; p. 208. [CrossRef]
Trogrlić, R.Š.; Rijke, J.; Dolman, N.; Zevenbergen, C. Rebuild by Design in Hoboken: A Design Competition as a Means for Achieving Flood Resilience of Urban Areas through the Implementation of Green Infrastructure. *Water* **2018**, *10*, 553. [CrossRef]
Cortinovis, C.; Grazia, Z.; Geneletti, D. Assessing Nature-Based Recreation to Support Urban Green Infrastructure Planning in Trento (Italy). *Land* **2018**, *7*, 112. [CrossRef]
Van Leeuwen, J.P.; Timmermans, H.J.P. *Innovations in Design & Decision Support Systems in Architecture and Urban Planning*, 1st ed.; Springer: Dordrecht, The Netherlands, 2006; p. 502. [CrossRef]
Niedźwiecka-Filipiak, I.; Rubaszek, J.; Potyrała, J. The Method of Planning Green Infrastructure System with the Use of Landscape-Functional Units (Method LaFU) and its Implementation in the Wrocław Functional Area (Poland). *Sustainability* **2019**, *11*, 394. [CrossRef]
Jansson, M. Green space in compact cities: The benefits and values of urban ecosystem services in planning. *Nord. J. Archit. Res.* **2014**, *2*, 139–160. Available online: https://arkitekturforskning.net/na/article/view/498 (accessed on 17 August 2022).
Kabisch, N.; Strohbach, M.; Haase, D.; Kronenberg, J. Urban green space availability in European cities. *Ecol. Indic.* **2016**, *70*, 586–596. [CrossRef]
Bhairappanavar, S.; Liu, R.; Coffman, R. Beneficial Uses of Dredged Material in Green Infrastructure and Living Architecture to Improve Resilience of Lake Erie. *Infrastructures* **2018**, *3*, 42. [CrossRef]
Gill, S.E.; Handley, J.F.; Ennos, A.R.; Pauleit, S. Adapting Cities for Climate Change: The role of the Green Infrastructure. In *Built Environment*, 1st ed.; Gill, S.E., Handley, J.F., Ennos, A.R., Pauleit, S., Eds.; Alexandrine Press: Marcham, UK, 2007; Volume 33, pp. 115–133. [CrossRef]
Infield, E.M.H.; Abunnasr, Y.; Ryan, R.L. *Planning for Climate Change: A Reader in Green Infrastructure and Sustainable Design for Resilient Cities*, 1st ed.; Routledge: Abingdon, UK, 2018; p. 392.
Loder, A. *Small-Scale Urban Greening: Creating Places of Health, Creativity, and Ecological Sustainability*, 1st ed.; Routledge: Abingdon, UK, 2020; p. 75. [CrossRef]
Mathey, J.; Rößler, S.; Lehmann, I.; Bräuer, A. Urban Green Spaces: Potentials and Constraints for Urban Adaptation to Climate Change. *Resilient Cities* **2011**, *1*, 479–485. [CrossRef]
Rubio-Bellido, C.; Pulido-Arcas, J.A.; Cabeza-Lainez, J.M. Adaptation Strategies and Resilience to Climate Change of Historic Dwellings. *Sustainability* **2015**, *7*, 3695–3713. [CrossRef]
Hagen, B. *Public Perception of Climate Change: Policy and Communication*; Routledge: London, UK, 2016; ISBN 978-138-79523-5. [CrossRef]
Kupka, J. *"Za města krásnější." Angažmá měšťanské společnosti vokrašlování měst. ("For more beautiful cities." The engagement of the bourgeois society with urban embellishment.) In Královéhradecko. Historický Sborník pro Poučenou Veřejnost*; Muzeum Východních Čech: Králové, Czech Republic, 2010; Volume 7, pp. 67–69. ISBN 978-80-85031-86-7. ISSN 1214-5211.
Indikátory Kvality Života v Oblasti Bezpečnosti (Indicators of Quality of Life in the Field of Safety). Úřad Vlády České Republiky. Available online: https://www.cr2030.cz/strategie/wpcontent/uploads/sites/2/2018/05/11_Bezpe%C4%8Dnost.pdf (accessed on 13 September 2022).
Lovasi, G.S.; Schwartz-Soicher, O.; Quinn, J.W.; Berger, D.K.; Neckerman, K.M.; Jaslow, R.; Lee, K.K.; Rundle, A. Neighborhood safety and green space as predictors of obesity among preschool children from low-income families in New York City. *Prev. Med.* **2013**, *57*, 189–193. Available online: https://pubmed.ncbi.nlm.nih.gov/23732240/ (accessed on 17 August 2022). [CrossRef]
Ruan, X.; Hogben, P. *Topophilia and Topophobia: Reflections on Twentieth Century Human Habitat*; Routledge: New York, NY, USA, 2007; 232p.

41. Cole, H.; Abel, T.D. Resisting green gentrification, Seattle's South Park neighborhood struggles for environmental justice. *Green City and Social Injustice: 21 Tales from North America and Europe*, 1st ed.; Anguelovski, I., Connolly, J.J.T., Eds.; Routled New York, NY, USA, 2021.
42. Bolger, N.; Zuckerman, A. A framework for studying personality in the stress process. *J. Pers. Soc. Psychol.* **1995**, *69*, 890–9 [CrossRef]
43. Tomášek, J. *Úvod do Kriminologie: JAK Studovat Zločin*; (Introduction to Criminology: How to Study Crime); Grada: Praha, Cze Republic, 2010; ISBN 978-80-247-2982-4.
44. Jíchová, J.; Temelová, J. Kriminalita a její percepce ve vnitřním městě: Případová studie pražského Žižkova a Jarova (Crime a its perception in the inner city: A case study of Prague's Žižkov and Jarov). *Geografie* **2012**, *117*, 329–348. [CrossRef]
45. Jíchová, J.; Temelová, J. Kriminalita a riziková místa centrálního a vnitřního města: Sonda do názorů obyvatel vybraný pražských čtvrtí (Crime and risky places of the central and inner city: A probe into the opinions of the inhabitants of select Prague districts). In *Nové Sociálně Prostorové Nerovnosti, Lokální Rozvoj a Kvalita Života*; Temelová, J., Pospíšilová, L., Ouředníč M., Eds.; Aleš Čeněk: Plzeň, Czech Republic, 2012; pp. 46–76.
46. Krulichová, E.; Buriánek, J. (Eds.) *Obavy ze Zločinu: Mýty a Realita (Fears of Crime: Myths and Reality)*; Karolinum: Praha, Cze Republic, 2020.
47. Toušek, L.; Hejnal, O. *Analýza Pocitu Bezpečí Občanů Vybraných Obcí Plzeňského Kraje (Analysis of the Feeling of Security of Citizen Selected Municipalities of the Pilsen Region)*; Krajský Úřad Plzeňského Kraje: Pilsen, Czech Republic, 2011.
48. Fisher, B.S.; Nasar, J.L. Fear of Crime in Relation to Three Exterior Site Features. Prospect, Refuge, and Escape. *Environ. Beh* **1992**, *24*, 35–65. [CrossRef]
49. Zoophobia (Fear of Animals). Cleveland Clinics. Available online: https://my.clevelandclinic.org/health/diseases/227 zoophobia-fear-of-animals (accessed on 13 September 2022).
50. Entomophobia (Fear of Insects). Cleveland Clinics. Available online: https://my.clevelandclinic.org/health/diseases/225 entomophobia-fear-of-insects (accessed on 13 September 2022).
51. Ornithophobia (Fear of Birds). Cleveland Clinics. Available online: https://my.clevelandclinic.org/health/diseases/225 ornithophobia-fear-of-birds (accessed on 13 September 2022).
52. Gastaldi, F.; Camerin, F. Brownfield infrastructures. In *The Elgar Companion to Urban Infrastructure Governance Innovation, Conce and Cases*; Finger, M., Yanar, N., Eds.; Edward Elgar Publishing Limited: Cheltenham, UK; Northampton, MA, USA, 20 [CrossRef]
53. Rink, D.; Schmidt, C. Afforestation of Urban Brownfields as a Nature-Based Solution. Experiences from a Project in Leip (Germany). *Land* **2021**, *10*, 893. [CrossRef]
54. Český Statistický Úřad (Czech Statistical Office). Available online: https://www.czso.cz/csu/czso/databaze-demograficky udaju-za-vybrana-mesta-cr (accessed on 18 April 2022).
55. Eckart, K.; Ehrke, S. *Social, Economic and Cultural Aspects in the Dynamic Changing Process of Old Industrial Regions: Ruhr Dist (Germany), Upper Silesia (Poland), Ostrava Region (Czech Republic)*; Lit Verlag Münster: Münster, Germany, 2003; Volume 1, p. 2
56. Jiřík, K. *Dějiny Města Ostravy. (History of Ostrava)*; Sfinga: Ostrava, Czech Republic, 1993.
57. Český Statistický Úřad (Czech Statistical Office). Available online: https://www.czso.cz/csu/xt/charakteristika_okresu_ostra mesto (accessed on 15 February 2021).
58. Vize Prostorového Rozvoje Ostravy. Available online: https://vize-prostoroveho-rozvoje-mappaova.hub.arcgis.com/ (acces on 12 March 2022).
59. ZdravaOstrava—Oficiální Web Města Ostravy k Životnímu Prostředí- Příroda a biodiverzita. (The Official Website of the City Ostrava Devoted to the Environment- Nature and Biodiversity). Available online: https://zdravaova.cz/category/priroda biodiverzita/ (accessed on 25 September 2022).
60. ZdravaOstrava—Oficiální Web Města Ostravy k Životnímu Prostředí -Adaptační Strategie na Dopady a Rizika, Vyplývajíc Změny Klimatu. (The Official Website of the City of OSTRAVA Devoted to the Environment—Adaptation Strategies to Clim Change Impacts and Risks). Available online: https://zdravaova.cz/adaptacni-strategie-na-zmeny-klimatu/ (accessed on September 2022).
61. FajnOva—Oficiální Web Města Ostravy k Strategickému Plánování. (The Official Website of the City of Ostrava for the Strateg Planning). Available online: https://fajnova.cz/pocitova-mapa (accessed on 1 October 2020).
62. Internal Documents ARR (MSID). Regional Development Agency Now Moravian Silesian Investment and Development terní Materiál k Přípravě Databáze Brownfieldů. (Internal Material for the Preparation of a Database of Brownfields)). 2 Unpublished work. p. 26.
63. Mekjavic, I.B.; Yogev, D.; Ciuha, U. Perception of Thermal Comfort during Skin Cooling and Heating. *Life* **2021**, *11*, 681. [Cross [PubMed]
64. Schweiker, M.; Huebner, G.M.; Kingma, B.R.M.; Kramer, R.; Pallubinsky, H. Drivers of diversity in human thermal perception- Review for Holistic Comfort Models. *Temperature* **2018**, *5*, 308–342. [CrossRef] [PubMed]
65. Orcígr, V.; Vidovićová, L.; Šerák, M.; Fernandová, E. *Pražští Senioři a Extrémní Horka* (Prague Seniors and Extreme Heat). Arr 2020. Available online: https://www.rds.org.co/apc-aa-files/205ec78c9cca6d1850bdca24e20e50bf/arnika_studie_seniori. (accessed on 25 September 2022).

Kalvach, Z.; Zadák, Z.; Jirák, R.; Sucharda, P.; Zavázalová, H. *Geriatrie a Gerontologie (Geriatrie and Gerontologie)*; Grada: Praha, Czech Republic, 2004; Volume 567.
Fudała, M. Urban environmental acupuncture for improving access to green spaces in cities—Example from urban region in Central Europe. In *Integrated Environmental Management of Land and Soil in European Urban Areas*; Starzewska-Sikorska, A., Ed.; Polish Academy of Sciences: Warsaw, Poland, 2021.
Český Statistický Úřad (Czech Statistical Office). Available online: https://www.czso.cz/staticke/animgraf/cz080/index.html?lang=cz (accessed on 12 April 2022).
Braungart, M.M.; Braungart, R.G.; Hoyer, W.J. Age, Sex, and Social Factors in Fear of Crime. *Sociol. Focus* **1980**, *13*, 55–66. [CrossRef]
Baruch, E.M.; Voss, K.A.; Blaszczak, J.R.; Delesantro, J.; Urban, D.L.; Bernhardt, E.S. Not all pavements lead to streams: Variation in impervious surface connectivity affects urban stream ecosystems. *Freshw. Sci.* **2018**, *37*, 673–684. [CrossRef]
Burden, D. *Urban Street Trees: 22 Benefits, Specific Applications*, 1st ed.; Glatting Jackson and Walkable Communities Inc.: Orlando, FL, USA, 2006; pp. 1–21.
Djedjig, R.; Belarbi, R.; Bozonnet, E. Green wall impacts inside and outside buildings: Experimental study. *Energy Procedia* **2017**, *139*, 578–583. [CrossRef]
Dolan, L.M.J.; Bohemen, H.; Whelan, P.; Akbar, K.F.; O'Malley, V.; O'Leary, G.; Keizer, P.J. Towards the sustainable development of modern road ecosystems. In *Environmental Pollution, The Ecology of Transportation: Managing Mobility for the Environment*, 1st ed.; Davenport, J., Davenport, J.L., Eds.; Springer: Dordrecht, The Netherlands, 2006; Volume 10, pp. 275–331. [CrossRef]
Drozd, W. Problems and benefits of using green roofs in Poland. *IOP Conf. Ser. Earth Environ. Sci.* **2019**, *214*, 012076. [CrossRef]
Elsadek, M.; Liu, B.; Zefeng, L. Green facades: Their contribution to street recovery and well-being in high-density cities. *Urban For. Urban Green.* **2019**, *46*, 126446. [CrossRef]
Lee, K.-W.W.; Vinka, O.-C.; Kohm, S.; Chango, H. Cool Pavements As a Sustainable Approach to Green Streets and Highways. In Proceedings of the Green Streets and Highways Conference 2010, Denver, CO, USA, 14–17 November 2010; American Society of Civil Engineers (ASCE): Reston, VA, USA, 2010; pp. 235–247. [CrossRef]
Leroy, M.-C.; Portet-Koltalo, F.; Legras, M.; Lederf, F.; Moncond'huy, V.; Polaert, I.; Marcotte, S. Performance of vegetated swales for improving road runoff quality in a moderate traffic urban area. *Sci. Total Environ.* **2016**, *566–567*, 113–121. [CrossRef]
Li, W.C.; Yeung, K.K.A. A comprehensive study of green roof performance from environmental perspective. *Int. J. Sustain. Built Environ.* **2014**, *3*, 127–134. [CrossRef]
Scott, M.; Lennon, M.; Haase, D.; Kazmierczak, A.; Clabby, G.; Beatley, T. Nature-based solutions for the contemporary city/Re-naturing the city/Reflections on urban landscapes, ecosystems services and nature-based solutions in cities/Multifunctional green infrastructure and climate change adaptation: Brownfield greening as an adaptation strategy for vulnerable communities?/Delivering green infrastructure through planning: Insights from practice in Fingal, Ireland/Planning for biophilic cities: From theory to practice. *Plan. Theory Pract.* **2016**, *17*, 267–300. [CrossRef]
Soares, A.L.; Rego, F.C.; McPherson, E.G.; Simpson, J.R.; Peper, P.J.; Xiao, Q. Benefits and costs of street trees in Lisbon, Portugal. *Urban For. Urban Green.* **2011**, *10*, 69–78. [CrossRef]
Zhang, L.; Zhichao, D.; Liang, L.; Zhang, Y.; Meng, Q.; Wang, J.; Mattheos, S. Thermal behavior of a vertical green façade and its impact on the indoor and outdoor thermal environment. *Energy Build.* **2019**, *204*, 109502. [CrossRef]
Clarke, A. *Evaluation Research: An Introduction to Principles, Methods and Practice*; SAGE: London, UK, 1999.
Stangel, M. *Action Plan for UEA in Chorzów, Ruda Śląska and Świętochłowice*; SALUTE4CE Project of CE Programme: Vienna, Austria, 2021.
Anguelovski, I.; Connolly, J.J.T. *Green City Social Injustice: 21 Tales from North America and Europe*, 1st ed.; Routledge: London, UK, 2021.
Kupka, J. Počátky Veřejné Zeleně ve Strakonicích (The Beginnings of Public Greenery in Strakonice). In *Vlastivědný Sborník Strakonice 3*; Kapitoly ze Života Města. Strakonice: Město Strakonice, Czech Republic, 2005; pp. 251–272. ISBN 80-239-4790-7.
Mathey, J.; Arndt, T.; Banse, J.; Rink, D. Public perception of spontaneous vegetation on brownfields in urban areas—Results from surveys in Dresden and Leipzig (Germany). *Urban For. Urban Green.* **2016**, *29*, 384–392. [CrossRef]
Southona, G.E.; Jorgensena, A.; Dunnetta, N.; Hoylea, H.; Evansba, K.L. Biodiverse perennial meadows have aesthetic value and increaseresidents' perceptions of site quality in urban green-space. *Landsc. Urban Plan.* **2017**, *158*, 105–118. [CrossRef]
Norton, B.A.; Bending, G.D.; Clark, R.; Corstanje, R.; Dunnett, N.; Evans, K.L.; Grafius, D.R.; Gravestock, E.; Grice, S.M.; Harris, J.A.; et al. Urban meadows as an alternative to short mown grassland: Effects of composition and height on biodiversity. *Ecol. Appl.* **2019**, *29*, e01946. [CrossRef]
Prieto Curiel, R.; Bishop, S.R. Fear of crime: The impact of different distributions of victimisation. *Palgrave Commun.* **2018**, *4*, 46. [CrossRef]

forests

ticle

:alculating and Mapping the Naturalness of eri-Urban Greenways

tor Àvila Callau [1,*], Yolanda Pérez-Albert [1] and Jesús Vías Martínez [2]

[1] Grup de Recerca d'Anàlisi Territorial i Estudis Turístics (GRATET), Department of Geography, Universitat Rovira i Virgili, 43480 Vila-seca, Spain; myolanda.perez@urv.cat
[2] Department of Geography, Universidad de Málaga, 29071 Málaga, Spain; jmvias@uma.es
* Correspondence: aitor.avila@urv.cat

Abstract: Peri-urban forests often have extensive greenway networks that allow for outdoor recreation. However, information associated with these greenways often does not include their degree of naturalness, which is usually reduced to descriptions of the flora and fauna and often overlooks the factors that reduce naturalness. Therefore, in some cases, the naturalness of these greenways is lower than expected. Quantifying their naturalness would be helpful, especially for hikers interested in appreciating and enjoying nature. Additionally, this information would help outdoor recreation managers to design trails or decide which ones to promote as "greenways". The objectives of this study are (1) to design a method to calculate and map the naturalness of greenways using two approaches, one based on perceptual fieldwork and the other on geographic information systems (GIS); (2) to apply the designed method to a specific greenway; and (3) to compare both methodological approaches. The results show that, for the greenway studied, the naturalness scores obtained are low in all three types of analyses used. Around 70% of the greenway sections in the GIS visibility analysis and 80% in the GIS proximity analysis have a low naturalness index. In comparison, this value is reduced to 40% with the fieldwork analysis. Although the results of the GIS approach (proximity and visibility) generate naturalness indices and spatial patterns that are very similar, they differ significantly from those derived from the fieldwork analysis. The discussion of the results suggests that the three methodologies used are valid for analyzing the degree of naturalness of the trails. However, if used together, it could add flexibility to the type of variables incorporated in the analysis.

Keywords: naturalness of greenways; nature-based recreation; trail assessment; peri-urban forests; GIS; fieldwork

:ion: Àvila Callau, A.;
z-Albert, Y.; Vías Martínez, J.
ulating and Mapping the
ıralness of Peri-Urban
nways. *Forests* **2023**, *14*, 1181.
s://doi.org/10.3390/f14061181

lemic Editors: Panayiotis
.trakopoulos, Mario A. Pagnotta,
as Scholz and Arshiya Noorani

ived: 9 May 2023
sed: 4 June 2023
pted: 5 June 2023
ished: 7 June 2023

1. Introduction

In the present day, urban populations are subjected, on the one hand, to an overload of work motivated by their work or academic activity, and on the other hand, to living in a degraded environment resulting, among other factors, from pollution. The confluence of these two situations can cause negative impacts on people's health: higher levels of physiological and psychological stress and a higher risk of chronic pathologies, including cardiovascular diseases and mental health problems [1].

In order to counteract these adverse effects on health, the World Health Organization (WHO) recommends regular physical activity as it is beneficial for preventing and treating non-communicable diseases (such as cardiovascular diseases) and for mental health [2]. Additionally, according to Kerr et al. [3], the amount of time spent outdoors is related directly to the improvement of individuals' physical and emotional states. The combination of these two elements, physical activity and outdoor exposure, gives rise to outdoor recreation, which provides psychological, sociological, educational, and physical benefits to people's health [4].

When physical activity is performed outdoors in natural environments, it is generally referred to as "green exercise". Several authors [1,5–7] consider that "green exercise", when carried out in pleasant natural environments, may positively affect mental heal and blood pressure, an important indicator of cardiovascular health. In addition, "gree exercise" in natural green spaces may have advantages over that carried out in urban gre spaces, especially concerning mental health [1,8]. This may be because natural vegetati provides a sense of privacy, encourages personal relationships and physical exercise, a allows one to distance oneself from routines, which implies isolating oneself from sourc of stress [9,10]. The COVID-19 pandemic has encouraged outdoor activities, and, therefo the use of hiking trails has increased. Obradović and Tešin's work [11] states that, as a res of the pandemic, there was an increase in the number of people who practiced hiking, th people became more involved in this activity, but that perhaps this situation was motivat because other types of activities could not be carried out. Despite this, a large percenta of those surveyed stated that they would continue to practice outdoor hiking when t situation normalized.

Peri-urban forests often feature extensive networks of greenways, paths, or tra that offer a unique opportunity to escape the stress of urban life through sport and rec ation [12–16]. It has also been found that the availability and accessibility of trails promo an active lifestyle, increasing the likelihood that users will meet physical activity recomm dations prescribed by experts [17–24]. Moreover, from a semi-rural or peri-urban locati perspective, trails allow for interaction with the natural environment, which, along wi the benefits of physical activity, can improve the quality of life for its users [18,21,25].

The satisfying experience of a hike is conditioned by whether the trail's characterist align with the user's motivations or preferences [11]. If a trail features elements preferr by the user, such as a pleasant landscape, their motivation to walk the trail will increase Several authors [13,18] have established that trails are mainly used for physical activ However, in addition to physical exercise, other motivations, such as culture, herita or contact with nature drive physical activity on trails [18,22,26–30]. For example, Farí Torbidoni [26] analyzed the motivations of hikers in natural areas and concludes that main reason for walking the trail is related directly to nature (connecting with natu relaxing and disconnecting, and enjoying the landscape). Cordeiro and Alves [31] obtain a similar result by establishing that hikers' most important motivating factors are relaxati in the first place and exploration of nature in the second place. A study by Mayer a Lukács [32] revealed the desire to escape from everyday life and be in a natural environm as the main motivational factors for hiking trails. The results of Keith et al. [13] establish that, after physical activity, relaxing and escaping from city life, spending time with frier or family, and discovering and experiencing nature are the main motivations for using a tr Thus, contact with nature is one of the recurring motivations for trail use. Therefore, hiki in forests or natural environments is one of the most sought-after hiking experiences [33]. shown, several studies confirm a high level of preference for environments with abund natural vegetation, with natural beauty and contact with nature being one of the m reasons people use trails [31,33–35]. Therefore, evaluating the degree of naturalness natural interest of the trails would allow potential users to plan their open-air activit based on their primary motivation. This way, their level of satisfaction would incre since their expectations would be adjusted to the experience along the trail.

Naturalness is one of the dimensions of human perception of the environment [In addition, naturalness is a subjective concept with aesthetic and biophysical dim sions [36,37]. However, it can be quantified through scientific assessment, for exam based on the number of native versus non-native species in a forest [37]. In outdoor re ation, naturalness relates to protected nature, nature-based recreational experiences, wildlife [38]. In terms of trails, naturalness is deeply linked to the degree of conservatio the natural environments around them [39].

According to Fredman et al. [38], naturalness is synonymous with predomina natural rather than built or artificial, and the degree of naturalness can increase depend on the distance from a human settlement. In addition, the impacts (such as visual, odor or acoustic) derived from the activities in urban settlements and their peripheries compromise a landscape's perceived degree of naturalness. For example, in the case of sound, it is an element of the landscape that, on the one hand, forms part of the place's identity and its attractiveness, both in natural and cultural terms. On the other hand, noise (for example, associated with road or rail traffic) not only masks the characteristic sounds of the region but also represents a threat to the nature and naturalness of the place, causing environmental fragmentation and general ecological stress [40].

Traditional methodologies that study the characteristics of trails use a series of factors such as the attractiveness of the environment, the type of ground, the conditions of the trail itself, or the natural monuments and attractions present [41]. This method is in line with those applied by hiking federations, such as the European Rambler's Association that issues trail quality seals, such as the "Leading Quality Trails—Best of Europe", and that, to certify them, relies on the quality seal acquisition guide proposed by the "Deutsche Wanderverband Service". The collection and evaluation of these factors have traditionally been based on fieldwork, relying mainly on observation and perception, but in recent decades, these methods have been supplanted by geographic information systems (GIS) [37,42–45].

In summary, not all hikers have the same interests and motivations, with some prioritizing the athletic aspect, others the ecological and naturalistic aspect, and others the recreational and cultural motivation [29,46]. Additionally, user experience improves if the trail characteristics align with their preferences, so evaluating and communicating the degree of naturalness of the trails is a priority for users motivated by contact with nature. Thus, a detailed understanding of the relationships between users' motivations and trail characteristics can contribute to the design of strategies that increase user satisfaction and provide greater health benefits [27,46]. However, in analyzing and evaluating trails, the traditional fieldwork method has been substituted by GIS tools. The research question posed by this article is whether GIS tools effectively replace traditional fieldwork methods or whether they are complementary.

This study aims to calculate the naturalness of a greenway, that is, its affinity for the "nature-minded hiker" profile [26], whose main motivation is to appreciate, enjoy, and have direct contact with nature. The objectives of the study are: (1) to design a method that allows the calculation and mapping of the naturalness of greenways based on two approaches, one based on fieldwork and the other on GIS; (2) to apply the designed method to the selected greenway and assess its degree of naturalness; and (3) to compare both methodological approaches to contrast the different results, highlighting their respective strengths and weaknesses.

The study is applied to a local and circular greenway called Anella Verda ("Green Ring"), which is 34 km long and mainly runs through the peri-urban forests of the city of Tarragona (Spain). This was chosen because it is one of the most used greenways in the area [47] and because the responsible administration (Tarragona city council) promotes it as "green", meaning that it has a strong ecological and naturalness component.

2. Materials and Methods

The methodology is divided into 4 steps (Figure 1). Except for the first step, the following steps are commonly used in methodologies based on multi-criteria analysis techniques in GIS environments. The first step consists of selecting a greenway and defining its sections to operationalize the analysis of its degree of naturalness. The second step is to determine a list of factors that influence the naturalness of the selected greenway, either positively or negatively. The third step refers to collecting data on the selected factors, and in this phase, two different approaches are proposed: GIS and perceptual fieldwork. We compare these two methodologies because they are some of the most widely used for collecting data on factors in multi-criteria analysis. In turn, the GIS method is subdivided into two approaches: proximity analysis and visibility analysis. It is considered that a certain factor, although not directly visible from the greenway, may be nearby and impact the user's perception in some way. For example, no direct visual contact between a greenway user and a nearby wastewater treatment plant may exist; however, this element will potentially produce a different negative impact (acoustic and/or odoriferous) t the user will perceive equally, compromising the degree of naturalness of the greenw Therefore, it is appropriate to use proximity and visibility analyses referring to the (method in this study. Finally, the fourth and last methodological step consists of calculat the naturalness index of each section of the greenway according to the three approac and mapping it.

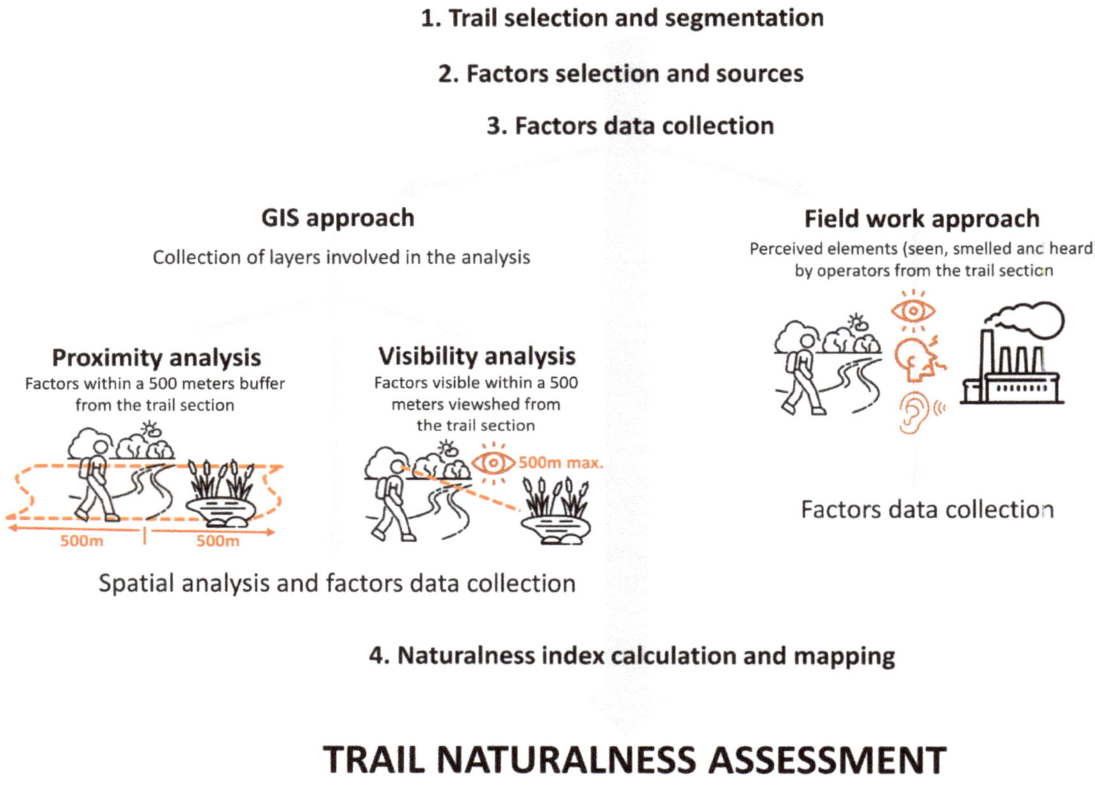

Figure 1. Outline of the methodological steps.

The segmentation of the greenway into sections (step 1 of the methodology) and it erence length as well as the selection of most of the factors that influence the naturalness greenway (step 2 of the methodology) are based on the Qualitätsweg Wanderbares (Qu Hiking Trail) guide (Deutscher Wanderverband Service GmbH, Kassel, Germany), moted by the Deutscher Wanderverband (German Hiking Association, DWV) which is a able online at https://www.wanderbares-deutschland.de/service/qualitaetsinitiati qualitaetswege (accessed on 4 February 2021). This guide details the requirements a trail in Germany must meet to be certified as Qualitätsweg Wanderbares Deutsch (Quality Hiking Trail of Wanderbares Deutschland). To achieve this, different paramete the trail are measured (materials, location, capacity to be enjoyed), the trail's develop as a means of communication, signage, the landscape and natural environment, cul aspects, and the humanization of the route, among others. This guide is widely recogn and used by prestigious outdoor recreation associations such as the European Ram Association (ERA), which awards a seal called Leading Quality Trails—Best of Eu

(ERA-EWV-FERP, Praha, Czech Republic) to the highest quality trails in Europe with the support of this guide (https://www.era-ewv-ferp.org/lqt/, accessed on 4 February 2021).

2.1. Study Area

The Anella Verda ("Green Ring") runs through the peri-urban area of Tarragona (Catalonia, Spain). This space is a highly fragmented traditional agricultural landscape due to the proliferation of industrial, logistical, commercial, and leisure areas and a dense network of infrastructures. The environment of this greenway is considered a chaotic area with many interstitial spaces, although with great potential for outdoor recreation as it has an extensive network of trails [48].

Through two variants, the selected greenway connects the rivers that flow into the municipality of Tarragona (the Francolí and the Gaià), resulting in a great excursion of 34 km in length. Its interior variant mostly runs through peri-urban forests, while the other variant runs along the coastline (Figure 2). The route covers a large part of the extensive historical and cultural heritage of the municipality, passing through a Roman aqueduct (Figure 2a), medieval farmhouses, defense towers (Figure 2c), Roman quarries, orchards, river mouths, villages of interest (Figure 2e), and coves or beaches of high natural value (Figure 2b,d). Although there are specific locations within the itinerary with the presence of tourists (such as the Roman aqueduct), the entire greenway is used mainly by the inhabitants of its surroundings [47].

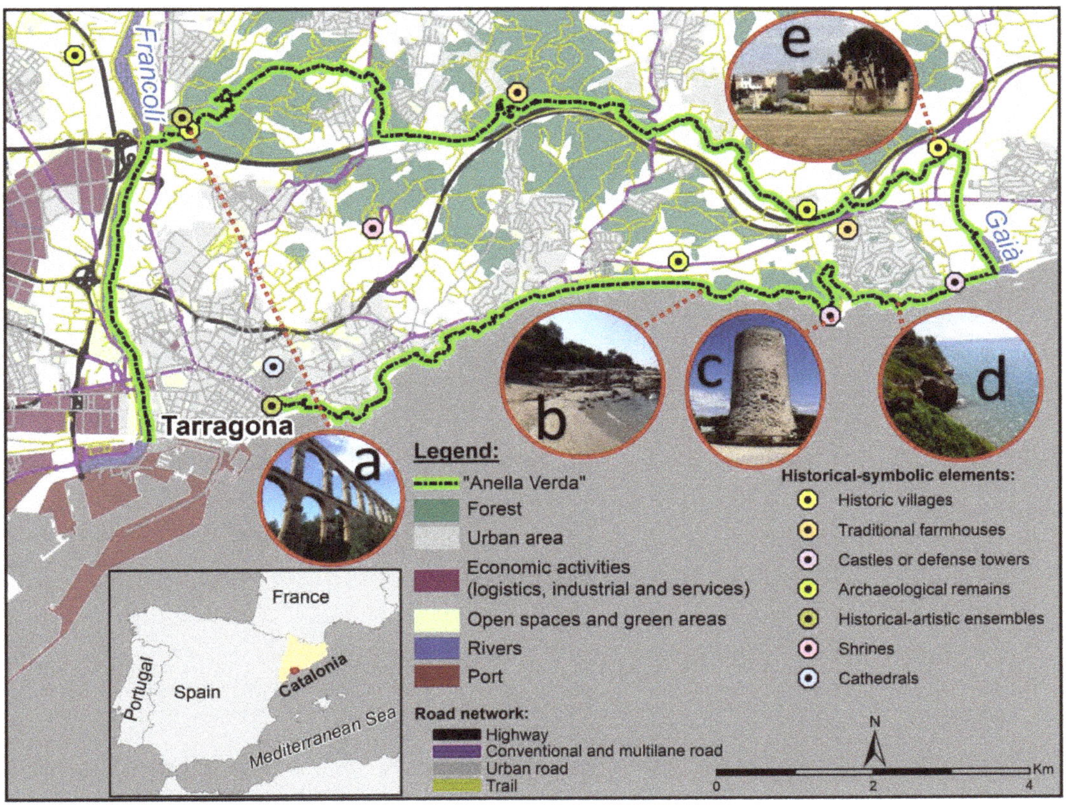

Figure 2. Study area with the representation of the analyzed greenway and some relevant photographs. (**a**) Roman aqueduct, (**b**) Platja Llarga beach, (**c**) La Mora defense tower, (**d**) La Mora coves and (**e**) Ferran village.

2.2. Trail Segmentation

The greenway layer has been downloaded from the website of the Tarragona Ci Council in GPX format and has been subdivided into different sections in order to c erationalize data collection. For this purpose, the greenway has been segmented usi satellite images through photointerpretation based on two criteria. The first (1) conside the morphology of the path and is based on detecting sudden changes, for example, wh its width or type of surface is radically modified. The second (2) establishes a travel distan of around 4 km, following the Qualitätsweg Wanderbares guide.

2.3. Factor Selection and Sources

For analyzing the naturalness of the sections of the Anella Verda, a set of factors h been selected that are mainly based on the Qualitätsweg Wanderbares guide and oth studies that evaluate naturalness using multi-criteria analysis techniques integrated in GIS, such as those by Vías and Ocaña [45] and Khazaee Fadafan et al. [43]. The select factors are grouped into two sets (Table 1), similar to the approach used by Khazaee Fadaf et al. [43]. On the one hand, there are factors related to the intensive use of space that m reduce the naturalness of the greenway, which could be considered negative factors. On t other hand, there are factors that add naturalness, i.e., those that shape natural landscap and their figures of protection that determine their relevance from a natural perspecti and all of them are positive factors.

Table 1. Factors included in the analysis.

Group	Factor Codes	Factor Description	Source (Geographic Information Systems—GIS)
Group 1: intensive use (negative factors)	F1_industry	Industrial areas	Urbanistic Map of Catalonia (MUC)
	F2_voltage	High-voltage towers and lines	Topographic Map (IGN)
	F3_treatment	Water treatment plants	Hypermap (Generalitat de Catalunya)
	F4a_transport	Roadways and railways	Urbanistic Map of Catalonia (MUC)
	F4b_transport *	Acoustic pollution from roads and railway	Noise Pollution Information System (SICA
	F5_urban	Urban centers	Urbanistic Map of Catalonia (MUC)
Group 2: natural landscapes (positive factors)	F6_forest	Woodland and forest	Hypermap (Generalitat de Catalunya)
	F7_n2000	Protected natural areas (Natura 2000 network)	Hypermap (Generalitat de Catalunya)
	F8_coast	Scenic coastal landscapes	Landscape Catalogue of the Camp de Tarragona (Landscape Observatory of Catalonia)
	F9_faunaflora	Areas of faunal and floral interest	Hypermap (Generalitat de Catalunya)
	F10_rivers	Watercourses (rivers, torrents, etc.)	Urbanistic Map of Catalonia (MUC)
	F11_sea	Sea	Hypermap (Generalitat de Catalunya)

In the factor codes column, the "x" of "Fx" means the number assigned to each factor. * For the analysis o transportation factor, with proximity analysis (F4b_transport), the layer indicating the influence area of noise f interurban roads and railway lines has been used (Ldia: average noise level during the day from 7 a.m. to 7 p.

Although there are many factors to use in the analysis, for this study, only th present in the study area have been considered. Specifically, through a previous explorat of the cartography associated with the Landscape Catalog of Camp de Tarragona from Landscape Observatory of Catalonia [49], only the factors within a 500 m influence a from the greenway are included. Those that cannot be measured by the three approac (GIS visibility, GIS proximity, and fieldwork) at the same time are excluded. In the c of approaches that require the use of GIS (proximity and visibility), the corresponding layers in ESRI shapefile format have been obtained from the different official organizations detailed in Table 1.

2.4. Factor Data Collection

The fieldwork was entirely perceptual or experiential. For each section, the presence of each factor was analyzed, considering a single scale of values from 1 to 5, with 1 indicating very little presence and 5 the opposite situation. The fieldwork was carried out in July 2021 by two of the authors of this article, having printed support cartography of all of the factors to be assessed. During data collection, the weather conditions were favorable, i.e., no episodes of wind or rain were experienced. The collected data were entered on analog fieldwork sheets by both operators independently and then the simple average of the evaluations made by each of them was calculated.

In the case of the GIS-based approaches, the analysis was restricted to an influence area of 500 m. According to Booth et al. [50], landscape characteristics can be separated into two categories based on their proximity to the viewer: aesthetic of view and aesthetic of landscape. The former is determined by the perceived characteristics at great distances, while the latter is limited to the perceived characteristics within a much closer range, typically by viewers within a natural area looking at a patch of landscape around them. Their landscape aesthetic model is limited to evaluating the characteristics of the 1000 m around the observer, based on their professional judgment and knowledge of the study area. However, they argue for the possibility of adapting this distance to the characteristics of the analyzed zone. Following the same criteria, and taking into account the characteristics of the study area, this work considers it appropriate to restrict both GIS analyses to the 500 m around each section of the greenway.

For the visibility analysis, visual basins were calculated using the "visibility analysis" plugin, accessible from the free and open-source software QGIS 3. As a previous step, a layer of viewpoints or observers was created for each of the greenway sections. To achieve this, the line layer, which represents the different sections, was transformed into a point layer based on its vertices, which generated 3716 observation points distributed, according to the section, as shown in Table 2. Attributes related to the observer's height (the default program value which is 1.60 m) and visibility radius (which is 500 m) are associated with this point layer.

Table 2. Number of observation points for each section (S).

Sections	S1	S2	S3	S4	S5	S6	S7	S8	S9	S10
Number of vertices	209	201	790	509	517	444	196	270	144	436

After configuring the observation points, the visual basins were calculated for each section using the default program values for the sphericity of the Earth and the atmospheric refraction index. In this study, due to its advantages in terms of precision and accuracy in modeling the terrain [51–53], a digital surface model (DSM) was used for calculating the visual basins, as they integrate the height of trees and buildings in the visibility calculations. This DSM was downloaded from the Spanish National Geographic Institute website.

This calculation process was performed 10 times, one for each set of points representing a section of the greenway. The resulting values were transformed into binary values: 0 (not visible) and 1 (visible). Then, the visible surface area in square meters was calculated for each of the 10 sections of the greenway. The visible surface area results from multiplying the number of pixels with a value of 1 by the surface area of each pixel, which in this study was 100 square meters. Finally, to determine the percentage of the visible surface area of each factor, its surface area was calculated by overlaying the layers representing the factors and the visual basin from each section.

In the case of the proximity analysis, the percentage of the surface area of each fac within a 500 m buffer around each section was calculated. For this, the "buffer" algorit of QGIS 3 was used to calculate the 10 buffers for each section of the greenway. Next, of the layers related to the different factors were clipped using each buffer as the clipp entity. Once the factors were clipped, they were overlaid with the buffers of each sectio calculate the proportion of the surface area of each factor relative to the total surface are each buffer. However, for the factor "F4b_transport", the percentage of the distance of e section that directly intersects with the layer of noise from interurban roads and railw was calculated. Therefore, unlike in the visibility analysis, proximity to the communica routes is not measured, but rather to the noise they generate, one of the negative imp these infrastructures can cause [40].

2.5. Naturalness Index Calculation and Mapping

To calculate the naturalness index, the ideal point method is used, adapted fr Malczewski [54] and expressed by the following equation:

$$Naturalness\ index = 1 - \left(\frac{\sqrt{\sum_1^n (x_i - p)^2}}{\sqrt{N(a-p)^2}} \right)$$

where x_i is the value of factor n in section i; p is the ideal point; a is the anti-ideal p (opposite to the ideal), and N is the total number of factors.

The ideal point method reflects the statistical distance of each case from the i situation. It is used in this study because, like in Vías and Ocaña [45], it is more appropr than a mere linear aggregation since the compensation derived from using the arithm mean balances situations with different dissimilarity values.

In the analyses derived from the GIS method, the value considered the ideal p for all factors is 100%, while for the data collected in the fieldwork, this value is 5. preliminary step to calculating the formula, the values of the factors belonging to gro (intensive use—negative factors) are inverted since, being factors that harm naturalness lower the original value collected, the closer they will be to the ideal point of natural For example, if the original value of factor 1 (F1_Industry) was 4.52%, it is inverted becomes 95.48%, considering 100% as the ideal point.

Considered as a measure of similarity, the result of calculating the part of the form included in the parentheses means that the value 0 is the ideal point, while 1 is its antith However, following the study of Vías and Ocaña [45], and in order to maintain a log order of scores (0 indicating lower degree of naturalness and 1 indicating higher degr naturalness), the value resulting from calculating the parentheses is inverted.

Once the naturalness index of each section was calculated, they were mapped u an equal interval classification to be comparable: low naturalness degree (from 0 to 0 medium naturalness degree (greater than 0.33 to 0.66), and high naturalness degree (gre than 0.66 to 1).

3. Results

3.1. Trail Sections' Characteristics

The segments that make up the analyzed greenway can be classified into three r groups: those that run along the river courses and are located at the eastern and wes ends (S1, S2, and partially S6); those in the interior (S3, S4, S5, and partially S6); and fi the coastal ones (S7, S8, S9, and S10) (Figure 3). In the first group, the presence o river is the main factor shaping the characteristics. However, there are some differe S1 is an entirely urban section that has been configured as a river park and is equip with urban furniture (fountains, benches, trash cans, etc.) and roads with different su characteristics (artificial, dirt, etc.) or users (bike lane or pedestrian lane); the S2 se corresponds to a wide riverbed, with little riparian vegetation and multiple trail and options; while the second part of S6 coincides with the final stretch of the Gaià river and has a dense and well-preserved riparian vegetation since it is included in the protected natural space of the mouth of the Gaià river.

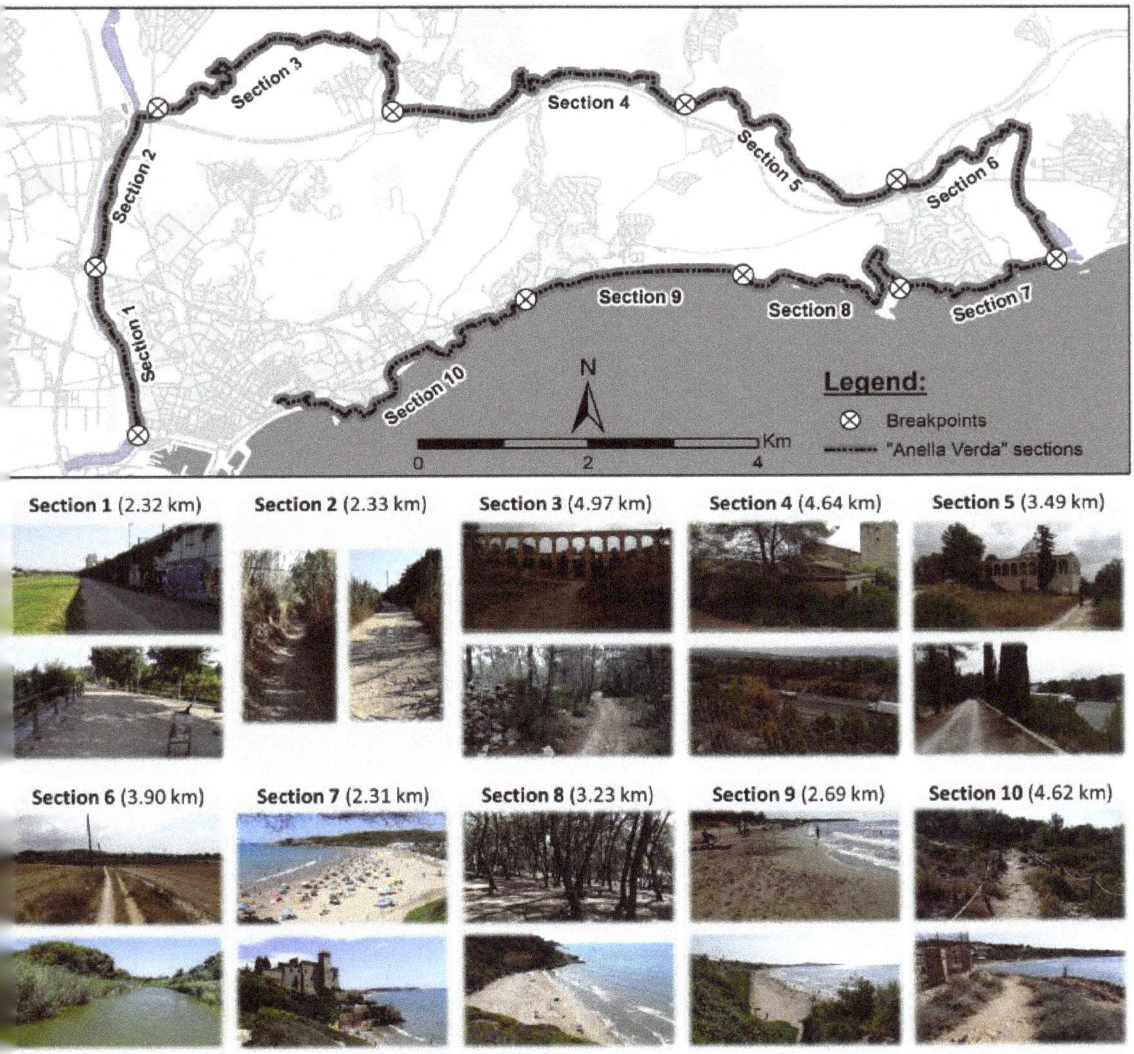

Figure 3. Mapping of the greenway sections and some representative photographs (own elaboration) of each one.

The interior sections display differences in land use, unique cultural elements, and t distance to fast traffic routes. Thus, in S3, there is the Roman aqueduct of Ferreres, and the first part of S6, the Roman quarry of Mèdol, both of which are first-rate historical a cultural attractions. In addition, examples of scattered traditional architecture (farmhous appear throughout the set of interior sections. S3 runs mostly through wooded areas a is located away from fast traffic routes, while in S4 and S5, and the first part of S6, the is a mosaic of forests, abandoned or non-abandoned agricultural areas, and low-intensi residential areas. Finally, it is noteworthy that the entirety of S4 runs parallel and adjace to fast-traffic routes.

The coastal sections are arranged from east (S7) to west (S10) and are characterized a greater or lesser degree of urbanization in their immediate hinterland and by the distan to the railway line. Thus, S7 starts at Tamarit Castle and corresponds to a beach surround by low-density urbanization. S8 runs through the protected natural space of the Bosc la Marquesa, which has been preserved thanks to the railway line acting as a barrier urbanization. S9 and S10 correspond to beaches, with the railway line parallel to the co and very close, with an urban or practically urban hinterland.

3.2. Factor Mapping

Figure 4a shows the factors used to assess the naturalness of the greenway, whi have been described previously. The factors that detract from the naturalness of t extensive surface landscape have different spatial distributions. Industry (F1_industry) concentrated in the western sector, while urban areas (F5_urban) have a large surface ar corresponding to the city of Tarragona adjacent to the industry and a series of random distributed patches in the entire study area (low-density residential area). The linear fact (F2_voltage, F4_transport in both variants) cross the territory horizontally, from east west, although the network of communication routes is denser in the western sector. should be noted that the transportation factor used in the proximity analysis (F4b_transp corresponds to a polygon that, in its western section, occupies a good part of the territ The factors that add naturalness to the landscape are mostly superficial in nature, from to F9, with a greater profusion in the eastern sector of the study area, most of them locat in the interior zone, except for the aesthetic coastal landscape located on the coast, also the east. Finally, it should be noted that in the river factor (F10_rivers), they are located the ends and have a vertical (N–S) disposition.

Since the GIS analysis is carried out using two approaches, visibility and proxim a visual basin has been generated for the first case (Figure 4b), which covers an area 17.53 km^2, and an influence area or buffer for the second case that delimits a polyg equidistant to 500 m from the Anella Verda and has an area of 37.98 km^2 (Figure 4c).

3.3. Trail Naturalness Assessment

The results of the naturalness index calculated for each section are different for three methodological approaches proposed (Figure 5). First, according to the GIS visibi analysis, 70% of the sections have a naturalness score of less than 0.33 and the remaini 30% (S7, S8, and S9) have an intermediate naturalness. No section with a high naturaln index is identified. Spatially, the three sections with an intermediate level of naturaln are contiguous and entirely on the coastal axis of the greenway.

Secondly, the proximity GIS analysis results follow a very similar pattern to previous one, with 80% of the sections having a low degree of naturalness. The remain 20% (S7 and S8) do not exceed a score of 0.66, placing them in an intermediate level naturalness. With this analysis, no sections appear to have a naturalness index higher th 0.66. The spatial pattern is identical to the previous one, except for S9, with a score of 0 which is located at a low level of naturalness.

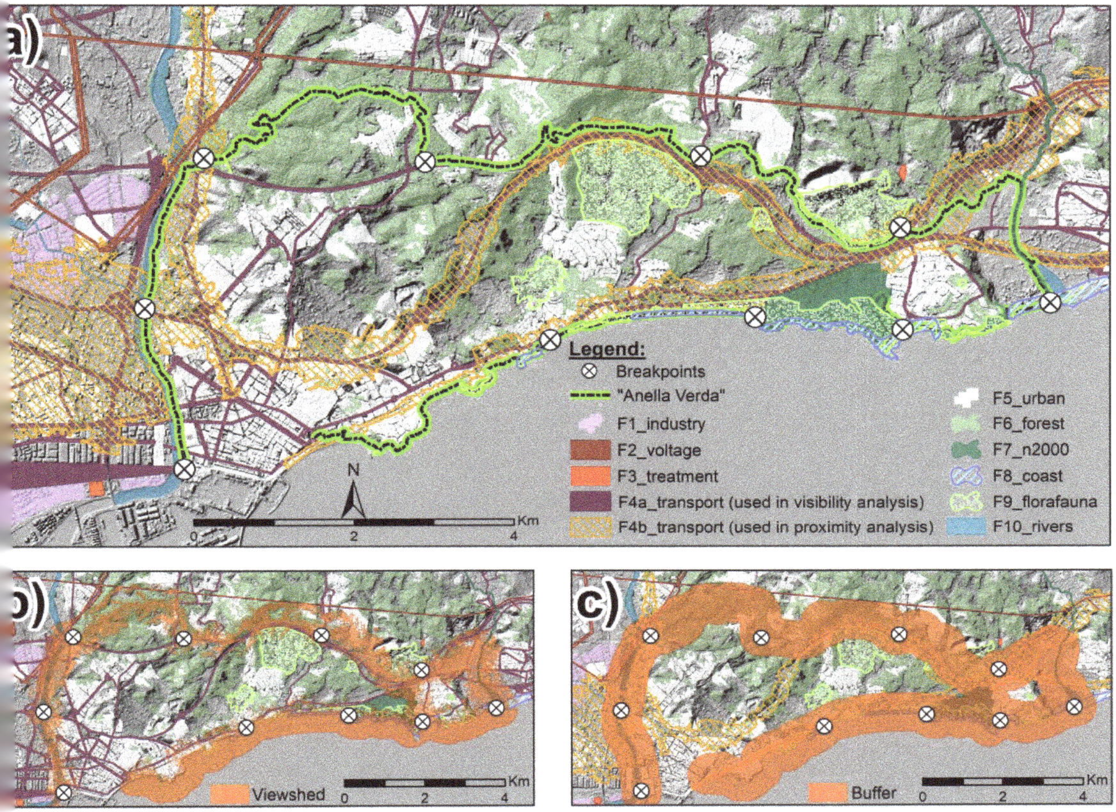

Figure 4. Mapping of the factors used (**a**), visual basin (**b**), and buffer (**c**) of all sections.

Thirdly, in the fieldwork, more significant variability is observed, both in the sections' scores and their spatial pattern. Specifically, less than half of the sections (40%) obtain a low naturalness score, half obtain an intermediate level of naturalness, and the remaining section (S8) is the only one that obtains a high naturalness score (0.70). From the spatial pattern, it can be deduced that the coastal sections have higher naturalness indices than the rest, with S8 standing out, having several positive factors (F6_forest, F7_n2000, F8_coast, F9_faunaflora, and F11_sea) that coincide spatially while lacking negative factors that reduce naturalness.

3.4. Comparison of Methods

The results obtained through the three methodological approaches to the naturalness of the greenway sections show, in general, low values of naturalness, which are consistent with the degree of human intervention in this territory. This area is located near the city of Tarragona, with many urbanized areas occupied by various human activities.

Out of the 10 sections analyzed, in half of them (S1, S2, S4, S5, and S7), the three methodological approaches coincide in the naturalness class (Figure 6). In contrast, at least two methodological approaches coincide in the other half of the sections (S3, S6, S8, S9, and S10). There was no section where the three approaches offered a different naturalness class in each one. In all sections, at least two approaches coincide.

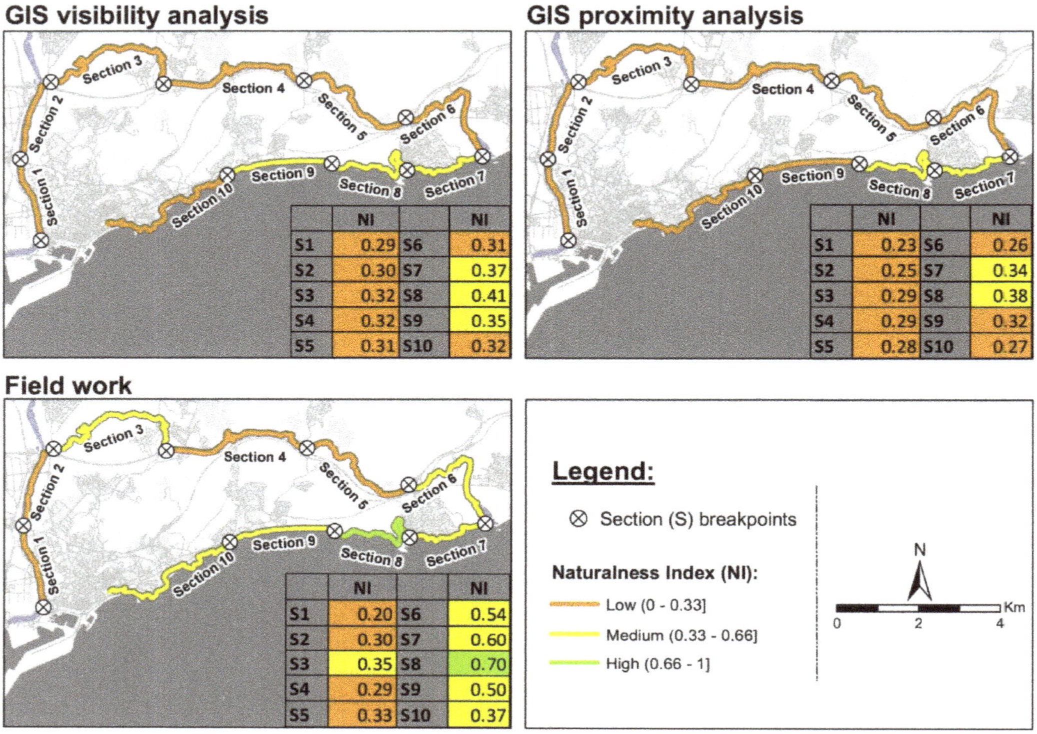

Figure 5. The naturalness index (NI) obtained for each section (S) according to the three methodological approaches used.

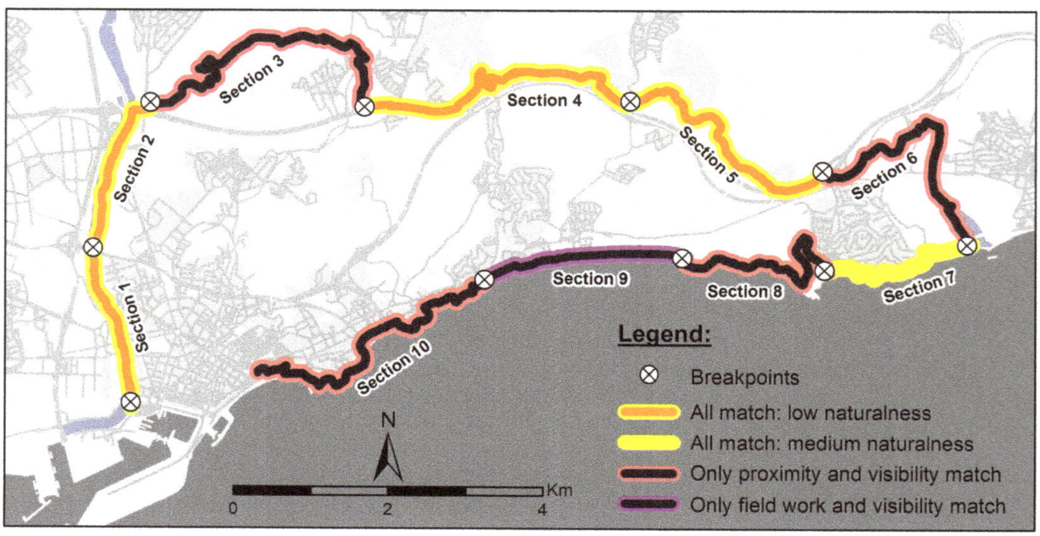

Figure 6. Coincidence in naturalness class of the sections according to the different methodological approaches used.

Sections S1, S2, S4, and S5 have the same value of naturalness, which is low, meaning they have little naturalness. The main feature that identifies them is the high presence of low-density residential areas, except in S2 where the low naturalness is due to industrial areas and power lines.

In section S7, the three methodological approaches also agree. However, this time the naturalness increases to the medium class due to the lower residential urbanization and the increase in positive elements such as the presence of forests, protected areas, coastal landscape, and relevant species of flora and fauna.

The coincidence of the degree of naturalness, according to the three approaches, has been observed primarily in the eastern area (highly urbanized), the northern area, and one section in the southern area. On the other hand, the southern area is the most heterogeneous in terms of naturalness results. Of the four sections in this area, only one is coincident in the three approaches, while the other three (S8, S9, and S10) only coincide in two. S10 and S8 are coincident in the methodological approaches with GIS (proximity and visibility), in which the values of natural and water landscapes dominate. In contrast, in S9, where the results of visibility and fieldwork coincide, the presence of the sea is significant.

4. Discussion

The information associated with an official trail usually does not provide an indicator that quantifies its degree of naturalness; the signs, brochures, and other available documentation typically only describe the characteristics of the fauna and flora of the greenway in question [55]. The availability of information on the degree of naturalness of a trail when choosing a greenway or a section of it to travel would be very useful for users, but especially for the profile of hiker interested in appreciating and enjoying nature. Despite the advantages that knowledge of the degree of naturalness of a trail could provide its users, a limited number of studies address this topic (e.g., [36,37,39,40,45,56]).

This study demonstrates that, although the administration promotes the analyzed greenway as "green" and "natural", it does not have the degree of naturalness that a nature-minded hiker would likely expect to find. In this regard, similarities have been found with the study by Vías and Ocaña [45], in which a multi-criteria evaluation model integrated into a GIS approach was used to assess the suitability of a network of trails for hiking. One of the criteria included in the generated model was natural interest. As a result of this criterion, they found that more than three-quarters of the analyzed sections had a below-average natural interest, which can be considered a low value in the context of that study area (the Sierra de las Nieves natural protected area in southern Spain). Similarly, in the work by Pavão et al. [37], the hypothesis was made that natural forests could have a relevant role as a tourist attraction for trails in the Azores (Portugal). To demonstrate this hypothesis, a quantitative analysis based on GIS was applied to determine the predominant types of cover found along these trails. They concluded that the predominant ground cover along the routes in this nature tourism destination correspond to anthropogenic landscapes of low naturalness (e.g., reforested forests, artificial pastures, or invasive forests).

Table 3 summarizes the applied methodologies in this study. Regarding the analysis speed, establishing the perception of naturalness through fieldwork is considered the slowest method. On the other hand, the fastest method is proximity to the trail using GIS tools. More layers are needed for the visibility analysis (the third method used), and the calculation of visual basins requires more processing time. The cost behavior also shows differences between fieldwork and GIS methods. The former requires a high and continuous budget for personnel to carry out data collection campaigns, while in GIS, the investment is initial and dedicated to the purchase of hardware, since free GIS software can be used, and the necessary information layers are usually available free of charge from institutions responsible for cartography. This cost difference is accentuated if we consider that, although computer equipment is not necessary during the collection of information in perceptual fieldwork, it is used in the subsequent treatment of the data.

Table 3. Comparison of the three analyses used.

	Perceptual Fieldwork	GIS Visibility	GIS Proximity
Implementation speed	Slow	Moderate	Fast
Costs	High and continuous	Medium and initial	Medium and initial
Conditioning factors in the selection of factors	Low	High	Medium
Risk of variation in the evaluation of the factors	High	Low	Low

One crucial aspect to consider is the selection of factors that will be part of the analy model for the degree of naturalness. In the case of perceptual fieldwork, the limitatio in selecting factors are very low since, based on perception, any factor can be collect such as bad smells or the presence of garbage. The only limitation detected is based on t time dedicated to this task, which is very demanding. In the case of GIS, the selection factors is based on technical criteria considering the availability and scale of cartograp which for some factors are difficult to find or there is an absence of information or it is n at an appropriate scale. Furthermore, in visibility analysis with GIS, these limitations a increased since, in addition to obtaining information layers, an additional layer, the digi surface model (DSM), is required to calculate the visual basins. In the selection of factors is considered that fieldwork is more versatile since it can incorporate factors easily sinc is only based on perception. Due to the characteristics of perception-based fieldwork, t risk of variations in the assessment of factors is very high since, despite having detail descriptors for each category of the range, external factors such as weather conditions the experience and judgment of the technicians performing the assessment can be differe This situation is diametrically opposed to the use of GIS, which obtains equivalent resu for any section and situation by applying the same parameters.

Regarding the initial research question, the results obtained from comparing t methods show that if a quick analysis is required, the best option is to use a GIS-bas proximity analysis. However, if the time and budget are available, the fieldwork techniq can be used, providing greater flexibility when selecting factors.

According to the results obtained from this study, it is not possible to affirm t GIS can be a complete substitute for fieldwork, but rather that both methodologies complementary. The three methods applied are valid independently when measuring degree of naturalness of a trail. However, GIS would provide a quick and objective analy in an integrated methodology. At the same time, fieldwork could introduce, on the c hand, factors that are not usually available in cartographic layers, such as the presence trash or foul odors and, on the other hand, information related to the user's emotions experiences, such as the attractiveness of a picturesque place or the tranquility of a for Other authors have already used this complementarity between GIS and fieldwork in ot contexts, leading to satisfactory results (e.g., [57–59]).

In any case, it is considered that the administration should expand the availa information on the degree of naturalness of the trails with the intention that when user makes their choice, it fits their motivations and preferences to increase their le of satisfaction [11]. In addition to increasing satisfaction, having this information encourage outdoor recreational and sports activities with potentially positive effects on health and well-being of people.

The main methodological limitations of this study are related to the segmentation the original greenway into sections and the area of influence used. In this work, crit related to the longitudinal homogeneity of the sections and their characteristics are u to segment the trail. Other works, such as Vías and Ocaña [45], use intersections w other trails to establish the different section breakpoints. Therefore, depending on how criterion is established to locate these section breakpoints, the segmentation of the itiner can directly influence the final calculations of naturalness for each section. Additionally, in the specific case of the GIS approach, determining the area of influence (both for visibility analysis and proximity analysis), which in this study was 500 m, can also affect the results.

Additionally, it is essential to consider that the adaptation of the factors to each specific case study limits the replicability of this method. That is, depending on the characteristics of the analyzed greenway and the study area in which it is located, the factors to include, both for GIS analysis and fieldwork analysis, may vary. This makes it difficult to replicate the method in another study area without first conducting a diagnosis of the factors present to customize the selection. For example, the same factors cannot be used to measure the degree of naturalness of a path that runs through a desert area, a forest, or an inland area versus a coastal area.

5. Conclusions

This study contributes to the international literature by adapting multi-criteria analysis techniques to calculate and map the level of naturalness of greenways, a field scarcely explored so far. It also provides a comparative analysis of the data collection methods used to detect and discuss their strengths and weaknesses.

The three analyses applied in this study (two through GIS and the other with fieldwork) are shown to be valid for calculating the degree of naturalness of a greenway. However, from the discussion of the results, it is concluded that integrating both techniques (GIS and fieldwork) can lead to more precise and detailed results in trail naturalness assessment.

The two GIS analyses obtain similar naturalness indices for this case study, although the values are consistently lower than those derived from the fieldwork analysis. Additionally, although the greenway analyzed in this study is promoted as "green", the resulting naturalness indices are generally low in all three methods. This suggests that this type of itinerary or trail only sometimes meets expectations, meaning it does not have the degree of naturalness that users would expect to find. Moreover, it would be interesting for the local administration to consider these results to, on the one hand, transfer the level of naturalness of each section to the users of this greenway (for example, with the implementation of informative panels or signs on the greenway). On the other hand, if appropriate, implement the necessary measures to increase the levels of naturalness of the sections that require it, primarily to mitigate the negative impacts potentially produced by the factors considered in this study as negative or intensive use.

The results obtained from both the analyzed greenway and the replication of this methodology in other study areas could be of great use for land and outdoor recreation managers. On the one hand, it could help them design trails by selecting those segments with the highest naturalness, and on the other hand, it could help them promote these trails. For hikers, it means having more and higher quality information about the naturalness of greenways for their decision-making process before using a particular greenway or section thereof.

Author Contributions: All authors (A.À.C., Y.P.-A. and J.V.M.) contributed equally to conceptualizing, writing, and reviewing this work. All authors have read and agreed to the published version of the manuscript.

Funding: This research was funded by MCIN/AEI/10.13039/501100011033 (grant number PID2020-114363GB-I00) and the GRATET Research Group, which is funded by the Catalan Government under code 2021-SGR-00657.

Data Availability Statement: Not applicable.

Acknowledgments: This article has been possible with the support of the grant PRE2018-084802 funded by MCIN/AEI/10.13039/501100011033 and by "ESF Investing in your future".

Conflicts of Interest: The authors declare no conflict of interest. The funders had no role in the design of the study; in the collection, analyses, or interpretation of data; in the writing of the manuscript; or in the decision to publish the results.

References

1. Li, H.; Zhang, X.; Bi, S.; Cao, Y.; Zhang, G. Psychological benefits of green exercise in wild or urban greenspaces: A meta-anal of controlled trials. *Urban For. Urban Green.* **2022**, *68*, 127458. [CrossRef]
2. WHO. *Guidelines on Physical Activity and Sedentary Behaviour*; World Health Organ: Genève, Switzerland, 2020; pp. 1–582.
3. Kerr, J.; Marshall, S.; Godbole, S.; Neukam, S.; Crist, K.; Wasilenko, K.; Golshan, S.; Buchner, D. The relationship between outc activity and health in older adults using GPS. *Int. J. Environ. Res. Public Health* **2012**, *9*, 4615–4625. [CrossRef] [PubMed]
4. Breitenstein, D.; Ewert, A. Health benefits of outdoor recreation: Implications for health education. *Health Educ.* **1990**, *21*, 16 [CrossRef]
5. Barton, J.; Pretty, J. The beneficial effects of green exercise on health. In *Forest Medicine*; Nova Science Publishers, Inc.: Hauppa NY, USA, 2013; pp. 201–219. ISBN 9781626184626.
6. Calogiuri, G.; Elliott, L.R. Why do people exercise in natural environments? Norwegian adults' motives for nature-, gym-, sports-based exercise. *Int. J. Environ. Res. Public Health* **2017**, *14*, 377. [CrossRef] [PubMed]
7. Hill, E.; Gómez, E. Perceived health outcomes of mountain bikers: A national demographic inquiry. *J. Park Recreat. Adm.* **202**C 31–44. [CrossRef]
8. Davies, N.J.; Lumsdon, L.M.; Weston, R. Developing Recreational Trails: Motivations for Recreational Walking. *Tour. Plan.* **2012**, *9*, 77–88. [CrossRef]
9. Fumagalli, N.; Maccarini, M.; Rovelli, R.; Berto, R.; Senes, G. An exploratory study of users' preference for different plan combinations along rural greenways. *Sustainability* **2020**, *12*, 2120. [CrossRef]
10. Senes, G.; Rovelli, R.; Bertoni, D.; Arata, L.; Fumagalli, N.; Toccolini, A. Factors influencing greenways use: Definition of a met for estimation in the Italian context. *J. Transp. Geogr.* **2017**, *65*, 175–187. [CrossRef]
11. Obradović, S.; Tešin, A. Hiking in the COVID-19 era: Motivation and post-outbreak intentions. *J. Sport Tour.* **2022**, *26*, 147- [CrossRef]
12. Chen, X.; de Vries, S.; Assmuth, T.; Dick, J.; Hermans, T.; Hertel, O.; Jensen, A.; Jones, L.; Kabisch, S.; Lanki, T.; et al. Rese challenges for cultural ecosystem services and public health in (peri-)urban environments. *Sci. Total Environ.* **2019**, *651*, 2118– [CrossRef]
13. Keith, S.J.; Larson, L.R.; Shafer, C.S.; Hallo, J.C.; Fernandez, M. Greenway use and preferences in diverse urban commun Implications for trail design and management. *Landsc. Urban Plan.* **2018**, *172*, 47–59. [CrossRef]
14. Komossa, F.; Wartmann, F.M.; Kienast, F.; Verburg, P.H. Comparing outdoor recreation preferences in peri-urban landsc using different data gathering methods. *Landsc. Urban Plan.* **2020**, *199*, 103796. [CrossRef]
15. Komossa, F.; van der Zanden, E.H.; Verburg, P.H. Characterizing outdoor recreation user groups: A typology of peri-u recreationists in the Kromme Rijn area, the Netherlands. *Land Use Policy* **2019**, *80*, 246–258. [CrossRef]
16. Žlender, V.; Ward Thompson, C. Accessibility and use of peri-urban green space for inner-city dwellers: A comparative s *Landsc. Urban Plan.* **2017**, *165*, 193–205. [CrossRef]
17. Brownson, R.C.; Housemann, R.A.; Brown, D.R.; Jackson-Thompson, J.; King, A.C.; Malone, B.R.; Sallis, J.F. Promoting phy activity in rural communities: Walking trail access, use, and effects. *Am. J. Prev. Med.* **2000**, *18*, 235–241. [CrossRef]
18. Corning, S.E.; Mowatt, R.A.; Chancellor, H.C. Multiuse trails: Benefits and concerns of residents and property owners. *J. U Plan. Dev.* **2012**, *138*, 277–285. [CrossRef]
19. Evenson, K.R.; Herring, A.H.; Huston, S.L. Evaluating change in physical activity with the building of a multi-use trail. *A Prev. Med.* **2005**, *28*, 177–185. [CrossRef] [PubMed]
20. Neff, L.J.; Ainsworth, B.E.; Wheeler, F.C.; Krumwiede, S.E.; Trepal, A.J. Assessment of trail use in a community park. *Community Health* **2000**, *23*, 76–84. [CrossRef]
21. Schasberger, M.G.; Hussa, C.S.; Polgar, M.F.; McMonagle, J.A.; Burke, S.J.; Gegaris, A.J., Jr. Promoting and developing a network across suburban, rural, and Urban communities. *Am. J. Prev. Med.* **2009**, *37*, S336–S344. [CrossRef] [PubMed]
22. Wan Omar, W.R.; Patterson, I.; Pegg, S. A Green Pathway for Future Tourism Success: Walking Trails in Kuala Lumpur. *Tour. Dev.* **2012**, *9*, 57–76. [CrossRef]
23. Wang, G.; Macera, C.A.; Scudder-Soucie, B.; Schmid, T.; Pratt, M.; Buchner, D. Cost effectiveness of a bicycle/pedestrian development in health promotion. *Prev. Med.* **2004**, *38*, 237–242. [CrossRef] [PubMed]
24. Xie, B.; Lu, Y.; Zheng, Y. Casual evaluation of the effects of a large-scale greenway intervention on physical and mental heal natural experimental study in China. *Urban For. Urban Green.* **2022**, *67*, 127419. [CrossRef]
25. Park, T.; Eyler, A.A.; Tabak, R.G.; Valko, C.; Brownson, R.C. Opportunities for Promoting Physical Activity in Rural Commu by Understanding the Interests and Values of Community Members. *J. Environ. Public Health* **2017**, *2017*, 8608432. [Cros [PubMed]
26. Farías-Torbidoni, E.I. Managing for recreational experience opportunities: The case of hikers in protected areas in Cata Spain. *Environ. Manag.* **2011**, *47*, 482–496. [CrossRef]
27. Farías-Torbidoni, E.I.; Grau, R.; Camps, A. Trail preferences and visitor characteristics in Aigüestortes i Estany de Sant M National park, Spain. *Mt. Res. Dev.* **2005**, *25*, 51–59. [CrossRef]
28. Farías-Torbidoni, E.I.; Mas-Alòs, S.; Gil-Moreno-de-Mora, G.; Lavega-Burgués, P.; Castañer, M.; Lorente-Catalán, E.; S Urbaneja, J.; Lacasa-Claver, E. Health and Well-Being in Protected Natural Areas—Visitors' Satisfaction in Three Diff Protected Natural Area Categories in Catalonia, Spain. *Int. J. Environ. Res. Public Health* **2020**, *17*, 6746. [CrossRef] [PubMe

Vías, J.; Rolland, J.; Castillo, S.; del Campo-Ávila, J.; Luque, A. Creación de rutas senderistas mediante análisis de redes y algoritmo multicriterio. Aplicación en el parque natural Sierra de las Nieves. In Proceedings of the XVI Congreso Nacional de Tecnologías de Información Geográfica, Alicante, Spain, 25–27 June 2014; pp. 947–953.
Watts, G.; Bauer, J. Tranquillity trails–design, implementation and benefits for healthy leisure. *World Leis. J.* **2021**, *64*, 156–165. [CrossRef]
Cordeiro, B.; Alves, L. Protected areas, tourism and hiking trails. The hiker's profile in Serra da Lousã (Portugal). *PASOS Rev. Tur. Y Patrim. Cult.* **2022**, *20*, 939–949. [CrossRef]
Mayer, K.; Lukács, A. Motivation and mental well-being of long-distance hikers: A quantitative and qualitative approach. *Heliyon* **2021**, *7*, e06960. [CrossRef]
Saayman, M.; Viljoen, A. Who are wild enough to hike a wilderness trail? *J. Outdoor Recreat. Tour.* **2016**, *14*, 41–51. [CrossRef]
Gaffar, V.; Yuniawati, Y.; Ridwanudin, O. A Study of Outdoor Recreation Motivation and Activity Preferences. *Xinan Jiaotong Daxue Xuebao/J. Southwest Jiaotong Univ.* **2019**, *54*, 3. [CrossRef]
Lee, J.; Lee, H.-S.; Jeong, D.; Shafer, C.S.; Chon, J. The relationship between user perception and preference of greenway trail characteristics in urban areas. *Sustainability* **2019**, *11*, 4438. [CrossRef]
Chon, J.; Shafer, C.S. Aesthetic responses to urban greenway trail environments. *Landsc. Res.* **2009**, *34*, 83–104. [CrossRef]
Pavão, D.C.; Porteiro, J.; Ventura, M.A.; Borges Silva, L.; Medeiros, A.; Moniz, A.; Moura, M.; Moreira, F.; Silva, L. Land cover along hiking trails in a nature tourism destination: The Azores as a case study. *Environ. Dev. Sustain.* **2021**, *23*, 16504–16528. [CrossRef]
Fredman, P.; Wall-Reinius, S.; Grundén, A. The Nature of Nature in Nature-based Tourism. *Scand. J. Hosp. Tour.* **2012**, *12*, 289–309. [CrossRef]
Oishi, Y. Toward the improvement of trail classification in national parks using the recreation opportunity spectrum approach. *Environ. Manag.* **2013**, *51*, 1126–1136. [CrossRef]
Malec, M.; Kędzior, R.; Ziernicka-Wojtaszek, A. The Method of Soundscape Naturalness Curves in the Evaluation of Mountain Trails of Diversified Anthropopressure—Case Study of Korona Beskidów Polskich. *Sustainability* **2023**, *15*, 723. [CrossRef]
Molokáč, M.; Hlaváčová, J.; Tometzová, D.; Liptáková, E. The Preference Analysis for Hikers' Choice of Hiking Trail. *Sustainability* **2022**, *14*, 6795. [CrossRef]
Beeco, J.A.; Hallo, J.C.; Brownlee, M.T.J. GPS Visitor Tracking and Recreation Suitability Mapping: Tools for understanding and managing visitor use. *Landsc. Urban Plan.* **2014**, *127*, 136–145. [CrossRef]
Khazaee Fadafan, F.; Soffianian, A.; Pourmanafi, S.; Morgan, M. Assessing ecotourism in a mountainous landscape using GIS MCDA approaches. *Appl. Geogr.* **2022**, *147*, 102743. [CrossRef]
Peterson, B.A.; Brownlee, M.T.J.; Marion, J.L. Mapping the relationships between trail conditions and experiential elements of long-distance hiking. *Landsc. Urban Plan.* **2018**, *180*, 60–75. [CrossRef]
Vías, J.; Ocaña, C. Multicriteria evaluation by GIS to determine trail hiking suitability in a natural park. *Bol. La Asoc. Geogr. Esp.* **2014**, *66*, 323–339.
Vías, J.; Rolland, J.; Gómez, M.L.; Ocaña, C.; Luque, A. Recommendation system to determine suitable and viable hiking routes: A prototype application in Sierra de las Nieves Nature Reserve (southern Spain). *J. Geogr. Syst.* **2018**, *20*, 275–294. [CrossRef]
Àvila Callau, A. *El Análisis de los Servicios Ecosistémicos Culturales A Partir de Información Geográfica Voluntaria. El Caso del Periurbano De Tarragona*; Universitat Rovira i Virgili: Tarragona, Spain, 2022.
Àvila Callau, A.; Pérez-Albert, Y.; Serrano Giné, D. Quality of GNSS Traces from VGI: A Data Cleaning Method Based on Activity Type and User Experience. *ISPRS Int. J. Geo-Inf.* **2020**, *9*, 727. [CrossRef]
Observatori del Paisatge. *Catàleg de Paisatge del Camp de Tarragona*; Observatori Del Paisatge: Olot, Spain, 2010.
Booth, P.N.; Law, S.A.; Ma, J.; Buonagurio, J.; Boyd, J.; Turnley, J. Modeling aesthetics to support an ecosystem services approach for natural resource management decision making. *Integr. Environ. Assess. Manag.* **2017**, *13*, 926–938. [CrossRef] [PubMed]
Klouček, T.; Lagner, O.; Šímová, P. How does data accuracy influence the reliability of digital viewshed models? A case study with wind turbines. *Appl. Geogr.* **2015**, *64*, 46–54. [CrossRef]
Noblejas, H.C.; Martínez, J.V.; Rodríguez, M.F.M. Relation between the Views and the Real Estate Application to a Mediterranean Coastal Area. *ISPRS Int. J. Geo-Inf.* **2022**, *11*, 365. [CrossRef]
Pellicer, I.; Estornell, J.; Martí, J. Application of airborne LiDA R data in viewshed analysis. *Rev. Teledetec.* **2014**, *41*, 9–18. [CrossRef]
Malczewski, J. *GIS and Multicriteria Decision Analysis*; John Wiley & Sons: New York, NY, USA, 1999; ISBN 0471329444.
Weis, K.; Hronček, P.; Tometzová, D.; Gregorová, B.; Přibil, M.; Jesenský, M.; Čech, V. Analysis of notice boards (Panels) as general information media in the outdoor mining tourism. *Acta Montan. Slovaca* **2019**, *24*, 269–283.
Getzner, M.; Meyerhoff, J. The benefits of local forest recreation in austria and its dependence on naturalness and quietude. *Forests* **2020**, *11*, 326. [CrossRef]
Babaie-Kafaky, S.; Mataji, A.; Sani, N.A. Ecological capability assessment for multiple-use in forest areas using GIS- based multiple criteria decision making approach. *Am. J. Environ. Sci.* **2009**, *5*, 714–721. [CrossRef]

58. Duran-Llacer, I.; Arumí, J.L.; Arriagada, L.; Aguayo, M.; Rojas, O.; González-Rodríguez, L.; Rodríguez-López, L.; Martín Retureta, R.; Oyarzún, R.; Singh, S.K. A new method to map groundwater-dependent ecosystem zones in semi-arid environmen A case study in Chile. *Sci. Total Environ.* **2022**, *816*, 151528. [CrossRef] [PubMed]
59. Gupta, M.; Srivastava, P.K. Integrating GIS and remote sensing for identification of groundwater potential zones in the hi terrain of Pavagarh, Gujarat, India. *Water Int.* **2010**, *35*, 233–245. [CrossRef]

forests

MDPI

cle

reen Infrastructure Network Identification at a Regional Scale: ıe Case of Nanjing Metropolitan Area, China

Liu [1], Hao Xu [1,*], Xiaotong Zhang [2,3] and Wenqi Jiang [1]

[1] College of Landscape Architecture, Nanjing Forestry University, Nanjing 210037, China; liuwei2019@njfu.edu.cn (W.L.); j13033051382@163.com (W.J.)
[2] Shanghai Key Lab for Urban Ecological Processes and Eco-Restoration, Shanghai 200241, China; xtzhang0711@outlook.com
[3] College of Ecological and Environmental Sciences, East China Normal University, Shanghai 200241, China
* Correspondence: hao_xu@njfu.edu.cn

Abstract: Clustered urban development has caused increasing fragmentation and islanding of regional ecological spaces. Creating a green infrastructure network (GIN) is a practical method of ensuring regional ecological security. This study proposed a method of GIN identification at the regional scale based on the Nanjing Metropolitan Area as an example. In this method, green hubs were identified using morphological spatial pattern analysis and connectivity indexes, green corridors were simulated based on the least-cost path model, and key optimization nodes were identified using circuit theory. The results indicated that green hubs covered an area of 5042.07 km^2, of which, 15.40% were cross-border, and the potential corridors were distributed in a network, with the key ecological nodes primarily narrowly situated. By comparing the hubs with the statutory green space protection area and the urban ecological control line, the identification results were more than 70% accurate, showing that the results were valid and reliable. This method not only made the identification of regional GIN more practical and replicable but also further identified key areas that need priority protection. This study provides a method for constructing regional GIN and serves as a strong guide for ecological and development planning of other urban clustered areas.

Keywords: green infrastructure; ecological network; regional scale; morphological spatial pattern analysis; circuit theory; Nanjing Metropolitan Area

on: Liu, W.; Xu, H.; Zhang, X.; W. Green Infrastructure ›rk Identification at a Regional The Case of Nanjing politan Area, China. *Forests* 3, 735. https://doi.org/)/f13050735

mic Editors: Panayiotis ıitrakopoulos, Mario notta, Miklas Scholz :shiya Noorani

ed: 22 March 2022
:ed: 6 May 2022
ıed: 9 May 2022

her's Note: MDPI stays neutral gard to jurisdictional claims in ıed maps and institutional affil-

1. Introduction

Green infrastructure (GI) refers to natural or semi-natural green spaces [1], including forests, grasses, and parks. As one of the main habitats for animals, GI plays an important role in maintaining biodiversity and preserving the integrity of ecosystems [2]. However, with massive urbanization, GI has been encroached upon, which brings about serious fragmentation, islanding, and declining connectivity [3]. Of particular concern is the rise in clustered urban development, such as urban agglomerations and metropolitan areas, where rapid urban expansion and intense human activities have led to increasing depletion of GI, severe loss of biodiversity, and serious threats to the structure and function of green ecosystems [4,5]. Building a green infrastructure network (GIN) is ecologically important for biodiversity conservation and is an effective means of enhancing landscape connectivity [6,7].

Being able to identify a GIN has attracted widespread attention, and a research paradigm of "hubs identification-ecological corridors construction-network optimization" has progressively emerged [8–10]. Hubs provide habitats for animal survival and migration, and they are normally identified based on the biological habitat needs in terms of function or scale dimensions through ecosystem service evaluation, ecological sensitivity evaluation, landscape connectivity evaluation, etc. [11–13]. Corridors are areas in ecological networks that provide passage for animal migration. They are mostly constructed based on focal species, whose spatial locations, orientations, and widths are identified using mo simulations such as the least-cost path, the minimum cumulative resistance, graph the and the ant colony algorithm [14–16]. Ecological nodes are key strategic points that h an important impact on ecosystem stability and connectivity, and most studies have lected ecologically sensitive areas or corridor narrows as ecological nodes through a spa overlay with landscape components, such as road networks and construction land [7]. evident that most of the existing studies on GINs are constructed with focal species as target, and the effectiveness of this construction technique was demonstrated [17,18].

However, the existing methods for identifying GINs are mostly applied at the ur scale and confined to specific administrative boundaries. Fewer studies have been car out at the regional scale. In fact, the construction of GINs is not entirely constrai by administrative boundaries. The spatial continuity of green space patterns and spatial mobility of ecosystem services dictate that the identification of a GIN requ the consideration of relevant influencing factors within a larger natural geograph context [16]. Traditional focal-species-based approaches to GIN construction are somew limited at the regional scale. First, the larger the area, the richer the species, and habitat requirements of focal species and dispersal paths may not be representative o species [19]. Second, animal dispersal behavior analysis is key to the use of this meth but the dispersal behavior of species is mostly uncertain [20]. The GI resource endowm regional physical geography, and socio-economic conditions at the regional scale are e more variable, which could cause species dispersal behavior to be more unpredicta Third, the method requires a large amount of data, including detailed species survey and habitat quality assessment data, and is complex to calculate. As a result, its applica to large-scale GIN construction takes a lot of time and effort [21].

The identification of hubs, corridors, and key ecological nodes at the regional s is conducive to building a GIN that breaks through administrative boundary restricti and can offer effective guidelines for building a GIN at the urban scale [17,22]. So scholars have realized the importance of regional GINs and investigated the meth for their construction. The morphological spatial pattern analysis (MSPA) approac based on graph theory [23] and uses image processing to identify GI elements that an important role in maintaining ecological network connectivity [24]. This approach fewer data, emphasizes the structural connectivity of ecological networks, and prese the continuity and integrity of landscape patterns; therefore, it has been applied in construction of national- and urban-scale GINs [13,25]. The circuit theory (CT) def the dispersal behavior of species as a stochastic behavior similar to the motion of ele charges, identifying ecological networks by assigning different ecological meaning physical quantities, such as resistance, current, and voltage [26]. Without a need for identification of focal species, CT could identify the corridors that meet the migratory n of multiple species, thus being more suitable for species dispersal characteristics [27] an broadly adaptable at several spatial scales [21,28].

The application of MSPA and CT has promoted the identification of GINs at regional scale, but they have advantages in hubs selection and corridors identifica respectively. The combination of these two methods to identify regional GINs not avoids the subjective interference of artificial hub selection but also obtains green corri with structural connectivity and functional connectivity [7,29]. Therefore, the MSPA CT method combined to study regional GINs is more scientific, which would provide practical and accurate references for regional ecological planning. However, few stu have combined both methods [30], and relevant studies at the regional scale are scarcer. Therefore, taking the Nanjing Metropolitan Area in China as an example, this s combined MSPA and CT to construct a methodological framework for identifying a regi GIN, including how to identify regional green hubs, how to create a resistance surface, how to construct green corridors and identify key points; furthermore, the accuracy c identified GIN is discussed. This study offers a detailed methodological framework fo construction of regional-scale GINs and a new perspective for regional-scale GIN plan

2. Materials and Methods

2.1. Study Area

Located in the lower reaches of the Yangtze River of China, the Nanjing Metropolitan Area (29°57′ N–34°06′ N, 117°09′ E–119°59′ E) spans the provinces of Jiangsu and Anhui and consists of Nanjing, Zhenjiang, Yangzhou, Huai'an, Chuzhou, Ma'anshan, Wuhu, Xuancheng, and Changzhou (including Jintan District and Liyang District only), with a total area of approx. 66,000 km^2 (Figure 1). Principally located in the subtropical monsoon climate zone, this region features hot summers, warm winters, and abundant precipitation. The terrain is dominated by low hills and alluvial plains, which are higher in the south and lower in the north. The region is endowed with an excellent natural background, with mountains such as Mt. Huangshan, Mt. Tianmu, and Mt. Jiuhua in the south, and Ningzhen, Laoshan, and Maoshan mountain ranges in the center.

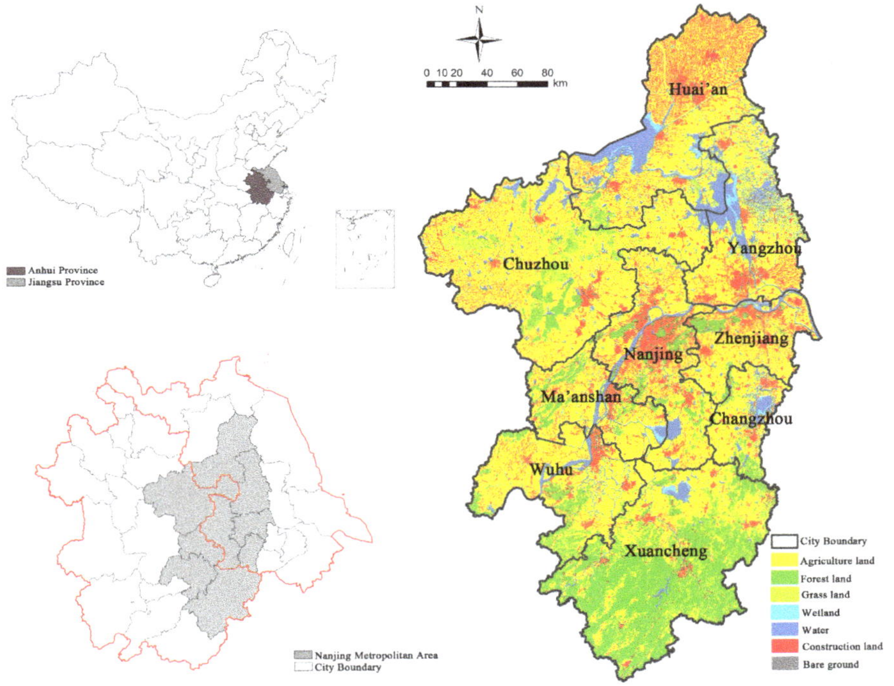

Figure 1. Location of the study area.

As a pivotal region linking eastern and central China, the Nanjing Metropolitan Area is an important part of China's Yangtze River Delta City Agglomeration and the first metropolitan area set up across provinces in China. By the end of 2019, it had a resident population of approx. 35 million and a gross domestic product per capita of RMB 113,000 [31], making it one of the most economically developed regions in China. In recent years, the vigorous development and construction of the Nanjing Metropolitan Area have brought about serious erosion of GI, a reduction in biodiversity, and increasing pressure on the regional ecological environment. Under the double pressure of clustered urban development and green spaces protection, Jiangsu and Anhui provinces jointly proposed to build a network of green corridors in Nanjing metropolitan area [31], providing a win–win solution for accelerating urbanization and protecting ecosystems. The results of this study can serve as a direct reference and data support for the construction of a regional GIN in the study area.

2.2. Data

The data sources used in this study principally included the 30 m resolution lar cover classification data from 2020 (from http://www.globallandcover.com/ (accesse on 10 October 2021)) and the 30 m resolution DEM digital elevation model data (fro http://www.gscloud.cn/ (accessed on 15 November 2021)). The study also referred to tl Regional Planning for Ecological Space Control in Jiangsu Province (from http://ww jiangsu.gov.cn (accessed on 15 November 2021)) and the Ecological Protection Control Li of Anhui Province (from https://www.ah.gov.cn/ (accessed on 15 November 2021)) extract the ecological protection control line within the study area.

The land cover classification data of Nanjing Metropolitan Area was divided into ba ground, shrub, grass, forest, agriculture, construction land, wetland, and water (Figure To facilitate identification, shrub, grass, forest, and wetland were defined as the GI. view of the fact that the vegetation type of agricultural land was relatively homogeneo mostly dominated by cash crops with low biodiversity maintenance value [32], and th agricultural land is subject to a separate and strict protection regime in China, the stu did not include agriculture within the GI.

2.3. Methods

In this study, a regional-scale GIN construction framework was proposed based the research mode of "hubs identification-corridors construction-network optimizatio by integrating approaches such as MSPA, least-cost path (LCP), and CT (Figure 2). T detailed process is described in the following sections.

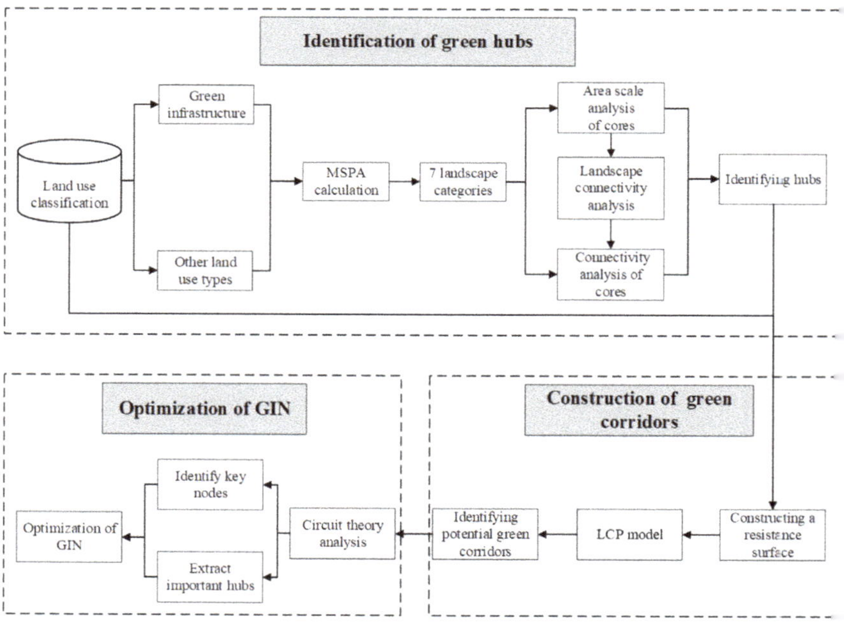

Figure 2. Flow chart of the study.

2.3.1. Identification of Landscape Elements Based on MSPA

Morphological spatial pattern analysis (MSPA) is an image-processing technique t measures, identifies, and segments the spatial pattern of a raster image based on mat matical morphological principles [33]. This method simplifies the process of determin landscape patterns and has been applied in the construction of regional-scale ecolog grids [25]. First, the land use data was reclassified and converted into binary maps us the ArcGIS software, with the foreground set to the GI and the background set to other types of land. Second, with the help of the GuidosToolbox software, the binary raster data was analyzed by using the eight-neighborhood rule (edge width set to 1) to extract 7 types of MSPA landscapes, i.e., core, bridging, branch, perforation, islet, edge, and loop, which did not overlap with each other. The cores were mostly large GI patches that offered larger habitats for species and inform the screening of green hubs.

2.3.2. Hubs Extraction Based on Scale and Connectivity

The extraction of hubs requires consideration of both the size and the connectivity of the patch. The size of the core is decisive for habitat heterogeneity and species carrying capacity and scattered small patches play a limited role in the maintenance of regional ecosystem function. Therefore, larger core patches need to be selected as candidate hubs.

Connectivity refers to the degree to which horizontal movement of organisms or ecological processes is inhibited between landscape elements [34]. Cores with high connectivity offer higher survival of organisms. The integral index of connectivity (IIC), the possibility of connectivity (PC), and the connectivity importance index (dI) were chosen to determine the connectivity of candidate hubs and to identify cores of high importance as hubs. The specific formulae used for computation are as follows:

$$\text{IIC} = \frac{\sum_{i=1}^{n}\sum_{j=1}^{n}\frac{a_i \times a_j}{1+nl_{ij}}}{A_L^2}, \tag{1}$$

where n refers to the total number of GI patches; a_i and a_j denote the area of core i and j, respectively; nl_{ij} represents the number of connections between cores i and j; and A_L stands for the total area of GI. The larger the IIC, the more suitable the core is as a habitat.

$$\text{PC} = \frac{\sum_{i=1}^{n}\sum_{j=1}^{n} P_{ij}^{*} \times a_i \times a_j}{A_L^2}, \tag{2}$$

where P_{ij}^{*} refers to the maximum product probability of all connections between cores i and j. The smaller the PC value, the lower connectivity between cores.

$$\text{dI} = \frac{I - I'}{I} \times 100\%, \tag{3}$$

where dI represents the importance of the core; I denotes the connectivity index value of the core, namely, the IIC and PC values in the study; and I' stands for the connectivity index value after the removal of a certain core. The larger the dI value, the more important the core [35].

2.3.3. Corridors Construction Based on LCP

Corridors are important pathways for species migration and are carriers of material cycles and energy flows between hubs. The least-cost path (LCP) reflects the relative minimum cost consumed by a species during their migratory dispersal, and the corridors constructed on the basis of the LCP are theoretically the best pathways for species migration [2]. In the study, the Linkage Mapper corridor simulation software was used to construct potential least-cost green corridors between hubs. The key to the LCP is the setup of landscape resistance. The ecological resistance varies depending on the nature of the landscape unit. In most cases, the higher the habitat suitability of a landscape unit, the lower the ecological resistance. In this study, based on the results of the MSPA analysis, the value of resistance was determined by considering the natural environment of the region, different land-use types, and the level of human interference. On this basis, different resistance values were assigned, where larger values represent greater resistance (Table 1). Based on 30 experts' questionnaire responses and using hierarchical analysis, the weight coefficients of four resistance factors (i.e., MSPA landscape type, land-use type, topography, and human activities) were identified. It should be notable that the resista values in Table 1 represent relative resistance only and not absolute resistance values the landscape.

Table 1. Assignments and weights of landscape resistance factors.

Resistance	Weight Coefficient	Resistance Factor		Resistance Value
MSPA landscape type	0.38	Core	Hubs	1
			Other cores	5
		Bridge		10
		Other six MSPA types		50
		Background		600
Land-use type	0.25	GI		5
		Water (km^2)	River width >100 m, and lakes area >50	1000
			River width 50–100 m, lakes area <50 and >10	400
			Lakes area <10	40
		Agriculture land		60
		Construction land		1000
		Other types of land		600

Resistance	Weight Coefficient	Resistance Factor		Resistance Value
Topography	0.15	Elevation (m)	<50	5
			50–175	20
			175–350	60
			350–650	200
			>650	600
		Slope (°)	<5	5
			5–15	20
			15–25	200
			>25	600
Human activities	0.22	Distance from built-up area (km)	>2.5	5
			1.5–2.5	100
			1.0–1.5	500
			0.5–1.0	800
			<0.5	1000
		Distance from traffic artery (km)	>3.0	5
			2.0–3.0	100
			1.5–2.0	500
			0.5–1.5	800
			<0.5	1000

2.3.4. GI Optimization Based on CT

The circuit theory (CT) abstracts organisms as randomly traveling charges, w the landscape pattern is viewed as a conducting surface, where the current reflects migration probability of organisms passing through corresponding pathways [26]. Ci theory has been extensively used for ecological point identification. Based on CT, the st employed the Linkage Mapper plug-in in GIS10.6 software to generate hub-to-hub cur densities to further identify important key areas that require priority protection or reco

1. Centrality identification

Centrality refers to the direct communication capability of the hub in the network can reflect the importance of the hub in maintaining the overall connectivity of the netw Specifically, each hub is considered a node, the least-cost path between any two hul considered a circuit, and the cost weighted distance of each path is considered a resista then, a current of 1 ampere is fed into one hub and another hub is connected, and the current value between the two hubs is calculated; finally, all hubs are iterated and final cumulative current value for each hub is the centrality of that hub [36]. The hi the centrality (current value), the more important the hub is for maintaining the ov connectivity of the GIN. The study employed the Centrality Mapper tool in the Link Mapper plug-in to work out the centrality of each hub, setting the green corridor w at a weighted cost distance of 5000 and using a weighted cost distance of 20,000 as corridor width.

2. Key ecological nodes identification

Key ecological points refer to the key nodes for material exchange between adja hubs [37], to which special attention should be given to strengthening the construction protection. In this study, key ecological points were identified using the Pinchpoint Ma tool. The areas with high currents, i.e., the areas with high degrees of ecological mob were identified by calculating the cumulative current value for each image element in the corridor and should be the key-point areas of corridors. Key-point areas are important for maintaining the connectivity of the entire GIN.

3. Results

3.1. Identification Analysis of Hubs

3.1.1. Landscape Pattern of the GI

The GI covered an area of 12,436 km^2, accounting for 19.00% of the total area of Nanjing Metropolitan Area. The core was about 7151.76 km^2, accounting for 57.51% of the GI (Table 2). Large cores were clustered in the south, while the central cores were distributed in strips. Sparse and fragmented cores were observed in the north, mostly wetlands, grasslands, or waterfront forest land (Figure 3). The bridge and branch are ecologically important for species migration and dispersal as structural linkages between cores. These two types accounted for only 14.98% of the GI, indicating that the cores were more scattered and independent of each other. Hence, it's necessary to carry out studies on the construction of regional GINs.

Table 2. Statistical description of MSPA landscape types.

Landscape Type of the GI	Area (km^2)	Proportion of the Area of the GI (%)	Proportion of the Area of the Study Region (%)
Core	7151.76	57.51	10.93
Bridge	1152.23	9.27	1.76
Islet	448.09	3.60	0.68
Perforation	148.96	1.20	0.23
Edge	2185.07	17.57	3.34
Loop	638.66	5.14	0.98
Branch	710.97	5.71	1.09

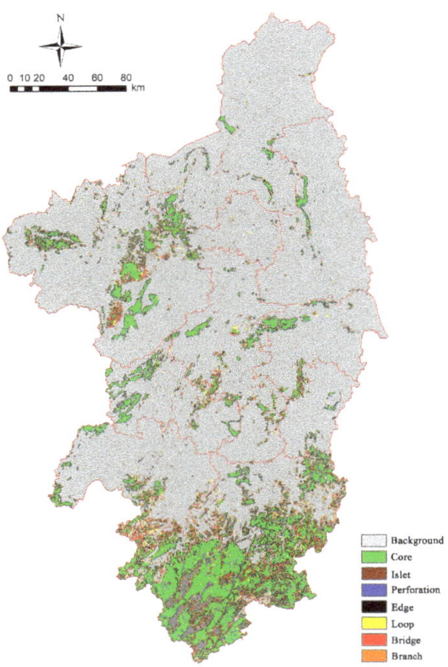

Figure 3. MSPA-based landscape classification map of the GI.

3.1.2. Identification of Hubs

There were 6208 cores, which varied greatly in size from 900 m^2 to 1811.70 km According to Figure 4, it can be seen that along with the reduction in the area of patche the number of patches increases rapidly, but the contribution to the core area decrease The threshold of 10 km^2 preserved more than 80% of the core areas and less than 1.5 of the total number. Therefore, 90 cores that were larger than 10 km^2 were chosen candidate hubs.

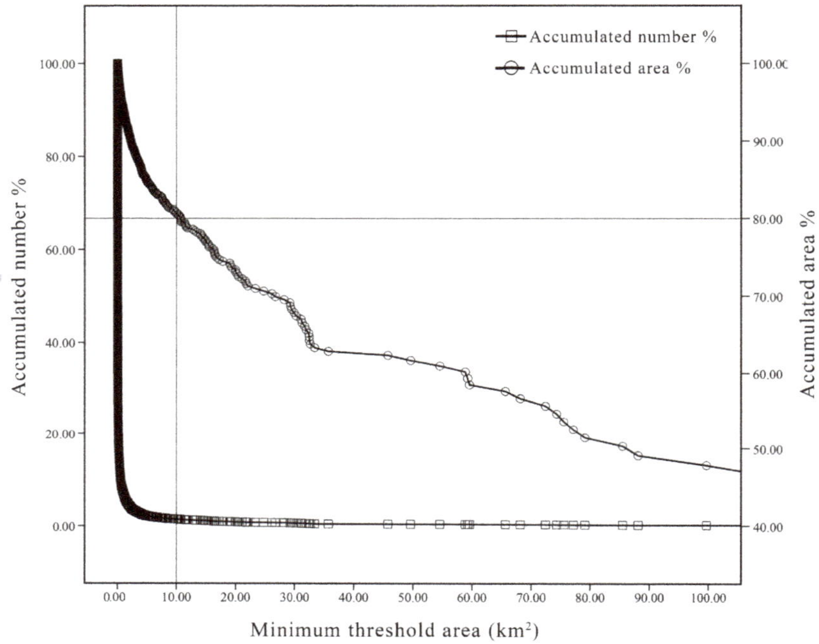

Figure 4. The results regarding candidate hub thresholds.

The landscape connectivity indexes among the 90 candidate hubs were calculat based on Confer 2.6 software. The distance threshold was set to 15 km and the connectiv probability was set to 0.5. The results were divided into three classes, i.e., high, medium, a low, using the natural breakpoint method. Then, 57 cores in the high and medium clas were chosen as green hubs. The largest hub covered an area of 1811.70 km^2, the minim was 10.01 km^2, and the total was 5042.07 km^2, which accounted for 40.54% of the GI. shown in Figure 5, the spatial pattern of hubs was similar to that of the core. The north hubs were small in size with higher fragmentation than the southern ones. The hubs w principally large-scale green ecological patches, and forest land accounted for 82.7% of total of the hubs, such as Qingliang Peak Nature Reserve, Banqiao Nature Reserve, a Laoshan National Forest Park, which can provide good habitats for living creatures. the whole, hubs were distributed in nine cities, but Xuancheng accommodated the m and Yangzhou and Changzhou the least. The high distribution of hubs was due to complicated mountain topography, abundant precipitation, and high vegetation covera in Xuancheng. In contrast, the flat terrain, less precipitation, and intensive human activi in Yangzhou and Changzhou resulted in fewer green hubs. It should be notable that 15.4 of hubs were located on administrative boundaries, with the total size being 1237.92 k

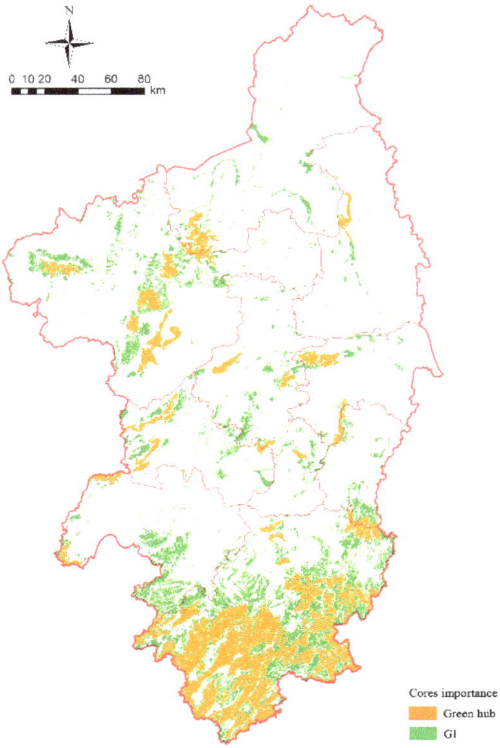

Figure 5. Spatial pattern of green hubs.

3.2. Construction of the GIN

In reference to each factor resistance surface, the final combined resistance surface was obtained using overlay analysis (Figure 6). Higher-resistance areas were principally located along the Yangtze River due to the fact that the dense towns, the developed economy, and the high level of land development along the river brought about high landscape resistance. In southern Xuancheng and central Chuzhou, low resistance was primarily due to a subtropical monsoon climate, which produced abundant rainfall and dense vegetation. Furthermore, human activities were constrained by the complicated geology and geomorphology, including mountains, hills, basins, tablelands, and plains. The hub-to-hub least-cost path was identified using an operational LCP model (Figure 7a). It was found that some of the hubs had two or more connecting paths to each other. The study ultimately identified 101 potential green corridors with a total length of 1961.85 km, with individual corridor lengths ranging from 120.45 km to 0.20 km (Figure 7b). Because of the scattered distribution of hubs, the green corridors were reticulated. From the green corridors distribution and number in each city (Figure 8), Xuancheng had the largest number of corridors (more than 70) and a shorter total length (only 270.32 km), which indicated that the GIN of Xuancheng was better with closely distributed hubs and shorter distance corridors. Nanjing had the longest total length of green corridors, over 500 km, but the number of corridors was only 32, indicating that its GIN was dominated by long-distance corridors. This may have been due to Nanjing being located in the center of the study area and the frequent material exchange between internal and external hubs, but with the high urbanization and the high value of landscape resistance, resulting in long corridor distances. There was only one green corridor in Huai'an. Huai'an was located in a flat area of Jianghuai Plain, which provided convenient conditions for urban development and agricultural production. As a result of dense construction and lack of green space was characterized by small sparsely green hubs, and the number and length of corrid were limited.

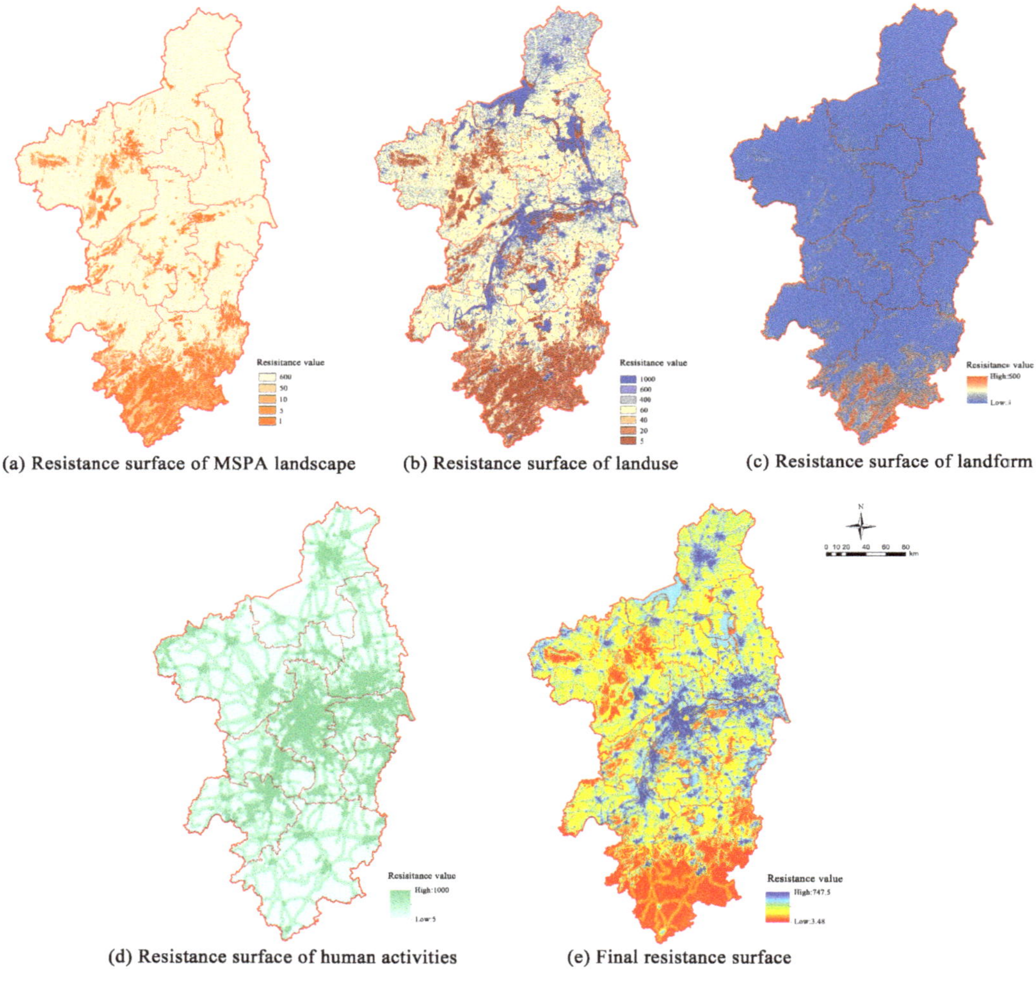

Figure 6. Landscape resistance surfaces.

To further learn about the importance of corridors, the ratio of the cost-weig distance to the least-cost path length was employed to measure the quality of corri The lower the ratio, the smaller the relative resistance of the path and the better the qu of the corridor. The corridors were divided into three classes, i.e., high, medium, low quality, using the natural breakpoint method. As shown in Figure 7b, more 85% of the corridors were of medium-to-high quality, indicating that the quality o corridors in the study area was better. Low-grade corridors were principally concentr in the central areas, which were characterized by intensive land development, develo cross-area transportation facilities, and great resistance to the migration of organisms t surrounding areas.

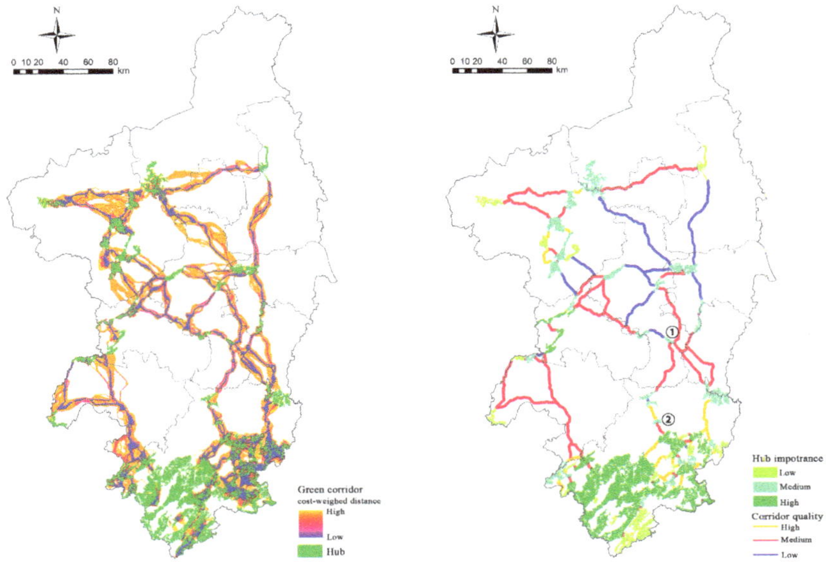

(a) The least-cost corridors in the landscape (b) Green infrastructure network

Figure 7. The GIN of the Nanjing Metropolitan Area. The number of the figure is the hub code.

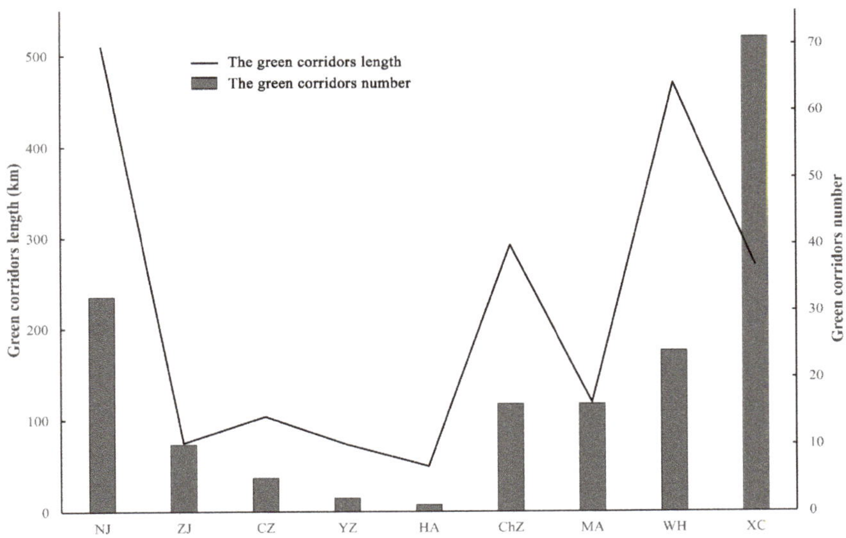

Figure 8. The length and number of green corridors in each city. NJ is Nanjing, ZJ is Zhenjiang, CZ is Changzhou, YZ is Yangzhou, HA is Huai'an, ChZ is Chuzhou, MA is Ma'anshan, WH is Wuhu, and XC is Xuancheng.

Based on the construction of GIN, the centrality of the hub was calculated to measure its importance. The results were also divided into the high, medium, and low classes. As shown in Figure 7b, there were only 10 high-class hubs, which were concentrated in the southern mountainous areas; the number of medium-class hubs was 26, which could be found in all cities except Yangzhou; the number of low-class hubs was 21 with a scattered distribution. In terms of size, the high-class hubs had the largest area, with a total 2829.70 km^2, accounting for 56.12% of the total area of the hubs; the low-class hubs had t smallest area, with a total area of 852.80 km^2, accounting for 16.91% of the total hubs. should be notable that some of the smaller hubs, such as hubs 1 and 2, were of the mediu class, indicating that they were highly capable of material exchange despite their small si

3.3. *Identification of Key Ecological Nodes*

The study identified the key ecological nodes in the network by calculating the curre values for each raster of the potential corridor. A total of 26 key nodes were extracte (Figure 9). Most of the nodes were predominantly long and narrow. The land-cover typ corresponding to key nodes were principally small forest, agricultural land, and grasslar which were located in small-sized, low-resistance sites. According to Figure 6d, traf networks passed through several ecological nodes. For instance, nodes 2 and 14 we passed through by expressways, while there were railways passing through nodes 12 ar 15 (Appendix A Figure A1). Linear transportation facilities had a restrictive effect on t flow of organisms.

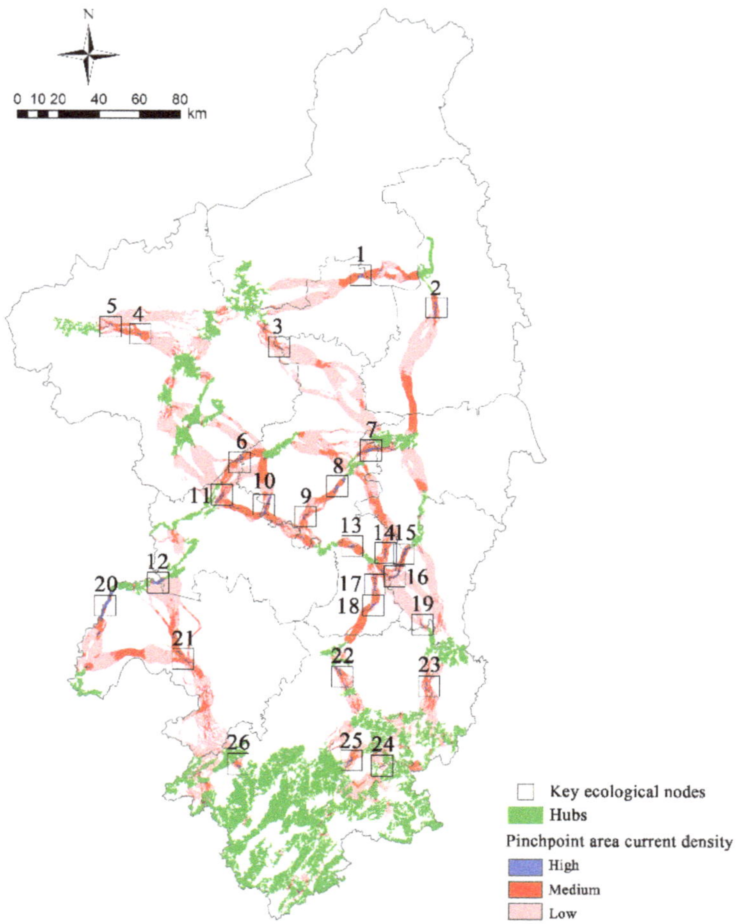

Figure 9. Result of the identification of the key ecological nodes. The number of the figure is node code.

Approx. 46.2% of the key nodes were located in Nanjing. As a key zone of connection, Nanjing had a high probability of communication with surrounding species. Nodes 22–26 were concentrated in Xuancheng. This area was characterized by many hubs, a high probability of biological dispersal, and a high current density, which brought about several key ecological nodes. However, there were fewer nodes in Yangzhou, Zhenjiang, and Huaian due to the lack of hubs and the large construction cost distance of corridors such that the biological dispersal and migration were more susceptible to threats. Hence, more attention should be paid to these nodes and their protection should be enhanced.

4. Discussion

4.1. Effectiveness of the Identification of the GIN

The Nanjing Metropolitan Area contains 76 statutory green space reserves, including provincial and higher-level nature reserves, forest parks, and wetland parks, which are the most important regional GI. After overlaying a GIN and the statutory green space reserves, it was observed that the location match between the hubs and the green space reserves of over 10 km^2 was 72.50% (Figure 10), which illustrated that it was highly reliable for selecting the green hubs. This view was consistent with the results of previous studies [7,30], having offered a further basis for MSPA and landscape connectivity indexes to identify the hubs of GI on a regional scale.

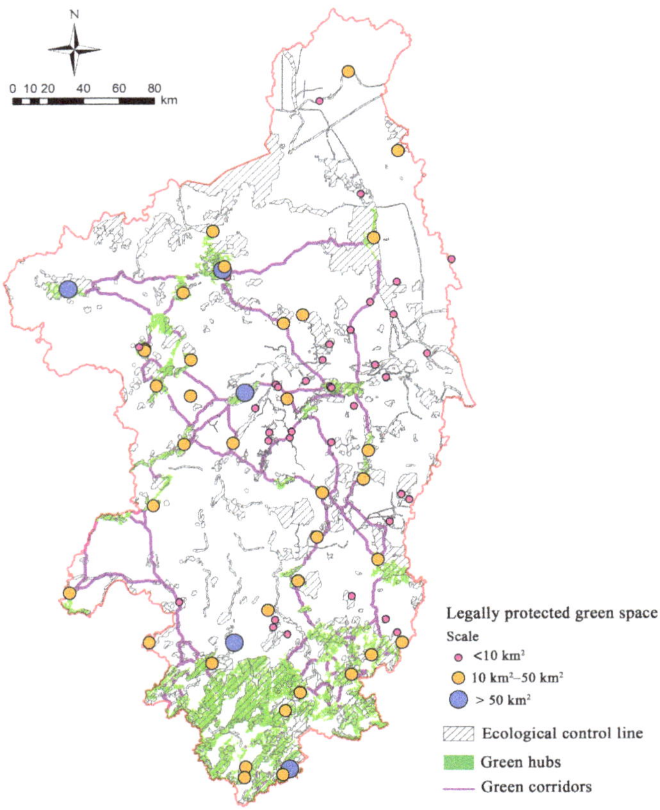

Figure 10. Overlay map of the GIN, statutory green space protection areas, and ecological control lines.

Urban ecological control lines are the bottom lines for each city to preserve its ecological security [38]. After overlaying the GIN and the ecological control lines of each member city, it was observed that the hubs within the control lines covered an area of 3653.57 k which accounted for 72.46% of all hubs; only less than 30% of hubs were outside the con lines. Almost hubs outside control lines were situated around such lines, perhaps beca the ecological control lines of China were normally composed of the core and conservat areas of ecological space, and the surrounding green spaces were not included. In contr the hubs identified in this study from spatial pattern and connectivity completely continuously depicted the scope of individual green spaces.

Since the study area was dominated by low mountains, hills, and plains where a culture was well-developed, agricultural land accounted for a large proportion of the gr corridors. However, agriculture was not included in the ecological control line progr of China; therefore, only 640.12 km of the corridors were located within the control li accounting for 32.63% of all corridors. In the ecological control lines and the statut green space reserves, the present linear GI acted as ecological corridors, but this existing alone may result in poor hub-to-hub connectivity [7]. The corridors identified in the st are playing an important role in the stability of the ecosystem and the connectivity of ecological network, and thus, they are a powerful supplement to ecological planning the Nanjing Metropolitan Area.

The GIN identified in this study is highly consistent with the ecological control li and statutory reserves. The hubs were favorably intact, while the corridors could effectiv supplement existing ecological control lines, and guarantee the stability and connecti of the ecosystem.

4.2. *Theoretical and Practical Implications of the Study*

4.2.1. Methodological Advantages

In this work, a regional-scale GIN identification method was proposed by combir methods such as MSPA, LCP, and CT (Figure 2). The method followed the parad of "hubs identification-corridors construction-network optimization". First, hubs w identified using land-use data as input data based on the scale and connectivity with help of MSPA and connectivity indexes; furthermore, landscape resistance surfaces w determined by considering landscape types, land-cover types, topography, and hur activities; then, the spatial locations of corridors were identified using LCP models; th the CT was introduced to determine the corridor and hub levels and identify the key po of optimization affecting the network connectivity.

The GIN identified in the study combined structural and functional connectivity however, when compared with the GIN constructed for certain protected species, identification results of our study may miss some small-scale quality habitats. Neverthe the identification method of the study still offers the following advantages: First, method principally employs land-cover data. It requires fewer data that are easier to ob and calculate, making it easier to build a large-scale GIN. Second, the method eliminates need to select focal species and the simulated ecological corridors can accommodate m species migration. In the absence of basic species survey data, it is still possible to const ecological networks with favorable connectivity [21]. Third, the method proposed in study can determine the location and pattern of key areas in the GIN that need pric protection. In summary, this study provides a practical and replicable methodolog framework for identifying a regional-scale GIN.

4.2.2. Policy Implications

China has entered its model of coordinated regional development with metropo areas as the mainstay. Compared to the urban scale, the identification of a GIN a regional scale breaks through the constraints of urban boundaries, helping to provi macro contextual grasp of regional ecological security and supporting the developme win–win ecological planning for individual member cities [39]. While regional GIN ning is not yet mandatory in China, more and more scholars and government agencie becoming aware of its importance. For example, Shi and Qing obtained the key elemen the Zhengzhou–Kaifeng metropolitan area GIN [40], and Zhang et al. attempted to identify landscape ecological security patterns in the Beijing–Tianjin–Hebei region [22]. Moreover, state authorities also advocated building regional GINs to resolve conflicts between land development and ecological protection. In 2019, the Guidance for Fostering the Development of Modern Metropolitan Areas was promulgated by the National Development and Reform Commission, which requires strengthening the regional green corridors' connectivity [41].

As the first inter-provincial metropolitan area in China, the Nanjing Metropolitan Area is faced with litmus tests in terms of regional environmental protection. The study serves as a reference for the Nanjing Metropolitan Area and its similar fast-growing urban clusters in formulating development policies to better deal with the contradictions between regional ecological protection and economic development. Approx. 15.40% of the hubs were found to span at least two cities in the study. This was because the administrative boundaries of Chinese cities are usually defined on the basis of natural elements, such as rivers and mountains, and there is plenty of GI at the junctions of cities. Such transboundary GI is generally large and has excellent ecological environments, making it suitable for containing large habitat patches, thereby turning into important components of GIN on a regional scale. However, during the development, GI erosion is often more serious in transboundary areas because of low land costs, weak government governance, and strong development vitality. The GI in these areas should be paid special attention, and targeted strategies should be provided from the perspective of regional coordination to gradually eliminate administrative control barriers and promote the protection of regional hubs.

There were relatively few green hubs, poor quality ecological corridors, and high cumulative resistance in the north of the Nanjing Metropolitan Area; therefore, it is necessary to set the identified hubs and corridors as no-build areas. Some hubs in particular play an important role in maintaining network connectivity out of proportion to their size. Hence, it is essential to enhance the investment in them and create buffer zones around them to minimize human activities. The key ecological nodes may be sensitive and vulnerable due to the lack of large types of low-resistance land cover, the disturbance by transport networks, etc. The narrowness of the ecological nodes indicates bottlenecks in network connectivity, where these nodes and their surroundings are critical to the connectivity of networks [42]. The identified key nodes should be investigated on the spot to learn the exact causes of each ecological point and formulate targeted protective and mitigating measures. For example, at nodes 12 and 15 (Figure 10), animal overpasses or underpasses could be established based on ecological assessments to provide more alternative pathways for species dispersal. For the high-quality corridors, it is essential to have strict control over the development intensity and pattern, reduce ecological stress, and optimize the regional GIN.

4.3. *Threshold Uncertainties and Study Limitations*

This study used MSPA and landscape connectivity indexes to identify green hubs and avoid the subjectivity of artificial selection that is frequently observed in previous studies. However, MSPA is very sensitive to the landscape scale [43], and thus the size of the input data image element may directly affect the recognition result. Since the study focused on a regional-scale GIN, a 30 m × 30 m scale was chosen to preserve the key elements of the GI [25,44]. Moreover, the edge width of MSPA represents the edge effect of a patch, and the setting of its value affects the sizes of core areas. Such a value is normally set in related studies based on protected species [13], but the high species richness at the regional scale complicates edge effects; hence, the default value of 1 was set to address the needs of most species [45].

When patch importance is calculated with the Confer software, the connectivity distance threshold needs to be defined. Patches are considered disconnected when the distance between them is greater than this threshold. As the GI in the study area consisted mainly of forest lands, wetlands, and grasslands, a threshold distance of 15 km with a connectivity probability of 0.5 was chosen [46], based on a consideration of the dispersal

distances of mammals, terrestrial birds, etc. [30,47]. In addition, Pinchpoint Mapper wa
used to identify the ecological points by setting a "width" threshold of the cost-weighte
corridor. This threshold merely affects the current density at pinch points and does n
affect the determination of locations of ecological points [26,48]. To identify the locations
ecological points more efficiently and conveniently, a threshold of 5000 m was chosen f
the study.

For the calculation in the study, most of the thresholds were defined based on previo
studies, but the applicability of such thresholds for the study area has not been verifie
These thresholds should be further studied and discussed in the subsequent steps. Th
selection of resistance surfaces and the assignment of weights for the landscape does n
refer to the resistance to the actual landscape, but rather indicates the relative resistan
which has flexibility in application and is adjustable depending on the characteristics
the region of interest. In addition, the way to optimize and implement the identified G
through concrete measures is a complex and worthwhile research topic. This study d
not offer an in-depth analysis of this topic, and future studies are expected to addre
it intensively.

5. Conclusions

Identification of regional GINs is essential for the sustainable development of regio
and the maintenance of regional ecological security. In the study, a method for identifyi
regional-scale GINs was established using MSPA, connectivity indexes, LCP, and CT, a
the identification results were validated with reference to statutory green space reserv
and urban ecological control lines by taking the Nanjing Metropolitan Area as an examp
The results of regional-scale hubs identified using MSPA integrated landscape connectiv
indexes are more reliable. The GIN of Nanjing Metropolitan Area was composed of gre
hubs that were mainly dominated by forest land, and reticulated green corridors we
dense in the south and sparse in the north. There were 57 green hubs, 101 potential gre
corridors, and 26 key ecological nodes in the Nanjing Metropolitan Area. The importa
hubs and quality corridors were concentrated in the south of the study area, while they we
unsatisfactory in the north. About 15.40% of hubs were across administrative boundari
Furthermore, most ecological nodes were long and narrow and were mainly located
small-sized and low-resistance sites.

As an exploration of the method for identifying regional GINs, this study provide
new way of thinking about construction at the regional scale. In addition, the study mak
suggestions for regional GIN conservation policy and planning formulation and provide
convincing reference for ecological planning and development planning of other rapi
developing city clusters. On this basis, the method of GIN identification proposed in t
study can be used to identify and manage the ecological spaces, including designati
expanding, upgrading, and strengthening management. In addition, it can indicate w
areas need priority protection and can be used in conjunction with habitat evaluation
determine the timing of the restoration of regional ecological resources. Moreover, t
method requires less data, is easier to replicate, and can be used as a long-term monitori
tool for the implementation of GIN planning. Once regional GIN planning has be
developed, realistic GINs can be identified in the future using the same approaches
fixed time intervals. By comparing the identified and planned GINs, it is possible
reveal the areas that comply with the plan, lag behind the plan, and violate the pl
which helps to judge the completion of GIN planning and management decisions can
promptly adjusted.

Author Contributions: Conceptualization and methodology, W.L. and H.X.; software, W.L. and X.Z.; data analysis, W.L. and W.J.; writing—original draft preparation, W.L.; writing—review and editing, W.L. and H.X.; supervision and project administration, H.X. All authors have read and agreed to the published version of the manuscript.

Funding: This research was funded by the Postgraduate Research & Practice Innovation Program of Jiangsu Province (KYCX20_0870) and a project funded by the Priority Academic Program Development of Jiangsu Higher Education Institutions (PAPD).

Institutional Review Board Statement: Not applicable.

Informed Consent Statement: Not applicable.

Data Availability Statement: The data presented in this study are available on request from the corresponding author.

Conflicts of Interest: The authors declare no conflict of interest.

Appendix A

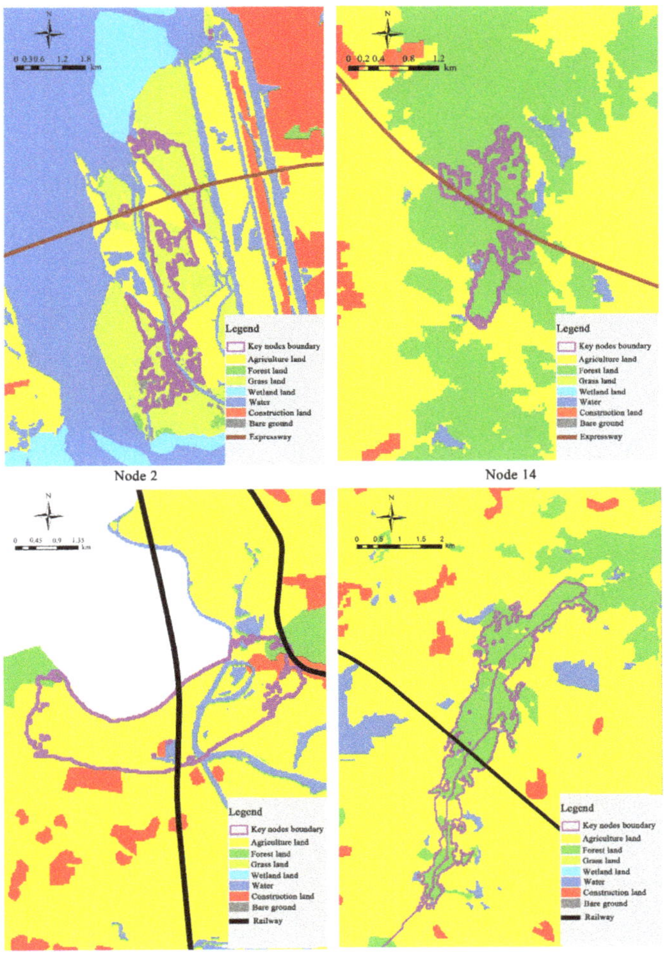

Figure A1. Land-use composition of key nodes.

References

1. Benedict, M.A.; McMahon, E.T. Green infrastructure: Smart conservation for the 21st century. *Renew. Resour. J.* **2002**, *20*, 12–
2. Kong, F.; Yin, H.; Nakagoshi, N.; Zong, Y. Urban green space network development for biodiversity conservation: Identifica based on graph theory and gravity modeling. *Landsc. Urban Plan.* **2010**, *95*, 16–27. [CrossRef]
3. Aminzadeh, B.; Khansefid, M. A case study of urban ecological networks and a sustainable city: Tehran's metropolitan a *Urban Ecosyst.* **2010**, *13*, 23–36. [CrossRef]
4. Jenerette, G.D.; Potere, D. Global analysis and simulation of land-use change associated with urbanization. *Landsc. Ecosyst.* **2** *25*, 657. [CrossRef]
5. Gao, Y.; Wu, Z.; Lou, Q.; Huang, H.; Cheng, J.; Chen, Z. Landscape ecological security assessment based on projection pursu Pearl River Delta. *Environ. Monit. Assess.* **2012**, *184*, 2307–2319. [CrossRef]
6. Nikolakaki, P. A GIS site-selection process for habitat creation: Estimating connectivity of habitat patches. *Landsc. Urban* **2004**, *68*, 77–94. [CrossRef]
7. Wang, S.; Wu, M.; Hu, M.; Fan, C.; Wang, T.; Xia, B. Promoting landscape connectivity of highly urbanized area: An ecolo network approach. *Ecol. Indic.* **2021**, *125*, 107487. [CrossRef]
8. Cui, L.; Wang, J.; Sun, L.; Lv, C. Construction and optimization of green space ecological networks in urban fringe areas: A study with the urban fringe area of Tongzhou district in Beijing. *J. Clean. Prod.* **2020**, *276*, 124266. [CrossRef]
9. Ma, Q.; Li, Y.; Xu, L. Identification of green infrastructure networks based on ecosystem services in a rapidly urbanizing *J. Clean. Prod.* **2021**, *300*, 126945. [CrossRef]
10. Peng, J.; Yang, Y.; Liu, Y.; Du, Y.; Meersmans, J.; Qiu, S. Linking ecosystem services and circuit theory to identify ecolo security patterns. *Sci. Total Environ.* **2018**, *644*, 781–790. [CrossRef]
11. Liquete, C.; Kleeschulte, S.; Dige, G.; Maes, J.; Grizzetti, B.; Olah, B.; Zulian, G. Mapping green infrastructure based on ecosy services and ecological networks: A Pan-European case study. *Environ. Sci. Policy* **2015**, *54*, 268–280. [CrossRef]
12. Cunha, N.; Magalhães, M. Methodology for mapping the national ecological network to mainland Portugal: A planning towards a green infrastructure. *Ecol. Indic.* **2019**, *104*, 802–818. [CrossRef]
13. Zhang, R.; Zhang, L.; Zhong, Q.; Zhang, Q.; Ji, Y.-W.; Song, P.; Wang, Q. An optimized evaluation method of an urban ecolo network: The case of the Minhang District of Shanghai. *Urban For. Urban Green.* **2021**, *62*, 127158. [CrossRef]
14. Ersoy, E.; Jorgensen, A.; Warren, P.H. Identifying multispecies connectivity corridors and the spatial pattern of the lands *Urban For. Urban Green.* **2019**, *40*, 308–322. [CrossRef]
15. Dong, J.; Peng, J.; Liu, Y.; Qiu, S.; Han, Y. Integrating spatial continuous wavelet transform and kernel density estimati identify ecological corridors in megacities. *Landsc. Urban Plan.* **2020**, *199*, 103815. [CrossRef]
16. Peng, J.; Zhao, S.; Dong, J.; Liu, Y.; Meersmans, J.; Li, H.; Wu, J. Applying ant colony algorithm to identify ecological sec patterns in megacities. *Environ. Model. Softw.* **2019**, *117*, 214–222. [CrossRef]
17. Modica, G.; Praticò, S.; Laudari, L.; Ledda, A.; Di Fazio, S.; De Montis, A. Implementation of multispecies ecological networ the regional scale: Analysis and multi-temporal assessment. *J. Environ. Manag.* **2021**, *289*, 112494. [CrossRef]
18. Li, J.; Weckworth, B.V.; McCarthy, T.M.; Liang, X.; Liu, Y.; Xing, R.; Li, D.; Zhang, Y.; Xue, Y.; Jackson, R. Defining prioriti global snow leopard conservation landscapes. *Biol. Conserv.* **2020**, *241*, 108387. [CrossRef]
19. Cushman, S.A.; Landguth, E.L.; Flather, C.H. Evaluating population connectivity for species of conservation concern i American Great Plains. *Biodivers. Conserv.* **2013**, *22*, 2583–2605. [CrossRef]
20. Dong, J.; Peng, J.; Xu, Z.; Liu, Y.; Wang, X.; Li, B. Integrating regional and interregional approaches to identify ecological sec patterns. *Landsc. Ecol.* **2021**, *36*, 2151–2164. [CrossRef]
21. Cao, Y.; Yang, R.; Carver, S. Linking wilderness mapping and connectivity modelling: A methodological framework for wil network planning. *Biol. Conserv.* **2020**, *251*, 108679. [CrossRef]
22. Zhang, L.; Peng, J.; Liu, Y.; Wu, J. Coupling ecosystem services supply and human ecological demand to identify lands ecological security pattern: A case study in Beijing–Tianjin–Hebei region, China. *Urban Ecosyst.* **2017**, *20*, 701–714. [CrossR
23. Soille, P.; Vogt, P. Morphological segmentation of binary patterns. *Pattern Recognit. Lett.* **2009**, *30*, 456–459. [CrossRef]
24. Vogt, P.; Riitters, K.H.; Iwanowski, M.; Estreguil, C.; Kozak, J.; Soille, P. Mapping landscape corridors. *Ecol. Indic.* **2007**, *7*, 48 [CrossRef]
25. Wickham, J.D.; Riitters, K.H.; Wade, T.G.; Vogt, P. A national assessment of green infrastructure and change for the conterm United States using morphological image processing. *Landsc. Urban Plan.* **2010**, *94*, 186–195. [CrossRef]
26. Mcrae, B.H.; Dickson, B.G.; Keitt, T.H.; Shah, V.B. Using circuit theory to model connectivity in ecology, evolution conservation. *Ecology* **2008**, *89*, 2712–2724. [CrossRef]
27. Koen, E.L.; Bowman, J.; Sadowski, C.; Walpole, A.A. Landscape connectivity for wildlife: Development and validati multispecies linkage maps. *Methods Ecol. Evol.* **2014**, *5*, 626–633. [CrossRef]
28. Lechner, A.M.; Doerr, V.; Harris, R.M.; Doerr, E.; Lefroy, E.C. A framework for incorporating fine-scale dispersal behaviou biodiversity conservation planning. *Landsc. Urban Plan.* **2015**, *141*, 11–23. [CrossRef]
29. Li, J.; Xu, J.; Chu, J. The construction of a regional ecological security pattern based on circuit theory. *Sustainability* **2019**, *11* [CrossRef]
30. An, Y.; Liu, S.; Sun, Y.; Shi, F.; Beazley, R. Construction and optimization of an ecological network based on morphological s pattern analysis and circuit theory. *Landsc. Ecol.* **2021**, *36*, 2059–2076. [CrossRef]

National Development and Reform Commission. *Nanjing Metropolitan Area Development Plan*; National Development and Reform Commission: Beijing, China, 2021.
Xie, G.; Zhang, C.; Zhang, L.; Chen, W.; Li, S. Improvement of the evaluation method for ecosystem service value based on per unit area. *J. Nat. Resour.* **2015**, *30*, 1243.
Vogt, P.; Ferrari, J.R.; Lookingbill, T.R.; Gardner, R.H.; Riitters, K.H.; Ostapowicz, K. Mapping functional connectivity. *Ecol. Indic.* **2009**, *9*, 64–71. [CrossRef]
Taylor, P.D.; Fahrig, L.; Henein, K.; Merriam, G. Connectivity is a vital element of landscape structure. *Oikos* **1993**, *68*, 571–573. [CrossRef]
Pascual-Hortal, L.; Saura, S. Comparison and development of new graph-based landscape connectivity indices: Towards the priorization of habitat patches and corridors for conservation. *Landsc. Ecol.* **2006**, *21*, 959–967. [CrossRef]
Carroll, C.; McRAE, B.H.; Brookes, A. Use of linkage mapping and centrality analysis across habitat gradients to conserve connectivity of gray wolf populations in western North America. *Conserv. Biol.* **2012**, *26*, 78–87. [CrossRef] [PubMed]
McRae, B.H.; Hall, S.A.; Beier, P.; Theobald, D.M. Where to restore ecological connectivity? Detecting barriers and quantifying restoration benefits. *PLoS ONE* **2012**, *7*, e52604. [CrossRef] [PubMed]
Hong, W.; Guo, R.; Su, M.; Tang, H.; Chen, L.; Hu, W. Sensitivity evaluation and land-use control of urban ecological corridors: A case study of Shenzhen, China. *Land Use Policy* **2017**, *62*, 316–325. [CrossRef]
Fabos, J.G. Greenway planning in the United States: Its origins and recent case studies. *Landsc. Urban Plan.* **2004**, *68*, 321–342. [CrossRef]
Shi, X.; Qin, M. Research on the optimization of regional green infrastructure network. *Sustainability* **2018**, *10*, 4649. [CrossRef]
National Development and Reform Commission. *The Guidance for Fostering the Development of Modern Metropolitan Areas*; National Development and Reform Commission: Beijing, China, 2019.
Pittiglio, C.; Skidmore, A.K.; van Gils, H.A.; McCall, M.K.; Prins, H.H. Smallholder farms as stepping stone corridors for crop-raiding elephant in northern Tanzania: Integration of Bayesian expert system and network simulator. *Ambio* **2014**, *43*, 149–161. [CrossRef]
Saura, S.; Vogt, P.; Velázquez, J.; Hernando, A.; Tejera, R. Key structural forest connectors can be identified by combining landscape spatial pattern and network analyses. *For. Ecol. Manag.* **2011**, *262*, 150–160. [CrossRef]
Pascual-Hortal, L.; Saura, S. Impact of spatial scale on the identification of critical habitat patches for the maintenance of landscape connectivity. *Lands. Urban Plan.* **2007**, *83*, 176–186. [CrossRef]
Wang, Y.; Lin, Q. The transformation of planning ideas and the exploration of planning methods of urban green space ecological network based on MSPA. *Chin. Landsc. Archit.* **2017**, *33*, 68–73.
Sutherland, G.D.; Harestad, A.S.; Price, K.; Lertzman, K.P. Scaling of natal dispersal distances in terrestrial birds and mammals. *Conserv. Ecol.* **2000**, *4*, 16. [CrossRef]
Saura, S.; Torne, J. Conefor Sensinode 2.2: A software package for quantifying the importance of habitat patches for landscape connectivity. *Environ. Modell. Softw.* **2009**, *24*, 135–139. [CrossRef]
Song, L.L.; Qin, M. Identification of ecological corridors and its importance by integrating circuit theory. *J. Appl. Ecol.* **2016**, *27*, 3344–3352. [CrossRef]

sustainability

MDPI

cle

‹amining the Economic Value of Tourism and Visitor ‹eferences: A Portrait of Sustainability Ecotourism in the ‹ngkahan Protection Area, Gunung Leuser National Park, ‹rth Sumatra, Indonesia

‹s Purwoko [1,*], Dodik Ridho Nurrochmat [2], Meti Ekayani [3], Syamsu Rijal [4] and Herlina Leontin Garura [1]

1 Faculty of Forestry, Universitas Sumatera Utara, Jl. Tri Dharma Ujung No.1 Kampus USU Medan, Medan City 20155, North Sumatra, Indonesia; herlinaleontinn@gmail.com
2 Faculty of Forestry and Environment, IPB University, Jl. Lingkar Akademik IPB Dramaga Campus, Bogor 16680, West Java, Indonesia; dnurrochmat@ipb.ac.id
3 Faculty of Economics and Management, IPB University, Jl. Agatis, IPB Dramaga Campus, Bogor 16680, West Java, Indonesia; meti@ipb.ac.id
4 Faculty of Forestry, Universitas Hasanuddin, Jl. Perintis Kemerdekaan No. Km.10, Tamalanrea Indah, Makassar 90245, South Sulawesi, Indonesia; jsyamsurijal@unhas.ac.id
* Correspondence: agus9@usu.ac.id; Tel.: +62-813-6051-3501

Abstract: North Sumatra Province has the Tangkahan Nature Tourism Area, which represents ecotourism managed by local communities, established in 2001, which has now become the leading tourism destination of North Sumatra both locally and internationally. Tangkahan ecotourism is an example of payment for environmental services for the Tangkahan community, which initially carried out illegal logging in the mount Leuseur national park and then agreed to preserve the national park through ecotourism. This study aims to analyze the economic value of tourism and the preferences of tourists to revisit, along with the factors that influence them, where these conditions can be an illustration of the sustainability of Tangkahan ecotourism. The travel cost method is used to calculate the economic value of Tangkahan Ecotourism environmental services. The factors that affect the economic value, intensity of visits, and interest in revisiting, were analyzed using multiple linear regression. The results showed that Tangkahan ecotourism has a relatively high economic value, supported by the intensity and interest of tourist visits. Factors that affect the economic value and preferences of tourist visits can be managed for the sustainability of Tangkahan ecotourism so as not to lose the economic value of the ecotourism environmental services.

Keywords: ecotourism; Tangkahan; economic value; intensity of visits; travel cost method; interest in revisiting

‹n: Purwoko, A.; Nurrochmat, ‹kayani, M.; Rijal, S.; Garura, ‹amining the Economic Value ‹ism and Visitor Preferences: A ‹t of Sustainability Ecotourism ‹angkahan Protection Area, ‹g Leuser National Park, North ‹a, Indonesia. *Sustainability* ‹4, 8272. https://doi.org/ ‹/su14148272

‹nic Editors: Miklas Scholz, ‹A. Pagnotta, Panayiotis ‹akopoulos and ‹ Noorani

‹d: 7 June 2022
‹ed: 2 July 2022
‹ed: 6 July 2022

‹er's Note: MDPI stays neutral ‹gard to jurisdictional claims in ‹ed maps and institutional affil-

1. Introduction

Forest resources have various interests that should be considered optimally. These interests are fragmented into interests of the community, law enforcement, conservation goals, and accelerated development [1]. Each region also faces the same challenges and is characterized by abundant natural resources on land and water. These natural resources are of interest and are included in the progress of a country, specifically the regions. Therefore, various forms of productive and multi-purpose use should be identified, planned, and developed. Varied use of forest resources is also important for conservation and resistance to pests and climate shocks [2]. This includes tourism activities as a more sustainable use of forest services. Thus, it is hoped that nature can be a solution for the community's economy and the conservation of nature itself.

As the main stakeholder, communities around forests are important parties to pay attention to in natural resource management. If the community obtains the benefits that can be felt, then various forms of participation can be developed [3]. If these conditions are according to [4], local governments can encourage community participation in activi such as trade, business exhibitions, various cultural festivals and museums, organiz sports and art attractions, and investing in various types of businesses based on ecotour Collaborative ecotourism management is also a demand in sustainable ecotourism m agement. The involvement of stakeholders both locally and internationally will hav positive impact on their concern about how ecotourism destinations can be developed managed sustainably [5]. Participation and involvement of local communities as wel the application of a responsible ecotourism model cannot be ignored, so that the goal sustainable ecotourism can be realized [3].

Indonesia, specifically North Sumatra Province, has Tangkahan Ecotourism mana by local communities and has been around for a long time [6]. The area has become of the leading tourism destinations in the province, locally, nationally, and internation This ecotourism was opened in 2001 and inaugurated in February 2004. The object i ecotourism area with excellent local community participation in nature conservation. T nature tourism activists are precisely communities that previously relied heavily on economy of the forest in harmful ways, such as illegal logging and hunting. The pres of ecotourism activities has become a solution for the community's economy that is in with the principles of sustainability. Additionally, the region demonstrates how ecotou growth may significantly protect the 17,000-hectare Gunung Leuser National Park (GL area in North Sumatra [7].

The intangible benefits of forests cannot be assessed using a market system, and sev users are unaware of these benefits. There is still a lack of appreciation for environme benefits in the form of scenic beauty. Landscape beauty can be enjoyed and used by hun through nature tourism [8]. Efforts should be made to develop the form and manager of its utilization to increase the value of the benefits. This study is necessary to ensure the planning for the development of Tangkahan Ecotourism in Gunung Leuser Nati Park can be truly effective and provide significant benefits for the welfare of the comm This is because, according to [4], the benefits to the surrounding community will gr affect the support of local communities for the development of sustainable tourism. T benefits should include both material and non-material domains.

People believe that the development of ecotourism can produce significant econ benefits for them [3]. The condition is that they must be able to play key roles in the decision-making process and in the formulation of the direction of ecotou management. For this reason, studies related to the economic benefits of Tangk ecotourism management are important to prove how much these economic benefit manifested. Aspects that are also important to study are the characteristics and beh of visitors. Therefore, this study analyzes the economic value of nature tourism ob the intensity of visits, the tourist interest in revisiting, and the affecting factors i Tangkahan Ecotourism Area, Langkat Regency, North Sumatra, Indonesia. Accor to [9], feedback from visitors is very helpful for ecotourism destination manage institutions to determine the priorities and directions of wisdom in the developme tourist attractions. The importance of visitor returns is also related to competition an application of the principles of sustainable ecotourism management.

2. Materials and Methods

The study was conducted in Tangkahan Ecotourism, Namo Sialang and Sei Se Villages, Batang Serangan Sub-district, Langkat Regency, North Sumatra Province, In sia (Figure 1).

Primary data were collected through questionnaires and field observations, secondary data were collected from various institutions, especially Tangkahan Ecoto managers. Quoted accidental sampling technique was used in collecting primary dat

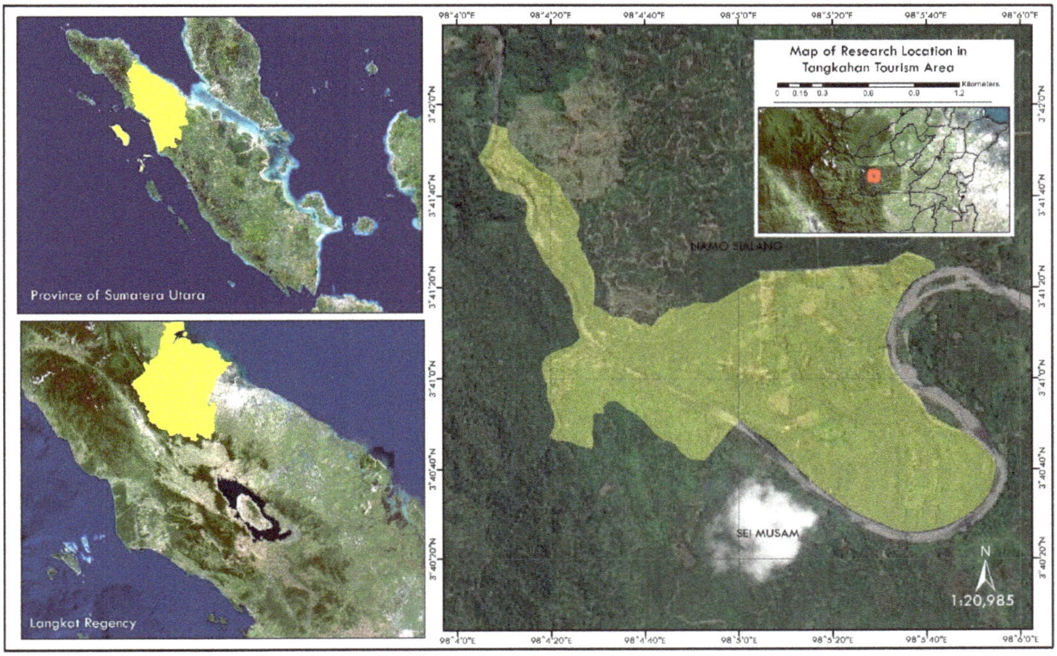

Figure 1. Tangkahan Ecotourism Location, Gunung Leuser National Park, North Sumatra, Indonesia (03°41′0″ N and 98°04′28.2″ E).

Secondary data collected are the characteristics of visitors and tourist attractions. The characteristics of visitors include the number of visitors each year, the number of inhabitants, the area of origin, as well as the number of inhabitants of the visitor's home zone. The characteristics of tourist attractions include location, area, biological conditions, tourism potential, accessibility, and recreational facilities.

The number of samples was determined by the Slovin formula, referring to the number of visitor populations of Tangkahan Ecotourism. The Slovin formula according to [11–13] is as follows:

$$n = \frac{N}{1 + N(e)^2}$$

Description:

n = number of samples
N = number of population
e = error tolerance (0, 1)

Refers to the total population of 31,200 people/year according to Tangkahan Tourism Institute. The sample taken was 99.68 (increased to 100 people). The Slovin formula in determining the sample for nature tourism was also used by [14] in Natsepa Beach, Maluku Province; [15] in Carocok Painan Beach, West Sumatra Province; [16] in Ciwidey, West Java; and [17] in Gunung Ciremai National Park.

2.1. Travel Cost Analysis

A zoned travel cost approach can be used to estimate the economic value of ecotourism [18–20]. The use of the travel cost method for the valuation of tourism objects is also used by [21] in Bozcaada (Turkey), [22] in Valencia (Spain), and [23] in Taman Tasik Cempaka (Malaysia). The stages in calculating the economic value of ecotourism [24–28] are as follows:

2.1.1. Calculating the Number of Visitors from Each Origin Area (Zone) Based on Interviews with Respondents

$$\text{Zi} = \text{Pi} \times \sum \text{Y}$$

Description:

Zi: Number of visitors zone i
Pi: Percentage of zone visits i
∑Y: Total visits

2.1.2. Determining the Average Travel Cost of the Total Travel Costs Incurred during Tra… or Recreational Activities

$$\text{BPR} = \text{TR} + \text{KA} + \text{TK} + \text{LL}$$

Description:

BPR: Average travel cost (IDR/person)
TR: Transportation cost (IDR/person)
KA: Consumption and accommodation cost during the trip (IDR/person)
TK: Ticket cost (IDR/person)
LL: Other costs (IDR/person)

2.1.3. Determining the Average Travel Cost of Zone *i*

$$\text{Xli} = \frac{\sum BPi}{Ni}$$

Description:

Xli: Average travel cost of zone i
Bpi: Travel cost of the sample
Ni: Total population of zone i

2.1.4. Determining the Visit Rate per 1000 People in Zone i in One Year

$$\text{LKi} = \frac{\sum \text{JPi}}{\sum \text{JPT}} \times 1000$$

Description:

LKi: Visit rate of visitors in zone i
JPi: Number of visitors in zone i
JPT: Total population in zone i

2.1.5. Determining the Total Economic Value (NET), Obtained from the Following Form…

$$\text{NET} = \text{Average Travel Cost} \times \text{Average Number of Visitors}$$

2.2. *Analysis of Factors Affecting Economic Value, Intensity of Visits, and Tourist Interest in Revisiting*

To determine the socioeconomic factors that influence the intensity of travel vi… multiple linear analysis is used. The multiple linear regression models used are:

$$Y_1 = \alpha_0 + \alpha_1 X_1 + \alpha_2 X_2 + \alpha_3 X_3 + \alpha_4 X_4 + \alpha_5 X_5 + \alpha_6 X_6 + e$$

$$Y_2 = \beta_0 + \beta_1 X_1 + \beta_2 X_2 + \beta_3 X_3 + \beta_4 X_4 + \beta_5 X_5 + \beta_6 X_6 + \beta_7 X_7 + \beta_8 Y_1 + e$$

$$Y_3 = \gamma_0 + \gamma_1 Y_1 + e$$

Description:
Y_1 = Travel Cost Value (individual)
Y_2 = Intensity of visits (frequency of visits up to the time of the study)
Y_3 = Tourist Interest in Revisiting
α_i, β_i, γ_i = Regression coefficient of independent variable
X_1 = Visitor Age
X_2 = Education Level
X_3 = Income Level
X_4 = Distance from Object
X_5 = Number of Members
X_6 = Travel Time
X_7 = Information Acquisition

In order to produce unbiased data, multiple regression analysis models are evaluated by econometric evaluation with classical assumption tests. Multicollinearity, heteroscedasticity, and autocorrelation tests were performed and met all assumptions. A Likert scale approach is used to measure various ordinal variables with a range of 1–5 [29]. It is used to measure visitors' conditions, attitudes, and opinions. The most positive opinion and in line with the theoretical assumptions are given a score of 5 (maximum), and the most negative opinion is given a score of 1 (minimum). After the data are obtained from the Likert scale, the validity and reliability tests are conducted to determine the validity and consistency of the received data.

2.3. *Overview of Study Location*

Tangkahan is developed as an ecotourism area located on Gunung Leuser National Park (GLNP) border. The area of the Tangkahan Ecotourism is ±103 hectares, which is divided into village and forest with an area of 18,526 ha and 17,653 ha, respectively [30]. Tangkahan is at an altitude of 130–200 m above sea level. The area's topography consists of hilly areas with varying slopes (45–90°). The Tangkahan area is located at the confluence of the Buluh and Batang Serangan rivers. This area has unique natural formations, beautiful landscapes, hot springs, waterfalls, caves, cliffs, high diversity of flora and fauna, and tropical rain.

2.4. *Socioeconomic Characteristics of Visitors*

An overview of the profile of visitors who travel to Tangkahan Ecotourism is obtained from the characteristics of respondents. The majority of visitors are domestic tourists from the area with a distance of 1–4 h, such as Stabat, Binjai, Medan, and several areas in North Sumatra. In certain seasons, it is also visited by many foreign tourists. The figures below show the distribution of respondents based on the type of tourists and their origin area (Figure 2). For the category of origin area, visitors consist of 20% and 80% of domestic and foreign tourists.

The characteristics of the visitors observed include age, gender, education level, occupation type, and income level. Visitors from the area are dominated by women (42% males and 58% females). A similar result was reported by [31], where 57% of Tangkahan visitors were female. This is because women prefer to spend recreational time with their friends. More women engage in tourism activities for various purposes [32–34]. The domination of women also occurs in families within the area. The average level of education of visitors is quite good. Most visitors have at least 12 years of education. Most visitors have received an education for at least 12 years. The majority of visitors' education levels are senior high school level (57%), 41% of visitors are undergraduate educated, and 2% have a master's degree. The occupation type of respondents is very diverse, with the largest proportion being students (33%), followed by entrepreneurs (17%) and private employees (15%).

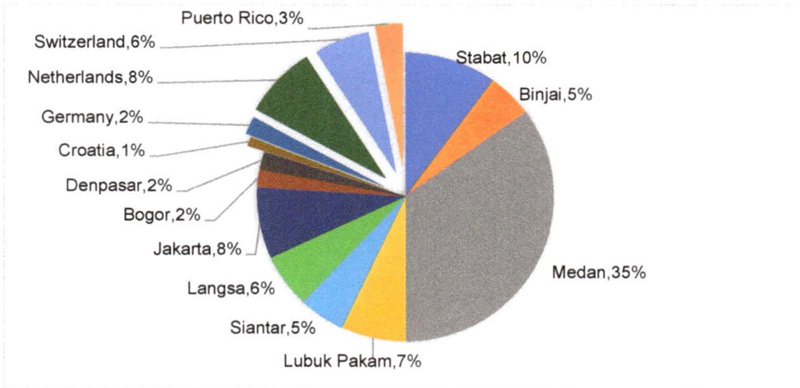

Figure 2. Distribution of Tangkahan Ecotourism respondents based on the origin of visitors.

The income level of visitors varies from the lowest to the highest segment (Figur
Theoretically, the level of visitor income will affect expenses during tourist visits.
allocation of expenditures includes transportation, consumption, accommodation,
other costs. Income is also expected to influence the choice of tourism objects to be vis
These data are consistent with the occupation type data, where most visitors are stud
Students do not have income at that age, and are still supported by their parents, inclu
budget allocations for tourism purposes.

Figure 3. Graphic of visitors' income level.

3. Results and Discussion

3.1. Valuation of Economic Value of Tangkahan Ecotourism

Calculating the value of the intangible benefits of a recreation area can be
through the approach of the travel cost method. Furthermore, the total value includ
the travel cost is the round trip cost plus the monetary value of the time spent on trave
recreational activities [35]. This method is widely used in various nature-based tou
objects such as Lake Limboto [36], Kalibiru [37], Parangtritis [38], Muara Angke [39],
Karas Beach, Pangandaran [40], Punti Kayu, Palembang [41,42], Thousand Islands
Ujung Genteng, Sukabumi [44], and others. The economic assessment of the Tangk
Ecotourism area collected includes information about the origin of visitors, the co
round-trip visitor travel, consumption costs during tourist visits, ticket fees for entrar
tourist objects, and other costs that must be paid (documentation, parking, storage, g
supporting equipment, and documentation fees).

Table 1 shows the highest average travel cost value comes from Zagreb at
14,600,000/visit, while the lowest comes from Stabat as the origin area of visitors clos

this tourism object at IDR 52,700/visit. The average travel costs that should be incurred from all visitors and all origin areas/countries of visitors is IDR 4,181,786/visit/person.

Table 1. Recapitulation of visitor data based on average travel cost.

No	Visitor Origin	$\bar{x}$ Transportation Cost (IDR)	$\bar{x}$ Consumption Cost (IDR)	Ticket Cost (IDR)	$\bar{x}$ Other Cost (IDR)	$\bar{x}$ Travel Cost (IDR)
1	Stabat	21,000	22,900	3000	5800	52,700
2	Binjai	38,000	28,000	3000	9000	78,000
3	Medan	56,971	31,343	5343	8257	101,914
4	Lubuk Pakam	42,857	58,857	3000	10,000	114,714
5	Siantar	51,000	49,000	3000	8800	111,800
6	Langsa	83,333	20,833	3000	7500	114,667
8	Jakarta	2,193,750	48,750	26,875	174,625	2,444,000
9	Bogor	3,000,000	200,000	100,000	250,000	3,550,000
10	Denpasar	3,494,000	80,000	100,000	107,500	3,781,500
11	Croatia	12,400,000	450,000	250,000	1,500,000	14,600,000
12	Germany	9,300,000	175,000	250,000	1,250,000	10,975,000
13	Netherlands	7,756,250	302,500	250,000	1,243,750	9,552,500
14	Switzerland	8,086,666	326,666	250,000	526,666	9,190,000
15	Puerto Rico	7,460,000	200,000	250,000	150,000	8,060,000
		Average after weighting each origin cluster of visitors				4,181,786

The costs incurred by respondents, according to the results of data recapitulation, in carrying out tourist activities (based on total travel costs), obtaining the economic value of the existence of Tangkahan Ecotourism, were IDR 72,708,168,000, -/year (equivalent to US $505,514.6). This value is obtained from the multiplication between the average value of visitor travel costs (IDR 4,181,786/visit) and the average number of tourist visits in 1 year. The average number of annual visits (31,200 people/year) was taken from data for the last three years from 2016 to 2018, based on data retrieved from the Tangkahan Tourism Institute in 2019.

The average travel cost of IDR 4,181,786 per visit is already high because Tangkahan is an ecotourism special interest tourism. Therefore, it selects visitors and attracts special groups of interested people willing to pay more to enjoy the specificity of a nature tourism area. In Bozkaada, Turkey, the economic value is TL 21,795,492.32 [21], and the travel cost value per person is TL 4.80 per travel or TL 110 per season [15]. Furthermore, the economic value of tourism in Kaziranga National Park is INR 773.45 million (INR 187.6 per visitor) [45]. Malaysia has a net economic value of MYR 6.2/visit/person [23].

This economic value is far greater than the revenue obtained from receiving an entrance ticket of IDR 3000/visit. The total revenue obtained by the manager from ticket sales is only IDR 71,814,000/year. This illustrates that the economic value of natural resources is from direct revenue and all benefits received by all parties related to the management/utilization of these natural resources. Therefore, this study uses a total travel cost approach to calculate the entire value received by the parties involved in Tangkahan Ecotourism.

In addition to material benefits, the existence of ecotourism also contributes to the welfare of communities around the forest in a non-material form. The important determinants of the quality of human life include the material and non-material domains [4]. The economic benefits of the existence of Tangkahan Ecotourism have an impact on material aspects, while changes in socio-cultural aspects such as increasing respect for the environment, strengthening community institutions in ecosystem management, changes in lifestyle and livelihoods, lack of potential disaster threats, and a sense of security are positive non-material impacts. These non-material benefits are inseparable from the development of Tangkahan Ecotourism as a form of active involvement of communities around national parks in managing forest ecosystems. Related to the above, as a result, [46] reported that the participating community in the development of Tangkahan Ecotourism has made a major contribution to the conservation of the Gunung Leuser National Park area. For more than 20 years the Tangkahan Ecotourism area has been running under the ausp:ces of a institution, namely the Tangkahan Tourism Institute (LPT) [11].

3.2. Factors Affecting Economic Value

An econometric evaluation with classical assumption tests, including multicollineari heteroscedasticity, and autocorrelation tests, was conducted before the regression analy The multicollinearity test results show that the VIF value is less than 10, and the toleran is less than 1 for all study variables. Based on the results of the multicollinearity test, t value of Variance Inflation Factor (VIF) was less than 10, and the value of Tolerance w less than 1 for all variables studied, so it was concluded that there was no multicollinear in the regression. The heteroscedasticity test using graphical aids also shows an ev distribution of points above and below the value 0. The autocorrelation test using t Durban Watson Test showed a DW value close to a value of 2. In general, the test resu stated that there was no violation of assumptions so it was feasible to proceed to the ne stage of testing [47–49].

Based on Tables 2 and 3, the regression equation obtained is $Y_1 = 8.295 + 0.192X_1$ $1.924X_2 + 1.518X_3 + 1.980X_4 - 0.733X_5 + 2.203X_6$. From the evaluation results of the t model, age, education, income, distance, number of members, and travel time significan affect individual visitors' travel costs with a coefficient of 57.1%. The factors that partia have a significant effect on the economic value of the Tangkahan Ecotourism area a education, income, distance of objects from the origin of visitors, and the travel time visitors to reach tourist sites.

Table 2. Simultaneous test results (F-test) for travel expenses value.

Model		Sum of Squares	Df	Mean Square	F	Sig.
1	Regression	324.812	6	54.135	38.260	0.000 [b]
	Residual	131.589	93	1.415		
	Total	456.402	99			

Dependent variable: the value of travel expenses; [b] Predictors: (constant), visitor age, education level, inco level, distance from area, number of members, length of travel, and acquisition of information.

Table 3. Partial test results (*t*-test) for travel expenses value.

Variable	B	Std. Error	Beta	T	Sig.
(Constant)	8.295	1.011		8.208	0.000
Age	0.192	0.223	0.051	0.860	0.392
Education	−1.924	0.875	−0.160	−2.199	0.030
Income	1.518	0.243	0.491	6.240	0.000
Distance	1.980	0.743	0.275	2.666	0.009
Number of Members	−0.733	0.562	−0.077	−1.305	0.195
Travel Time	2.203	0.760	0.309	2.899	0.005
R Adjusted	0.844				
R^2	0.712				

Education has a positive effect with a negative coefficient on the travel cost. There visitors with higher levels of education spend less on travel costs. This is not consis with the theoretical assumption that groups with higher education should be willin spend more [50,51]. Based on field observations, the lower travel cost of the more educa group is due to their better ability to organize visits and reduce travel costs individu These efforts include using more mass transportation facilities, more planned managem of visit activities, and better access to information technology. Therefore, transactions be conducted more efficiently through online-based ordering and transaction services

Income has a significant and positive effect on travel costs. This is consistent w the theoretical assumption that higher-income visitors will spend more to enjoy na tourism areas, as reported by [52] in Kodagu District, India, [53] concerning Nature-Based Tourism (NBT), and [54] in Kalam Valley of Khyber Pakhtunkhwa, Pakistan. Distance and travel time are synchronous factors, where the distance is directly proportional to the travel time from the visitor's origin to the location of a tourism site. These two variables have a significant and positive effect; therefore, the distance and the travel time are directly proportional. Visitors need to spend more to enjoy Tangkahan nature tourism areas [55].

3.3. Factors Affecting Intensity of Visits

According to descriptive statistical data, tourists visited Tangkahan 1.6 times on average. This illustrates that many visitors repeat visits to Tangkahan tourism areas due to the good impression obtained from various aspects. The intensity of visits to Tangkahan is low/moderate/high compared to other tourism areas studied.

The intensity of tourist visits in Tangkahan illustrates that there are still demands to improve the impression of visitors by increasing aspects that significantly affect the intensity of tourists. Improving service quality through important elements is expected to be more effective [56–59].

Table 4 explains that the F-count value is greater than the F-table. Therefore, the independent variable simultaneously has a significant effect on the dependent variable (intensity of tourist visits). This shows that travel cost, age, education level, income, distance from residence to tourism objects, number of members in the group, travel time to be taken, and the acquisition of information related to tourism sites affect the intensity of visiting Tangkahan nature locations. The high coefficient of determination in this regression model (93.2%) indicates that the eight variables above can simultaneously explain almost all changes/variations in the intensity of visits.

Table 4. Simultaneous test results (F-test) for visit intensity.

	Model	Sum of Squares	Df	Mean Square	F	Sig.
1	Regression	14.489	8	1.811	17.440	0.000 [b]
	Residual	9.450	91	0.104		
	Total	23.940	99			

Dependent variable: intensity of visits; [b] Predictors: (constant), information, number of members, education, age, travel time, income, distance, travel cost.

The F-test has not shown which independent variables directly and significantly affect the dependent variable (intensity of visits). Therefore, it is continued with multiple linear regression analysis to determine which variables significantly affect the dependent. Based on Table 5, the regression equation obtained is $Y_2 = 1.570 + 0.072X_1 - 0.140X_2 + 0.415X_3 - 0.185X_4 - 1.457X_5 - 0.244X_6 - 0.026X_7 - 0.066X_8$.

Table 5. Partial test results (*t*-test) for visit intensity.

Variable	B	Std. Error	Beta	T	Sig.
(Constant)	1.570	0.361		4.352	0.000
Travel Cost	0.072	0.029	0.315	2.475	0.015
Age	−0.140	0.064	−0.162	−2.194	0.031
Education	0.415	0.244	0.151	1.705	0.092
Income	−0.185	0.079	−0.262	−2.357	0.021
Distance	−1.457	0.209	−0.882	−6.953	0.000
Number of Members	−0.244	0.154	−0.111	−1.587	0.116
Travel Time	−0.026	0.216	−0.016	−0.120	0.904
Information	−0.006	0.069	−0.006	−0.087	0.931
R Adjusted	0.778				
R^2	0.605				

Simultaneously, these predictor variables have a significant effect on the inten variables of visits. These independent variables significantly affect the dependent (inten of visits). However, the partial test shows that not all independent variables affect intensity of visits. The test results showed that of the seven socioeconomic varial observed in this study, there were only two variables that had a significant effect (us $\alpha = 0.05$) on the intensity of tourist visits to the Tangkahan Ecowista area. The two varial are the distance and the number of members with a negative sign. It shows that vi will increase as the distance decreases and the number of members decreases. Dista is very influential in the selection of tourism objects. Another study stated that 4 ou 10 independent variables tested directly affected tourist revisits, such as safety and secu description of destinations, infrastructure, and price [60].

This study found that the farther the tourism site from the residence, the less inte the tourist visits. Visitors prefer tourist destinations that are closer to their homes indicating distance significantly affects tourist visits, specifically costs and benefits [61- People closer to tourism areas are more supportive of tourism activities than those far a The distance factor is often a barrier to tourist visits to nature tourism areas; therefor is necessary to support adequate regional transportation infrastructure to minimize distance factor with good access quality to shorten travel time.

The greater the number of members in the visiting group, the less the intensity of v to Tangkahan. This shows the tendency of tourists to enjoy visits with fewer mem The major attractions of nature tourism are rivers and landscapes; thus, people p to experience them in smaller groups. Therefore, it is necessary to have facilities tourism object designs that support small and personal group-based activities to enjoy privacy more.

The other five variables, such as travel cost, age, education, income, travel time, information acquisition, have no significant effect (using $\alpha = 0.05$) on the intensity of to visits. In general, traveling can be conducted on weekends and national holidays. foreign tourists, visits are made during seasonal holidays. During that period most pe will plan trips to tourist attractions that present recreational attractions. The facto per-street cost, age, education, income, length of travel, and acquisition of information not to be significant considerations for tourists, so that it has a negligible influence o intensity of tourist visits.

3.4. Tourist Interest in Revisiting

The average score of interest in revisiting the Tangkahan Ecotourism area is 4.21 scale of 5). This total score is in the very high category (slightly past the very high- limit); therefore, most tourists express interest in revisiting. This answer does not de on the intensity of visits to describe the visitors' impression of the area. The intere revisiting is increases when it is close to a score of 5 (total score of 500), where all visito willing to come back for a tour according to Pareto terms. The interest score in revisitir 4.21 is very high compared to other tourism attractions. According to [64], there is a to 2948 people willing to revisit the Ciwangun Indah Camp. In Banyuwangi, the Effectiv of Tourism Destination Advertisements on Interest in Revisiting had a score of 3.66 [

This study shows that satisfaction is a factor that has a direct significant influ on short-term return visit intentions, while the novelty of tourist attractions is a f that has a significant effect on medium- to long-term return visit intentions [66]. I United Arab Emirates, satisfaction affects the interest in revisiting [67]. The impres of visitors which are positive but have not reached the highest score illustrate that though Tangkahan nature tourism areas are quite attractive to visitors, some aspe service need to be addressed. Improvements carried out effectively are expected to inc the average score of the interest in revisiting to close to 5. The aspects developed shou prioritized to affect the interest/willingness/intention to revisit significantly.

A simple model of estimating the tourist interest in revisiting (Y_3) is obtained by travel cost (Y_1) as the independent variable (Table 6). The regression analysis results that travel cost significantly affects the tourist interest in revisiting (Table 7). However, this regression model is simple, with a sufficient coefficient of determination at 43%. This also tends to be different from the theoretical assumption that the cost is a factor inhibiting tourist interest in revisiting. Empirically, this is possible because Tangkahan natural tourism is a special interest tourism, so the cost factor tends to be in elastic. The average travel cost to enjoy this tourist area is relatively low; hence, it has not been considered a factor that becomes a negative factor for visiting. However, it is still necessary to conduct a special study to explain this matter further.

Table 6. F-test results for tourist interest in revisiting.

	Model	Sum of Squares	Df	Mean Square	F	Sig.
	Regression	0.156	1	0.156	5.400	0.022 [b]
1	Residual	2.835	98	0.029		
	Total	2.991	99			

Dependent variable: interest in revisiting; [b] Predictors: (constant), travel cost.

Table 7. Linear regression results for tourist interest in revisiting.

Variable	B	Std. Error	Beta	T	Sig.
(Constant)	2.288	0.102		22.406	0.000
Travel Cost	0.019	0.008	0.229	2.324	0.022
R Adjusted	0.229				
R^2	0.052				

4. Conclusions

The economic value of the Tangkahan Ecotourism area with the zoned travel cost method is IDR 72,708,168,000/year. On average, tourists have visited Tangkahan 1.6 times. Simultaneously, the factors of age, education, income level, distance, number of members, and travel time significantly affect the value of the individual travel costs of visitors, with a regression model $Y_1 = 8.295 + 0.192X_1 - 1.924X_2 + 1.518X_3 + 1.980X_4 - 0.733X_5 + 2.203X_6$. Education, income, distance, and travel time partially affect the economic value. This is a reference for the right promotional segmentation policy in order to increase the economic value of the existence of Tangkahan Ecotourism. The tourist interest in revisiting Tangkahan nature tourism objects is very high (score 4.21). Generally, travel cost, age, education level, income level, distance, number of members, travel time, and information acquisition significantly affect the intensity of visits with regression model $Y_2 = 5.975 + 1.040 \times 10^{-8}X_1 - 0.097X_2 + 0.267X_3 + 0.121X_4 - 0.723X_5 - 0.515X_6 - 0.116X_7 - 0.190X_8$. Distance, number of members, and travel cost significantly affect the intensity of tourist visits to the Tangkahan Ecotourism area. Ecotourism managers must improve accessibility infrastructure, increase comfort for visitors with large groups, and minimize travel costs to increase tourist interest in visiting again. Socio-cultural variables and visitors' assessment of ecotourism sustainability aspect are recommended to be involved in the next research.

Author Contributions: Drafting concepts, A.P. and D.R.N.; data analysis, A.P., D.R.N. and H.L.G.; format adjustment, H.L.G.; fund-raising, A.P.; clarification of data, A.P., M.E. and H.L.G.; method design, A.P.; field equipment, H.L.G.; software, S.R.; supervision, S.R. and M.E.; visual display, A.P. and H.L.G.; writing and editing drafts, A.P. and H.L.G. All authors have read and agreed to the published version of the manuscript.

Funding: This study received no external funding.

Institutional Review Board Statement: The study was conducted in accordance with the Declaration of Helsinki, and approved by the Ethics Committee of Universitas Sumatera Utara (No. 396/KEPK/USU/2022, date of approval 22 April 2022).

Informed Consent Statement: Informed consent was obtained from all subjects involved in the study.

Data Availability Statement: Not applicable.

Acknowledgments: The authors would like to thank the Universitas Sumatera Utara for fundi this research. We also extend our gratitude to The Manager of Gunung Leuser National Park (TNC and the Tangkahan Tourism Institute (LPT) for access, data, and explanations in the field. We a express our gratitude to visitors to Tangkahan Ecotourism for their willingness to be respondents

Conflicts of Interest: The authors declare no conflict of interest.

References

1. Fisher, M.R.; Dhiaulhaqi, A.; Sahide, M.A.K. The politics, economies, and ecologies of Indonesia's third generation of soc forestry: An introduction to the special section. *For. Soc.* **2019**, *3*, 152–170. [CrossRef]
2. Affandi, O.; Zaitunah, A.; Batubara, R. Potential Economic and Development Prospects of Non-Timber Forest Products Community Agroforestry Land around Sibolangit Tourism Park. *For. Soc.* **2017**, *1*, 68–77. [CrossRef]
3. Chan, J.K.L.; Marzuki, K.M.; Mohtar, T.M. Local Community Participation and Responsible Tourism Practices in Ecotouri Destination: A Case of Lower Kinabatangan, Sabah. *Sustainability* **2021**, *13*, 13302. [CrossRef]
4. Eslami, S.; Khalifah, K.; Mardani, A.; Streimikiene, D.; Han, H. Community attachment, tourism impacts, quality of life a residents' support for sustainable tourism development. *J. Travel Tour. Mark.* **2019**, *36*, 1061–1079. [CrossRef]
5. Roxas, F.M.Y.; Rivera, J.P.R.; Gutierrez, E.L.M. Mapping stakeholders' roles in governing sustainable tourism destinations. *J. H Tour. Manag.* **2020**, *45*, 387–398. [CrossRef]
6. Rachmawati, E.; Joanna, F. Role of External Stakeholders in Tourism Development and Community Empowerment. *For.* **2017**, *1*, 25. [CrossRef]
7. Yusnikusumah, T.R.; Sulystiawati, E. Evaluation of Ecotourism Management in Tangkahan Ecotourism Area of Gunung Leu National Park, North Sumatra. *J. Reg. City Plan* **2016**, *27*, 173–189.
8. Hayati, N.; Wakka, A.K. Valuasi Ekonomi Manfaat Air Di Taman Nasional Bantimurung Bulusaraung, Sulawesi Selatan. *J. Per Sos. Dan Ekon. Kehutan.* **2016**, *13*, 47–61. [CrossRef]
9. Ponte, J.; Couto, G.; Pimentel, P.; Sousa, A.; Oliveira, A. Tourism planning in the Azores and feedback from visitors. *Tour. Ma Stud.* **2021**, *17*, 7–15. [CrossRef]
10. Ernawan, F.N.; Harini, R. Economic valuation of Blue Lagoon Tourism Village Widodomartani Sleman. *E3S Web Conf.* **2** *200*, 03003. [CrossRef]
11. Tejada, J.J.; Joyce, R.B.P. On the Misuse of Slovin's Formula. *Philipp. Stat.* **2012**, *61*, 129–136.
12. Ellen, S. Slovin's Formula Sampling Techniques. 2012. Available online: http://www.ehow.com/way_5475547_slovins-form sampling-techniques.html (accessed on 3 May 2020).
13. Sugiyono, D. *Metode Penelitian Kuantitatif, Kualitatif dan R&D*; Penerbit Alfabeta: Bandung, Indonesia, 2017.
14. Sulistio, T.D.; Fitriana, R.; Melisa, L. The Influence of Words of Mouth and Sapta Pesona on the Decision to Visit Natsepa Be *Enrich. J. Manag.* **2021**, *11*, 334–337.
15. Salfadri; Hadya, R. Determining Factors for Tourists Visiting Carocok Painan Beach. *Int. J. Multi Sci.* **2020**, *1*, 67–81.
16. Rahmafitria, F.; Nurazizah, G.R. Community Based Tourism: A Correlation Between Knowledge and Participation in Moun Based Destination. In Proceedings of the 1st UPI International Conference on Sociology Education (UPI ICSE 2015), Band Indonesia, 12 October 2015.
17. Syamsudina, T.; Kosmaryandi, N.; Nopiansyah, F. Evaluation of Natural Tourism Objects Potential in Nature Tourism Re Kuningan National Park Management Section I Gunung Ciremai National Park. *Int. J. Sci. Basic Appl. Res. (IJSBAR)* **202** 44–54.
18. Das, S. Travel Cost Method For Environmental Valuation. Center of Excellence in Environmental Economics. *Madras Sch.* *Chennai* **2013**, *600*, 025.
19. Brown, G., Jr.; Mendelsohn, R. The hedonic travel cost method. *Rev. Econ. Stat.* **1984**, *66*, 427–433. [CrossRef]
20. Lamtrakul, P.; Teknomo, K.; Hokao, K. Public Park Valuation Using Travel Cost Method. *Proc. East. Asia Soc. Transp. Stud.* **20** 1249–1264.
21. Çay, R.D.; Taşli, T.C. Determination of Recreation and Tourism Use Value of Bozcaada by Travel Cost Analysis Methods *J. Environ. Stud.* **2021**, *30*, 35–45. [CrossRef]
22. Gimenez, F.V.; Mas, C.V. The Valuation of Recreational Use of Wetlands and the Impact of the Economic Crisis. *Int. J. Environ Public Health* **2020**, *17*, 3228. [CrossRef]
23. Othman, J.; Jafari, Y. Economic Valuation of an Urban Lake Recreational Park: Case of Taman Tasik Cempaka in Bandar Bangi, Malaysia. *Sustainability* **2019**, *11*, 3023. [CrossRef]
24. Douglas, A.J.; Taylor, J.G. A New Model for The Travel Cost Method: The Total Expenses Approach. *Environ. Model Softw.* *14*, 81–92. [CrossRef]
25. Sulistiyono, N. *Pemanfaatan Jasa Lingkungan Hutan sebagai Kawasan Ekotourisme*; USU e-Repository: Medan, Indonesia, 200
26. Mayor, K.; Scott, S.; Tol, R.S.J. Comparing the travel cost method and the contingent valuation method—An applicati convergent validity theory to the recreational value of Irish forests. In *ESRI Working Papers No 190*; Environmental Econo Institute for Environmental Studies: Daejeon, Korea, 2014.

Latinopoulos, D. The impact of economic recession on outdoor recreation demand: An application of the travel cost method in Greece. *J. Environ. Plann. Manag.* **2014**, *57*, 254–272. [CrossRef]
Ezebilo, E.E. Economic value of a non-market ecosystem service: An application of the travel cost method to nature recreation in Sweden. *Int. J. Biodivers. Sci. Ecosyst. Serv. Manag.* **2016**, *12*, 314–327. [CrossRef]
Croasmun, J.T.; Lee, O. Using Likert-Type Scales in the Social Sciences. *J. Adult Educ.* **2011**, *40*, 19–22.
Lembaga Pariwisata Tangkahan. *Database Lembaga Pariwisata Tangkahan: Annual Report of Tangkahan Tourism Institute. Stabat*; Tangkahan Tourist Institute: Tangkahan, Indonesia, 2019.
Nasution, H.R.; Herwanti, S.; Febryano, I.G.; Winarno, G.J. Characteristics of Business Actors and Visitors in the Tangkahan Ecotourism Area during The COVID-19 Pandemic. *Gorontalo J. For. Res.* **2021**, *4*, 61–71. [CrossRef]
McGuckin, N.; Murakami, E. Examining Trip-Chaining Behavior: Comparison of Travel by Men and Women. *Transp. Res. Rec. J. Transp. Res. Board* **1999**, *1693*, 79–85. [CrossRef]
Collins, D.; Tisdell, C. Gender and Differences in Travel Life Cycles. *J. Travel Res.* **2002**, *41*, 133–143. [CrossRef]
Hasan, R.; Shaw, C.; Russell, M.; Keall, M.; Wild, M.; Woodward, A.; MacBride-Stewart, S. How do Female Patterns of Utility Cycling, and Travel More Generally, Differ From Men? Cross Sectional Survey in New Zealand. *J. Transp. Health* **2020**, *14*, 100635. [CrossRef]
Suzana, B.O.L.; Timban, J.; Kaunang, R.; Ahmad, F. Economic Valuation of Mangrove Forest Resources in Palaes Village, West Likupang Subdistrict, North Minahasa Regency. *Agri-Sosioekonomi* **2011**, *7*, 9–38. [CrossRef]
Noho, Y.; Wijaya, R.; Anugrah, K. Economic Value Analysis of Ecotourism Development of Lake Limboto Area Using Travel Cost Method Approach. *J. Ilmu Pendidik. Nonform.* **2020**, *6*, 1–8. [CrossRef]
Arsalan, A.; Gravitiani, E.; Irianto, H. Valuasi Ekonomi Ekowisata Kalibiru dengan Individual Travel Cost Method. In *National Seminar on Biology Education*; Sebelas Maret University: Surakarta, Indonesia, 2018; pp. 110–118. ISBN 978-602-61265-2-8.
Khoirudin, R.; Khasanah, U. Valuasi Ekonomi Objek Wisata Pantai Parangtritis, Bantul Yogyakarta Economic Valuation of Parangtritis Beach, Bantul Yogyakarta Pendahuluan. *J. Ekon. Dan Pembang. Indones.* **2018**, *18*, 152–166. [CrossRef]
Mahardhika, S.M.; Saputra, S.W.; Ain, C. Economic Valuation of Fisheries Resources and Mangrove Ecotourism in Muara Angke, Jakarta. *J. Maquares* **2018**, *7*, 458–464.
Zulpikar, F.; Prasetiyo, D.E.; Shelvatis, T.V.; Komara, K.K.; Pramudawardhani, M. Economic Valuation of Environmental Service-Based Tourism Object in Batu Karas Beach-Pangandaran Using The Travel Cost Method. *J. Reg. Rural Dev. Plan.* **2017**, *1*, 53–63.
Subardin, M.; Yusuf, M.K. Valuasi Ekonomi Menggunakan Metode Travel Cost pada Taman Wisata Alam Punti Kayu Palembang. *J. Ekon. Pembang.* **2011**, *9*, 81–89.
Premono, B.T.; Kunarso, A. Economic Valuation on Punti Kayu Recreation Park Palembang. *J. Penelit. Hutan Dan Knservasi Alam* **2010**, *7*, 13–23. [CrossRef]
Marjuka, M.Y. Econometric Valuation with Travel Cost Method on The Ecotourism Objects of Small Islands (Case of Kepulauan Seribu). *Bina Ekonomi* **2007**, *11*, 80–100. [CrossRef]
Aryanto, R.; Mardjuka, M.Y. Valuasi Ekonomi Dengan Travel Cost Method pada Obyek Ekowisata Pesisir (Kasus Kawasan Ujung Genteng, Sukabumi). *J. Ilm. Pariwisata* **2005**, *10*, 58–76.
Bharali, A.; Mazumder, R. Application of Travel Cost Method to Assess The Pricing Policy of Public Parks: The Case of Kaziranga National Park. *J. Reg. Dev. Plan.* **2012**, *1*, 44–52.
Anggraini, R.I.; Budhi, G. Ecotourism Development in National Parks: A New Paradigm of Forest Management in Indonesia. *E3S Web Conf.* **2021**, *249*, 03010. [CrossRef]
Qudratullah, M.F. *Analisis Regresi Terapan: Teori, Contoh Kasus, dan Aplikasi Dengan SPSS*; Penerbit Andi: Yogyakarta, Indonesia, 2012.
Hamid, M.; Sufi, I.; Konadi, W.; Akmal, Y. *Analisis Jalur Dan Aplikasi Spss Versi 25*, 1st ed.; Sefa Bumi Persada: Lhokseumawe, Indonesia, 2019.
Ghozali, I. *Aplikasi Analisis Multivariate Dengan Program SPSS*; Badan Penerbit Universitas Diponegoro: Semarang, Indonesia, 2011.
Rivas, J.C.; Sánchez, R.M. Willingness to Pay for More Sustainable Tourism Destinations in World Heritage Cities: The Case of Caceres, Spain. *Sustainability* **2020**, *11*, 5880. [CrossRef]
Cai, Y.; Jin, L.; Du, B. Effects on Willingness to Pay for Marine Conservation: Evidence from Zhejiang Province, China. *Sustainability* **2018**, *10*, 2298. [CrossRef]
ardana, K. Tourists' Willingness to Pay for Restoration of Traditional Agro-forest Ecosystems Providing Biodiversity: Evidence from India. *Ecol. Econ.* **2019**, *159*, 362–372. [CrossRef]
McCreary, A.; Fatoric, S.; Seekamp, E.; Smith, J.W.; Kanazawa, M.; Davenport, M.A. The Influences of Place Meanings and Risk Perceptions on Visitors' Willingness to Pay for Climate Change Adaptation Planning in a Nature-Based Tourism Destination. *J. Park Recreat. Adm.* **2018**, *36*, 121–140. [CrossRef]
anaullah, F.; Rabbi, S.A.; Khan, Z.; Zamin, M. Visitors' Willingness to Pay for Conservation of the Biodiversity and Tourism in alam Valley of Khyber Pakhtunkhwa, Pakistan. *Sarhad J. Agric.* **2020**, *36*, 81–94. [CrossRef]
uea, L.; Zhang, Y. The effect of distance on tourist behavior: A study based on social media data. *Ann. Tour. Res.* **2020**, *82*, 102916. [CrossRef]

56. Narayan, B.; Rajendran, C.; Sai, L.P.; Gopalan, R. Dimensions of service quality in tourism—An Indian perspective. *Total Q Manag.* **2009**, *20*, 61–89. [CrossRef]
57. Haghkhah, A.; Nosratpour, M.; Ebrahimpour, A.; Hamid, A.B.A. The impact of service quality on tourism industry. In Proceed of the 2nd International Conference on Business and Economic Research Proceeding, Langkawi, Malaysia, 14–16 March 201
58. Ţîţu, M.A.; Răulea, A.S.; Ţîţu, Ş. Measuring service quality in tourism industry. *Procedia-Soc. Behav. Sci.* **2016**, *221*, 294– [CrossRef]
59. Kachwala, T.; Bhadra, A.; Bali, A.; Dasgupta, C. Measuring customer satisfaction and service quality in tourism industry. *SM J. Bus. Manag. Stud.* **2018**, *14*, 42–48. [CrossRef]
60. Ngoc, K.M.; Nguyen, T.T. Factors Affecting Tourists' Return Intention towards Vung Tau City, Vietnam-A Mediation Analys Destination Satisfaction. *J. Adv. Manag. Sci.* **2015**, *3*, 292–298. [CrossRef]
61. Soukotta, L.M.; Willem, T.; Eygner, G.T. Relationship of Marketing Mix Factors and Interest in Returning to Lubang Buaya B Tourism. In Proceedings of the National Seminar on Marine and Fisheries, Nusaniwe, Indonesia, 18–19 December 2019; Fac of Fisheries and Marine Sciences, Universitas Pattimura Ambon: Nusaniwe, Indonesia, 2019. [CrossRef]
62. Wiseza, F.C. Faktor-faktor yang Mendukung Pengembangan Obyek Wisata Bukit Khayangan di Kota Sungai Penuh Prov Jambi. *J. Nur. El-Islam* **2017**, *4*, 89–106.
63. Jurowski, C.; Gursoy, D. Distance Effects on Residents' Attitudes Toward Tourism. *Ann. Tour. Res.* **2004**, *31*, 296–312. [Cross
64. Wiradipoetra, F.A.; Erlangga, B. Analysis of Travelers' Perceptions of The Decrease in The Quality of Tourist Attractio Visiting Interests. *J. Pariwisata* **2016**, *3*, 129–137. [CrossRef]
65. Suryaningsih, I.B.; Kristian, S.W.N. Epic Model: Effectiveness of Banyuwangi Regency Tourist Destination Ads Against Dom Tourists' Revisiting Interest. *Manag. Insight* **2018**, *13*, 8–26. [CrossRef]
66. Jang, S.C.; Feng, R. Temporal destination revisit intention: The effects of novelty seeking and satisfaction. *Tour. Manag.* **200** 580–590. [CrossRef]
67. Jimenez, R.G.; Reyes, Z.M.; Garcíab, J.M.P.; Margalida, A. Economic Valuation of Non-Material Contributions to People Prov by Avian Scavengers: Harmonizing Conservation and Wildlife-based Tourism. *Ecol. Econ.* **2021**, *187*, 107088. [CrossRef]

forests

MDPI

ticle

nhancing Regional Tourism Development in the Protected reas Using the Total Economic Value Approach

aaz Kabil [1,2], Rahaf Alayan [1,*], Zoltán Lakner [1] and Lóránt Dénes Dávid [1]

1 Doctoral School of Economic and Regional Science, Hungarian University of Agriculture and Life Sciences (MATE), 2100 Godollo, Hungary; moaazkabil@cu.edu.eg (M.K.); lakner.zoltan.karoly@uni-mate.hu (Z.L.); david.lorant.denes@uni-mate.hu (L.D.D.)

2 Faculty of Urban and Regional Planning, Cairo University, Giza 12613, Egypt

* Correspondence: alayan.rahaf.ahmad@phd.uni-szie.hu

Abstract: This research aims to boost tourism development in natural protected areas through the classification of the tourism hiking trails based on biodiversity ecosystem services values. The Total Economic Value (TEV) approach was used as the main research method to estimate the monetary value of the various ecosystem resources in Abu Qubies Syrian protected area. Five main tourism hiking trails in Abu Qubies were identified in order to be classified based on the economic value of their ecosystem resources. The study findings highlighted the importance of protected areas in enhancing tourism activities, especially natural-based ones. Additionally, this research identified the most economically valuable tourism hiking trails in the reserve, thus providing a supporting tool for decision-makers regarding tourism development in protected areas. This study presents the importance of the conservation perspective of natural resources in protected areas without ignoring their physical monetary value that can help governments in boosting local communities as well as the national, regional and local economies.

Keywords: nature reserves; nature-based tourism; biodiversity; TEV; hiking; ecosystem services

ion: Kabil, M.; Alayan, R.; er, Z.; Dávid, L.D. Enhancing nal Tourism Development in rotected Areas Using the Total omic Value Approach. *Forests* 13, 727. https://doi.org/ 90/f13050727

emic Editors: Panayiotis mitrakopoulos, Mario gnotta, Miklas Scholz and ya Noorani

ved: 27 March 2022
ted: 4 May 2022
shed: 6 May 2022

1. Introduction

Protected areas represent significant importance in our world. According to The International Union for Conservation of Nature (IUCN), approximately 25% of the Earth planet (10% of the land and 15% of the territorial waters) are designated as various forms of nature reserves and national parks [1]. Additionally, the number and size of these areas have increased dramatically in the last 20 years [2]. Protected areas also have rich economic value due to their biological, cultural and environmental resources. All these aspects related to economic value have made these nature reserves the focus of various global economic sectors and industries, including the tourism [3]. Therefore, many protected areas have gained prominence as tourism destinations, especially with the expansion of nature-based tourism forms such as ecotourism, as well as the growing demand for outdoor tourism activities [4,5].

The relationship between protected areas and tourism as an economic sector can be described as a "complex" relationship [6]. The definition of the protected areas according to IUNC is "An area of land and/or sea especially dedicated to the protection and maintenance of biological diversity, and of natural and associated cultural resources, and managed through legal or other effective means" [7]. Within this previous definition, the concept of establishing the nature reserves seeks to achieve sustainable conservation and management of the rare ecosystem resources in these areas such as the natural landscape, unique ecosystem services, rare fauna and flora, local communities and cultural heritage [8,9].

Therefore, tourism and protected areas are mutually beneficial. For one, tourism seeks to create new tourism patterns/markets, cover the various needs of different tourists, increase the economic value of this important economic sector and increase its contribution to the GDP of the nations. Additionally, the tourism industry seeks to use the protect areas' potential to boost economic opportunities by providing jobs for local communiti improving their living standards, increasing possible sources of funding for nature reserv and local societies and encouraging local goods and SMEs [4].

On the other hand, the continuing momentum that tourism creates for protected are by exporting them as promising tourism destinations highlights the interest in conservati and maintenance of the natural values that nature reserves seek [10]. For example, p tecting natural ecosystems and watersheds, creating economic values for underrated lo natural resources, supporting environmental principles for visitors and locals, enhanci intercultural understanding, helping locals to learn foreign languages and improving th educational skills, transmitting conservation value and protecting the natural and cultu heritage of the countries; all these listed issues are considered as the most important bene that protected areas reap from being considered tourist attractions [11–13].

This relationship between protected areas and tourism development has also hi lighted the importance of achieving sustainability during using the various natural sources in the different kinds of nature reserves. With the emergence of the United Natic Sustainable Development Goals (SDGs) in 2015 [14], tourism has been a major activity sustainable development. Additionally, two of the main SDGs have been associated w protected areas: SDG No. 14 "Life below water" and SDG No. 15 "Life on land". Th the relationship between nature reserves and tourism activities has gained ground on global stage in terms of realizing the goals of the sustainability approach.

Moreover, it is worth mentioning the recent impact of COVID-19 on the tourism sec which is one of the most affected economic sectors. According to the Organisation Economic Co-operation and Development (OECD) statistics, the global tourism sec deteriorated by 45% due to the pandemic in 2020 [15]. Moreover, the World Tourism O nization (UNWTO) reported that 100% of the countries involved imposed travel restricti on tourism destinations, and 45% of global tourism destinations were completely close tourists because of this pandemic [16]. Accordingly, world tourism organizations and ag cies have sought different tourism patterns to enhance tourism development [17]. Theref outdoor activities in nature reserves highlighted the importance of protected areas as glo tourism destinations, which play a significant role in achieving the sustainability a prosperity of the global tourism industry [18].

Natural environmental resources in protected areas support numerous tourism tivities, particularly outdoors ones. For example, walking, cycling, camping, hiking a wildlife observation, as well as diving and kayaking in the case of marine nature reserves Tourism hiking trails are one of the most famous recreational and tourism activities t support the sustainability paradigm in the tourism industry [19]. Tourism hiking trails a link between nature and people [20]. Although hiking trails are considered recreatic and leisure activities for people, they are currently one of the most important tour products, especially after the COVID-19 pandemic [21]. Additionally, it is worth no that tourism hiking trails have gained importance on the global stage because of t inexpensive infrastructure commensurate with the current deterioration and declir growth of the tourism industry globally, as well as the current tourism demand [20,22

Despite the aforementioned importance of the relationship between protected a and tourism as an economic activity, the negative impact that tourism may have on th areas must not be overlooked. Economically, the various tourism activities in the prote areas increase the demand for tourism products and services such as restaurants, ho and other attractions. Thus, tourism activities lead to providing many basic serv such as policing, safety, electricity, and health care, which in turn lead to higher taxe these local areas that may be difficult for the locals to live and cope with [4]. In addi the fact that the tourism sector is a fragile and seasonal one and the protected a show a large dependence on this sector may lead to a reduction in their ability to co certain external factors such as natural disasters, climate change, changing tourist beha and other economic and political crisis [4]. Consequently, this can affect the econc stability of local communities in these protected areas. Socially, the tourism industry may have a negative impact on nature reserves by depriving the locals of their rights and freedom to make their own decisions and control their lives [9]. Additionally, tourism activities can cause a proliferation of many problems that disturb the locals' lives, such as overcrowding, vandalism, littering, crime, unpredictability and disrespect for local customs and traditions [23]. Environmentally, intensive tourism activities can have serious consequences for protected areas such as habitat loss, changing land uses, air pollution, increasing fire risk, flora and fauna damage, etc. [24,25].

Because of all these possible negative impacts that may be caused by the tourism development of nature reserves, good development plans and policies (economically, socially and environmentally) must be developed, pursued and accomplished, thereby reducing the impact of these consequences and enhancing the concept of the sustainable tourism development in protected areas [26].

Accordingly, this paper seeks to amplify the conservation of the environmental resources in the nature reserves with achieving their most economic use, through the classification of the tourism hiking trails in Abu Qubies. The Total Economic Value (TEV) was used as the main approach to classify the tourism hiking trails in Abu Qubies by valuing the ecosystem services. Thus, economic tools are given to decision makers to facilitate the adoption of the best development decisions that will enhance the concept of sustainable tourism development in Abu Qubies and the economic optimization of its natural resources.

2. Materials and Methods

2.1. Study Area

Abu Qubies natural protected area is located on the coastal eastern mountain range of the Syrian governorate Hama (see Figure 1). Moreover, Abu Qubies has a strategic location linking three Syrian governorates (Latakia, Hama and Tartus), with an area of 11,000 hectares and an altitude ranging from 230 to 1370 m above sea level. Abu Qubies was declared as an environmental protected area by the Syrian Ministry of Agriculture and Agrarian Reform (MoAAR) in 1999, with the aim of protecting biodiversity in the Syrian western region and for conducting environmental scientific research projects and studies [27]. According to the United Nations Development Programme (UNDP) in Syria and the Global Environment Facility (GEF), only three Syrian protected areas (out of 115) play a significant role in preserving biodiversity in Syria, and Abu Qubies is one of them, in addition to Al Frunloq and Jabal Abdul-Aziz [28].

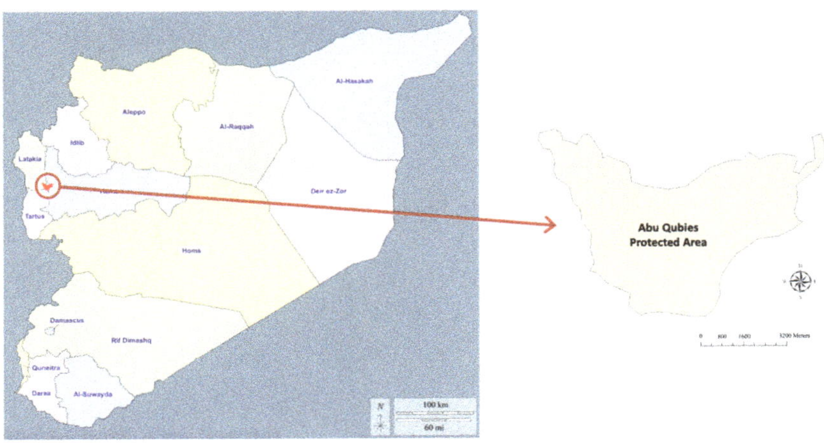

Figure 1. Abu Qubies protected area, Syria (case study).

A range of factors were considered before choosing the study area, both national local. Regarding the national level, the 115 Syrian wilderness reserves cover a surfac 1,132,286 hectares, making up a total of 10.2% of Syria's territory [29,30]. Addition these Syrian protected areas are highly valuable in terms of biodiversity, being a habitat about 70% of the country's unique endangered plant and animal species [31]. Although importance of the protected areas in terms of both biodiversity and economic value, have not contributed effectively to the Syrian economy. According to the Syrian Natic Bureau of Statistics for 2002, all different kinds of Syrian protected areas contribute 0.01% of the Syrian GDP, taking into account all economic sectors. Such a scarce econor relevance is often associated with a notable neglect of environmental conservation iss

On the local level, Abu Qubies has rare different features that make it suitable for research proposal and gives it a qualitative significance away from other Syrian prote areas. For example, Abu Qubies has a distinct geographical location, being an impor area in the migratory trails of birds and a habitat for many birds included on the of endangered birds by BirdLife International [32]. Furthermore, Abu Qubies has m cultural values that can enhance tourist development within it, such as Roman monum the remains of a fortress called the Abu Qubies Castle at 550 m above sea level, a gr of local villages and several religious shrines. Additionally, Abu Qubies well repres the western Syrian landscape and the Mediterranean basin region. Where the nat vegetation cover of the Abu Qubies is characterized by valuable biodiversity and distin ecosystems, with a series of different species of plant communities, making it a rich na heritage [33,34].

The key ecological features of Abu Qubies include birds (e.g., the endangered Sy eagle and other 102 species of birds for which the protected area represents a bottle during their migration) [35], natural plants (e.g., more than 300 tree, shrub and species) [34], mammals (e.g., artiodactyls and hyenas) [36], reptiles and amphibians unique land uses (see Figure 2).

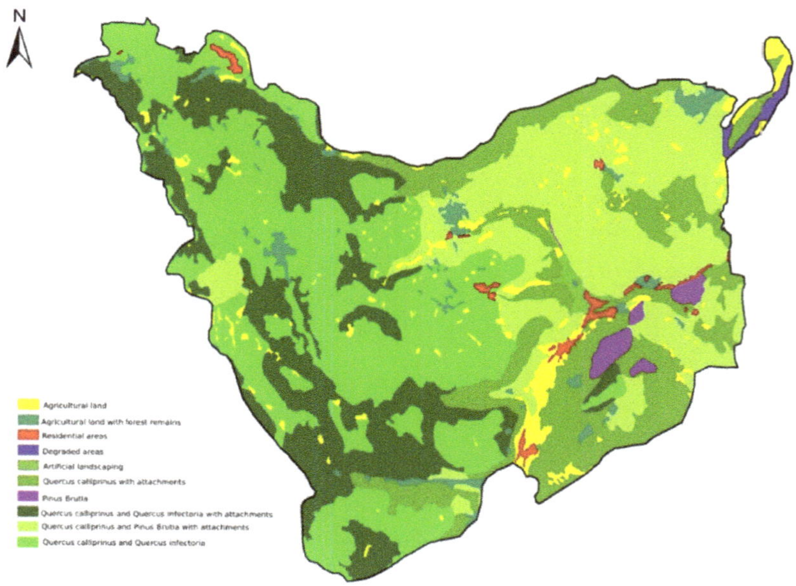

Figure 2. Land use and land cover in Abu Qubies as a unique ecological feature.

Consequently, this research selected Syria in general—and the Abu Qubies pro area in particular—to present the idea of how to optimize the economic dimensi ecosystem services' value in the tourism development sector while preserving the area's unique natural resources.

2.2. Research Method

This research basically aims to classify the tourism hiking trails in the Abu Qubies by valuing the ecosystems services, therefore providing helpful tools for decision-makers that will promote the sustainable management of the protected area's resources and optimize the economic value of its ecosystem services provided by biodiversity. The research method combined two main stages: (I) identify the tourism hiking trails in Abu Qubies and (II) calculate/estimate the ecosystem services' values in the Abu Qubies protected area by using the Total Economic Value (TEV) approach. Different software and analytical tools are used in order to apply this proposed research method such as on-site observation, Geographic Information System (GIS) and Earth Resources Data Analysis System (ERDAS).

2.2.1. Stage (I): Identify the Tourism Hiking Trails in Abu Qubies Protected Area

At this stage, the tourism hiking trails of Abu Qubies were identified by relying on two major data sources: the Syrian Ministry of Agriculture and Agrarian Reform (MoAAR), as well as the researchers' on-site observation and interview with the director of the protected area. According to MoAAR, the Abu Qubies has five main tourism hiking trails. The first trail is "Sheikh Ali Majdal-Beer Eljabal", located in the south of Abu Qubies and considered one of the longest tourism hiking trails with a length of 7.7 km. The second trail is "Abu Qubies-Mshta Elbeer", located in the north of the protected area and considered the second longest trail with a length of 6.8 km. The third trail is "Ras Alshareh-Sheikh Hatem", located in the west of Abu Qubies and is 6.3 km long. The fourth trail is "Sheikh Abdullah-Beer Eljabal", located in the middle of the protected area, which is 3.9 km long. Finally, the fifth trail is "Abu Qubies-Beer Eljabal", which is located in the middle of Abu Qubies and considered the shortest tourism hiking trail in the area with an average length of 1.4 km. Additionally, all information related to the five tourism hiking trials in Abu Qubies was confirmed by the researchers' on-site observations. Moreover, the researchers conducted a comprehensive interview with the Abu Qubies director to identify more about the hiking trails in Abu Qubies, especially with his expertise that was based on his scientific knowledge and his organization of approximately 390 trips in these hiking trails with different groups such as environmental experts, locals, tourists and international organization delegations from 2012 to 2020. Figure 3 shows the five tourism hiking trails in Abu Qubies.

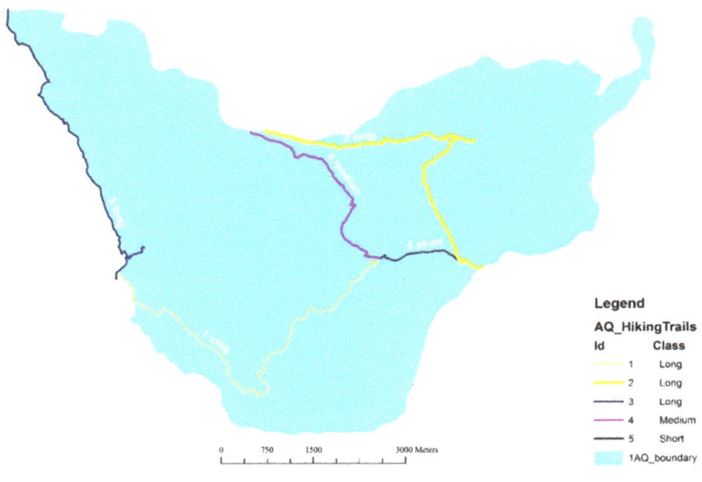

Figure 3. Tourism hiking trails in Abu Qubies protected area.

2.2.2. Stage (II): Valuation of Abu Qubies Protected Area Ecosystem Services

After identifying the five main tourism hiking trails in Abu Qubies in the first sta this stage focuses on calculating the ecosystem services value in Abu Qubies using the To Economic Value (TEV) approach, which included determining the most valuable hiki trail among these five tourism hiking trails. TEV approach is a framework that seeks identify the monetary value of various natural resources in any ecosystem [37]. The To Economic Value approach's concept is based on the idea that every service or good h unique attributes, some of them are physical and easy to measure, while others are mo difficult to define. Therefore, TEV is the summation of all attributes of different goods a services, not only the quantifiable ones [38]. Additionally, it is worth mentioning that TEV is interested in measuring both sides of different goods and services in any ecosyste willingness to pay (WTP) and willingness to accept (WTA). WTP represents the moneta value of obtaining an ecosystem service, while WTA represents the monetary value avoiding loss [39].

In order to conduct the Total Economic Value approach in this research, its ma components should be identified first. TEV's main components vary slightly from c economist to another, but generally comprise the following: use-value (UV) and non- value (NUV) (see Figure 4).

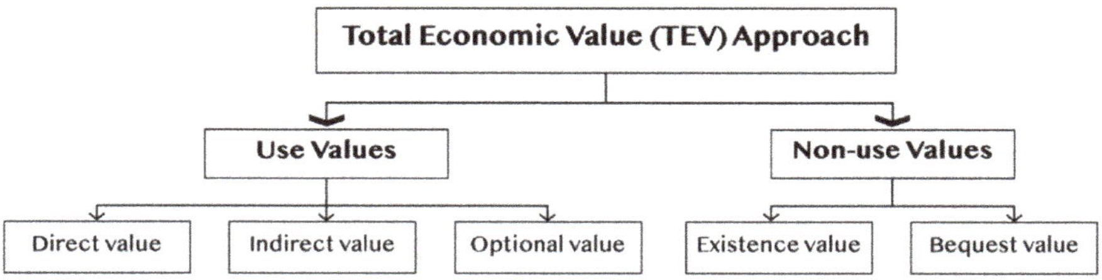

Figure 4. Main components of the Total Economic Value (TEV) Approach. Source: adapted f (Nature Conservancy et al., 2004; Pascual et al., 2015; Tinch et al., 2019; UNEP, 2010).

Use value (UV) includes three main components: direct use (monetary), indirect and optional value (OV). Direct use, also known as consumptive use-value, originates f services or goods that can be extracted and consumed directly. In addition, direct use v refers to the different products of any ecosystem that are easily quantified and recogni by analyzing and monitoring market interactions [40]. For example, timber and har crops such as fruits and herbs are considered as direct use value products in forests other natural reserves. Furthermore, other non-consumptive goods can be classifie direct use value such as hiking or camping, even if people who enjoy these activities do consume any of the natural resources in these forests or reserves [41]. All these goods belong to direct use value are tangible, can be quantified and have real value. Compa to indirect use products, direct ones commonly have observable quantities and their p can be also traced, which facilitates their measurement and calculation. Regarding indi use, which is known as a functional value, it originates from ecosystem ecological so products and services. Unlike direct use products, indirect use services are difficu identify clearly and explicitly in different marketplaces. Thus, measuring these indirec services is difficult compared to the direct ones [42]. For example, stunning landscap natural reserves are considered as indirect use values, where people can enjoy their vi aesthetic without diminishing the pleasure of others and without having to pay any kin material or monetary value for it. Finally, the optional value (OV) goods or services, w resemble the concept of an insurance policy, represents the future use of certain goods services (whether direct or indirect) at a time when they may not seem important. In c words, the optional value refers to the potential of creating a future value from using tangible or intangible natural goods or services [40,41].

The non-use value (NUV), which is known as passive use, originates from the idea that any natural ecosystem must be preserved and protected. This non-use value consists of two main types: existence value and bequest value. Existence value is obtained from the fact that an ecosystem resource exists, even if people do not utilize it or intend to use it [43]. For example, in marine environments, many organizations and aquatic conservation associations seek to protect some marine species from extinction. In order to do that, they raise donations from people, who in one way or another help even if they may never see these marine species. Regarding bequest value, it comes from the fact that people should ensure the potential of using any current ecosystem resources for future generations [44]. Therefore, what people pay or act to achieve this conservation of natural resources is considered as a bequest value. Generally, it is difficult to measure the non-use value which keeps it a challenging step when valuation the ecosystem resources and applying the Total Economic Value (TEV) approach.

Subsequently, the Total Economic Value (TEV) approach is the summation of use value (UV) and non-use value (NUV) in any existed ecosystem (see Equation (1)). Thus, TEV's output is frequently considered as natural capital since it includes the whole value of the ecosystem resources [40,42,43].

$$TEV = UV + NUV \tag{1}$$

It is worth mentioning that the practical application of the TEV approach relies heavily on the available data. Thus, there are many practical methods and measures to calculate the ecosystem services such as Market Price Method, Social Cost Method and contingent Valuation Method (CVM) [45]. The market price is a method that determines the value of ecosystem services by looking at the prices of different commercial goods and services that are purchased and sold in the markets [46]. According to the US National Academies, social cost of carbon (SCC) estimated the monetized value of different damages caused by one-ton increases in CO_2 levels [47]. These damages include damage from natural disasters such as flooding, human health, agriculture productivity changes and other nonmarket damages such as that happened in the ecosystem services [47]. The SCC monetized value is estimated to be in the range of USD 10/tCO_2 to USD 1000/tCO_2 [48]. In our case, the used value was USD 75/tCO_2 according to the average global price used by the Syrian Ministry of Local Administration and Environment. Additional details about the calculation process of SCC can be found in [49]. Finally, the contingent valuation method (CVM), which is a survey-based technique that expressed the idea of willing to pay for sustainable use of natural resources or the avoidance of losing them [50]. All these methods are used to measure the ecosystem services values in the Abu Qubies protected area. Regarding the used method to classify and rank the tourism trails, the Zonal Statistics tool was used. It is a spatial analysis toolbox that calculates specific statistics values in a certain zone (cell) which are represented as inputs for raster data and creates a table of outputs and statistical features such as mean, median, minority and majority [51]. Figure 5 presents the concept of using the zonal statistics tool in this research.

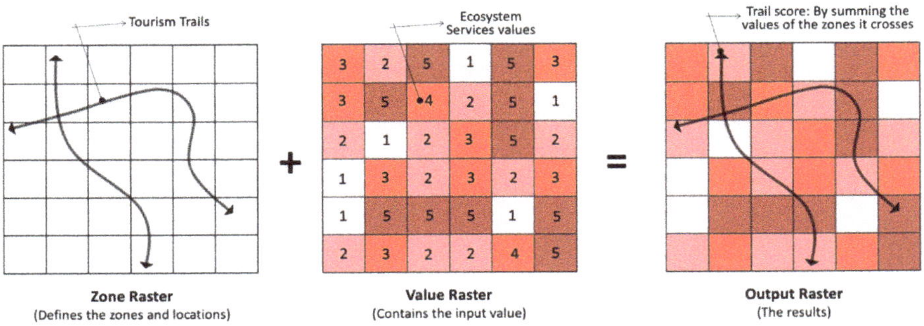

Figure 5. Zonal Statistics Tool concept. Source: based on [52] by researcher.

3. Results

The TEV approach adopted and conducted depends on the different goods a services in Abu Qubies protected area. Regarding the use value, the direct use va is represented by various goods and services such as wood products (wood products—f trees), other non-timber products (honey—thymus plant), while the indirect use valu represented by carbon reduction. In addition, the optional value which represente pharmaceutical industries from different medicinal plants such as Laurus and orch Regarding the non-use value in Abu Qubies, this is represented by two types of goods services that represent natural-based tourism and community-based tourism (CBT) suc endangered species, landscapes, heritage and cultural resources and local and agricult products. All these goods and services are introduced in detail as follows.

3.1. Use Value in Abu Qubies Protected Area

3.1.1. Direct Use Value

The direct use value in Abu Qubies is represented by two main goods: wood p ucts and non-wood products. Regarding the wood products, these are composed of elements: hardwood/firewood and fruit trees. The market price was the selected m surement method to estimate the valuation of these two elements as they are consid commercial goods and transformative services. At the market price of USD 70.67/ton total value of wood was estimated to be USD 201,890 [52–54]. While the main exampl the fruit trees in Abu Qubies are *Malus trilobata* (wild apple), *Crataegus azarolus* (hawth olive, *Pyrus syriaca* (Syrian pear) and *Prunus ursina*. The production of these fruit tre estimated at approximately 4500 kg/year at a price of up to USD 0.38/kg, with a total v of about USD 1730.76. The non-wood products include honey, thymus and plants tha used in the preparation of medicines and pharmaceutical industries such as Laurus nc and orchids. The honey production from Abu Qubies was about 500 kg/year at an ave price of USD 9.61 per kg, bringing the total value of the honey produced from the prote area to about USD 4807.69. Thymus production was 200 kg/year at an average price of 5.76/kg for a total value of about USD 1153.84. Laurus is a forest plant and its impor comes from its use in the medical and industrial industries (soap), its production in Qubies is estimated to be 50,000 kg/year at an average price of USD 0.43/kg, so its value was USD 21,634.61. For orchids the total production was around 100–150 kg/ye an average price of USD 28.84/kg, with a total value of USD 19,038. This orchid produ is commonly sold to nearby medicine laboratories. By adding these monetary valu honey, thymus, Laurus and orchids, the total economic price of non-wood products in Qubies protected area was USD 5961.44.

3.1.2. Indirect Use Value

The indirect use value in Abu Qubies is represented by Carbon reduction a environmental dimension of natural resources in this protected area. By using Social of Carbon (SCC) as a measurement method to estimate the expected damage caus releasing one additional ton of CO_2 into the atmosphere, the total value of Carbon redu in Abu Qubies was USD 2,450,000 [27,47,49,53].

3.1.3. Optional Value

The optional value in Abu Qubies protected area represents the willingness to p guarantee the availability of the ecosystem services for future use by the individu fact, option value cannot be accurately estimated for ecosystem services in the Abu Q Reserve. The option value represents the economic difference between the net bene optimal and non-optimal decision-making because it ignores the gains that can be ach by delaying the decision and learning during this delay period. Therefore, estim the option values requires the presence of two strategies, and each of them is eval with the aim of making the best decision. Consequently, option values cannot eas calculated separately and added to the benefit-cost equation [55].

Table 1 represents the estimated values of goods and services of use value in Abu Qubies, while Figure 6 presents a graded color map of the estimated values of goods and services calculated in (Table 1).

Table 1. Monetary values of ecosystem use value in Abu Qubies protected area.

TEV Approach Classification (Use Value)				Valuation			Measurement Method
Value Classification		Category	Good/Service	Monetary Value ($/year)	Total ($/year)	%	
Use Value	Direct Value	Wood Products	Hardwood/Firewood and fruit trees	203,620.76	235,543.81	8.77	Market Price
		Non-wood Products	Honey and Thymus syriacus	5961.44			
			Laurus nobilis and Orchids (plants)	25,961.61			
	Indirect Value	Environmental Dimension	Carbon reduction	2,450,000	2,450,000	91.23	Social Cost of Carbon (SCC)
	Optional Value			Not estimated			
Total Valuation of Use Value in Abu Qubies Syrian Protected Area				2,685,543.81		100	

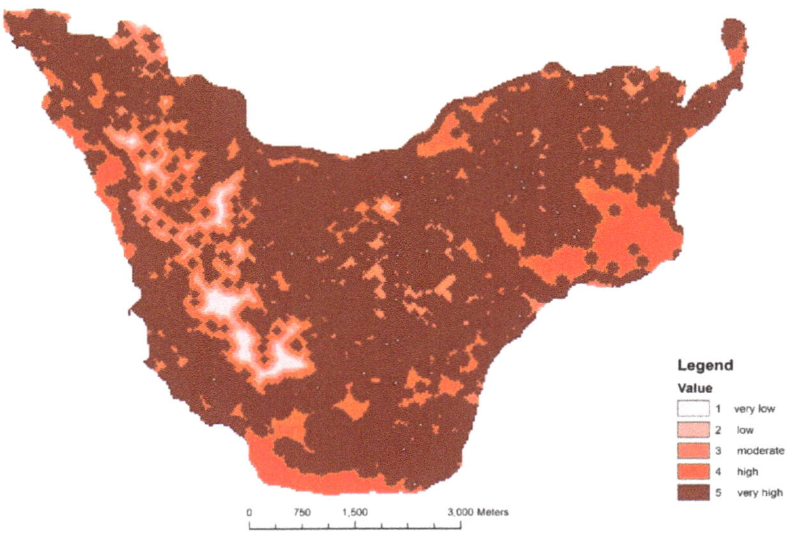

Figure 6. Graded color map representing Abu Qubies' use values (direct and indirect).

3.2. Non-Use Value in Abu Qubies Protected Area

3.2.1. Existence Value

The existence value in Abu Qubies protected area is represented in costs to protect the existing ecosystem resources. Abu Qubies has many endangered animals that are an attraction for many visitors and tourists, such as the *Salamandra infraimmaculata* (Arouss Al-Ayn), *Green Toad* and the *Greek Tortoise*. This is in addition to many birds, such as the *Egyptian Vulture* and the *Greater Spotted Eagle*, for which Abu Qubies is important, due to its geomorphological diversity, and is seen as a major stop in their migration paths. For natural and geomorphological diversity, Abu Qubies has many climates and geographical environmental zones that make it the focus of many visitors in various seasons of the year. Consequently, the locals in Abu Qubies were asked how much they would be willing to pay for protecting the existence of these unique animals and birds. The average willingness to pay per family was around USD 11/month (USD 132/year). When this value is multiplied by the number of families in the Abu Qubies (585 families), the estimated monetized val of the existence value was approximately USD 77,220/year.

3.2.2. Bequest Value

The bequest value in Abu Qubies protected area was represented by the monetiz value that locals are willing to pay to preserve the rights of future generations to bene from the ecosystem services and resources in Abu Qubies. To estimate the bequest valu the locals were asked through a questionnaire about their willingness to pay to comb poaching, excessive logging and land degradation. The results expressed strong supp from the locals to protect the ecosystem resources. The average willingness to pay w estimated at USD 17/month/family. Therefore, the total monetized value of the bequ value was approximately USD 119,340 per year.

Table 2 represents the estimated values of goods and services of non-use value in A Qubies. While Figure 7 presents a graded color map of the estimated values of goods a services calculated in Table 2.

Table 2. Monetary values of ecosystem non-use value in Abu Qubies.

TEV Approach Classification (Non-Use Value)				Valuation			Measurement Metho
Value Classification		Category	Good/Service	Monetary Value (Per Year)	Total ($/Year)	%	
Non-use Value	Existence Value	Willingness to pay	The existence of unique animals and birds species	77,220	77,220	39.2	CVM
	Bequest Value	Willingness to pay	Combat poaching, excessive logging and soil degradation	119,340	119,340	60.8	
Total Valuation of Non-use Value in Abu Qubies Syrian Protected Area				196,560		100	

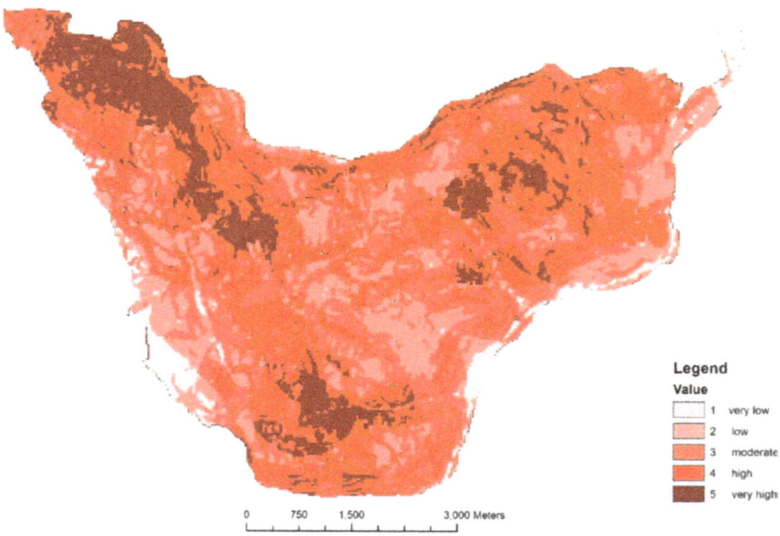

Figure 7. Graded color map representing Abu Qubies non-use values (existence—bequest).

Before moving to the main result of this analysis, which is the classification o tourism hiking trails based on the valuation of total ecosystem services value in Abu Qu it is also worth noting that there were some negative values found in Abu Qubies, w affected the total net monetary value of the natural resources in Abu Qubies. These neg values include soil degradation, fires and mismanagement. Figure 8 shows the differen vegetation in Abu Qubies protected area between 2019 and 2021, through which vegetation is significantly reduced. The monetary value of these negative values mentioned before was estimated at USD 704,410.92 [52].

FID	Shape	Grid Code	Class Name	Area (ha)	%
0	Polygon	1	Urban areas	641.89	17%
1	Polygon	2	Degraded areas (Roke-Soil-Jaobat)	919.6	24%
2	Polygon	3	Forests (low density)	1217.04	32%
3	Polygon	4	Forests (high density)	984.11	26%

FID	Shape	Grid Code	Class Name	Area (ha)	%
0	Polygon	1	Urban areas	761.11	20%
1	Polygon	2	Degraded areas (Roke-Soil-Jaobat)	1116.64	30%
2	Polygon	3	Forests (low density)	1173.77	31%
3	Polygon	4	Forests (high density)	710.69	19%

2019

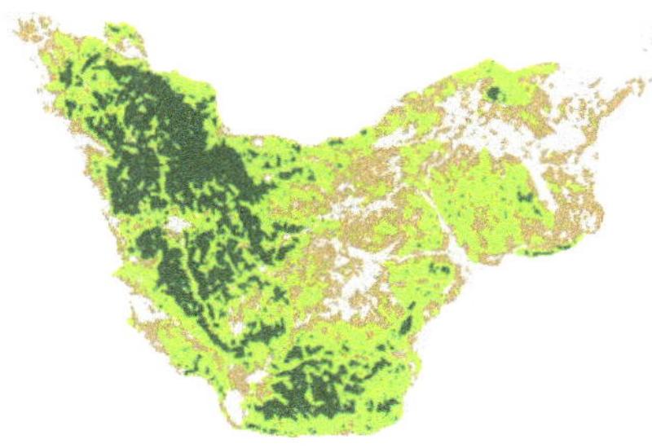

2021

Figure 8. Change of vegetation in Abu Qubies protected area in 2019 and 2021.

3.3. Classification of Tourism Hiking Trails in Abu Qubies Protected Area Based on TEV Approach

In order to determine the most important tourism hiking trial in Abu Qubies based on the gradient map (see Figure 9), the Zonal Statistics tool is used. In Abu Qubies the input values of the cells were determined based on the biodiversity resources values. The total inputs value for each tourism hiking trail was estimated by calculating all cell values that the trail passed through. The final classification and sorting of the tourism hiking trials in Abu Qubies are presented in Table 3. In addition to other statistical parameters predetermined by authors such as majority, minority, median, mean and path length. The most economic value tourism hiking trail in Abu Qubies was the second trail "Abu Qubies—Mshta Elbeer" with a 1463-point score.

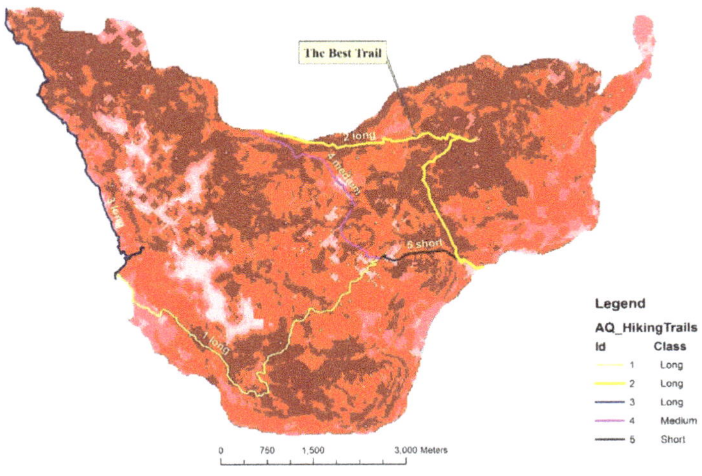

Figure 9. The best tourism hiking trails in Abu Qubies protected area based on the valuation of ecosystem services (use value—non-use value).

Table 3. Ranking of the tourism hiking trails in Abu Qubies based on ecosystem services values

ID	Tourism Hiking Trail	Length (km)	Mean	Median	Minority	Majority	Total Sco
1	Sheikh Ali Majdal—Beer Eljabal	7.77	4.4	4	2	4	1339
2	Abu Qubies—Mshta Elbeer	6.89	4.7	5	3	5	1463
3	Ras Alshareh—Sheikh Hatem	6.34	3.1	4	2	3	591
4	Sheikh Abdullah—Beer Eljabal	3.93	3.9	4	1	4	703
5	Abu Qubies—Beer Eljabal	1.44	2.5	3	3	4	273

4. Discussion

By reviewing the Total Economic Value (TEV) analysis results and identifying expected monetary value of the biodiversity resources in Abu Qubies, it was emphasi that there are enormous values in Abu Qubies that can be economically relied upo develop the resources of the protected area on the one hand and to encourage tou industry with its various activities on the other. The findings of the analysis show tha most valuable biodiversity resource in Abu Qubies is indirect values, specifically ca reduction, for a total of USD 2,450,000. That is because Abu Qubies is characterize its high elevations and high-density trees, which makes it function environmental reducing carbon efficiently. Wood comes in second place in terms of its total economic as an ecosystem good in Abu Qubies, with at a value of USD 203,620.76. The import of wood in Abu Qubies comes from the fact that about 90% of local households rel it for heating, cooking and food preparation. The locals' willingness to pay to co poaching, excessive logging and soil degradation ranked third in the monetary val Abu Qubies (USD 119,340). For other values, it had the lowest share in terms of econ valuation, starting from USD 77,220 for the willingness to pay to protect the existen unique animals and bird species, and ending with USD 5691.44 for non-wood prod such as honey and Thymus Syriacus.

Despite the diversity of tourism hiking trails at Abu Qubies protected area, identi the most significant trails in terms of ecosystem services values is an important dec support tool [22]. The tourism hiking trail "Abu Qubies—Mshta Elbeer" ranked first a trail containing the highest economic and monetary ecosystem services values with cell input points of 1463 (see Table 2). "Abu Qubies—Mshta Elbeer" trail is characteriz its diverse environmental zones, making it the center of attention for Abu Qubies tou visitors and locals as well (see Figure 10). This tourism hiking trail also runs thr some of the main touristic areas in the reserve, such as ancient Roman monuments some religious shrines, which represent the identity of the reserve [31], in addition to steep mountain letters along the trail which are considered ideal sites for establis environmental observatories to monitor wildlife in the area.

Figure 10. The different environmental zones in "Abu Qubies—Mshta Elbeer" tourism hikin in Abu Qubies. Source: author, 2021.

Accordingly, many activities can be proposed in these tourism hiking trails in order to be sources of income for all human elements in Abu Qubies, whether management or community. For example, trail entry and use fees, the establishment of a visitors' database, providing tourist guides services, on-site subsistence allowance, rental allowance for certain equipment (e.g., camping), sales of local products and artefacts, fees for local product markets and shops, and finally fees for using the protected area resources as material for academic research and studies.

In addition, some recommendations could be set to help decision-makers make optimal use of the protected area's resources, especially in the tourism sector. First, the government should concentrate the distribution of the available investments in specific development zones with future economic and social returns, which will help all Abu Qubies system elements (administrative, locals, tourists and visitors). Second, the results of this study confirmed the possibility of estimating an ecological landscape or ecosystem services in terms of monetary value [56]. Third, the study emphasizes the importance of achieving all environmental standards during establishing various tourism projects in the protected areas, in order to preserve its unique ecosystem services. Fourth, the study supports the idea of flexible management in Abu Qubies by emphasizing continuous follow-up of its ecosystem services and monitoring of irregularities that may result from mass tourism.

Finally, it is also worth noting that choosing the best tourism hiking trails in Abu Qubies in terms of its economic values does not in any way mean that the other trails will be neglected. In particular, the five tourism hiking trails in Abu Qubies have their own natural resources and different attractions for tourists and visitors. But the importance of this research and its results comes from the limited investments in the current Syrian situation, the massive neglect of nature protected areas and the mismanagement of the reserve's resources [27]. Accordingly, the process of concentrating tourism investments in one hiking trail is one of the most realistic and effective solutions in the light of these facts on the ground [18].

5. Conclusions

Generally, this study is aimed at enhancing the importance of nature reserves and protected areas as tourist destinations, especially in the current context of the importance of natural resources and the sustainability approach in the COVID-19 era. The specific main objective of this research focuses on the classification of the tourism hiking trails in Abu Qubies (study area) in order to find out which of these trails contains the most beneficial economic and monetary ecosystem services. The Total Economic Value (TEV) approach has been used to estimate the monetary value of the different ecosystem services of Abu Qubies.

The relationship between nature-protected areas and tourism development has become strong in recent years, especially after the COVID-19 pandemic and the growing demand for outdoors tourism activities. Hence, many tourism patterns appeared to be seeking to optimize the use of nature reserves' resources in various tourism activities, such as ecotourism. Consequently, this complex relationship between nature reserves and tourism development has been an encouraging factor for researchers to conduct this research. Regarding the selection of Syria in general and Abu Qubies in particular as a study area for this research was reliant on several reasons. In addition to the previously analyzed relationship between tourism and protected areas, the current Syrian situation is also one of the main criteria for choosing this case study. For example, political volatility, high inflation rates and the continuous international economic fines from global society promoted the selection of Syria and Abu Qubies. This highlights the importance of this research on how to use and develop domestic resources efficiently. As an example, valuing the market and monetary value of the ecosystem services of Abu Qubies, which in turn leads to identifying the most economically desirable tourism hiking trails, helps decision-makers to channel "limited" investments in optimal places to maximize their potential. Additionally, the current Syrian situation has obliged the local community and development partners to ensure the success of the proposed tourism and economic development schemes, whi will have a direct impact on all activities that are part of their daily lives.

6. Limitations and Future Research

Most of the limitations in this research revolve around the central question of "H much is an ecosystem worth?". Although the diversity of natural resources in ecosyste (e.g., protected areas) creates major importance for these ecosystems (economically, en ronmentally or socially), it causes a burden when trying to transform all these vital valu some of which are intangible, into easily measurable values. In this context, this stu adopted the Total Economic Value (TEV) approach to estimate the monetary values of A Qubies biodiversity resources by using different measurement methods that are subject to certain limitations; for example, the constant change in data available on natural sources, which often results in a change in production. In addition, it is difficult to estim some intangible resources, such as the value of life and the value of looking at a sce landscape. Moreover, some overestimations of the actual values of natural resources ecosystems can occur. Finally, many of these measures need enormous amounts of accur and sensitive data. Therefore, we must be extremely careful when calculating the monet values of biodiversity resources in any ecosystem, especially the intangible ones. In t research, the researchers have tried to take all these limitations into account and often on on-site observations to verify the authenticity of the data and thus reach more relia and robust results.

Future work needs to be oriented to study tourism development in protected ar and their implications for the environment, accordingly achieving the conservation a sustainable use of natural resources, as they represent capital for these nature reserves their local communities. Additionally, there is a need for future research to measure ot tourism dimensions related to tourism development on natural reserves such as tou satisfaction, changes in tourist behavior and preferences, etc.

Author Contributions: Conceptualization, M.K. and R.A.; methodology, M.K.; software, R.A.; idation, M.K., Z.L. and L.D.D.; formal analysis, M.K.; investigation, M.K.; resources, R.A.; curation, R.A.; writing—original draft preparation, M.K.; writing—review and editing, M.K sualization, M.K. and R.A.; supervision, Z.L. and L.D.D.; project administration, Z.L. and L.I funding acquisition, Z.L. and L.D.D. All authors have read and agreed to the published versio the manuscript.

Funding: This research received no external funding.

Data Availability Statement: The data presented in this study are available on request from corresponding author.

Acknowledgments: This research is supported by the Hungarian University of Agriculture and Sciences (MATE), Cairo University and the Egyptian Ministry of Higher Education and Scier Research (MHESR). Additionally, the researchers want to thank Yasser Nassour, the director of Qubies protected area, for his help to organize several on-site observations in order to collec required data for this study.

Conflicts of Interest: The authors declare no conflict of interest.

References

1. IUCN. The World Now Protects 15% of Its Land, but Crucial Biodiversity Zones Left Out. Available online: https://www.iucn news/secretariat/201609/world-now-protects-15-its-land-crucial-biodiversity-zones-left-out (accessed on 22 December 2021).
2. UNEP-WCMC; IUCN. *Protected Planet Report 2016*; Protected Planet Reports; UNEP: Cambridge, UK; Gland, Switzerland, 2016;
3. WCPA; IUCN. *Economic Values of Protected Areas: Guidelines for Protected Area Managers*; Best Practice Protected Area Guidelines S IUCN Publications Services Unit: Cambridge, UK; Gland, Switzerland, 1998; ISBN 2-8317-0461-8.
4. Eagles, P.; McCool, S.; Haynes, C. *Sustainable Tourism in Protected Areas: Guidelines for Planning and Management*; Best Practice Prot Area Guidelines Series; IUCN—The World Conservation Union: Gland, Switzerland; Cambridge, UK, 2002; ISBN 2-8317-0648-3.
5. Whitelaw, P.A.; King, B.E.M.; Tolkach, D. Protected Areas, Conservation and Tourism—Financing the Sustainable D *J. Sustain. Tour.* **2014**, 22, 584–603. [CrossRef]

Wilson, E.; Nielsen, N.; Buultjens, J. From Lessees to Partners: Exploring Tourism Public–Private Partnerships within the New South Wales National Parks and Wildlife Service. *J. Sustain. Tour.* **2009**, *17*, 269–285. [CrossRef]
IUCN. *Guidelines for Protected Area Management Categories*; IUCN: Cambridge, UK; Gland, Switzerland, 1994; ISBN 2-8317-0201-1.
Bushell, R.; Bricker, K. Tourism in Protected Areas: Developing Meaningful Standards. *Tour. Hosp. Res.* **2017**, *17*, 106–120. [CrossRef]
Kc, B. Complexity in Balancing Conservation and Tourism in Protected Areas: Contemporary Issues and Beyond. *Tour. Hosp. Res.* **2021**, *22*, 146735842110158. [CrossRef]
Lanier, P. *The Positive Impacts of Ecotourism in Protected Areas*; WIT Press: Opatija, Croatia, 8 July 2014; pp. 199–209.
Kim, M.; Xie, Y.; Cirella, G.T. Sustainable Transformative Economy: Community-Based Ecotourism. *Sustainability* **2019**, *11*, 4977. [CrossRef]
Koens, J.F.; Dieperink, C.; Miranda, M. Ecotourism as a Development Strategy: Experiences from Costa Rica. *Environ. Dev. Sustain.* **2009**, *11*, 1225–1237. [CrossRef]
Mondino, E.; Beery, T. Ecotourism as a Learning Tool for Sustainable Development. The Case of Monviso Transboundary Biosphere Reserve, Italy. *J. Ecotour.* **2019**, *18*, 107–121. [CrossRef]
United Nations. The 2030 Agenda for Sustainable Development. Available online: https://sdgs.un.org/goals (accessed on 13 January 2022).
OECD. *OECD Tourism Trends and Policies 2020*; OECD Tourism Trends and Policies; OECD: Paris, France, 2020; ISBN 978-92-64-70314-8.
UNWTO. 100% of Global Destenations Now Have COVID-19 Travel Restrictions, UNWTO Reports. Available online: https://www.unwto.org/news/covid-19-travel-restrictions (accessed on 13 January 2022).
Spenceley, A.; McCool, S.; Newsome, D.; Báez, A.; Barborak, J.R.; Blye, C.-J.; Bricker, K.; Sigit Cahyadi, H.; Corrigan, K.; Halpenny, E.; et al. Tourism in Protected and Conserved Areas amid the COVID-19 Pandemic. *PARKS* **2021**, 103–118. [CrossRef]
Balmford, A.; Green, J.M.H.; Anderson, M.; Beresford, J.; Huang, C.; Naidoo, R.; Walpole, M.; Manica, A. Walk on the Wild Side: Estimating the Global Magnitude of Visits to Protected Areas. *PLoS Biol.* **2015**, *13*, e1002074. [CrossRef]
Rodriges, A.; Kastenholz, E. Hiking as a Recreational and Tourist Activity—Comparing Portuguese Hikers with Those from Other Nationalities. *Rev. Tur. Desenvolv.* **2007**, *7*, 83–91.
Li, W.; Ge, X.; Liu, C. Hiking Trails and Tourism Impact Assessment In Protected Area: Jiuzhaigou Biosphere Reserve, China. *Environ. Monit. Assess* **2005**, *108*, 279–293. [CrossRef] [PubMed]
Newsome, D. The Collapse of Tourism and Its Impact on Wildlife Tourism Destinations. *J. Tour. Futures* **2021**, *7*, 295–302. [CrossRef]
Mnguni, D.E.M. Community-Based Tourism Development: A Hiking Trails Perspective. *Tour. Leis.* **2017**, *6*, 17.
Cole, D.N.; Petersen, M.E.; Lucas, R.C. *Managing Wilderness Recreation Use: Common Problems and Potential Solutions*; U.S. Department of Agriculture, Forest Service, Intermountain Research Station: Ogden, UT, USA, 1987; p. INT-GTR-230.
Finnessey, L. The Negative Effects of Tourism on National Parks in the United States. Honors Theses. Bachelor's Thesis, Johnson & Wales University, Wales, UK, 2012.
Kavallinis, I.; Pizam, A. The Environmental Impacts of Tourism— Whose Responsibility Is It Anyway? The Case Study of Mykonos. *J. Travel Res.* **1994**, *33*, 26–32. [CrossRef]
Stronza, A.L.; Hunt, C.A.; Fitzgerald, L.A. Ecotourism for Conservation? *Annu. Rev. Environ. Resour.* **2019**, *44*, 229–253. [CrossRef]
MoAAR; The Ministry of State for Environment Affairs—MSEA. *Abu Qubies Protected Area Management Plan "Biodiversity Conservation and Protected Area Management Project" SYR/05/010*; The Ministry of State for Environment Affairs—MSEA: Damascus, Syria, 2010.
GEF. *GEF Country Portfolio Evaluation: Syria (1994–2008)*; Global Environment Facility Evaluation Office: Washington, DC, USA, 2009.
Ministry of Local Administration and Environment. Accomplishments and Actions Taken in Connection with the Work of the Directorate of Biodiversity and Land. Available online: http://www.mola.gov.sy/mola/ (accessed on 18 April 2022).
Martini, G. *Forest Sector Policy and Institutional Development*; Food and Agriculture Organization (FAO): Damascus, Syria, 2009.
MoAAR. The Syrian Ministry of Agriculture and Agrarian Reform. Available online: http://moaar.gov.sy/main/archives/19575 (accessed on 13 January 2022).
UNDP; GEF. *Biodiversity Conservation & Protected Area Management (PIMS 227)*; UNDP: Damascus, Syria, 2014.
Ilieș, D.C.; Hodor, N.; Indrie, L.; Dejeu, P.; Ilieș, A.; Albu, A.; Caciora, T.; Ilieș, M.; Barbu-Tudoran, L.; Grama, V. Investigations of the Surface of Heritage Objects and Green Bioremediation: Case Study of Artefacts from Maramureş, Romania. *Appl. Sci.* **2021**, *11*, 6643. [CrossRef]
Yazbek, A. A Study of the Distribution of Forest Cover in (Abu Qubais) Reserve Using Remote Sensing Techniques (RS) and Geographic Information Systems (GIS). Unpublished. Master Thesis, Damascus University, College of Agriculture, Department of Renewable Natural Resources and Environment, Damascus, Syria, 2011.
Syrian Society for the Conservation of Wildlife (SSCW). Wildlife for People. Available online: https://sscw-syr.org/ (accessed on 8 April 2022).
GEF; UNDP; Ministry of State for Environmental Affairs. *The Fourth National Report on Biodiversity in the Syrian Arab Republic*; UNDP: Damascus, Syria, 2009.

37. Lette, H.; de Boo, H. *Economic Valuation of Forests and Nature—A Support Tool for Effective Decision-Making*; International Agricult Centre (IAC): Ede, The Netherlands, 2002.
38. Dushin, A.V.; Yurak, V.V. Total Economic Value Concept: Essence, Evolution and Author's Approach. In Proceeding the International Scientific Conference "Far East Con" (ISCFEC 2018), Vladivostok, Russia, 2–4 October 2018; Atlantis P Vladivostok, Russia, 2019.
39. Defra. *An Introductory Guide to Valuing Ecosystem Services*; Department for Environment, Food and Rural Affairs: London, 2011; p. 68.
40. Aanesen, M.; Armestrong, C.; Kahui, V. TEV (Total Economic Value) Analysis of a Marine Environment in Norway. In Proceed of the Montpellier Proceedings, Montpellier, France, 13–16 July 2010; International Institute of Fisheries Economics and T (IIFET): Montpellier, France, 2010; p. 11.
41. Tinch, R.; Beaumont, N.; Sunderland, T.; Ozdemiroglu, E.; Barton, D.; Bowe, C.; Börger, T.; Burgess, P.; Cooper, C.N.; Fac M.; et al. Economic Valuation of Ecosystem Goods and Services: A Review for Decision Makers. *J. Environ. Econ. Policy* **20** 359–378. [CrossRef]
42. Pascual, U.; Termansen, M.; Hedlund, K.; Brussaard, L.; Faber, J.H.; Foudi, S.; Lemanceau, P.; Jørgensen, S.L. On the Value o Biodiversity and Ecosystem Services. *Ecosyst. Serv.* **2015**, *15*, 11–18. [CrossRef]
43. *Mainstreaming the Economics of Nature: A Synthesis of the Approach, Conclusions and Recommendations of Teeb*; UNEP (Ed.) Economics of Ecosystems & Biodiversity; UNEP: Geneva, Switzerland, 2010; ISBN 978-3-9813410-3-4.
44. Croci, E.; Lucchitta, B.; Penati, T. Valuing Ecosystem Services at the Urban Level: A Critical Review. *Sustainability* **2021**, *13*, [CrossRef]
45. Nature Conservancy; World Bank; IUCN. *How Much Is an Ecosystem Worth?—Assesssing the Economic Value of Conservation*; W Bank: Washington, DC, USA, 2004.
46. Ghermandi, A.; Nunes, P.A.L.D.; Portela, R.; Rao, N.; Teelucksingh, S.S. 12.11—Recreational, Cultural, and Aesthetic Ser from Estuarine and Coastal Ecosystems. In *Treatise on Estuarine and Coastal Science*; Wolanski, E., McLusky, D., Eds.; Acad Press: Waltham, MA, USA, 2011; pp. 217–237, ISBN 978-0-08-087885-0.
47. National Academies of Sciences; Engineering; Medicine. *Valuing Climate Changes: Updating Estimation of the Social Cost of C Dioxide*; National Academies Press: Washington, DC, USA, 2017; p. 24651, ISBN 978-0-309-45420-9.
48. Ricke, K.; Drouet, L.; Caldeira, K.; Tavoni, M. Country-Level Social Cost of Carbon. *Nat. Clim. Change* **2018**, *8*, 895–900. [Cros
49. Anthoff, D.; Tol, R.S.J. The Uncertainty about the Social Cost of Carbon: A Decomposition Analysis Using Fund. *Clim. C* **2013**, *117*, 515–530. [CrossRef]
50. Venkatachalam, L. The Contingent Valuation Method: A Review. *Environ. Impact Assess. Rev.* **2004**, *24*, 89–124. [CrossRef]
51. ArcGIS Pro. How the Zonal Statistics Tools Work. Available online: https://pro.arcgis.com/en/pro-app/2.8/tool-refer spatial-analyst/how-zonal-statistics-works.htm (accessed on 20 April 2022).
52. Nassour, Y. *Interview with the Director of Abu Qubies Protected Area*; Latakia, Syria, 2021.
53. Ministry Of Agriculture and Agrarian. Reform Statistical Group—Price List. Available online: http://moaar.gov.sy/r archives/24005 (accessed on 18 April 2022).
54. Abu Qubies. *Protected Area—Field Observation*, 2021.
55. OECD Quasi Option Value. In *Cost-Benefit Analysis and the Environment*; OECD: Paris, France, 2006; pp. 145–154, ISBN 978- 01004-8.
56. Russo, G.; Beritognolo, I.; Bufacchi, M.; Stanzione, V.; Pisanelli, A.; Ciolfi, M.; Lauteri, M.; Brush, S.B. Advances in Biocu Geography of Olive Tree (Olea Europaea L.) Landscapes by Merging Biological and Historical Assays. *Sci. Rep.* **2020**, *10*, [CrossRef] [PubMed]

sustainability

MDPI

ticle

an Nature-Based Solutions (NBSs) for Stress Recovery in Green Hotels Affect Re-Patronage Intention?

nmi Yun [1] and Taeuk Kim [2,*]

[1] Department of Hospitality and Tourism Management, Sejong University, Seoul 05006, Korea; sunmiyun80@gmail.com
[2] Department of Hotel & Restaurant Management, Kyonggi University, Seoul 03746, Korea
* Correspondence: teokim1305@naver.com; Tel.: +82-10-4307-3903

Abstract: Our research framework in this paper investigated natural-based solutions (NBSs) at green hotels. We employed attention restoration theory (ART) to test the mediating effect of perceived stress (PS), psychological wellness (PW), satisfaction (SA), and the moderating effect of health consciousness (HC) on re-patronage intentions (RI). Data were collected through a survey of 544 customers who frequently visited green hotels in Korea, and structural equation modeling (SEM) was used to test the research hypotheses. The findings generally supported the hypothesized associations of the study variables within our proposed theoretical framework (PS, PW, SF) in order of the mediating effect on RI and confirmed the moderating effect of HC. In addition, the study's results have important theoretical and practical implications for the environment. In the former case, our results demonstrate the application of ART and NBS by explaining the effect of the relationship among PS, PW, and SF on RI and confirm the mediating effect of the ART (PS, PW, SF) on RI, as demonstrated in previous studies. Moreover, in the latter case our results may encourage green hotels to participate in the prevention of environmental problems.

Keywords: natural-based solutions (NBSs); attention restoration theory (ART); perceived stress (PS); psychological wellness (PW); mental health (MH); well-being (WB); satisfaction (SA); health consciousness (HC); re-patronage intention (RI); green hotels

on: Yun, S.; Kim, T. Can re-Based Solutions (NBSs) for Recovery in Green Hotels : Re-Patronage Intention?. nability 2022, 14, 3670. https:// g/10.3390/su14063670

emic Editors: Miklas Scholz, iotis G. Dimitrakopoulos, A. Pagnotta and ya Noorani

ved: 11 February 2022
ted: 13 March 2022
hed: 21 March 2022

her's Note: MDPI stays neutral egard to jurisdictional claims in ned maps and institutional affil-.

1. Introduction

Research shows that the official COVID-19 casualty should be double the number that has actually appeared. The actual casualties should be multiplied by 1.5× from the official report for the USA; for Japan only one tenth of casualties have been counted [1]. The Centers for Disease Control and Prevention (CDC) reported a higher lethality rate for COVID-19 during 2021 compared to 2020. This indicates that multiple elements, such as low vaccination rate, low mask usage rate, eased social distance, and the highly contagious Delta variant are in effect [2]. Nations equipped with effective anti-COVID prevention measures are now transitioning to normal life under the "With Corona" program. However, as the Omicron variant appears, we are under an increasing period of COVID-19 patients as the hazard of COVID-19 rises. COVID-19 is damaging the global industry. The tourist industry (i.e., hotel, travel agency, airlines) suffered the most, as COVID-19-related travel attenuation caused USD 1.2 trillion to USD 3.3 trillion in losses; this is close to 2.8–4.2% of global GDP [3]. The global tourist industry took three steps under COVID-19: Challenge → Resilience → Transformation. During the resilience stage, there were positive effects from regional affection. Credits from consumer and related businesses now include the sustainability, well-being of society, climate action, and local communities, which expect to take an imperative role during the transformation phase [4].

A warning about climate change in the World Economic Global Risk Report 2021, which began publishing in 2013, supports this issue. They report that the past two decades saw natural capital decrease by 40% [5]. The natural capital is being systematically c pleting. However, at the same time, nature and its ecosystem services are at the cent of the hotel industry business proposition, from food and beverage offers to guests' e joyment of the natural landscape at the destination. Nonetheless, nature is a "capit component available to hotels and a natural-based solution (NBSs) to mitigate and ada to climate change and protect biodiversity while ensuring human mental health (M and well-being (WB) [6]. NBS represents an activity to protect, manage, and restore t natural and deformed ecosystem, sustainably providing the WB of humanity, benef of biodiversity, and adaptive, efficient, and rapid solutions to mend social issues sim taneously [7]. Eco-friendly environments such as WB parks, green cities, gardens, a urban forests can reduce stress, depression, attention deficit hyperactivity disorder, a cardiovascular disease, and improve pregnancy and respiratory symptoms in physiologi and psychological ways [8,9]. NBS is expanding to be an important social phenomen Thus, an efficient realization of NBSs can improve MH and WB along with other vari environmental and social tasks [10,11]. Green hotels can positively influence custom satisfaction (SA) with their eco-friendly structure, spaces, and items. Influences such reduced stress and depression are a representative sample [12,13].

Attention restoration theory (ART) has been actively studied in environmental p chology due to the benefits NBSs provide to people. In previous studies, people stay in a natural environment and green space to which it is applied to relieve stress [14, According to that, people follow NBSs to enjoy leisure in nature, reducing their stress a restoring their attention [16,17], which are affected by the stress that comes from dwell in urban society [18,19]. MH, emotional WB, SA, and stress are the core factors to h explain customers' post-purchase activity. Thus, NBSs and ART take key roles in custo behavior study. Improving customers' MH and WB is now the head issue in the glo customer service and tourism business [20,21]. However, understanding the eco-frien concepts in inducing customers' good MH and WB is not deep enough, even though N is taking a pivotal role in the area, as is the study of inducing customers to green ho With that said, promising customer maintenance and SA can be a possible output of eff to effectively maintain customers' MH and WB effectively [21,22]. Previous studie the tourist industry based on NBSs [13,23,24] have proven that NBSs can enhance hur MH and reduce stress though direct and indirect influences. This can be highly relate people's behavior, intention, or attitude.

During the COVID-19 pandemic, travelers collected health knowledge and maintai strict policies and norms to protect their health. To prevent the spreading of COVID among future travelers, health consciousness (HC) will take a vital role in the tou industry [25]. Research shows that customers with high HC are more likely to avoi cancel a tour instead of joining health-related activities to make travel [25].

Examining victims of the COVID-19 pandemic from their stress status to percei wellness will enhance the current study of NBS and ART, and may provide prac implications as well. However, the study of psychological wellness (PW), perceived st (PS), and re-patronage intention (RI) grafted with NBS and ART is not deep eno currently. This study aims to create a merged model of NBS and ART, sampling gues South Korea's green hotel to identify the green hotels' structure of PS, PW, SA, and RI shed light on the mediating effects of PW, PS, and SA on the relationship between N green hotels, and RI. At the end of the study, we verify the HC within society under conditions of the COVID-19 crisis.

2. Conceptual Framework

2.1. Green Hotels as Natural-Based Solutions

NBSs are not a new concept. It is close to an umbrella concept, including na based approaches such as ecosystem-based adaptation or ecosystem-based mitigatio Ultimately, NBS aims to help achieve sustainable development goals (i.e., long-term security, climate change, water security, human health, disaster management, society, economic development). While the definition of NBS is not agreed upon under universal consensus yet, it is known as a helpful approach in various studies.

In a service-related NBS study [26,27], NBS is defined as "a method of improving/enhancing ecosystems to protect nature and solve various social/environmental/economic issues by performing sustainable management and restorations" [28]. From the view of the IUCN and EU, NBSs can be interpreted as a solution to improve natural environments' vulnerable points with social and environmental benefits to maximize the positive effect. The OECD's perception of NBSs emphasizes better healing from COVID-19's affects. They insist on valuing natural resources to secure the health of nature. In their concept, ecosystem services are indispensable for human health; they are essential to healing our society from pandemic damage. Mainly, they are focused on forbidding lumber harvesting and land utilization that can be related to the spread of COVID-19.

NBSs fall into three categories: (1) NBSs related to green infrastructure [26,29]; (2) NBSs related to climate change mitigation and adaptation [30]; and (3) NBSs related to ecosystem services [26,27]. These human health studies related to NBSs share one thing: they are all superficial. Most of the studies focus on the three concepts mentioned above and discuss basic ideas, which indicates that NBS is mainly about environmental and ecosystem issues. Moreover, using the NBS in city planning is essential to improve residents' mental health and welfare and resolve various social tasks [11,31].

Embodying NBS in the hotel industry means improving customer MH and WB. Thus, it has great salience [24]. Green-infra-related studies such as green spaces, green surfaces, green items, and nature play an imperative role in eco-friendly hotels, and structures where the service performance and character can directly lead to the product quality evaluation [32–34]. Human health can be affected by the surrounding social and physical environment; when someone collaborates with nature, nature can improve their MH [24]. Therefore, we can confirm the importance of NBS in human WB. Further studies to determine the benefits of NBSs and their effects should be performed.

2.2. Attention Restoration Theory Linked to Natural-Based Solutions Such as Green Hotels

Early studies of stress focused on physical harm and stress applied by accident or by any change [35]. However, recent studies have found that stress could be one reason for various mental sicknesses, including neurosis, depression, anxiety, suicidal tendency, or disruptive mood dysregulation disorder [36]. Half of the world's population dwells in cities. That residential condition causes stress to negatively affect city-dwellers' mental health. For that reason, people seek nature-based leisure activities to reduce stress and restore their attention [15]. Nature can reduce stress for those who contact it, allowing them to recover their brain and mental fatigue [37,38]. There is evidence that nature can offer positive emotion to improve MH or restore attention [15,31,39].

ART draws on nature's attention-recovering potential [40,41]. ART is an environmental psychology theory explicating nature as a restorative environment to heal the fatigue that results from directional attention [40]. RE refers to the environmental trait of attention restoration. When one loses directional attention capability due to the stress of city dwelling, it can be recovered by recognizing the natural environment's absolution effect, which requires no other extra effort [42]. Moreover, restorative experience results from actual restoration experiences such as distress, attention restoration, mental peace, recharging energy, and vigor [41]. The interest in environmental psychology has been present in nature-based activity for a long time. Active research based on ART is commonly conducted, which is expanding to tourism as healing tourism intended for mental healing increases [15,41,42].

Green hotels use NBSs through a nature-based approach (i.e., ecosystem-based adaptation, ecosystem-based relaxation) [12,13] to build a restorative environment, as nature can decrease PS of people and create a restorative environment by imbuing positive effects on PW (i.e., MM, WB) [24,42,43]. In the green-infra-related aspect of NBSs, the physical green environment such as green spaces, green surfaces, and green items may be essential for green hotels [33,34]. Therefore, the physical green environment inside the green hotel and the natural environment of the exterior are crucial component of NBSs [11–13]. green physical environment of green hotels can provide stress reduction, body acti enhancement, health inequality relief, emotion improvement, and increased WB and It will bring various positive effects on the health of individuals and society [1–3,31 Previous studies claim that people who reside in the natural environment and green sp can be distressed [14,15].

Furthermore, research shows PW and PS of NBSs can affect SA [12,13]; Han et al. affected customer satisfaction by increasing MH and WB perception and reducing stres NBSs in a study of customer maintenance strategy and green hotel-related NBSs. In et al. [12], it was shown that NBSs could affect customers' SA and RI by perceiving change. The previous studies mentioned above developed the hypothesis of our stud

Hypothesis 1 (H1). *Green hotels as NBSs have a significant effect on psychological wellness*

Hypothesis 2 (H2). *Green hotels as NBSs have a significant effect on perceived stress.*

Hypothesis 3 (H3). *Perceived stress has a significant effect on psychological wellness.*

Hypothesis 4 (H4). *Perceived stress has a significant effect on satisfaction.*

2.3. Re-Patronage Intention

Intentions are subjective judgments against individual activity, and re-patronage i tion (RI) usually functions as an outcome variable in the service business. RI can be de as "An individual judgment to use the assigned service of the same place after consid one's current and possible situation" [45]. If a customer receives RI from a place, o likely to re-patronize and recommend the place to their acquaintances [46]. This cor assumes that a customer's choice interrelates to loyalty based on user experience [4 Butcher [48], RI is claimed to be a measurable service result. On the other hand, Soder and Ohman [49] took RI as a decisive intention based on expectation. The study was key to understanding travelers' behavior intentions. Moreover, it evolved as t patronage intention derived from the sum of the travelers' satisfaction, behavior inte and evaluation of travel destination. Naturally, further research into recommendin sharing behavior of favored travel destination became the linchpin [50,51]. Belo related previous studies: Lin [52] showed that customers' mental well-being can aff on cruise travel; Kim et al. [21] showed that the well-being of the lounge experienc affect satisfaction and RI; Han et al. [12] examined customers' SA affecting the effect o by increasing MH and WB perception and reducing stress. Previous studies [11,12,21, developed the hypothesis of our study.

Hypothesis 5 (H5). *Psychological wellness has a significant effect on re-patronage intentio*

Hypothesis 6 (H6). *Satisfaction has a significant effect on re-patronage intention.*

2.4. The Mediating Effect of Perceived Stress, Psychological Wellness, and Satisfaction

Previous studies related to stress reduction [12,13,53,54] show that green spaces, tures, and NBS have a high correlation with customer RI by improving their MH an In a study of green hotel visitors, Han et al. [13] found a significant mediating eff customer WB and self-rated MH on the relationship between NBSs and customer be This shows that a nature-based restorative environment can reduce the stress leve users [14]. Moreover, Grahn and Stigsdotter [55] sampled eight environments to prov restorative environment has the highest stress-reducing potential.

The detailed analysis between average nature-based activity period and emo WB mediated by the restoration experience and visit shows that restoration experien affect the WB. This is due to exposure to natural environments and mediating the emc WB under specific nature-based activities [15]. In Qiu et al. [56], it was shown th natural environment could directly affect tourists' attention restoration and quality of life. The sampled Chinese mountaineers and Australian coast visitors showed that, based under ART, examining ART can apply to various natural environments. Moreover, ART can be a valuable tool to improve people's mental health post COVID-19.

Hypothesis 7 (H7). *Perceived stress, psychological wellness, and satisfaction will be parallel mediation on the relationship between green hotels as NBS and re-patronage intention.*

2.5. Health Consciousness as Moderator

People with HC are highly interested in private health. They are more conscious about health and are likely to improve and join health activities to remain healthy [57]. People with HC are active in doing health-related research. They will act quickly on the information they collect, performing more than one related activity to secure a healthy life [58]. HC can be treated as a psychological state, predicting possible outcomes from health attitude and activity [59]. This can define HC as a preparatory stage to perform a health activity, making people join the health improvement and maintenance activities [60]. Many theories claim that HS can promote disease prevention, and is the first health improvement action [61].

In this study, we will examine HC as a prevention and protection against COVID-19 derived infection and wariness. Each individual reacts differently to recognized health hazards depending on their level of HC [62]. This can lead to a gap in HC depending on their response to health-related messages [63]. People with high HC take the COVID-19 situation as a high health risk and strengthen their health attitude and norms when traveling [64]. Additionally, tourists with high HC are very likely to join the health-related activity under COVID-19 pandemic. This leads them to cancel or avoid their travel plans [25].

The following articles are previous studies of HC and the tourist industry. In Zhang et al. [65], tourists from rural, eco-friendly areas were sampled, showing that travelers with high HC have a high desire to travel for their health and to protect it. In Wu et al. [64], the authors verified COVID-19 awareness and the regulating effect of HC on social distancing among the sampled Chinese hotel employees. This study set the following assumptions based on the previously mentioned studies:

Hypothesis 8a (H8a). *Health consciousness plays a significant moderating role in the relationship between green hotels as NBSs and perceived stress.*

Hypothesis 8b (H8b). *Health consciousness plays a significant moderating role in the relationship between perceived stress and psychological wellness.*

Hypothesis 8c (H8c). *Health consciousness plays a significant moderating role in the relationship between perceived stress and satisfaction.*

Hypothesis 8d (H8d). *Health consciousness plays a significant moderating role in the relationship between green hotels as NBSs and psychological wellness.*

Hypothesis 8e (H8e). *Health consciousness plays a significant moderating role in the relationship between psychological wellness and satisfaction.*

Hypothesis 8f (H8f). *Health consciousness plays a significant moderating role in the relationship between psychological wellness and re-patronage intention.*

Hypothesis 8g (H8g). *Health consciousness plays a significant moderating role in the relationship between satisfaction and re-patronage intention.*

3. Materials and Methods

3.1. Measures and Questionnaire Development

This study focused on (1) expected re-patronage intention of green hotel custom (2) adding ART to base NBSs though the green hotel model, (3) the utility of NBSs as verifi through previous studies of environmental science [7,11,26,30,44,66] and ART throu environmental psychology [38,40,44,56,67–69]. With these, we intended to accumul studies of hotel industry hypotheses in the previous studies mentioned above wit research model based on preset hypothesis.

Measurement variables of previous studies examined the change of customer tentions against green hotels' NBSs, mediated by PS, PW and SA. However, there w insufficient studies related to the hotel industry [12], as most of the variables were fr studies of the tourist industry [13,15,23–25,42]. Thus, we developed a measurement v able based on modified previous study material and in-depth interviews made by the h manager and CEO.

In detail, we first selected five hotel managers with over ten years of experie and five professors of hotel management who were willing to join the in-depth intervi Second, we made detailed explanations of the purpose of the study and interviews. Th to check the suitability of the questionnaire, the questionnaire used in the previous st was shown to the experts through a one-on-one interview, and it was verified whet the questionnaire was suitable for green hotels research through the interview. Fourth combining five interviews with experts, it was modified and supplemented to fit the gr hotels study. Lastly, we deployed the survey to be distributed to twenty customers v had visited the Grand Walkerhill Seoul Hotel in the time span from 11 November 202 13 November 2021 as a pretest under final review, providing transparency to the sur The Grand Walkerhill Seoul Hotel (independent hotel, 799 rooms) surveyed is green h which is located overlooking the Han River and the slopes of the Acha Mountain; plants and flowers inside the hotel are all natural, and the hotel has large glass wind for natural lighting.

Measurement items shown below (Appendix A) include (1) eight items extrac from [12,13] for "green hotel as NBS"; in this study, an NBS is defined as "a special con or method; developed to induce social and environmental outputs from COVID-19 provide a solution for various social and environmental challenges by utilizing nat Two factors were derived, indoor green hotels (IGH) and outdoor green hotels (OGH), (2) nine items are extracted from [12,13,70] for PW.

In this study, PW is "an individual's self-assessment of their ability to deal wi mental crisis in order to return to a pre-crisis state." Items stem from two factors (MH, by following the previous studies. (3) Four items were extracted (i.e., I fear I cannot control of the important matters) from [71] for PS. This study defined PS as "a subjec consciousness against environmental stimulants, variable to individual traits". (4) items are extracted (i.e., 'as a whole, I have really enjoyed myself at this hotel as expec from [21,72] for SA. Customers had to NBSs after visiting the green hotel. (5) Four i were extracted (i.e., I have a strong intention to visit the service provider again) from [7 RI, which is the most reliable factor of behavior intention against the green hotel. This s used 34 items to evaluate six variables, using a Likert five-point scale: 1. Not importa all, 2. Not important, 3. Normal, 4. Important, 5. Very important.

3.2. Data Collection Process and Data Analysis

Targets were adults aged 20 and older who visited green hotels within the Before the primary survey, we performed a pilot test to find any structural issues wit paper. From 8 November 2021 to 4 November 2021, fifty students under the doctor's management course were applied. We used the online research company embrain between 15 November 2021 and 28 November 2021, and received 600 surveys. Fift undependable responses were removed, leaving 544 (90.7%) valid samples for data ana The tools for the analysis were SPSS 20.0 for exploratory factor analysis and AMOS 24.

confirmatory factor analysis (CFA), structural equation modeling (SEM), mediator effect, and moderator effect.

4. Results

4.1. Characteristics of Respondent

Of the 544 respondents, 353 respondents (64.9%) were female and 191 respondents (35.1%) were male. Present marital status was highest for married people (307 respondents, 56.4%), followed by single people (232 respondents, 42.6%), and other (five respondents, 0.9%). Among the participants, 37.7% were 30–39 years old (205 respondents), 11.9% were 20–29 years old (65 respondents), 9.0 were 40–49 years old (49 respondents), and 3.7% were 50 years old or over (20 respondents). The education level was generally high; 328 respondents (60.3%) were college and university graduates, and 146 respondents (26.8%) were in graduate school or higher. Additionally, 70 respondents (12.9%) had attended high school and below. The annual household income of USD 20,000–29,999 (238 respondents, 43.8%) was the highest, followed by USD 40,000 and more (109 respondents, 20.0%), USD 30,000–39.999 (99 respondents, 18.2%), and under USD 20,000 (98 respondents, 18.0%). By job, employed (219 respondents, 40.3%) showed the highest, followed by professionals (113 respondents, 20.8%), housewives (80 respondents, 14.7%), self-employed (74 respondents, 13.6%), and students (43 respondents, 7.9%). The number of hotel visits for one year was as high as 46.5% (253 respondents) for 3–5, followed by 1–2 times 24.6% (134 respondents), 6–9 times 20.2% (110 respondents), and ten and more times 8.6% (47 respondents). The purpose of visiting the green hotel was leisure 69.5% (378 respondents), business 16.9% (92 respondents), and other 13.6% (74 respondents).

4.2. Reliability and Validity Assessments and Confirmatory Factor Analysis

The measurement unit was developed from previous studies, to offer content validity. To analyze the measurements' reliability, we performed exploratory factor analysis; If Cronbach's α value is above 0.6, we can say it has internal consistency. During the analysis of all measurement items (NBSs, PW, SA, RI, and HC), one of the PW items ("I feel healthy and happy when staying at this hotel") was removed, as it had a Cronbach's α value below 0.5. Reverse-coding was performed to PS for analysis.

To verify the construct reliability and convergent validity, we performed confirmatory factor analysis (CFA) on the constructs (NBSs, PS, PW, SA, RI, and HC). The IPS helped to increase the scale's reliability when performing the CFA. Parceling is a procedure for computing average scores or sums through multiple items. The average scores or sum instead of the individual item scores serve as indicators of latent factors in the SEM analysis. In this study, items with high correlation or similar factor loadings in factor analysis were grouped using Cattle's radial parceling [74].

We measured IPS based on factor loading for study, derived from exploratory factor analysis of NBSs items (IGH, OGH) and PW items (MH, WB). When the average variance extracted (AVE) and construct reliability (CR) are above 0.5 and 0.7, we can say there is convergent validity [75]. We found an AVE between 0.589 and 0.861 and CR between 0.749 and 0.927; these values satisfy the convergent validity, as displayed in Table 1.

Normally, the goodness-of-fit index should have CFI, IFI, and TLI above 0.9 and below 1.0 [76]. RMSEA should remain below 0.05, and is acceptable between 0.05~0.1 [77]. Our study showed a well-established model to fit the data satisfactorily (goodness-of-fit statistics for measurement models: $\chi2 = 289.844$, df = 137, $p < 0.000$, $\chi2/\text{df} = 2.116$, RMSEA = 0.092, CFI = 0.929, IFI = 0.931, TLI = 0.902). We present correlation to verify discriminant validity.

After the analysis, all variables showed a square price below AVE value and all correlation coefficients did not include 1 in standard error estimation. Thus, all variables obtained discriminant validity. To verify the discriminant validity, we performed Pearson Correlation Analysis. All variables (NBSs, PS, PW, SA, RI, and HC) showed a significance level of $p < 0.001$, index of the correlation coefficients.

Table 1. The measurement model and correction.

Construct and Scale Item		Standardized Loading	MEAN (SD)	AVE (CR)	NBSs	PW	PS	SA	RI	H(
NBSs	IGH OGH	0.836 0.709	4.323 (0.567)	0.601 (0.749)	1					
PW	MH WB	0.961 0.893	4.259 (0.749)	0.861 (0.925)	0.758 *** (0.575)	1				
PS	PS1 PS2 PS3 PS4	0.737 0.841 0.857 0.609	4.393 (0.581)	0.589 (0.849)	0.636 *** (0.404)	0.597 *** (0.356)	1			
SA	SA1 SA2 SA3	0.883 0.940 0.939	4.487 (0.622)	0.848 (0.944)	0.730 *** (0.533)	0.827 *** (0.684)	0.615 *** (0.378)	1		
RI	RI1 RI2 RI3 RI4	0.837 0.863 0.901 0.681	4.227 (0.641)	0.650 (0.878)	0.650 *** (0.423)	0.857 *** (0.734)	0.704 *** (0.496)	0.842 *** (0.709)	1	
HC	HC1 HC2 HC3 HC4	0.878 0.781 0.952 0.873	3.882 (0.951)	0.762 (0.927)	0.404 *** (0.163)	0.486 *** (0.236)	0.404 *** (0.163)	0.567 *** (0.321)	0.557 *** (0.310)	

Note 1. NBSs = natural-based solutions, IGH = indoor of green hotels, OGH = outdoor of green hotels, PS ceived stress, PW = psychological wellness, MH = mental health, WB = well-being, SA = satisfaction, R patronage intention, HC = health consciousness. Note 2. Goodness-of-fit statistics for the measurement r $\chi2$ = 289.844, df = 137, $p < 0.000$, $\chi2$/df = 2.116, RMSEA = 0.092, CFI = 0.929, IFI = 0.931, TLI = 0.902). Note factor loadings are significant at $p < 0.001$ (***). Correlations between variables are below the diagonal. Sq correlations between variables are within parentheses

4.3. Research Hypotheses Testing and Structural Equation Modeling

Variables of the research model were the following items: Inner variable, NBSs; neous variable, RI; intervening variables, PW, PS, and SA. We used a covariance matri maximum likelihood estimation for the analysis. Goodness-of-fit statistics were favo ($\chi2$ = 141.909, df = 77, $p < 0.001$, $\chi2$/df = 1.843, RMSEA = 0.080, CFI = 0.960, IFI = TLI = 0.938). Additionally, SEM showed high prediction power for RI n general (R^2 = C The standardized path coefficients and t-values are shown in Table 2. The hypot test results are provided in Figure 1. The path estimates show that NBSs had a po direct effect on ascription of PW (β = 0.648, $p < 0.001$) and PS (β = 0.541, $p < 0.001$); H1 and H2 were supported. The result of estimation indicated that PS had a signi positive effect on PW (β = 0.327, $p < 0.01$) and PS (β = 0.291, $p < 0.01$); thus, H3 and H4 supported. As expected, PW had an impact on SA (β = 0.639, $p < 0.001$) and RI (β = $p < 0.001$); H5 and H6 were supported. It was found that SA had a significant positive on RI (β = 0.292, $p < 0.05$); thus, H7 was supported.

We found direct and indirect effects of NBS on RI related to PW, PS, and th diating effect of SA. To verify the significance of indirect effect, we used the paran bootstrapping, and the result is displayled in Table 2. The findings revealed that significance affected PCBI (β NBSs → PW and PS and SA → RI = 0.692, $p< 0.001$) indi through PW, PS, and SA, thus confirming them as partial mediating variables. Ther H8 was supported.

Table 2. Structural model results and hypotheses testing.

Hypothesized Paths	Coefficients	*t*-Values
H1: NBSs → PW	0.648	5.607 ***
H2: NBSs → PS	0.541	4.156 ***
H3: PS → PW	0.327	2.842 **
H4: PS → SA	0.291	3.098 **
H5: PW → SA	0.639	6.921 ***
H6: PW → RI	0.427	3.960 ***
H7: SA → RI	0.292	2.467 *
Explained variable:	R^2(PW) = 0.629 R^2(SA) = 0.745	R^2(PS) = 0.420 R^2(RI) = 0.819
Indirect effect:	β (NBSs → PW& PS & SA → RI) = 0.692 ***	
Total effect on RI:	β(RI) = 0.692 ***	
The results:	H8 supported	

Note 1. NBSs = natural-based solutions, PS = perceived stress, PW = psychological wellness, MH = mental health, WB = well-being, SA = satisfaction, RI = re-patronage intention. Note 2. Goodness-of-fit statistics for the structural model: $\chi2$ = 141.909, df = 77, $p < 0.001$, $\chi2$/df = 1.843, RMSEA = 0.080, CFI = 0.960, IFI = 0.961, TLI = 0.938. Note 3. * $p < 0.5$, ** $p < 0.01$, *** $p < 0.001$.

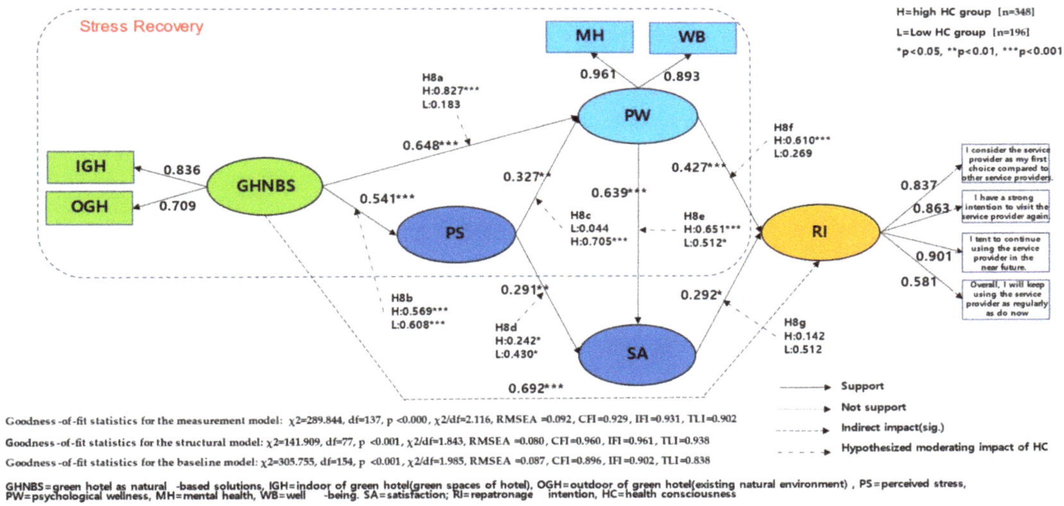

Figure 1. Structural equation model estimation and test for structural metric invariance.

4.4. Moderating Effect of Health Consciousness

We performed K-mean cluster analysis on moderating variable HC to check whether NBSs build a significant HC gap on RI. The result was in two groups, one with high HC (n = 322) and one with low HC (n = 207). The baseline model showed an acceptable level for the data suitability (goodness-of-fit statistics for the baseline model: $\chi2$ = 305.755, df = 154, $p < 0.001$, $\chi2$/df = 1.985, RMSEA = 0.087, CFI = 0.896, IFI = 0.902, TLI = 0.838).

If the constrained model and unconstrained model's chi-square value gap exceed 3.84 and reach a significance level ($p < 0.5$), we can say there is a regulation effect (as shown in Table 3 and Figure 1). The chi-square test results with measurement invariance revealed that GHNBS-PW path ($\Delta\chi2$ [1] = 11.398, $p > 0.05$) between high and low HC groups and PS-PW path ($\Delta\chi2$ [1] = 1O.624, $p > 0.05$) between high and low HC groups did differ significantly, and more than 3.83 for chi-square difference. Hence, this result supports Hypothesis H9a and H9c. However, H9b, H9d, H9e, and H9F could not be supported, as they did not show significant results.

Table 3. Result of moderating effect.

Paths	High-HC Group (n = 348)		Low-HC Group (n = 196)		Nested Model Constraine to Be Equal
	Coefficients	t-Values	Coefficients	t-Values	
H8a: NBSs-PW	0.827	3.394 ***	0.183	1.266	$\chi2$ (155) = 317.153 [a]
H8b: NBSs-PS	0.569	3.936 ***	0.608	3.642 ***	$\chi2$ (155) = 305.941 [b]
H8c: PS-PW	−0.044	−0.280	0.705	4.267 ***	$\chi2$ (155) = 316.379 [c]
H8d: PS-SA	0.242	2.179 *	0.430	2.071 *	$\chi2$ (155) = 306.185 [d]
H8e: PW-SA	0.651	5.793 ***	0.512	2.576 *	$\chi2$ (155) = 305.882 [e]
H8f: PW-RI	0.610	4.626 ***	0.269	1.179	$\chi2$ (155) = 306.602 [f]
H8g: SA-RI	0.142	1.094	0.512	1.742	$\chi2$ (155) = 306.412 [g]

Chi-square difference test:	Test results:	
[a] $\Delta\chi2$ (1) = 11.398, $p > 0.05$	H8a: Supported	
[b] $\Delta\chi2$ (1) = 0.186, $p > 0.05$	H8b: Not support	
[c] $\Delta\chi2$ (1) = 10.624, $p > 0.05$	H8c: Supported	Goodness-of-fit statistics for the baseline model:
[d] $\Delta\chi2$ (1) = 0.430, $p > 0.05$	H8d: Not supported	$\chi2$ = 305.755, df = 154, $p < 0.001$, $\chi2$/df = 1.985,
[e] $\Delta\chi2$ (1) = 0.127, $p > 0.05$	H8e: Not support	RMSEA = 0.087, CFI = 0.896, IFI = 0.902, TLI = 0.838
[f] $\Delta\chi2$ (1) = 0.847, $p > 0.05$	H8f: Not supported	
[g] $\Delta\chi2$ (1) = 0.657, $p > 0.05$	H8g: Not supported	

Note 1. NBSs = natural-based solutions, PS = perceived stress, PW = psychological wellness, MH = mental he WB = well-being, SA = satisfaction, RI = re-patronage intention, HC = health consciousness. Note 2 * p < *** $p < 0.001$.

5. Results and Discussion

5.1. Conclusions

As the COVID-19 pandemic increases interest in social and environmental issues, importance of NBSs has come to the forefront [13,78]. The COVID-19 pandemic can ca stress, while green hotels can reduce stress; thus, we performed this study to find gr hotel NBSs' customer mental care ability and effect on SA and RI by restoring PS.

This study had three goals. First, to merge NBS and ART to investigate a mode eco-friendly hotel and PS, PW (MH, WB), SA, and RI. This study can provide stand data for a solid practical implication, exceeding the limits of previous studies. Second, shaped the verification results of previous studies by examining the multiple medi effects of PW, PS, and RI drawn from green hotel NBSs and RI. Third and last, we trie shed light on the mediating role of HC against NBSs, PW, PS, SA, and RI. After COVIL occurred, people showed behavioral and perceptual changes. Customers who had CO 19 developed relatively high HC compared to those who had not. Thus, monitoring behavior change was considered vital to us.

Our empirical analysis shows that NBSs can significantly impact PW and PS. result supports previous research [37,70]. The significant impact of PS on PW and matches the results of previous studies [79]. Additionally, the impact of PW on SA an matches previous studies [12,13]. The partial mediation effect of PS, PW, and SA betw NBSs and RI matches previous studies as well [79]. Moreover, we found a mediating e of HC on NBSs, PW, PS, SA, and RI. This supports H9a and H9b and partially supp previous studies [64,65].

5.2. Theoretical Implication

The research implications of this study identified the positive effect of green h NBS. We applied ART to supplement PS, expanding the academic field of NBSs to s restoration. In the past, NBSs focused on recognizing improvement of human MH and This study applied ART to NBSs, re-examining NBSs' stress restoration aspect of redu PW and PS. It is relevant that we managed to examine what NBSs does to PS, SA, RI via applying ART. It was relevant to examine what restorative experience can d customer satisfaction and RI.

After the COVID-19 pandemic, people recognized health risks more than ever w raised HC. Additionally, people with high HC are likely to make healthier lifestyle decis Thus, there is a theoretical implication around examining HC as a moderator between PS, SA, and RI, as this was the first empirical research about the topic.

5.3. Practical Implication

The practical implications of this study include first, that green hotels' NBSs can help PW (MH, WB) and reduce PS. From this, green hotels' NBSs can reduce stress caused by COVID-19. Moreover, stress is the core factor of various diseases. Thus, hotel managers should secure an eco-friendly environment and nature accessibility for customers looking for the stress-reducing effect of a green hotel. Deploying roads and signs to mountains, rivers, and oceans near green hotels can increase nature accessibility. Additionally, installing an eco-friendly bench will offer a fine rest in nature.

Second, customers with perceived stress visiting green hotels affect psychological wellness (mental health, well-being) and satisfaction, although not as high. In order to alleviate perceived stress, healing through a rehabilitation program using nature may be helpful. In addition to the healing and recovery environment provided by nature, stress can be managed through leisure activities that can be carried out in nature. For example, in a green hotel in nature, stress management can be performed by healing the tired physical conditions and mind by creating programs such as healing yoga in the forest.

Third, customers with high SA visiting green hotels had PW which was high by the characteristics of NBSs, and low on RI. As a result of this study, it can be seen that the variable of customer's SA at green hotels does not have significant effect on RI. On the other hand, green hotel customers seem to be satisfied through MH and WB from rest. However, the purpose of hotel customers' visits (restaurant visit, casino, leisure (swimming pool, tennis court, kids' activity, etc.) should be considered in various ways. It is important to ensure that these things and the advantages of the Green Hotel can be accompanied at the same time as RI. Specifically, green hotels should supplement suitable facilities so that customers can feel attracted to the green physical environment. For example, green hotels should make a hotel swimming pool close to nature for customers who feel like swimming in a river or sea and provide the comfort of an indoor swimming pool.

Fourth, while NBSs of green hotels did not directly affect RI, it was found to have a significant effect on PW, PS, and SA as indirect effect. These results show that the characteristics of green hotels' NBSs help customers recover from stress, increasing their willingness to revisit intentions. By increasing the accessibility of green hotels, the NBS characteristics of green hotels can have a positive effect on customers' stress recovery. For example, in order to compensate for the geographical shortcomings of green hotels that must be close to nature, transportation that makes them easily accessible to green hotels located outside should be provided. Such a pickup service (i.e., limousine or hotel shuttle bus) is provided for convenient access from airports or train stations located around green hotels.

Fifth and last, the moderation effect of HC was valid only on NBSs and PW and PS and PW. It was ineffective on HC and other combinations between NBSs and PW, PS, and SA on RI. Furthermore, if the HC is not high enough, it is not practical for any group. COVID-19 incursion raised the HC of people generally. Additionally, high HC can affect the relationship between NBSs and PW. We found that a group with high HC on the relationship between PS and PW (MH, WB) has significant effect. Plus, a previous study by [25] defined HC as a critical factor in preventing COVID-19 contagion. A group with high HC is likely to cancel or change their travel plans when the destination is under high infection risk. Thus, it seems that anxiety towards society is the catalyst that boosts the effect of green hotels' NBSs and PS on PW. Nevertheless, as the COVID-19 situation is not yet resolved, a secure quarantine system should be considered to reduce customers' anxiety.

5.4. Limitations

This study has several limitations. First, the research method was limited to cross-section; measurements were made only from recognizing a limited period and aspect. A longitudinal study of change over time should be carried out, i.e., comparing pre-COVID and post-COVID data to check the effect of green hotel NBS on PW, PS, SA, and RI in order to collect diverse data. Second, the target was limited. Continuing research may target the food service industry (i.e., eco-friendly restaurants or coffee houses) to find more valu factors. Third and last, this study focused on the green hotel visiting experience c Continuing research may differentiate the sample and examination by target area (i.e., r and urban areas) to find any significant gaps between them. Lastly, although this st was conducted on customers who visited green hotels located in Korea, future rese needs to expand on this and compare customers from Asia, Europe, and the America addition to Korea.

Author Contributions: Conceptualization, S.Y. and T.K.; methodology, S.Y.; formal analysis, investigation, T.K.; resources, T.K.; data curation, T.K.; writing—original draft preparation, S.Y T.K.; writing—review and editing, S.Y. and T.K. All authors have read and agreed to the publi version of the manuscript.

Funding: This research received no external funding.

Conflicts of Interest: Authors declare no conflict of interest.

Appendix A. Measurement Items

Indoor of the green hotel [12,13]
I easily see green interior decorations and diverse living plants in the lobby area of this hotel.
I easily see a variety of green items and light through glass windows in this hotel's restaurants.
I easily see diverse flowers, trees, and potted plants in the hotel's coffee lounge.
green space can easily be seen everywhere in this hotel.
Outdoor of the green hotel [12,13]
This hotel has easy access to the natural environment (i.e., mountains, forests, rivers, seas, lakes, natural parks).
The region surrounding this hotel has good weather (i.e., temperature, humidity, and precipitation).
The region surrounding this hotel has good and fresh quality air.
The region around this hotel is safe from natural disasters (i.e., earthquakes, typhoons, tsunamis, floods).
Mental health [12,13]
Staying at this hotel plays an important role in relieving my mental stress and anxiety.
Staying at this hotel helps me rising my confidence in everyday life.
Staying at this hotel is worthwhile as it makes me think that I am a valuable and important person.
Staying at this hotel is worthwhile as it helps me turn any anxiety and worry into confidence
Well-being [12,13]
I feel healthy and happy during my stay in this hotel.
I feel emotionally secure during my stay at this hotel.
This hotel plays an important role in making my mind calm and peaceful.
Thanks to this hotel, I was able to relax comfortably.
Thanks to this hotel, I was able to refresh my mood.
Perceived stress [35]
For the past 30 days, I have been unable to control the important things in my life.
For the past 30 days, I have not been able to handle my personal problems.
In the past 30 days, my work has not gone my way.
In the past 30 days, I have felt difficulties that I cannot overcome.
Satisfaction [72]
Overall, I am satisfied with my experience at this hotel.
My decision to stay at this hotel was a wise one.
As a whole, I have really enjoyed myself at this hotel as expected.
Health consciousness [64]
I often reflect on my health a lot and try to protect it.
I'm awfully self-conscious about my health.
I am generally watchful to my inner feelings about my health.
I am incessantly checking my health.
Re-patronage intention [73]
I consider the nature-based solution of this hotel is better compared to other hotels.
I have a strong intention to visit this hotel again.
I intend to continue using this hotel in the near future
Overall, I will keep using this hotel as regularly as I do now.

References

1. Institute for Health Metrics and Evaluation. Available online: https://www.healthdata.org/sites/default/files/files/Pr COVID/briefing_US_092320.pdf (accessed on 14 December 2021).

Centers for Disease Control and Prevention. Available online: https://www.cdc.gov/coronavirus/2019-ncov/community/schools-childcare/child-care-guidance.html (accessed on 12 January 2022).
United Nations Conference on Trade and Development. Available online: https://unctad.org/system/files/official-document/tn_unctad_ict4d19_en.pdf (accessed on 13 January 2022).
United Nations. Available online: https://www.un.org/sustainabledevelopment/blog/2020/11/internet-governance-forum-calls-for-bridging-digital-divides-harnessing-the-internet-to-support-human-resilience-and-build-solidarity-amid-covid-19/ (accessed on 11 January 2022).
The World Economic Forum. Available online: https://www3.weforum.org/docs/WEF_The_Global_Risks_Report_2021.pdf (accessed on 13 December 2021).
Seddon, N.; Chausson, A.; Berry, P.; Girardin, C.A.; Smith, A.; Turner, B. Understanding the value and limits of nature-based solutions to climate change and other global challenges. *Philos. Trans. R. Soc. B* **2020**, *375*, 20190120. [CrossRef] [PubMed]
Cohen-Shacham, E.; Walters, G.; Janzen, C.; Maginnis, S. *Nature-Based Solutions to Address Global Societal Challenges*; IUCN: Gland, Switzerland, 2016; Volume 97.
Panno, S.; Matić, S.; Tiberini, A.; Caruso, A.G.; Bella, P.; Torta, L.; Stassi, R.; Davino, S. Loop mediated isothermal amplification: Principles and applications in plant virology. *Plants* **2020**, *9*, 461. [CrossRef] [PubMed]
Chaudhury, P.; Banerjee, D. Recovering with nature: A review of ecotherapy and implications for the COVID-19 pandemic. *Front. Public Health* **2020**, *8*, 604440. [CrossRef] [PubMed]
Pietilä, M.; Neuvonen, M.; Borodulin, K.; Korpela, K.; Sievänen, T.; Tyrväinen, L. Relationships between exposure to urban green spaces, physical activity and self-rated health. *J. Outdoor Recreat. Tour.* **2015**, *10*, 44–54. [CrossRef]
Van den Bosch, M.; Sang, Å.O. Urban natural environments as nature-based solutions for improved public health–A systematic review of reviews. *Environ. Res.* **2017**, *158*, 373–384. [CrossRef]
Han, H.; Yu, J.; Hyun, S.S. Effects of nature-based solutions (NBS) on eco-friendly hotel guests' mental health perceptions, satisfaction, switching barriers, and revisit intentions. *J. Hosp. Mark. Manag.* **2019**, *29*, 592–611. [CrossRef]
Han, H.; Jongsik, Y.; Hyun, S.S. Nature based solutions and customer retention strategy: Eliciting customer well-being experiences and self-rated mental health. *Int. J. Hosp. Manag.* **2020**, *86*, 102446. [CrossRef]
Beil, K.; Hanes, D. The influence of urban natural and built environments on physiological and psychological measures of stress—A pilot study. *Int. J. Environ. Res. Public Health* **2013**, *10*, 1250. [CrossRef]
Korpela, K.; Borodulin, K.; Neuvonen, M.; Paronen, O.; Tyrväinen, L. Analyzing the mediators between nature-based outdoor recreation and emotional well-being. *J. Environ. Psychol.* **2014**, *37*, 1–7. [CrossRef]
Baur, J.W.; Tynon, J.F. Small-scale urban nature parks: Why should we care? *Leis. Sci.* **2010**, *32*, 195–200. [CrossRef]
Heintzman, P. Nature-based recreation, spirituality, and persons with disabilities. *J. Disabil. Relig.* **2014**, *8*, 97–116. [CrossRef]
Kebbede, G. *Living with Urban Environmental Health Risks: The Case of Ethiopia*; Routledge: New York, NY, USA, 2017; pp. 1–251. [CrossRef]
Lederbogen, F.; Kirsch, P.; Haddad, L.; Streit, F.; Tost, H.; Schuch, P.; Meyer-Lindenberg, A. City living and urban upbringing affect neural social stress processing in humans. *Nature* **2011**, *474*, 498–501. [CrossRef] [PubMed]
Chua, B.L.; Lee, S.; Han, H. Consequences of cruise line involvement: A comparison of first-time and repeat passengers. *Int. J. Contemp. Hosp. Manag.* **2017**, *29*, 1658–1683. [CrossRef]
Kim, H.C.; Chua, B.L.; Lee, S.; Boo, H.C.; Han, H. Understanding airline travelers' perceptions of well-being: The role of cognition, emotion, and sensory experiences in airline lounges. *J. Travel Tour. Mark.* **2016**, *33*, 1213–1234. [CrossRef]
Hwang, J.; Han, H. Examining strategies for maximizing and utilizing brand prestige in the luxury cruise industry. *Tour. Manag.* **2014**, *40*, 244–259. [CrossRef]
Olya, H.; Taheri, B. Introduction to the Special Issue: Nature-Based Solutions in Hospitality and Tourism Management. *J. Hosp. Tour. Res.* **2021**, *46*, 415–417. [CrossRef]
Yu, J.; Ariza-Montes, A.; Hernández-Perlines, F.; Vega-Muñoz, A.; Han, H. Hotels' eco-friendly physical environment as nature-based solutions for decreasing burnout and increasing job satisfaction and performance. *Int. J. Environ. Res. Public Health* **2020**, *17*, 6357. [CrossRef]
Pahrudin, P.; Chen, C.T.; Liu, L.W. A modified theory of planned behavioral: A case of tourist intention to visit a destination post pandemic Covid-19 in Indonesia. *Heliyon* **2021**, *7*, e08230. [CrossRef]
Chen, J.; Zhu, L.; Fan, P.; Tian, L.; Lafortezza, R. Do green spaces affect the spatiotemporal changes of PM 2.5 in Nanjing? *Ecol. Process.* **2016**, *5*, 7. [CrossRef]
Santiago Fink, H. Human-nature for climate action: Nature-based solutions for urban sustainability. *Sustainability.* **2016**, *8*, 254. [CrossRef]
Organisation for Economic Co-operation and Development. Available online: https://unfccc.int/sites/default/files/resource/OECD.pdf (accessed on 12 December 2021).
Derkzen, M.L.; van Teeffelen, A.J.; Verburg, P.H. Green infrastructure for urban climate adaptation: How do residents' views on climate impacts and green infrastructure shape adaptation preferences? *Landsc. Urban Plan.* **2017**, *157*, 106–130. [CrossRef]
Bennett, G.; Cassin, J.; Carroll, N. Natural infrastructure investment and implications for the nexus: A global overview. *Ecosyst. Serv.* **2016**, *17*, 293–297. [CrossRef]

31. Richardson, M.; Maspero, M.; Golightly, D.; Sheffield, D.; Staples, V.; Lumber, R. Nature: A new paradigm for well-being a ergonomics. *Ergonomics* **2017**, *60*, 292–305. [CrossRef]
32. Liang, H.H.; Chen, C.P.; Hwang, R.L.; Shih, W.M.; Lo, S.C.; Liao, H.Y. Satisfaction of occupants toward indoor environm quality of certified green office buildings in Taiwan. *Build. Environ.* **2014**, *72*, 232–242. [CrossRef]
33. Moon, H.; Yoon, H.J.; Han, H. Role of airport physical environments in the satisfaction generation process: Mediating the imp of traveler emotion. *Asia Pac. J. Tour. Res.* **2016**, *21*, 193–211. [CrossRef]
34. Trang, H.L.T.; Lee, J.S.; Han, H. How do green attributes elicit pro-environmental behaviors in guests? The case of green hotel Vietnam. *J. Travel Tour. Mark.* **2019**, *36*, 14–28. [CrossRef]
35. McEwen, B.S.; Seeman, T. Protective and damaging effects of mediators of stress: Elaborating and testing the concepts of allost and allostatic load. *Ann. N. Y. Acad. Sci.* **1999**, *896*, 30–47. [CrossRef]
36. Polanco-Roman, L.; Miranda, R. Culturally related stress, hopelessness, and vulnerability to depressive symptoms and suic ideation in emerging adulthood. *Behav. Ther.* **2013**, *44*, 75–87. [CrossRef]
37. Ulrich, R.S.; Simons, R.F.; Losito, B.D.; Fiorito, E.; Miles, M.A.; Zelson, M. Stress recovery during exposure to natural and ur environments. *J. Environ. Psychol.* **1991**, *11*, 201–230. [CrossRef]
38. Kaplan, R.; Kaplan, S. *The Experience of Nature: A Psychological Perspective*; Cambridge University Press: Cambridge, UK, 198
39. Martin, L.; White, M.P.; Hunt, A.; Richardson, M.; Pahl, S.; Burt, J. Nature contact, nature connectedness and associations w health, wellbeing and pro-environmental behaviors. *J. Environ. Psychol.* **2021**, *68*, 101389. [CrossRef]
40. Kaplan, S. The restorative benefits of nature: Toward an integrative framework. *J. Environ. Psychol.* **1995**, *15*, 169–182. [Cross
41. Hartig, T.; Korpela, K.; Evans, G.W.; Gärling, T. A measure of restorative quality in environments. *Hous. Theory Soc.* **1997** 175–194. [CrossRef]
42. Tyrväinen, L.; Ojala, A.; Korpela, K.; Lanki, T.; Tsunetsugu, Y.; Kagawa, T. The influence of urban green environments on st relief measures: A field experiment. *J. Environ. Psychol.* **2014**, *38*, 1–9. [CrossRef]
43. Van den Berg, M.; Wendel-Vos, W.; Van Poppel, M.; Kemper, H.; Van Mechelen, W.; Maas, J. Health benefits of green spaces in living environment: A systematic review of epidemiological studies. *Urban For. Urban Green.* **2015**, *14*, 806–816. [CrossRef]
44. Vujcic, M.; Tomicevic-Dubljevic, J.; Grbic, M.; Lecic-Tosevski, D.; Vukovic, O.; Toskovic, O. Nature based solution for impro mental health and well-being in urban areas. *Environ. Res.* **2017**, *158*, 385–392. [CrossRef]
45. Hellier, P.K.; Geursen, G.M.; Carr, R.A.; Rickard, J.A. Customers repurchase intention: A general structural equation model. *E Mark.* **2003**, *37*, 1762–1800. [CrossRef]
46. Pollack, B.L. Linking the hierarchical service quality model to customer satisfaction and loyalty. *J. Serv. Mark.* **2009**, *23*, 42 [CrossRef]
47. Kusumawati, A.; Rahayu, K.S. The effect of experience quality on customer perceived value and customer satisfaction an impact on customer loyalty. *TQM J.* **2020**, *32*, 1525–1540. [CrossRef]
48. Butcher, K. Differential impact of social influence in the hospitality encounter. *Int. J. Contemp. Hosp. Manag.* **2005**, *17*, 125- [CrossRef]
49. Soderlund, M.; Ohman, N. Behavioral intentions in satisfaction research revisited. *J. Consum. Satisf. Dissatisf. Complain. B* **2003**, *16*, 53–66.
50. Ryu, K.; Han, H.; Jang, S.S. Relationships among hedonic and utilitarian values, satisfaction and behavioral intentions ir fast-casual restaurant industry. *Int. J. Contemp. Hosp. Manag.* **2010**, *22*, 416–432. [CrossRef]
51. Som, A.P.M.; Marzuki, A.; Yousefi, M. Factors influencing visitors' revisit behavioral intentions: A case study of Sabah, Mala *Int. J. Mark. Stud.* **2012**, *4*, 39–50. [CrossRef]
52. Lin, C.H. Effects of cuisine experience, psychological well-being, and self-health perception on the revisit intention of hot spr tourists. *J. Hosp. Tour. Res.* **2014**, *38*, 243–265. [CrossRef]
53. Sohaib, M.; Wang, Y.; Iqbal, K.; Han, H. Nature-based solutions, mental health, well-being, price fairness, attitude, loyalty, evangelism for green brands in the hotel context. *Int. J. Hosp. Manag.* **2022**, *101*, 10312. [CrossRef]
54. Han, H.; Hyun, S.S. Impact of hotel-restaurant image and quality of physical-environment, service, and food on satisfaction intention. *Int. J. Hosp. Manag.* **2017**, *63*, 82–92. [CrossRef]
55. Grahn, P.; Stigsdotter, U.A. Landscape planning and stress. *Urban For. Urban Green.* **2003**, *2*, 1–18. [CrossRef]
56. Qiu, M.; Jin, X.; Scott, N. Sensescapes and attention restoration in nature-based tourism: Evidence from China and Australia. *Manag. Perspect.* **2021**, *39*, 100855. [CrossRef]
57. Mai, R.; Hoffmann, S. Taste lovers versus nutrition fact seekers: How health consciousness and self-efficacy determine the consumers choose food products. *J. Consum. Behav.* **2012**, *11*, 316–328. [CrossRef]
58. Moorman, C.; Matulich, E. A model of consumers' preventive health behaviors: The role of health motivation and health abil *Consum. Res.* **1993**, *20*, 208–228. [CrossRef]
59. Karn, S.; Amarkantak, A.M.; Swain, S.K. Health consciousness through wellness tourism: A new dimension to new age trav *Afr. J. Hosp. Tour. Leis.* **2017**, *6*, 1–9.
60. Kraft, F.B.; Goodell, P.W. Identifying the health-conscious consumer. *Mark. Health Serv.* **1993**, *13*, 18.
61. Newsom, J.T.; McFarland, B.H.; Kaplan, M.S.; Huguet, N.; Zani, B. The health consciousness myth: Implications of the independence of major health behaviors in the North American population. *Soc. Sci. Med.* **2005**, *60*, 433–437. [CrossRef]

Gidengil, C.A.; Parker, A.M. Zikmund-Fisher, B.J. Trends in risk perceptions and vaccination intentions: A longitudinal study of the first year of the H1N1 pandemic. *Am. J. Public Health* **2012**, *102*, 672–679. [CrossRef]
Iversen, A.C.; Kraft, P. Does socio-economic status and health consciousness influence how women respond to health related messages in media? *Health Educ. Res.* **2006**, *21*, 601–610. [CrossRef]
Wu, J.; Zhang, X.; Zhu, Y.; Yu-Buck, G.F. Get Close to the Robot: The Effect of Risk Perception of COVID-19 Pandemic on Customer–Robot Engagement. *Int. J. Environ. Res. Public Health* **2021**, *18*, 6314. [CrossRef]
Zhang, Y.; Wong, I.A.; Duan, X.; Chen, Y.V. Craving better health? Influence of socio-political conformity and health consciousness on goal-directed rural-eco tourism. *J. Travel Tour. Mark.* **2021**, *38*, 511–526. [CrossRef]
Nesshöver, C.; Assmuth, T.; Irvine, K.N.; Rusch, G.M.; Waylen, K.A.; Delbaere, B.; Wittmer, H. The science, policy and practice of nature-based solutions: An interdisciplinary perspective. *Sci. Total Environ.* **2017**, *579*, 1215–1227. [CrossRef]
Lee, T.J.; Kim, J.S. *Natural Toursim: The Impact of Korea's Nature-Based Recreation Settings on Tourists' Emotions, Satisfaction and Subjective Happiness*; Routledge: New York, NY, USA, 2017; pp. 46–57. [CrossRef]
Mody, M.; Suess, C.; Dogru, T. Restorative servicescapes in health care: Examining the influence of hotel-like attributes on patient well-being. *Cornell Hosp. Q.* **2020**, *61*, 19–39. [CrossRef]
Wang, T.C.; Tsai, C.L.; Tang, T.W. Restorative quality in tourist hotel marketing pictures: Natural and built characteristics. *Curr. Issues Tour.* **2019**, *22*, 1679–1685. [CrossRef]
Kolokotsa, D.; Lilli, A.A.; Lilli, M.A.; Nikolaidis, N.P. On the impact of nature-based solutions on citizens' health & wellbeing. *Energy Build.* **2020**, *229*, 110527. [CrossRef]
VanKim, N.A.; Nelson, T.F. Vigorous physical activity, mental health, perceived stress, and socializing among college students. *Am. J. Health Promot.* **2013**, *28*, 7–15. [CrossRef]
Oliver, R.L. A cognitive model of the antecedents and consequences of satisfaction decisions. *J. Mark. Res.* **1980**, *17*, 460–469. [CrossRef]
Min, J.; Yang, K.; Kim, J. The role of perceived vulnerability in restaurant customers' creation behavior and re-patronage intention during the COVID-19 pandemic. *J. Vacat. Mark.* **2021**, *28*, 38–51. [CrossRef]
Cattell, R.B. Second-order personality factors in the questionnaire realm. *J. Consult. Psychol.* **1956**, *20*, 411–418. [CrossRef]
Fornell, C.; Larcker, D.F. Structural equation models with unobservable variables and measurement error: Algebra and statistics. *J. Mark. Res.* **1981**, *18*, 382–388. [CrossRef]
Byrne, B.M. *Structural Equation Modeling with LISREL, PRELIS, and SIMPLIS: Basic Concepts, Applications, and Programming*; Psychology Press: Hove, East Sussex, UK, 2013; pp. 259–286.
Turner, L.W.; Reisinger, Y. Shopping satisfaction for domestic tourists. *J. Retail. Consum. Serv.* **2001**, *8*, 15–27. [CrossRef]
Soga, M.; Evans, M.J.; Tsuchiya, K.; Fukano, Y. A room with a green view: The importance of nearby nature for mental health during the COVID-19 pandemic. *Ecol. Appl.* **2021**, *31*, e2248. [CrossRef]
Valikhani, A.; Ahmadnia, F.; Karimi, A.; Mills, P.J. The relationship between dispositional gratitude and quality of life: The mediating role of perceived stress and mental health. *Pers. Individ. Differ.* **2019**, *141*, 40–46. [CrossRef]

icle

esearch Framework Built Natural-Based Solutions (NBSs) as reen Hotels

uk Kim [1] and Sunmi Yun [2,*]

[1] Department of Hotel & Restaurant Management, Kyonggi University, Seoul 03746, Korea; teokim1305@naver.com
[2] Department of Hospitality and Tourism Management, Sejong University, Seoul 05006, Korea
* Correspondence: sunmiyun80@gmail.com; Tel.: +82-10-4055-2066

Abstract: In this study, value-belief-norm (VBN) theory and the social exchange theory (SET) were applied to predict hotel customers' pro-environmental responsibility behavior intention (PRBI) for the characteristics of NBSs in green hotels—specifically, to investigate the relationship between NBSs as green hotel and PRBI, and to test its mediating effect on pro-environmental perceived (PPV), pro-environmental perceived belief (PPVBE), personal pro-environmental norms (PPN), attitude toward environmental behavior (ATEB), mental health (MH), well-being (WB), and satisfaction (SA) and the moderating effect of locations (urban, rural) among these variables toward pro-environmental responsibility behavior intention (PRBI). Data were collected using a survey of 440 customers who had visited green hotels in the Republic of Korea within the last 12 months. We used to test the research hypotheses by structural equation modeling (SEM). The findings generally supported the hypothesized associations between variables within our proposed theoretical framework and confirmed the moderating effect of location. The study's results have important theoretical and practical implications for the environment. We investigated the relationship between the characteristics of NBSs and PRBI of green hotels, and we investigated the relationship between psychological state, attitude, and behavior of green hotel customers by applying variables suitable for ART, SET, and VBN. In addition, we verified the moderating effect of customers' green behavior and attitudes toward green hotels located in urban and rural areas. Moreover, these findings herein may encourage green hotels to participate in preventing environmental problems. It provides primary data on customers' perception of ecofriendliness in establishing corporate management strategies.

Keywords: natural-based solutions (NBSs); pro-environmental value (PPV); pro-environmental perceives belief (PPVB); personal pro-environmental norms (PPN); attitude toward environmental behavior (ATEB); mental health (MH); well-being (WB); satisfaction (SA); pro-environmental responsibility behavior intention (PRBI); green hotels

ion: Kim, T.; Yun, S. Research ework Built Natural-Based ions (NBSs) as Green Hotels. *nability* **2022**, *14*, 4282. //doi.org/10.3390/su14074282

emic Editors: Panayiotis nitrakopoulos, Mario gnotta, Miklas Scholz rshiya Noorani

ved: 24 February 2022
ted: 28 March 2022
hed: 4 April 2022

1. Introduction

As COVID-19 spread rapidly around the world and the number of confirmed cases and deaths continued to increase, the World Health Organization (WHO) declared the disease a pandemic on 11 March 2020 [1], the third pandemic in history after Hong Kong flu in 1968 and novel swine-origin influenza in 2009. For this reason, it has been hit hard in various fields such as politics, economy, and culture [1].

The continued effects of COVID-19 are engendering a strong sense of crisis in terms of the environment and the impact of changes to consumer life [2]. The world is not only facing the COVID-19 pandemic but is also struggling with the increase in waste mainly from plastics affecting both society and healthcare sectors [2–4]. For instance, there is unintentionally and rapidly generated plastic waste from single-use plastics such as facial masks, hand sanitizer bottles, surgical gloves, takeout food packaging, food and polyethylene goods packaging, protective medical suits, and medical test kits [4]. This is because of disposable products and materials for health and safety reasons, whi has worsened the plastic pollution problem [5]. According to 'The Global Risks Rep 2020' published by the World Economic Forum, environmental problems present the m significant risk today.

As such, interest in global environmental problems has increased due to the impact COVID-19, and in response, natural-based solutions (NBSs) have emerged as an essent global phenomenon [2]. The concept of NBSs is a growing global phenomenon that c help societies address various social and environmental challenges sustainably [6,7]. D to sustainable characteristics and the pursuit of social and environmental benefits, NI have attracted attention in various contexts in recent years [6]. NBSs are presently bei applied to the design of residential areas, workplaces, buildings, and urban spaces enhance the quality of human life [6–9]. Hotels are increasing their NBSs endeavors create green spaces inside the building and protect the natural outdoor environment This creates what NBSs call green building that result in eco-environmental physi environment that influences human responses and behaviors [8,10]. Thus, hotel gr spaces, together with pro-environmental physical environments, often play a crucial r in positively affecting the well-being (WB) and mental health (MH) of the customer encourage continued retention [2,8,9].

This study focused on Value-Belief-Norm (VBN) theory, attitude toward sustainabi and psychological health as NBSs to explicate customer pro-environmental responsi behavior intention processes. We took this approach because social psychologists sugg that values, beliefs, and norms play a principal role in motivating pro-environmer behaviors [10–15]. Thus, consumers' personal characteristics such as social factors (i.e., ticipation in society, normative influence, and information influence), values, and perso emotions can influence patronage intention. In particular, social factors play a critical in environmental problems [16].

Second, it was utilized as it plays an important role in understanding NBSs ecofriendliness and customer attitudes toward the sustainable natural environment ecofriendly behavioral intentions (responsible behavioral intentions for the environment) According to SET, the structure of society is formed through the interchange of tangible intangible valuable things [18,19]. For example, tangible values might refer to green sp of ecofriendly hotels, and intangible values might refer to actions that help ecofriendly tels, such as values, beliefs, norms, and attitudes toward the environment [17]. When th tangible and intangible things are exchanged in social phenomena, they react condition in proportion to their relative values [20]. Applying SET to ecofriendly hotel compa can be an interchange between NBS of ecofriendly hotels and customers using these ho

Third, according to attention restoration theory (ART), NBSs of eco-friendly ho play an essential role in inducing satisfaction and eco-friendly behavioral intentions wi positive effect on customer MH and WB [8,9,21–25]. ART is a concept of the psycholog recovery effect natural spaces provide to humans [20]. In environmental psychology, n ral space research on individual happiness and recovery has been conducted, revealing nature provides positive emotional outcomes such as improving MH and restoring a tion [26]. For instance, Korea is currently considering possible travel (visiting ecofrie hotels) to recover from fatigue caused by COVID-19 and daily life. Because overseas tr has been restricted due to the COVID-19 pandemic, hotel stays are enjoying popularit an alternative travel option; moreover, people prefer green hotels when choosing whe stay [27], when you feel negative emotions in your daily life, your desire to feel happi through leisure and travel activities increases [28].

To summarize the above, the first purpose of this study is to verify the impact rela ship on PRBI of customers who have stayed at green hotels as NBSs. The second pur is to verify the mediating effect by applying VBN to explain the customer's ecofrie behavior and SET to explain the customer has perceived costs and benefits. Finally, t amine changes in perceptions and behaviors at green hotels in an urban and rural loca Therefore, this study aims to provide academic and practical implications for the influ relationship of NBS as green hotels on customers' PRBI through various variables applied in the theories of VBN, ART, and SET.

2. Conceptual Framework

2.1. Value-Belief-Norm Theory

VBN is a theory that explains people's ecofriendly behavior as three factors—value, belief, and norm—and is an extension of the norm activation model [13]. VBN theory is most widely used along with the theory of planned behavior when analyzing factors related to environmental behavior [15]. VBN theory was developed with a focus on environmentalism [15]. It integrates the perspective of value theory [14], the norm activation model of [29], and the new environmental paradigm of [13,30]. Thus, VBN theory identifies ecofriendly behavior by organizing the factors influencing ecofriendly behavior into values, new paradigms, perception of results, ascribed responsibility, and personal norms. In particular, personal norms (such as being willing to pay premium prices for ecofriendly food [31]) are noted in previous studies, such as [32], which focus on ecofriendly organic wine purchase intention. It was focused that personal norms having a positive effect on ecofriendly travel [33].

In addition, [15] proposes an action based on norms according to [29]'s theory of activation of norms. This process refers to the belief that an individual accepts a specific value, and if that value is threatened, actions are taken to help alleviate the threat and restore its value to activate personal norms. Moreover, the VBN theory explains how an individual's moral obligation leads to action. The value perceived by an individual leads to a norm that forms a belief and determines an individual's behavior [13]. The mindset factor most related to the actual behavior of VBN theory is a personal norm, which can be defined as creating a moral obligation to practice as an action and a willingness to engage in environmentally friendly behavior. Thus, personal norms are essential variables that can predict the behavior of ecofriendly consumers in various environments [13].

In this context, previous studies in the tourism industry using VBN are described as follows. In a study by [34] targeting users of national parks in India, analysis of the impact of the relationship with ecofriendly behavior of users applying the VBN model verified that ecofriendly values, attitudes, and norms are important variables in predicting the ecofriendly behavior of nature-based tourists. In addition, in a study by [35] targeting restaurant users with experience in using delivery food services, the VBN model was applied, confirming the significant positive effect in intention to use, word of mouth, and additional cost payment. In a study conducted by [36] on Vietnamese travelers, the analysis of ecofriendly behavioral intentions on Vietnamese agricultural tourism showed a difference in ecofriendly behavior between Vietnamese locals and overseas travelers.

These studies verified that these characteristics positively affect the ecofriendly behavior of users and travelers. In this context, studies using the concept of NBSs are gradually increasing about ecofriendly behavior, a study by NBSs applied to the service industry [3,37]. The successful implementation of NBS in the hotel industry can improve both MH and WB awareness of customers and employees [9]. Green spaces, green surfaces, green items, and natural environments are all part of environmentally friendly hotels and buildings whose service performance or characteristics are the main feature of product quality evaluation [38–40], which can provide high psychological and emotional value to customers.

Specifically, previous studies in the tourism industry applying VBN are as follows. In a study by [41] targeting customers who chose ecofriendly hotels (linen reuse program, recycling gray water from sinks and shows, low-flow water fixtures), a significant influence on the relationship between attitudes toward ecofriendly hotels and eco-friendly responsibility behavior was verified. In addition, by verifying the behavioral results in a study by [42] that applied VBN, it was confirmed that the attitudes of those who demonstrated eco-friendly values increased with a commitment to the environment, intention green hotels, and willingness to pay a premium price.

Here, the VBN theory can explain how the eco-friendly elements of NBSs (tha
the characteristics provided by eco-friendly hotels that build eco-friendly environme
can increase perceived value for customers. Therefore, based on the prior studies,
hypothesis is established that there is a relationship between eco-friendly values, bel
and norms in eco-friendly hotels.

Hypothesis 1 (H1). *Green hotels as NBSs significant effect on pro-environmental perceived v*

Hypothesis 2 (H2). *Pro-environmental perceived value significantly effects on pro-environme perceived belief.*

Hypothesis 3 (H3). *Pro-environmental perceived belief has a significant effect on personal no*

Hypothesis 4 (H4). *Satisfaction has a significant effect on pro-environmental responsible be ior intention.*

2.2. Social Exchange Theory

SET is a general sociological theory related to understanding resource exchang tween individuals and groups in interaction situations [43]. In the process of excha members of society rely on each other for the consequences of what they value. Hence, try to obtain more positive values and reduce negative values, assuming that all individ participating in the interaction process are based on subjective cost–benefit analysis alternative comparison. In this process, individuals continue to interact when they more benefits are generated [44]. According to social exchange, human behavior be in exchanging costs and rewards with others. In the social exchange theory develo by [18], it is said that when humans receive benefits such as compensation from the c party of exchange, they form an exchange relationship that makes them feel obligat repay it someday in the future. In this context, SET-applied tourism suggests a sen obligation among consumers to pay back in the future when an 'exchange relationsh' established through tourism activities when compensation or benefits are obtained the other party. Based on the tangible and intangible resources of tourism effects thro tourism experiences, tourists and residents form an attitude toward tourism by evalu and comparing the benefits and costs they earned through the exchange process [4 other words, healthy life is affected by the surrounding social and physical environ especially when people and the natural environment form a complementary relation accordingly, the environment can improve people's mental and physical health [6].

Lovaglia explained social exchange relations by dividing them into economic social exchanges [17]. It was explained that social exchange theory is a relationship tha a sense of obligation to pay someday when receiving benefits such as compensation the other party, and that economic exchange is based on the response method of the c to be exchanged, such as the transaction of goods. Customers who use the hotel cho situation where they can benefit a lot by maximizing the benefits of using the hote minimizing the cost. In this context, when customers stay at green hotels, they can how to do ecofriendly behavior, enjoy fresh and healthy food, and benefit from funct and emotional benefits such as making you feel good by others, while they perceive p relatively high and pay for financial and nonmonetary costs.

Moreover, SET predicts that a person will leave a relationship when they perceiv costs of the relationship outweigh the benefits [46]. This framework is considered su for examining residents' perceptions of tourism [47,48]. Residents evaluate touris terms of social exchange (expected benefits or costs in return for the services they su Perceived high costs may stimulate negative attitudes from the host— for example, th over-exploitation [47]. SET has not been applied in the content of green hotels, some studies suggest that customers perceive functional and emotional value during stay at green hotels [49], although the price was the only cost item included in the measurement scale [50]. Although it is unclear what green hotel customers will sac many studies related to green product consumption or daily eco-friendly activities suggest high monetary and nonmonetary costs, and risks associated with making lifestyle changes hinder customers from going green [51].

Examining prior research in the tourism industry reveals a significant effect of eco-tourism on mental health, satisfaction, and behavior. One study [9] found positive effects from a survey of ecofriendly surfaces hotel users in Korea. NBSs were explained in the order of naturally ecofriendly, ecofriendly surfaces, ecofriendly spaces, and ecofriendly items. After analyzing the impact of user's revisit intention by applying the expanded theory of planned behavior on customers who attended a green hotel that had received a Leadership in Energy and Environmental Design (LEED) award, [51] a significant positive influence on revisit intention was revealed according to attitude, subjective norms, and perceived behavioral control. A further study by [52] surveyed customers using ecofriendly hotels in Malaysia and found that the values and emotions derived from the physical environment in green hotels positively affected customers' attitudes and perceived behavioral control. Therefore, the following hypothesis was established. In addition, in a study focusing on green hotels conducted by [53], significant partial mediating effects were confirmed after verifying the mediating effects of MH, emotional WB, and green brand attitude in the relationship between NBSs and green brand evangelism. Therefore, the following hypothesis was established based on previous studies. A study by [54] confirmed that the perceived value of eco-friendly hotels has a significant positive effect on customers' attitudes toward green hotels. Therefore, the following hypothesis was established based on the prior studies.

Hypothesis 5 (H5). *Green hotels as NBSs have a significant effect on attitude toward environmental behavior.*

Hypothesis 6 (H6). *Attitude toward environmental behavior has a significant effect on satisfaction.*

Hypothesis 7 (H7). *Attitude toward environmental behavior has a significant effect on pro-environmental responsible behavior intention.*

2.3. Green Hotels as Natural-Based Solutions

"NBSs are acts that are inspired by, supported by, or imitated from nature" and are intended to handle a variety of environmental concerns in a cost-effective and adaptable manner while also bringing economic, social, and environmental advantages [6,8,9]. NBSs are clearly described in the final report of the Horizon 2020 Expert Group [55]. The definition of NBS is not agreed upon worldwide but is recognized and used as a helpful approach in various fields. NBSs cover all natural-based approaches, such as ecosystem-based adaptation and ecosystem-based mitigation [56]. The International Union for Conservation of Nature (IUCN) defines NBSs as a measure for protecting, continuing management, and restoring natural or transformed ecosystems that provide human well-being and biodiversity benefits [56].

NBSs are characterized by their ability to solve various social problems simultaneously as providing natural benefits. They have significant advantages in improving resilience and increasing biodiversity and sustainability in the face of climate change. Because NBSs are compatible with the human environment, they are considered an essential solution for human life and activity over time [57]. Moreover, since 2015, NBSs have been considered an optimized solution that is resilient, adaptable, resource-efficient, and locally adjustable for improving and maximizing the quality of life for urban residents [58]. Ultimately, NBSs contribute to achieving sustainable development goals such as long-term food security, climate change, water security, human health, disaster risk management, and social and economic development [59]. The OECD's approach to implementing NBSs emphasizes better recovery in overcoming the impact of COVID-19. In terms of human health, ecosystem services are essential to human economic activities, and health, deforestation, and land use are all related to the spread of diseases [2]. Therefore, the successful implementation NBSs to improve MH and WB awareness of both customers and employees is also ve important in the hotel industry [2]. Emphasizing the importance of NBSs, psychologic health and WB are also subject to study by various researchers in the tourism industry [9, In terms of NBSs related to green infrastructure, green spaces, green surfaces, green ite and natural environments are essential for environmentally friendly hotels and buildin where service performance and characteristics are the main features of product qual evaluation [38–40]. The ecofriendly physical environments of hotels that accept the asp of NBSs can have very positive results for MH and WB beyond overall customer satisfacti and satisfaction with their stay.

People's psychological aspects and attitudes are highly correlated with whether hc users can easily access ecofriendly spaces (rivers, lakes, mountains), and whether varic nature-friendly spaces are available. In a survey of hotel users [6], verifying a signific influence on ecofriendly space and psychological elasticity, customer attitudes, and havioral intentions that explain ecofriendly accessibility, availability, and diversity, it v found that attitude has no significant effect on behavioral intention. In addition, in a stu by [2] that targeted hotel customers, the specific role of the NBSs-based hotel ecofrien physical environment was identified as a significant influencing relationship on WB, M and satisfaction. Therefore, NBSs play an important role in people's psychological w being, and the following hypotheses were established by requiring in-depth and dive studies on the benefits and effects of NBSs.

Hypothesis 8 (H8). *Green hotels as NBSs have a significant effect on mental health.*

Hypothesis 9 (H9). *Green hotels as NBSs have a significant effect on well-being.*

Hypothesis 10 (H10). *Mental health has a significant effect on satisfaction.*

2.4. Pro-Environmental Responsible Behavior Intention

In marketing, the concept of post-action greatly influences potential consumer cisions, such as repurchase intention and recommendation intention [60]. It can be s that after consumers form an attitude toward an object, it is their personal will and lief to express a specific future behavior, revisit intention or positive recommenda intention. In this context [61] on eco-friendly behavioral intentions, people who are b perpetrators of environmental problems and subjects of the issues to be solved partici in environmental protection, defined as a responsible environmental behavior divided educational, economic, civic, physical, and legal and persuasive actions. Moreover, defined environmentally friendly behavior as contributing to environmental conserva and conservation, and [63] stated that eco-friendly behavior is an action to reduce negative impact on the environment. As such, many researchers define ecofriendly be ior as an ecological behavior to refrain from personal behavior and manage and pres the natural environment and living environment to reduce and prevent environme pollution damage [64]. As environmental awareness has increased, efforts have been m to clarify what antecedents of consumers' interests and attitudes toward the environn can lead to actual ecofriendly behavior in existing studies to solve environmental probl Most studies have shown that environmental consciousness and ecofriendly beha influence consumers' willingness to purchase environmental products [65]. In addi studies have recently been conducted to emphasize the characteristics of NBSs in stu such as [9,53] applying NBS and to verify the influence of perceived value and satisfac with ecofriendly behavior.

Eco-friendly service products increase satisfaction with customers, and it can be that such ecofriendly satisfaction has a high correlation with moral standards and beha toward ecofriendly action. Therefore, the preceding studies in the hotel industry that a ecofriendly behavior as a result variable are as follows. A study by [8] surveyed ecofrie hotel customers in Vietnam and found that empirical satisfaction in ecofriendly hotels had a significant positive effect on moral norms and ecofriendly behavioral intentions. In addition, as a result of an experimental study on eco-friendly behavior [33] targeting Swiss-based four-star eco-friendly hotel customers, it was found that customers with high awareness of ecofriendly norms had high towel reuse. In other words, users with high norms show high ecofriendly behavior. In addition, in a study by [66] targeting participants in the Macau Eco-friendly Festival, it was verified that subjective norms, a subfactor of the theory of planning behavior, had a significant influence on eco-friendly values and ecofriendly behavior. As an environmental characteristic being developed by green hotels, previous studies empirically analyzed customer satisfaction and norms on eco-friendly behavioral intentions [8,33,67].

Thus far, studies on eco-friendly behavior, cost and benefits approach, and studies based on moral and normative interests have been examined. VBN theory, an extended model of the theory of planned behavior and norm activation model, has been a commonly used model in research that explains ecofriendly behavior. Therefore, this study established a model by combining the VBN, SET, and NBS models, and the following hypotheses were developed.

Hypothesis 11 (H11). *Well-being has a significant effect on satisfaction.*

Hypothesis 12 (H12). *Personal norms have a significant effect on pro-environmental responsible behavior intention.*

2.5. The Moderating Effect on the Locations (Urban, Rural)

While some studies argue that environmental concern is higher in cities [68], that urban residents are more concerned about the over-exploitation of natural resources [69], and that the perception of environmental problems increases as the size of place of residence increases [70], others suggest (after controlling for other sociodemographic variables) that there are no attitudinal or behavioral differences between the two types of samples [66]. A study by [71] compared urban (Madrid) and rural (Non-Madrid) residents living in Spain with an explanation of the social structure relationship of people's ecofriendly values, attitudes, and behaviors. As a result of the analysis, both urban and rural residents showed great environmental concern and low pro-environmental behavior. Urban residents have high environmental responsible values but low pro-environmental orientation. However, the more rural individuals are aware of environmental responsibilities, the more consistent they are in expressing behavioral intentions that are environmentally friendly. In the context of previous studies, examining the differences in perceptions, attitudes, and behaviors of urban and rural people in tourist destinations is considered very meaningful to figure out the structural relationship of value-belief-norms, attitudes, and well-being to predict green behavior according to the urban and rural green hotels. Therefore, the following hypothesis was established.

Hypothesis 13 (H13). *Location (urban, rural) plays a significant moderating role in the relationship between green hotels as NBSs and pro-environmental responsibility behavior intention.*

3. Methods

3.1. Conceptual Framework

In a situation where the COVID-19 continues for a long time, as the global environmental problem becomes more severe due to COVID-19, people are increasingly interested in global environmental problems [2–4]. Because of this, NBS has emerged as an essential issue in the hotel industry [2,6–9]. Therefore, the research framework has built NBSs into green hotels. The research framework of this study is composed of the following three main theories, as shown in Figure 1. First, this framework was centered on VBN theory, attitude towards sustainability, and psychological health as NBSs to explain the process of pro-environmental responsible behavior intention. Second, SET was used becaus plays an important role in understanding NBSs for eco-friendliness as well as custom ATEB and PRBI. Last, NBSs of green hotels have a positive effect on customers' MH WB and play an essential role in inducing SA and PRBI; ART was used as the framew for this study.

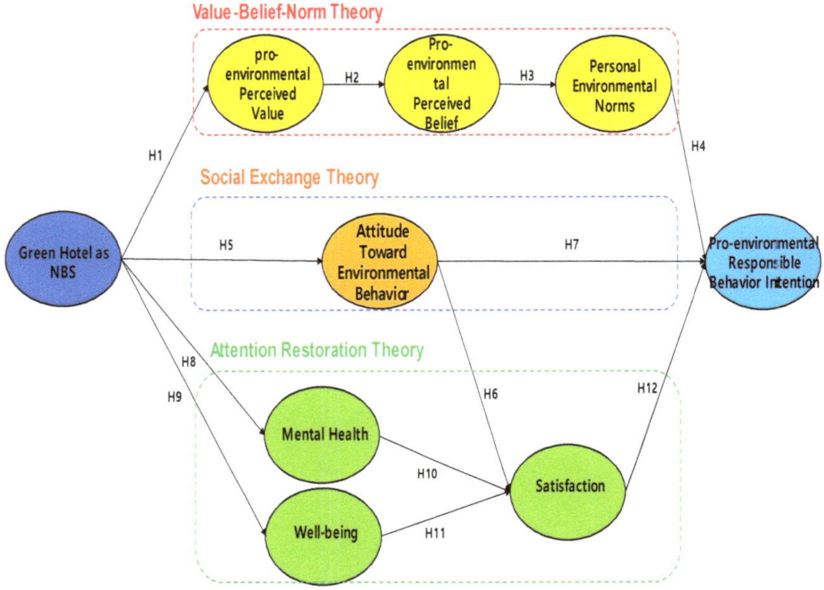

Figure 1. Proposal research model.

3.2. Measures and Questionaire Development

In order to investigate how green hotel customers' PRBI change according to N previous studies in the extant literature [2,8,9,14,38–40,53,54] provide measurement ables. The measures were modified to be adequate in the present research context. questionnaire is presented in Appendix A. Multiple items with a five-point scale ("Stro disagree" [1] to "Strongly agree" [5]) were used for the assessment of all study variable particular, four times for indoor green hotels (e.g., "I easily see green interior decora and diverse living plants in the lobby area of green hotels") and four items for out of green hotels (e.g., "The region surrounding green hotels has good and fresh qu air.]") were used. WB was evaluated with five items (e.g., "I feel healthy and happy du my stay in green hotels."). MH was measured with four items (e.g., "Staying at g hotels plays an important role in relieving my mental stress and anxiety"). In additio used three items for each measurement to measure PPV (e.g., "Visiting the green h was a pleasant experience."), PPVB (e.g., "Staying in green hotels makes you feel diff from others."), PPN (e.g., "I feel that it is important to make green hotels environmer sustainable, reducing the harm to the wider environment"), and SA (e.g., "Overall, satisfied with my experience at green hotels."). Moreover, four items were utiliz evaluate ATEB (e.g., "Green hotels are concerned about the environment"). Lastly, was measured with four items (e.g., "I try to participate in the activities pursued by g hotels"). The survey questionnaire containing these measures, a research description questions for demographic information was pre-tested by experts of two hotel practiti and two hospitality academics. A slight modification was made by these experts bas their feedback, then we further improved and finalized the questionnaire.

3.3. Data Collection Process and Data Analysis

For data collection, an online research methodology used the system of an internet research company. The survey was randomly selected from the database of an internet research company, and only customers who had visited the green hotels within the last 12 months were asked to participate in the survey. A survey invitation e-mail with a research description was delivered to general hotel customers in South Korea. The survey was conducted for about three weeks, which was from 13 December 2021 to the last day of December 2021. The survey took approximately 15 min, the average time to complete. This survey was used for data analysis and the purpose of this study only. A total of 450 questionnaires were completed through this procedure. The final sample size after removing unusable responses that were used for the data analysis was 440 cases. For data analysis of this study, using SPSS 20.0 program and AMOS 20.0 program, demographic characteristics, correlation analysis, confirmatory factor analysis (CFA), structural equation modeling (SEM), and moderator effect were analyzed.

4. Results

4.1. Characteristic of Respondents

Of the 440 respondents, 292 respondents (66.4%) were female, and 148 respondents (33.6%) were male. Present marital status was highest for married people (252 respondents, 57.3%) and followed by single people (188 respondents, 42.7%). Survey participants were 30–39 years old (38.2%, 169 respondents), 40–49 years old (35.5%, 156 respondents), 50 years old or older (13.6%, 60 respondents), and 20–29 years old (12.7%, 56 respondents) appearing in order. As for the level of education, college graduates or higher (252 respondents, 57.3%), graduate school students or higher (128 respondents, 29.1%), and high school graduates (60 respondents, 13.6%) were in order, and overall, the level of education was high. The annual household income of $20,000–$39,999 (204 respondents, 46.2%) was the highest, followed by $60,000 or over (80 respondents, 18.2%), under $20,000 (80 respondents, 18.2%), and $40,000–$59,999 (76 respondents, 17.3%). In terms of the number of visits to the Green Hotel in the past year, 3~4 times (204 respondents, 46.4%), 5~6 times (84 respondents, 19.1%), 1–2 times (112 respondents, 25.2%), 10 times or more visits (7 people, 9.1%) were in that order, the purpose of visiting the green hotels was leisure (240 respondents, 72.7%), business (72 respondents, 16.4%), and other purposes (12 respondents, 10.9%). The people who visited green hotel together appeared in the order of family (164 respondents, 41.0), friends (100 respondents, 22.7%), lover (72 respondents, 14.5%), alone (72 respondents, 14.5%), and colleague (32 respondents, 7.3%). The respondents who visited the green hotel located were 168 respondents (42.7%) in the urban areas and 252 respondents (57.3%) in the rural areas.

4.2. Confirmatory Factor Analysis

The confirmatory factor analysis (CFA) was performed to verify the reliability and validity. As a result of the measurement model, Goodness-of-fit statistics for the structural model ($\chi2$ = 711.297, df = 398, $p < 0.000$, $\chi2$/df = 1.787, RMSEA = 0.085, CFI = 0.921, IFI = 0.922, TLI = 0.908) were judged to be overall excellent [72]. Factor loadings, significance probability of t−value, average variance extracted (AVE), and construct reliability (CR) were checked to check the convergent validity of the latent variables of the measurement model. The confidence coefficients (Cronbach's alpha) of factor loading were shown between 0.706 and 0.952, which was more significant than the 0.6 suggested by [73]. Moreover, AVE values and CE values were respectively constructed ranged from 0.707 to 0.855 and from 0.833 to 0.957. These values were all greater than that level of 0.5 and 0.7 suggested by [74]. In addition, correlation analysis was performed as shown in Table 1 to verify discriminant validity. As a result of Pearson's correlation analysis, all variables of NBSs, PPV, PPVB, PPN, ATEB, MH, WB, SA, and PRBI were $p < 0.01$, indicating a significant correlation association [75]. Thus, discriminant validity was confirmed.

Table 1. The measurement model and correlation.

Construct and Scale	Item	Standardization Loading	NBSs	MN	WB	PPV	PPVB	PPN	ATEB	SA	PRBI
NBSs	NBSIO	0.952	1								
	NBSOH	0.823									
MH	MH1	0.866	0.639 **	1							
	MH2	0.878									
	MH3	0.851									
	MH4	0.902									
WB	WB1	0.901	0.658 **	0.858 **	1						
	WB2	0.889									
	WB3	0.915									
	WB4	0.922									
	WB5	0.888									
PPV	PPV1	0.938	0.608 **	0.745 **	0.671 **	1					
	PPV2	0.927									
	PPV3	0.750									
PPVB	PPVB1	0.868	0.618 **	0.743 **	0.703 **	0.888 **	1				
	PPVB2	0.843									
	PPVB3	0.866									
PPN	PPN1	0.882	0.633 **	0.762 **	0.695 **	0.821 **	0.895 **	1			
	PPN2	0.780									
	PPN3	0.870									
ATEB	ATEB1	0.880	0.671 **	0.745 **	0.722 **	0.821 **	0.848 **	0.807 **	1		
	ATEB2	0.817									
	ATEB3	0.908									
	ATEB4	0.877									
SA	SA1	0.947	0.693 **	0.755 **	0.779 **	0.691 **	0.710 **	0.715 **	0.751 **	1	
	SA2	0.948									
	SA3	0.878									
PRBI	PRBI1	0.866	0.530 **	0.651 **	0.624 **	0.740 **	0.727 **	0.732 **	0.749 **	0.665 **	1
	PRBI2	0.706									
	PRBI3	0.889									
	PRBI4	0.888									
MEAN(SD)			4.321 (0.531)	4.252 (0.699)	4.356 (0.707)	4.242 (0.660)	4.185 (0.692)	4.079 (0.686)	4.302 (0.658)	4.476 (0.646)	4.20 (0.65
AVE (CR)			0.792 (0.833)	0.762 (0.929)	0.816 (0.957)	0.769 (0.907)	0.738 (0.894)	0.714 (0.882)	0.759 (0.926)	0.855 (0.947)	0.70 (0.90

Note 1. NBSs = natural-based solutions, PPV = pro-environmental perceived value, PPVB = pro-environme perceived belief, PPN = personal pro-environmental norms, ATEB = attitude toward environmental behavior, = mental health, WB = well-being, SA = satisfaction, PRBI = pro-environmental responsibility behavior inten Note 2. Goodness-of-fit statistics for the structural model: $\chi 2$ = 711.297, df = 398, $p < 0.000$, $\chi 2$/df = 1.787, RM = 0.085, CFI = 0.921, IFI = 0.922, TLI = 0.908. Note 3. All factor loadings are significant at * $p < 0.5$, ** $p <$ *** $p < 0.001$.

4.3. Structural Equation Modeling

In this study, the PRBI of green hotels customers was investigated based on the V theory, NBSs theory, ART theory, and social exchange theory. The structural equat model (SEM) analysis was generated by using the maximum likelihood estimation meth as an estimation method for both model and procedures' evaluation [72]. Goodnes Fit of Structural Model ($\chi 2$ = 741.761, df = 410, $p < 0.000$, $\chi 2$/df = 1.809, RMSEA = 0. CFI = 0.917, IFI = 0.918, TLI = 0.908) was satisfactorily higher than the standard va Moreover, SEM had shown high prediction power for PPV (R^2 = 0.861), ATEB (R^2 = 0.8 PPN (R^2 = 0.973), PPVB (R^2 = 0.921), MH (R^2 = 0.763), WB (R^2 = 0.837), SA (R^2 = 0.7 and PRBI (R^2 = 0.684) and t-values and standardized path coefficient were shown as result in Table 2. The path estimates show that NBSs had a significantly positive di effect on the PPV (β = 0.928, $p < 0.001$), PPV had a significantly positive direct ef on the PPVB (β = 0.963, $p < 0.001$), PPVB had a significantly positive direct effect on PPN (β = 0.989, $p < 0.001$), and PPN had a significantly positive direct effect on the P (β = 0.664, $p < 0.001$). Thus, H1, H2, H3, and H4 were supported. Moreover, NBSs ha significant positive on ATEP (β = 0.925, $p < 0.001$), and ATEB had a significant posi on SA (β = 0.332, $p < 0.01$); thus, H5 and H6 were supported. Moreover, the influenc the NBSs on MH and WB was a significant positive MH (β = 0.893, $p < 0.001$) and WB (β = 0.915, $p < 0.001$); thus, H8 and H9 were supported. Additionally, MH had a significant positive effect on SA (β = 0.389, $p < 0.01$), thus H10 was supported. However, WB had not been significantly associated with SA, and SA had not been positively significantly associated with PRBI; thus, H11 and H12 were not supported.

Table 2. The structural model results and hypotheses testing.

Hypothesized Paths	Coefficients	t-Values
H1: NBSs → PPV	0.928	9.394 ***
H2: PPV→ PPVB	0.963	12.570 ***
H3: PPVB → PPN	0.989	9.794 ***
H4: PPN → PRBI	0.664	5.971 ***
H5: NBSs → ATEB	0.925	8.197 ***
H6: ATEB → SA	0.332	2.333 *
H7: ATEP → PRBI	0.280	1.858
H8: NBSs → MH	0.873	8.616 ***
H9: NBSs → WB	0.915	8.895 ***
H10: MH → SA	0.389	3.370 ***
H11: WB → SA	0.186	1.858
H12: SA → PRBN	0.130	1.125
Explained variable: R^2 (PPV) = 0.861 R^2 (ATEB) = 0.856 R^2 (PPVB) = 0.921	R^2 (MH) = 0.763 R^2 (PPN) = 0.973 R^2 (WB) = 0.837	R^2 (SA) = 0.726 R^2 (PRBI) = 0.684

Note 1. NBSs = natural-based solutions, PPV = pro-environmental perceived value, PPVB = pro-environmental perceived belief, PPN = personal pro-environmental norms, ATEB = attitude toward environmental behavior, MH = mental health, WB = well-being, SA = satisfaction, PRBI = pro-environmental responsibility behavior intention. Note 2. Goodness-of-fit statistics for the measurement model: $\chi2$ = 741.761, df = 410, $p < 0.000$, $\chi2$/df = 1.809, RMSEA = 0.086, CFI = 0.917, IFI = 0.918, TLI = 0.908. * $p < 0.5$, *** $p < 0.001$.

4.4. The Moderating Effect on the Location (Urban, Rural)

In this study, an analysis was conducted to determine whether there was a moderating effect between NBSs and PRBI according to locations (urban, rural). If the customer had visited green hotels in the questionnaire, they were asked if they visited an urban or a rural location. As a result, out of 440 respondents, 255 respondents were the rural locations, and 188 respondents of an urban locations. The results of empirical comparisons are displayed in Table 3 and Figure 2. The goodness-of-fit statistics for the baseline model ($\chi2$ = 1746.640, df = 820, $p < 0.000$, $\chi2$/df = 2.130, RMSEA = 0.92, CFI = 0.891, IFI = 0.896., TLI = 0.862) was showed an acceptable level for the data suitability. Subsequently, to examine the effectiveness of the conditioning effect, a constraint model was established by dividing the modulating variables into the urban group and the rural group, and the difference in χ^2 between the free model and the constrained model was verified. There is a difference between the groups only if the $\Delta\chi^2$ of the baseline model (freely estimated) and the nested model (constrained to be equal) is 3.84 or more. As a result of the analysis, there was no difference between groups among PPV-PPVB, PPVB-PPN, ATEB-SA, WB-SA, and SA-PRBI. Therefore, H13b, H13c, H13g, H13k, and H13l were not supported.

Table 3. The result of the moderating effect.

Paths	Urban Group (n = 188)		Rural Group (n = 252)		Baseline Model (Freely Estimated)	Nested Model (Constrained to Be Equal)	Chi-Square Different Test	Test Resul
	Coefficients	t-Value	Coefficients	t-Value				
H13a: NBSs-PPV	0.874	6.279 ***	0.924	5.640 ***	$\chi 2(820) = 1746.640$	$\chi 2(821) = 1752.877$	$\Delta\chi 2(1) = 6.237$, $p > 0.01$	Support
H13b: PPV-PPVB	0.928	7.759 ***	0.969	9.158 ***	$\chi 2(820) = 1746.640$	$\chi 2(821) = 1746.683$	$\Delta\chi 2(1) = 0.043$, $p > 0.01$	Not suppo
H13c: PPVB-PPN	0.990	7.342 ***	0.972	6.087 ***	$\chi 2(820) = 1746.640$	$\chi 2(821) = 1747.525$	$\Delta\chi 2(1) = 0.855$, $p > 0.01$	Not suppo
H13d: PPN-PRBI	0.057	0.453	0.848	3.728 ***	$\chi 2(820) = 1746.640$	$\chi 2(821) = 1754.304$	$\Delta\chi 2(1) = 7.664$, $p > 0.01$	Support
H13e: NBSs-ATEB	0.816	5.477 ***	0.964	5.151 ***	$\chi 2(820) = 1746.640$	$\chi 2(821) = 1752.955$	$\Delta\chi 2(1) = 6.315$, $p > 0.01$	Support
H13f: ATEB-PRBI	0.964	5.027 ***	−0.189	−0.841	$\chi 2(820) = 1746.640$	$\chi 2(821) = 1756.143$	$\Delta\chi 2(1) = 9.503$, $p > 0.01$	Support
H13g: ATEB-SA	0.244	1.422	0.311	1.485	$\chi 2(820) = 1746.640$	$\chi 2(821) = 1746.676$	$\Delta\chi 2(1) = 0.036$, $p > 0.01$	Not suppo
H13h: NBSs-MH	0.934	6.378 ***	0.832	5.109 ***	$\chi 2(820) = 1746.640$	$\chi 2(821) = 1750.827$	$\Delta\chi 2(1) = 4.187$, $p > 0.01$	Support
H13i: NBSs-WB	0.999	7.199 ***	0.880	5.230 ***	$\chi 2(820) = 1746.640$	$\chi 2(821) = 1750.298$	$\Delta\chi 2(1) = 3.658$, $p > 0.01$	Support
H13j: MH-SA	−0.125	−0.394	0.509	3.376 ***	$\chi 2(820) = 1746.640$	$\chi 2(821) = 1751.520$	$\Delta\chi 2(1) = 4.880$, $p > 0.01$	Support
H13k: WB-SA	0.795	2.096	0.042	0.239	$\chi 2(820) = 1746.640$	$\chi 2(821) = 1749.052$	$\Delta\chi 2(1) = 2.412$, $p > 0.01$	Not suppo
H13l: SA-PRBI	−0.129	−0.744	0.208	1.535	$\chi 2(820) = 1746.640$	$\chi 2(821) = 1748.920$	$\Delta\chi 2(1) = 0.280$, $p > 0.01$	Not suppo

Note 1. NBSs = natural-based solutions, PPV = pro-environmental perceived value, PPVB = pro-environm perceived belief, PPN = personal pro-environmental norms, ATEB = attitude toward environmental beha MH = mental health, WB = well-being, SA = satisfaction, PRBI = pro-environmental responsibility beha intention. Note 2. Goodness-of-fit statistics for the baseline model: $\chi 2 = 1746.640$, df = 820, $p < 0.000$, $\chi 2/\text{df} =$ RMSEA = 0.92, CFI = 0.891, IFI = 0.896, TLI = 0.862. *** $p < 0.001$.

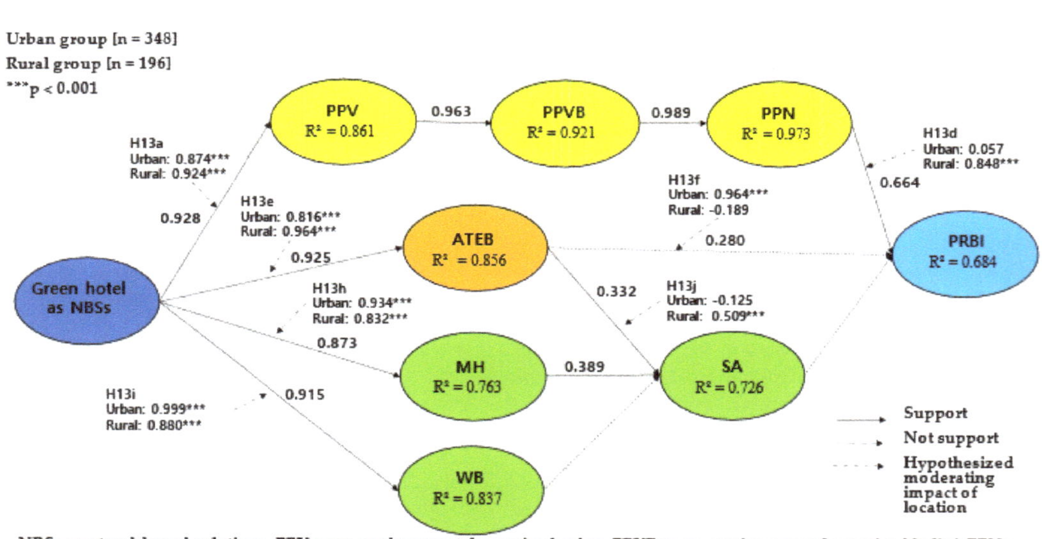

NBSs = natural-based solutions, PPV = pro-environmental perceived value, PPVB = pro -environmental perceived belief, PPN = personal pro-environmental norms, ATEB = attitude toward environmental behavior, MH = mental health, WB = well -being, SA = satisfaction, PRBI = pro-environmental responsibility behavior intention

Figure 2. Structural equation model estimation and test for structural metric invariance.

5. Discussion

In recent years, many hotels around the world have contributed to the acceler of global warming and continue to suffer natural disasters such as floods and typh due in part to the increase in plastic use in daily life (such as packaging containers plastic cups) and medical supplies (such as face mask) [2–4]. Therefore, hotels are grad implementing a Green Hotel as NBSs Policy for sustainable management [2,9], ma great efforts to raise the customer's PRBI. However, despite marketing activities to consider and create a sustainable environment, studies that can provide academic and practical implications for the personal relationship between their decision-making process and consumer behavior are lacking. For this reason, this study attempted to understand the PRBI of visitors to green hotels through VBN theory, SET, and ART. Moreover, the effect of green hotels as NBSs on PPV, PPVB, PPN, ATEB, MH, WB, SA, and PRBI was identified, and their relationship was verified. In addition, the difference between urban and rural green hotels was investigated by applying the moderating effect of location (urban, rural) in the relational model.

The summary of the results of this study is as follows. Green hotels as NBSs have a significant effect on PPV, BE, PEN, and PRBI and the result of these studies [3,15,34,35,37]. Other green hotels as NBSs have a significant effect on ATEB and SA, although ATEB has no significant effect on PRBI; therefore, previous studies [15,34] were partially supported. Green hotels and NBSs significantly affect MH and WB, which support previous studies [38–40]. However, WB has no significant effect on SA, not the same result as previous studies [76]. Moreover, SA has no significant effect on PRBI, not supporting previous studies [8,9].

Our result from the baseline model assessment and invariance test revealed that location (urban, rural) had a significant moderating role within the proposed conceptual framework. The paths from green hotels as NBSs to PPV (urban group: $\beta = 0.874$, $p < 0.01$ vs. rural group: $\beta = 0.964$, $p < 0.01$), from green hotels as NBSs to MH (urban group: $\beta = 0.934$, $p < 0.01$ vs. rural group: $\beta = 0.832$, $p < 0.01$), and from green hotels as NBSs to WB (urban group: $\beta = 0.909$, $p < 0.01$ vs. rural group: $\beta = 0.880$, $p > 0.01$), and ATEB to PRBI (urban group: $\beta = 0.964$, $p < 0.01$ vs. rural group: $\beta = -0.189$, $p > 0.05$) were significantly greater in the urban group than in the rural group. However, paths from PPN to PRBI (urban group: $\beta = 0.057$, $p > 0.05$ vs. rural group: $\beta = 0.848$, $p < 0.01$) and from ATEB to SA (urban group: $\beta = -0.125$, $p > 0.05$ vs. rural group: $\beta = 0.509$, $p < 0.01$) were significantly greater in the rural group than in the urban group.

5.1. Theoretical Implications

Based on the results of this study, theoretical and practical implications are deduced. The theoretic implications of this study are as follows. First, the PRBN of customers visiting green hotels was identified based on the VBN theory. Previous studies have identified general ecofriendly tourism behavior based on recognizing the importance of environmental behavior results of VBN theory [8,9,34,35,38–40]. In this study, the existing research was extended in that green hotels as NBSs identified PRBI through PPV, PPVB, and PPN. Second, green hotels are implementing NBSs policies to protect and preserve the environment to create a sustainable society and enterprises. Based on SET, it was determined that these corporate efforts influence customers' ATEB on PRBI. Third, the research framework has built NBSs into green hotels, and the researchers actively utilized the concepts of NBS theory, ART, and SET to construct this study. Our findings are theoretically significant in that they are the first empirical studies using SET and ART based on NBSs theory to determine PRBI. Last, in the study of NBSs, researchers need to understand this moderating nature of locations (urban, rural). They should actively utilize how they differ according to locations (urban, rural) as moderating when developing theory or conceptual frameworks related to NBS. Our findings have theoretical significance in that they are the first empirical studies to illuminate the interactions between NBSs and locations (urban and rural) that determine PRBI.

5.2. Practical Implication

The practical implications of this study are as follows. First, the green hotels' ecofriendly policy has a high impact on PPV and ATEB. These results show that caring for and protecting nature makes consumers feel more responsible for global warming, depletion of natural resources, and energy problems, and that hotel NBSs have a significant influence on customer perceptions. Therefore, hotels should feel responsible for the environment a manage sustainable businesses. For example, Marriott International Inc announced in 2021 SERVE 360 REPORT that it continues to produce positive effects to reduce waste 45% by the end of 2030, reduce food waste by 50%, and water and carbon dioxide by 1 and 30%, respectively [77]. In addition, The Athenee Hotel, Bangkok's Luxury Collecti Hotel, offers a Green Meeting Package of refreshments made of ecofriendly materials reduce meeting and event waste. As a result, the Athenee Hotel achieved the world's fi event sustainability management system standard (ISO: 20121) [78].

Second, although green hotel NBSs showed significant influence on PPV and AT PPV had a significant effect on PRBN through the mediation of BE and PPN, while ATEB c not affect PRBI. In other words, it is believed this is because there is an intense psychologi desire to copy others regardless of ATEB. Therefore, for green hotels' customers' PRBI increase, it is necessary to recognize the customer's environment, such as PPV, PPVB, a PPN. As more people see the cause of COVID-19 being due to climate change caused global warming [79], people's awareness of nature protection and sustainability is a increasing. Therefore, the demand for ecofriendly and alternative consumption is a increasing. For example, the French government has mandated the use of wood or natu materials for more than 50% of new public buildings built after 2022. Therefore, both ho companies and government should increase the PRBI of green hotel customers to creat sustainable society.

Third, in the present research, we have indicated that MH derived from NBSs fundamental concepts exerting a considerable influence on SA. However, WB had significant effect on SA and MH had a significant effect on SA with lower values. Analy of the hotel trends in 2022, including sustainable hotels [80]. Sustainable hotels are bu using natural products while protecting nature, using trees or ecofriendly materials construction, and restraint with accessories necessary for the hotel. In addition, it provi a pleasant atmosphere with relaxing music and a good aroma so that customers can h comfortably in green hotels, promoting MH and WB and reducing mental fatigue. Howe this has become a matter of course in green hotels. Therefore, to achieve high custon satisfaction, marketing is required one step ahead. For example, in 2022, another hotel tre virtual reality, augmented reality, and sustainable hotels will be combined to help custom create and enjoy virtual spaces according to their needs without destroying nature.

Finally, because differences were analyzed between locations (urban, rural), it v found that an urban location had a more significant effect on the psychological welln variable (MH, WB). In other words, green hotels located in urban areas are primarily n residential areas and can be easily accessed for rest and private time away from busy d life. Thus, green hotels located in cities need to make a proper resting place for consum In addition, it was found that green hotels located in rural areas were more affected socially required variables (PPV, ATEB) than those in urban locations. In other wo most customers of green hotels located in rural locations visited urban hotels beca they wanted to find nature and heal. Because these customers want to find peace a stability in nature, they feel the importance of nature more and want to protect nat Therefore, it is necessary to study how local green hotels can identify the needs of th customers, protect nature, and coexist with humans without destroying the environm For example, the Banyan Tree restored Bang Tao Bay, which was little more than ru presenting it as Banyan Tree Phuket [81]. Phuket's Vantao Bay restoration project be of Banyan Tree's sustainable project, using natural materials for interior design and lo unique architectural techniques. In addition, it preserves the community's environm where the Banyan Tree is located and supports social and economic development to h residents develop their abilities.

5.3. Limitations

The study used VBN, SET, and NBSs theories (theoretically certified for reliabi and validity). Although it is significant in that it was intended to verify green h users with various variables such as the characteristics of NBSs of green hotels and the psychological state of users, there are several limitations in this research process. First, as of December 2021, customers who visited green hotels in previous years' hotel stay experience were selected as samples, due to the specific situation of COVID-19; thus, visit surveys or interviews with sufficient explanation were not preceded. Instead, because we surveyed online, not only was the user's understanding of the survey low, but it was impossible to verify the change in attitude. It is hoped that this will lead to follow-up studies that can supplement future studies and solve limitations. Therefore, it is expected that standardized and generalized result analysis will be possible through regular and long-term surveys.

Second, the survey was targeted at visitors to ecofriendly hotels in Korea. Like Europe, citizens' awareness of ecofriendly hotels is still insufficient. Before the survey response, the characteristics of the green hotels were explained, although perceptions of hotels with ecofriendly marks and certification may be shallow. Thus, there may be gaps in survey responses through the experience of staying in the green hotels. In the aftermath of COVID-19, it will be necessary to explain to respondents directly with sufficient information using photos or pictures of ecofriendly hotels.

Third, the study analyzed the moderation effect of green hotels located in urban and rural areas, and a clear distinction between urban and rural areas may be unclear. The reason for this is that due to the geographical characteristics of Korea, the distinction between green hotels in the city center and the outskirts needs to be more apparent as the areas around lakes, rivers, seas, and mountains, where green hotels are located, are urbanized.

Fourth, this study did not consider hotel grades or global chains and local hotels. In future studies, it would be meaningful to identify the ecofriendly behavior characteristics of customers according to three-star, four-star, and five-star ratings and compare and analyze differences in the behavior of users according to the hotel star level. In addition, the comparative analysis of global chains and local hotels is also an essential part of predicting the ecofriendly behavior of users, so it is necessary to examine them in future studies.

Author Contributions: Conceptualization, S.Y. and T.K.; methodology, S.Y.; formal analysis, S.Y.; investigation, T.K.; resources, T.K.; data curation, T.K.; writing—original draft preparation, S.Y. and T.K.; writing—review and editing, S.Y. and T.K. All authors have read and agreed to the published version of the manuscript.

Funding: This research received no external funding.

Conflicts of Interest: Authors declare no conflict of interest.

Appendix A. Measurement Items

Indoor of green hotels [8,9]
I quickly see green interior decorations and diverse living plants in the lobby area of green hotels
I quickly see various green items and light through glass windows in green hotels' restaurants.
I quickly see mixed flowers, trees, and potted plants in green hotels' coffee lounges.
Green space can easily be seen everywhere in green hotels.
Outdoor of green hotels [8,9]
Green hotels have easy access to the natural environment (i.e., mountains, forests, rivers, seas, lakes, natural parks).
The green hotels region has good weather (i.e., temperature, humidity, and precipitation).
The region surrounding green hotels has excellent and fresh quality air.
The region around green hotels are safe from natural disasters (i.e., earthquakes, typhoons, tsunamis, floods).
Pro-environmental perceived value [14]
Visiting the green hotels was worth the money paid.
Visiting the green hotels was a pleasant experience.
Visiting the green hotels was to improve self-esteem.
Pro-environmental perceives belief [14]
Staying in green hotels is a good idea.
Staying in green hotels makes you feel different from others.
I tend to think of Staying in the green hotels.
Personal environmental norm [14]
I feel obligated to visit green hotels.
Green hotels help the natural environment.
I use green hotels as my values and principles.
Attitude toward environmental behavior [14,40,54]
I think it has enough electricity, water, and trees in the Republic of Korea
I think recycling is necessary.

I think recycling is very important to conserve natural resources.
Green hotels try to preserve the natural environment.
Mental health [2,8,9,53]
Staying at green hotels plays an important role in relieving my mental stress and anxiety.
Staying at green hotels gave me confidence in my daily life.
Staying in green hotels is worth it because it makes me feel that I am a precious and important person.
Staying in green hotels is worth it because it helps turn all your anxiety and worries into confidence.
Well-being [2,8,9,53]
I feel healthy and happy during my stay in green hotels.
I feel emotionally secure during my stay at green hotels.
Green hotels play an important role in making my mind calm and peaceful.
Thanks to green hotels, I was able to relax comfortably.
Thanks to green hotels, I was able to refresh my mood.
Satisfaction [8,38,40]
Overall, I am satisfied with my experience at green hotels.
My decision to stay at green hotels was a wise one.
As a whole, I have enjoyed myself at green hotels as expected.
Pro-environmental responsibility behavior intention [43]
I tend to participate in the pro-environmental actions required by green hotels (use disposable items upon request, recycling, reusing linen, etc.).
I tend to try to be ecofriendly in green hotels.
I try to participate in the activities pursued by green hotels
I try to abide by the rules and regulations in green hotels.

References

1. Cucinotta, D.; Vanelli, M. WHO declares COVID-19 a pandemic. *Acta Biomed.* **2020**, *91*, 157. Available online: https://www. nlm.nih.gov/pmc/articles/PMC7569573/ (accessed on 23 February 2022). [PubMed]
2. Soga, M.; Evans, M.J.; Tsuchiya, K.; Fukano, Y. A room with a green view: The importance of nearby nature for mental h during the COVID-19 pandemic. *Ecol. Appl.* **2021**, *31*, e2248. [CrossRef]
3. Chen, J.; Zhu, L.; Fan, P.; Tian, L.; Lafortezza, R. Do green spaces affect the spatiotemporal changes of PM 2.5 in Nanjing. *Processes* **2016**, *5*, 1–13. [CrossRef]
4. Khoo, K.S.; Chew, K.W.; Yew, G.Y.; Manickam, S.; Ooi, C.W.; Show, P.L. Integrated ultrasound-assisted liquid biphasic flo for efficient extraction of astaxanthin from Haematococcus pluvialis. *Ultrason. Sonochem.* **2020**, *67*, 105052. [CrossRef]
5. Benson, N.U.; Bassey, D.E.; Palanisami, T. COVID pollution: Impact of COVID-19 pandemic on global plastic waste foot *Heliyon* **2021**, *7*, e06343. [CrossRef] [PubMed]
6. Yu, J.; Ariza-Montes, A.; Hernández-Perlines, F.; Vega-Muñoz, A.; Han, H. Hotels' eco-friendly physical environment as n based solutions for decreasing burnout and increasing job satisfaction and performance. *Int. J. Environ. Res. Public Health* **20** 6357. [CrossRef] [PubMed]
7. Nesshöver, C.; Assmuth, T.; Irvine, K.N.; Rusch, G.M.; Waylen, K.A.; Delbaere, B.; Wittmer, H. The science, policy and pract nature-based solutions: An interdisciplinary perspective. *Sci. Total Environ.* **2017**, *579*, 1215–1227. [CrossRef]
8. Han, H.; Yu, J.; Hyun, S.S. Effects of nature-based solutions (NBS) on eco-friendly hotel guests' mental health percep satisfaction, switching barriers, and revisit intentions. *J. Hosp. Mark. Manag.* **2019**, *29*, 592–611. [CrossRef]
9. Han, H.; Jongsik, Y.; Hyun, S.S. Nature based solutions and customer retention strategy: Eliciting customer well-being experi and self-rated mental health. *Int. J. Hosp. Manag.* **2020**, *86*, 102446. [CrossRef]
10. Bitner, M.J. Servicescapes: The impact of physical surroundings on customers and employees. *J. Mark.* **1992**, *56*, 57–71. [Cro
11. Dietz, T.; Stern, P.C.; Guagnano, G.A. Social structural and social psychological bases of environmental concern. *Environ.* **1998**, *30*, 450–471. [CrossRef]
12. Dietz, T.; Fitzgerald, A.; Shwom, R. Environmental values. *Annu. Rev. Environ. Resour.* **2005**, *30*, 335–372. [CrossRef]
13. Stern, P. Toward a coherent theory of environmentally significant behavior. *J. Soc. Issues* **2000**, *56*, 407–424. [CrossRef]
14. Kim, T.; Yun, S. How will changes toward pro-environmental behavior play in customers' perceived value of environn concerns at coffee shops? *Sustainability* **2019**, *11*, 3816. [CrossRef]
15. Stern, P.C.; Dietz, T.; Abel, T.; Guagnano, G.A.; Kalof, L. A value-belief-norm theory of support for social movements: Th of environmentalism. *Hum. Ecol. Rev.* **1999**, *6*, 81–97. Available online: https://www.jstor.org/stable/24707060 (access 23 February 2022).
16. Pickett, G.M.; Kangun, N.; Grove, S.J. Is there a general conserving consumer? A public policy concerns. *J. Public Policy* **1993**, *12*, 234–243. [CrossRef]
17. Lovaglia, M.J. Social exchange theory. *Blackwell Encycl. Sociol.* **2007**. [CrossRef]
18. Homans, G.C. The humanities and the social sciences. *Am. Behav. Sci.* **1961**, *4*, 3–6. [CrossRef]
19. Vanya, H.; Erin, M.; BA, M.; Veda, T. Pe diatric Psychotropic Medication Initiation and Adherence: A Literature Review Ba Social Exchange Theory. *J. Child Adolesc. Psychiatr. Nurs.* **2010**, *23*, 151–172. [CrossRef]
20. Van Dyne, L.; Ang, S. Organizational citizenship behavior of contingent workers in Singapore. *Acad. Manag. J.* **1998**, *41*, 692–703. [Cr
21. Lin, C.H. Effects of cuisine experience, psychological well-being, and self-health perception on the revisit intention of hot s tourists. *J. Hosp. Tour. Res.* **2014**, *38*, 243–265. [CrossRef]
22. Pietilä, M.; Neuvonen, M.; Borodulin, K.; Korpela, K.; Sievänen, T.; Tyrväinen, L. Relationships between exposure to urban spaces, physical activity and self-rated health. *J. Outdoor Recreat. Tour.* **2015**, *10*, 44–54. [CrossRef]

Van den Bosch, M.; Sang, Å.O. Urban natural environments as nature-based solutions for improved public health–A systematic review of reviews. *Environ. Res.* **2017**, *158*, 373–384. [CrossRef] [PubMed]
Vujcic, M.; Tomicevic-Dubljevic, J.; Grbic, M.; Lecic-Tosevski, D.; Vukovic, O.; Toskovic, O. Nature based solution for improving mental health and well-being in urban areas. *Environ. Res.* **2017**, *158*, 385–392. [CrossRef]
Kaplan, R.; Kaplan, S. *The Experience of Nature: A Psychological Perspective*; Cambridge University Press: Cambridge, UK, 1989.
Southon, G.E.; Jorgensen, A.; Dunnett, N.; Hoyle, H.; Evans, K.L. Biodiverse perennial meadows have aesthetic value and increase residents' perceptions of site quality in urban greenspace. *Landsc. Urban Plan.* **2017**, *158*, 105–118. [CrossRef]
Kim, M.H. A Woman's Heart Completely Changed by the COVID-19, a Vacation in a Hotel instead of an Overseas Trip? Consumer Insight. 2020. Available online: https://www.consumerinsight.co.kr/voc_view.aspx?no=3130&id=pr10_list&PageNo=2&schFlag=0 (accessed on 10 November 2020).
Bemstein, B. Class and pedagogies: Visible and invisible. *Educ. Stud.* **1975**, *1*, 23–41. [CrossRef]
Schwartz, B. Administrative law: The third century. *Admin. Law Rev.* **1977**, *29*, 291–319. Available online: https://www.jstor.org/stable/40709018 (accessed on 23 February 2022).
Dunlap, R.E.; Van Liere, K.D. The new environmental paradigm. *J. Environ. Educ.* **1978**, *40*, 19–28. [CrossRef]
Wiidegren, Ö. The new environmental paradigm and personal norms. *Environ. Behav.* **1998**, *30*, 75–100. [CrossRef]
Thøgersen, J.; Ölander, F. Human values and the emergence of a sustainable consumption pattern: A panel study. *J. Econ. Psychol.* **2002**, *23*, 605–630. [CrossRef]
Terrier, L.; Marfaing, B. Using social norms and commitment to promote pro-environmental behavior among hotel guests. *J. Environ. Psychol.* **2015**, *44*, 10–15. [CrossRef]
Sharma, R.; Gupta, A. Pro-environmental behavior among tourists visiting national parks: Application of value-belief-norm theory in an emerging economy context. *Asia Pac. J. Tour. Res.* **2020**, *25*, 829–840. [CrossRef]
Hwang, J.; Kim, W.; Kim, J.J. Application of the value-belief-norm model to environmentally friendly drone food delivery services: The moderating role of product involvement. *Int. J. Contemp. Hosp. Manag.* **2020**, *32*, 1775–1794. [CrossRef]
Le, T.H.; Wu, H.C.; Huang, W.S.; Liou, G.B.; Huang, C.C.; Hsieh, C.M. Evaluating Determinants of Tourists' Intentions to Agrotourism in Vietnam using Value–Belief–Norm Theory. *J. Travel Tour. Mark.* **2021**, *38*, 881–899. [CrossRef]
Santiago Fink, H. Human-nature for climate action: Nature-based solutions for urban sustainability. *Sustainability* **2016**, *8*, 254. [CrossRef]
Liang, H.H.; Chen, C.P.; Hwang, R.L.; Shih, W.M.; Lo, S.C.; Liao, H.Y. Satisfaction of occupants toward indoor environment quality of certified green office buildings in Taiwan. *Build. Environ.* **2014**, *72*, 232–242. [CrossRef]
Moon, H.; Yoon, H.J.; Han, H. Role of airport physical environments in the satisfaction generation process: Mediating the impact of traveler emotion. *Asia Pac. J. Tour. Res.* **2016**, *21*, 193–211. [CrossRef]
Trang, H.L.T.; Lee, J.S.; Han, H. How do green attributes elicit pro-environmental behaviors in guests? The case of green hotels in Vietnam. *J. Travel Tour. Mark.* **2019**, *36*, 14–28. [CrossRef]
Choi, H.; Jang, J.; Kandampully, J. Application of the extended VBN theory to understand consumers' decisions about green hotels. *Int. J. Hosp. Manag.* **2015**, *51*, 87–95. [CrossRef]
Rahman, I.; Reynolds, D. The influence of values and attitudes on green consumer behavior: A conceptual model of green hotel patronage. *Int. J. Hosp. Tour. Adm.* **2019**, *20*, 47–74. [CrossRef]
Min, J.; Yang, K.; Kim, J. The role of perceived vulnerability in restaurant customers' co-creation behavior and re-patronage intention during the COVID-19 pandemic. *J. Vacat. Mark.* **2021**, *28*, 38–51. [CrossRef]
Nunkoo, R.; Ramkissoon, H. Developing a community support model for tourism. *Ann. Tour. Res.* **2011**, *38*, 964–988. [CrossRef]
Perdue, R.R.; Long, P.T.; Allen, L. Resident support for tourism development. *Ann. Tour. Res.* **1990**, *17*, 586–599. [CrossRef]
Emerson, R.M. Power-dependence relations. *Am. Sociol. Rev.* **1962**, *27*, 31–41. [CrossRef]
Ap, J. Residents' perceptions on tourism impacts. *Ann. Tour. Res.* **1992**, *19*, 665–690. [CrossRef]
McGehee, N.G.; Andereck, K.L. Factors predicting rural residents' support of tourism. *J. Travel Res.* **2004**, *43*, 131–140. [CrossRef]
Hu, B.; Fu, Y.; Wang, Y. An empirical study on the dimensions of consumer perceived value in green hotels. In Proceedings of the International Conference on Management and Service Science LEEE, Wuhan, China, 12–14 August 2011; pp. 1–4. [CrossRef]
Manaktola, K.; Jauhari, V. Exploring consumer attitude and behaviour towards green practices in the lodging industry in India. *Int. J. Contemp. Hosp. Manag.* **2007**, *19*, 364–377. [CrossRef]
Han, H.; Kim, Y. An investigation of green hotel customers' decision formation: Developing an extended model of the theory of planned behavior. *Int. J. Hosp. Manag.* **2010**, *29*, 659–668. [CrossRef]
Mohd Suki, N.; Mohd Suki, N. Consumption values and consumer environmental concern regarding green products. *Int. J. Sustain. Dev. World Ecol.* **2015**, *22*, 269–278. [CrossRef]
Sohaib, M.; Wang, Y.; Iqbal, K.; Han, H. Nature-based solutions, mental health, well-being, price fairness, attitude, loyalty, and evangelism for green brands in the hotel context. *Int. J. Hosp. Manag.* **2022**, *101*, 103126. [CrossRef]
Verma, V.K.; Chandra, B.; Kumar, S. Values and ascribed responsibility to predict consumers' attitude and concern towards green hotel visit intention. *J. Bus. Res.* **2019**, *96*, 206–216. [CrossRef]
EC. Results of the Stakeholder Consultation for the Horizon 2020 Societal: Challenge 5 (Climate Action, Environment, Resource Efficiency and Raw Materials) Advisory Group Report. 2014. Available online: https://ec.europa.eu/programmes/horizon2020/en/print/1063 (accessed on 18 July 2014).

56. Cohen-Shacham, E.; Walters, G.; Janzen, C.; Maginnis, S. *Nature-Based Solutions to Address Global Societal Challenges*; IUCN: Gla Switzerland, 2016; p. xiii-97. [CrossRef]
57. Rockstrm, J.; Steffen, W.; Noone, K.; Persson, A.; Chapin, F.; Lambin, E.A. safe operation space for humanity. *Nature* **2009**, 472–475. [CrossRef] [PubMed]
58. Faivre, N.; Fritz, M.; Freitas, T.; de Boissezon, B.; Vandewoestijne, S. Nature-Based Solutions in the EU: Innovating with natur address social, economic and environmental challenges. *Environ. Res.* **2017**, *159*, 509–518. [CrossRef] [PubMed]
59. Acharya, P.; Gupta, A.K.; Dhyani, S.; Karki, M. New pathways for NbS to realise and achieve SDGs and post 2015 targ Transformative approaches in resilience building. In *Nature-Based Solutions for Resilient Ecosystems and Societies*; Sprin Singapore, 2020; pp. 435–455. [CrossRef]
60. Anderson, E.W.; Fornell, C.; Lehmann, D.R. Customer satisfaction, market share and profitability: Findings from Sweden. *J. M* **1994**, *58*, 53–66. [CrossRef]
61. Hungerford, H.R.; Volk, T.L. Changing learner behavior through environmental education. *J. Environ. Educ.* **1990**, *21*, 8–21. [CrossRef]
62. Axelrod, L. Balancing personal needs with environmental preservation: Identifying the values that guide decisions in ecolog dilemmas. *J. Soc. Issues.* **1994**, *50*, 85–104. [CrossRef]
63. Kim, Y.J.; Kim, W.G.; Choi, H.M.; Phetvaroon, K. The effect of green human resource management on hotel employees' eco-frier behavior and environmental performance. *Int. J. Hosp. Manag.* **2019**, *76*, 83–93. [CrossRef]
64. Kaiser, F.G.; Fuhrer, U. Ecological behavior's dependency on different forms of knowledge. *Appl. Psychol.* **2003**, *52*, 598–613. [CrossRe
65. Schwepker Jr, C.H.; Cornwell, T.B. An examination of ecologically concerned consumers and their intention to purch ecologically packaged products. *J. Public Policy Mark.* **1991**, *10*, 77–101. [CrossRef]
66. Wong, I.A.; Wan, Y.K.P.; Huang, G.I.; Qi, S. Green event directed pro-environmental behavior: An application of goal syst theory. *J. Sustain. Tour.* **2021**, *29*, 1948–1969. [CrossRef]
67. Han, H. Consumer behavior and environmental sustainability in tourism and hospitality: A review of theories, concepts, latest research. *J. Sustain. Tour.* **2021**, *29*, 1021–1042. [CrossRef]
68. Van Liere, K.D.; Dunlap, R.E. Environmental concern: Does it make a difference how it's measured? *Environ. Behav.* **1981** 651–676. [CrossRef]
69. Arcury, T.A.; Christianson, E.H. Rural-urban differences in environmental knowledge and actions. *J. Environ. Educ.* **1993** 19–25. [CrossRef]
70. Samdahl, D.M.; Robertson, R. Social determinants of environmental concern: Specification and test of the model. *Environ. B* **1989**, *21*, 57–81. [CrossRef]
71. Berenguer, J.; Corraliza, J.A.; Martin, R. Rural-urban differences in environmental concern, attitudes, and actions. *Eur. J. Psy Assess.* **2005**, *21*, 128–138. [CrossRef]
72. Byrne, B.M. *Structural Equation Modeling with LISREL, PRELIS and SIMPLIS: Basic Concepts, Applications and Programm* Psychology Press: New York, NY, USA, 2013; pp. 259–286.
73. Bagozzi, R.P.; Yi, Y.; Nassen, K.D. Representation of measurement error in marketing variables: Review of approaches extension to three-facet designs. *J. Econom.* **1998**, *89*, 393–421. [CrossRef]
74. Anderson, J.C.; Gerbing, D.W. Structural equation modeling in practice: A review and recommended two-step approach. *Psy Bull.* **1988**, *103*, 411–423. [CrossRef]
75. Fornell, C.; Larcker, D.F. Structural equation models with unobservable variables and measurement error: Algebra and stati *J. Mark. Res.* **1981**, *18*, 382–388. [CrossRef]
76. Valikhani, A.; Ahmadnia, F.; Karimi, A.; Mills, P.J. The relationship between dispositional gratitude and quality of life: mediating role of perceived stress and mental health. *Personal. Individ. Differ.* **2019**, *141*, 40–46. [CrossRef]
77. Marriott International. SERVE 360 REPORT: Environmental, Social and Governance Pogress at Marriott International. Available online: http://serve360.marriott.com/wp-content/uploads/2021/09/2021_Serve_360_Report.pdf (accessed 10 February 2022).
78. The Athenea Hotel. Certified and Supervised Grande-A Organic Rice from Upcountry Farm to Sensorial Table. Avail online: https://www.marriott.co.uk/hotels/hotel-information/details-3/bkkla-the-athenee-hotel-a-luxury-collection-h bangkok/ (accessed on 10 February 2022).
79. News Medical Life Science. Climate Change and COVID-19. Available online: https://www.news-medical.net/health/Clir Change-and-COVID-19.aspx (accessed on 24 March 2021).
80. Site Minder. Hotel Industry Trends to Watch Out For. Available online: https://www.siteminder.com/r/hotel-trends-h hospitality-industry/ (accessed on 10 February 2022).
81. Banyan Tree Global Foundation. Sustainable Development; Building a Brighter Future. Available online: https://w banyantreeglobalfoundation.com/sustainable-development/ (accessed on 10 February 2022).

MDPI
St. Alban-Anlage 66
4052 Basel
Switzerland
www.mdpi.com

MDPI Books Editorial Office
E-mail: books@mdpi.com
www.mdpi.com/books

www.ingramcontent.com/pod-product-compliance
Lightning Source LLC
Chambersburg PA
CBHW061607220926
51850CB00005B/441
* 9 7 8 3 0 3 6 5 9 2 0 6 0 *